INSTRUMENTATION

Center for the Advancement of Process Technology

Prentice Hall

Boston Columbus Indianapolis New York San Francisco Upper Saddle River Amsterdam
Cape Town Dubai London Madrid Milan Munich Paris Montreal Toronto Delhi
Mexico City Sao Paulo Sydney Hong Kong Seoul Singapore Taipei Tokyo

Editor in Chief: Vernon Anthony
Acquisitions Editor: David Ploskonka
Editorial Assistant: Nancy Kesterson
Director of Marketing: David Gesell
Senior Marketing Coordinator: Alicia Wozniak
Marketing Assistant: Les Roberts
Associate Managing Editor: Alexandrina Benedicto Wolf
Project Manager: Alicia Ritchey
Senior Operations Supervisor: Pat Tonneman
Operations Specialist: Laura Weaver
Art Director: Diane Ernsberger
Cover Designer: Jayne Conte
Cover Art: Center for the Advancement of Process Technology and Eyewire
Lead Media Project Manager: Karen Bretz
Full-Service Project Management: Lisa Garboski, bookworks
Composition: TexTech
Printer/Binder: RR Donnelley/Kendallville
Cover Printer: Lehigh-Phoenix Color
Text Font: Times Ten Roman

Credits and acknowledgments borrowed from other sources and reproduced, with permission, in this textbook appear on appropriate page within text.

Library of Congress Cataloging-in-Publication Data
Instrumentation / CAPT.
p. cm.
ISBN 0-13-700413-3
1. Chemical process control—Instruments. I. CAPT (Organization)
TP155.75.I5345 2010
660'.2815—dc22 2009003567

15 14 13
V011

Prentice Hall
is an imprint of

www.pearsonhighered.com

ISBN-10: 0-13-700413-3
ISBN-13: 978-0-13-700413-3

Contents

Preface

The Process Industries Challenge

In the early 1990s, the process industries recognized that they would face a major manpower shortage due to the large number of employees retiring. Industry partnered with community colleges, technical colleges, and universities to provide training for their process technicians, recognizing that substantial savings on training and traditional hiring costs could be realized. In addition, the consistency of content and exit competencies of process technology graduates could be ensured if industry collaborated with education.

To achieve this consistency of graduates' exit competencies, the Gulf Coast Process Technology Alliance and the Center for the Advancement of Process Technology identified a core technical curriculum for the Associate Degree in Process Technology. This core, consisting of eight technical courses, is taught in alliance member institutions throughout the United States. This textbook is intended to provide a common standard reference for the *Instrumentation* courses that serve as part of the core technical courses in the degree program.

Purpose of the Textbook

Instructors who teach the process technology core curriculum, and who are recognized in the industry for their years of experience and their depth of subject matter expertise, requested that a textbook be developed to match the standardized curriculum. Reviewers from a broad array of process industries and education institutions participated in the production of these materials so that the widest audience possible would be represented in the presentation of the content.

The textbook is intended for use in community colleges, technical colleges, universities, and corporate settings in which process instrumentation is taught. However, educators in many disciplines will find these materials a complete reference for both theory and practical application. Students will find this textbook to be a valuable resource throughout their process technology career.

Organization of the Textbook

This textbook has been divided into 24 chapters with Chapters 1-12 covering the material in the Instrumentation I course and Chapters 13-24 covering the material in the Instrumentation II course. Each chapter is subdivided in the following progression:

- Learning Objectives
- Key Terms
- Introduction

- Key Topics
- Summary
- Checking Your Knowledge
- Student Activities

The **Learning Objectives** for a chapter may cover one or more sessions in a course. For example, Chapter 2 may take two weeks (or two sessions) to complete in the classroom setting.

The **Key Terms** are a listing of important terms and their respective definitions that students should know and understand before proceeding to the next chapter.

The **Introduction** may be a simple introductory paragraph, or may introduce concepts necessary to the development of the content of the chapter itself.

Any of the **Key Topics** can have several subtopics. Although these topics and subtopics do not always follow the flow of the learning objectives, all learning objectives are addressed in the chapter.

The **Summary** is a restatement of the learning outcomes of the chapter.

The **Checking Your Knowledge** questions are designed to help students self-test on potential learning points from the chapter.

The **Student Activities** section contains activities that can be performed independently or with other students in small groups, and activities that should be performed with instructor involvement.

Chapter Summaries

CHAPTER 1: INTRODUCTION TO INSTRUMENTATION

In this chapter, there is a progressive description of the evolution of instrumentation in the process industries from earlier versions through the many technological advances, and then to how they are used today. An introduction to various common terms is discussed with a preview of in-depth material on instrumentation. In addition, various relationships between process variables are introduced.

CHAPTER 2: PROCESS VARIABLES, ELEMENTS, AND INSTRUMENTS: PRESSURE

In this chapter, a review is given of the concepts and definitions of pressure and, in general terms, how to calculate the amount of pressure for various applications. Also discussed are the various types of measurement scales used to determine the amount of pressure in a given situation and how to convert the pressure reading from one scale to another. Different types of pressure-measuring devices are identified as to their purpose and operation, including their respective locations where they may be used in a processing environment.

CHAPTER 3: PROCESS VARIABLES, ELEMENTS, AND INSTRUMENTS: TEMPERATURE

In this chapter, a description is given of temperature and heat energy as they relate to molecular movement. Boiling and freezing points on commonly used temperature measuring scales are discussed along with the various temperature-measuring devices to include their specific purpose and operation. Conversions between the various temperature-measuring scales are also covered.

CHAPTER 4: PROCESS VARIABLES, ELEMENTS, AND INSTRUMENTS: LEVEL

In this chapter, the process variable identified as level is defined as well as the most common terms associated with level measurement. After this, the most common types of level measuring devices are discussed as to their respective purpose and operation. Then, a discussion of the relationships between level, temperature, density, and volume brings everything together to show the effects of each one on the other.

CHAPTER 5: PROCESS VARIABLES, ELEMENTS, AND INSTRUMENTS: FLOW

In this chapter, flow (fluids in motion) and the various types of flow-measurement devices and their respective flow-measurement units (e.g., mass or volume over time) are discussed. Differential pressure (d/p) is correlated to flow measurement, as are other principles of how flow rate or total flow is determined and impacted.

CHAPTER 6: PROCESS VARIABLES, ELEMENTS, AND INSTRUMENTS: ANALYTICAL

In this chapter, an introduction is given to the many different types of analytical instrumentation found within process facilities and the various types of physical properties that can be analyzed. The functionality of each type of analyzer is discussed with an association of the types of physical properties each type analyzes. In addition, the role of process technicians is covered as it pertains to the use of analytical instrumentation.

CHAPTER 7: MISCELLANEOUS MEASURING DEVICES

In this chapter, a discussion is given on other devices that do not fall in the main categories or types of instrumentation used in process facilities as described in Chapters 2–6. The two types that are discussed in this chapter include devices that measure vibration and speed with a discussion on the various applications and applicable type variations for each.

CHAPTER 8: INTRODUCTION TO CONTROL LOOPS: SIMPLE LOOP THEORY

In this chapter, the concept is introduced of how processes are controlled by combining various instruments to function together in a control loop. A discussion is presented of how each component of a control loop acts individually and together within a control loop to sense, measure, compare, and control or convert signals between the various components. Various types of signal transmission used within control loops are also introduced.

CHAPTER 9: CONTROL LOOPS: PRIMARY SENSORS, TRANSMITTERS, AND TRANSDUCERS

In this chapter, the individual control loop components called primary sensors, transmitters, and transducers are discussed in more detail. This discussion includes their function of sensing, measuring, and transmitting as well as their purpose and operation in each portion of the process. Also covered is how each component relates to others in the loop based on the process variable (PV).

CHAPTER 10: CONTROL LOOPS: CONTROLLERS AND FINAL CONTROL ELEMENT OVERVIEW

In this chapter, a closer look is taken at controllers to include how they function, what type of responses that they may take to various signals, and what types of signals they output. The last and final element of the control loop is also introduced to include what types of final control elements are used and how they relate to the signals produced by the controller.

CHAPTER 11: CONTROL LOOPS: CONTROL VALVES AND REGULATORS

In this chapter, the various components of control valves as well as their purpose and operation are covered. Simple troubleshooting tips are given as well as what happens during valve failure. Also included is a discussion on various actuators and valve positioners, input/output signal differences, fail-safe positions, and special valve types in various arrangements and configurations. In addition, a discussion about instrument air regulators and how they are used in the operation of control valves is also included.

CHAPTER 12: SYMBOLOGY: PROCESS DIAGRAMS AND INSTRUMENT SKETCHING

In this chapter, the types of drawings that a process technician may encounter are discussed including their variations and characteristics. Then, an introduction to symbology is given to include a discussion of the standard for symbols and codes used for denoting the types of instrumentation they represent. Basic equipment symbols are shown as well as standard symbols for lines, instruments (balloon representations) as they apply to process variables, and then samples of how these symbols are used together along with the appropriate interpretation for how to read them.

CHAPTER 13: INSTRUMENTATION TROUBLESHOOTING

In this chapter, the concept of troubleshooting, calibration, and the practices related to process technicians are introduced. Typical equipment malfunctions are identified and proper tool selection is discussed.

CHAPTER 14: SWITCHES, RELAYS, AND ANNUNCIATORS

In this chapter, various types of switches, their functions, their uses, and symbology associated with them are explained. In addition, various types of relays and their applications listed and explained. And finally, annunciator systems, their purposes, and terms associated with them are also explained.

CHAPTER 15: SIGNAL TRANSMISSION AND CONVERSION

In this chapter, the concept of signal transmission and the devices that protect the integrity and reliability of signal transmission are introduced. In addition, various signal transmission types and conversions are listed and explained.

CHAPTER 16: CONTROLLERS

In this chapter, process dynamics and control are introduced. In addition, different types of controllers and their actions (modes) are listed and explained. Process recording instrumentation and signatures are identified and explained, as well as tuning modes and a brief section on troubleshooting tuning errors.

CHAPTER 17: CONTROL SCHEMES

In this chapter, on/off, lead/lag, feedback, and feedforward control schemes are explained along with the following control modes: local manual, local automatic, remote manual, remote automatic, and cascading of the remote automatic.

CHAPTER 18: ADVANCED CONTROL SCHEMES

In this chapter, terminology associated with advanced control schemes is identified and explained. In addition, the purpose and functions of cascaded, ratio (fractional), split-range, and multivariable control schemes are discussed, as well as the steps required to change instrument controllers without bumping the process.

CHAPTER 19: INTRODUCTION TO DIGITAL CONTROL

In this chapter, the purpose of digital control is explained and the differences between analog and digital controllers are highlighted. Key terms associated with process control schemes are identified and explained. In addition, how analog and digital controllers transmit output signals is explained.

CHAPTER 20: PROGRAMMABLE LOGIC CONTROL

In this chapter, key terms associated with programmable logic controls (PLCs) are identified and explained. In addition, the purpose of a PLC is explained, as well as ladder logic and how it applies to PLCs.

CHAPTER 21: DISTRIBUTED CONTROL SYSTEMS (DCSs)

In this chapter, the purpose and general operation of a distributed control system (DCS) is explained, as well as the advantages of a DCS over an analog control system. In addition, the function of multiplexers and demultiplexars is also discussed along with the components of a typical DCS.

CHAPTER 22: INSTRUMENT POWER SUPPLY

In this chapter, the purpose of an uninterruptible power supply (UPS) system is explained and the components of a basic UPS are identified. In addition, key terms associated with instrumentation power supplies are identified and explained.

CHAPTER 23: EMERGENCY SHUTDOWN (ESD), INTERLOCKS, AND PROTECTIVE DEVICES

In this chapter, key terms associated with emergency shutdown systems, interlocks, and alarms are identified and explained. The use of permissives, interlocks, bypasses, emergency shutdown systems, and alarms are discussed, along with the various components and techniques that are implemented to help prevent unnecessary shutdowns.

CHAPTER 24: INSTRUMENTATION MALFUNCTIONS

In this chapter, various types of instrument failure modes are identified and explained along with the methods used to determine if the device is malfunctioning. In addition, how a control loop responds to malfunctions in primary sensing elements, transmitters, controllers, and final control elements is also explained.

Acknowledgments

The following organizations and their dedicated personnel voluntarily participated in the production of this textbook. Their contributions to making this a successful project are greatly appreciated. Perhaps our gratitude for their involvement can best be expressed by this sentiment:

> The credit belongs to those people who are actually in the arena . . . who know the great enthusiasms, the great devotions to a worthy cause; who at best, know the triumph of high achievement; and who, at worst, fail while daring greatly . . . so that their place shall never be with those cold and timid souls who know neither victory nor defeat. —Theodore Roosevelt

Process technicians both current and future will utilize the information within this textbook as a resource to more fully understand the instrumentation commonly found within the process industries. This knowledge will strengthen these paraprofessionals by helping to make them better prepared to meet the ever challenging roles and responsibilities within their specific process industries.

INDUSTRY CONTENT DEVELOPERS AND REVIEWERS

Lisa Arnold, Marathon Ashland, Texas
Ted Borel, Equistar Chemical, Texas
Gerald Canady, The Dow Chemical Company, Louisiana
Tim Carroll, BASF Corporation, Louisiana
Lester Chin, Hovensa, LLC, U.S. Virgin Islands
Sheldon Cooley, Westlake Group, Louisiana
Ed Couvillion, Sterling Chemicals, Texas
Greg Curry, The Dow Chemical Company, Texas
Lloyd Davis, Halliburton, Oklahoma
Steve Erickson, GCPTA, Texas
Steve Ernest, ConocoPhillips, Oklahoma

Billy Fridelle, ExxonMobil, Texas
Richard Honea, The Dow Chemical Company, Texas
Lee Hughes, BASF Corporation, Texas
Jennifer Martin, OG&E Horseshoe/Mustang Plant, Oklahoma
Mike McBride, BASF, Texas
Martha McKinley, Eastman Chemical, Texas
Jason Oxley, OG&E Horseshoe/Mustang Plant, Oklahoma
Aracelli Palomo, The Dow Chemical Company, Texas
Allen Parker, Crompton Corporation, Louisiana
Shawn Parker, Marathon Ashland, Texas
Clemon Prevost, BP, Texas
Robert Rabalaise, The Dow Chemical Company, Louisiana
Raymond Robertson, PPG, Louisiana
Matt Scully, BP, Alaska
Roy St. Romain, Eastman Chemical, Texas
Paul Summers, The Dow Chemical Company, Texas
Dennis Thibodeaux, Citgo, Louisiana
Robert Toups, Lyondell Chemical, Louisiana

EDUCATION CONTENT DEVELOPERS AND REVIEWERS

Louis Babin, ITI Technical College, Louisiana
Carrie Braud, Baton Rouge Community College, Louisiana
Tommie Ann Broome, Mississippi Gulf Coast Community College, Mississippi
Larry Callaway, Bellingham Technical College, Washington
Chuck Carter, Lee College, Texas
Lou Caserta, Alvin Community College, Texas
Ken Clark, Perry Technical Institute, Washington
Michael Connella III, McNeese State University, Louisiana
Richard Cox, Baton Rouge Community College, Louisiana
Mark Demark, Alvin Community College, Texas
Ned Duffey, South Arkansas Community College, Arkansas
Jerry Duncan, College of the Mainland, Texas
Tommy Edgar, Texas State Technical College, Texas
Steve Ernst, Northern Oklahoma College, Oklahoma
Charles Gaffen, Elizabeth High School, New Jersey
Gary Hicks, Brazosport Community College, Texas
Kevin Holmstrom, Bismarck State College, North Dakota
Raymond Johns, Northern Oklahoma College, Oklahoma
Larry Johnson, Nova Scotia Community College, Nova Scotia
Mike Kukuk, College of the Mainland, Texas
Tony Kuphaldt, Bellingham Technical College, Washington
Linton LeCompte, Sowela Technical Community College, Louisiana
Michael Murray, Copiah-Lincoln Community College, Mississippi
Richard Ortloff, Bellingham Technical College, Washington
Denise Rector, Del Mar College, Texas
Robert Robertus, Montana State University - Billings, Montana
Paul Rodriguez, Lamar Institute of Technology, Texas
Dean Schwarz, Southwestern Illinois College, Illinois
Mike Speegle, San Jacinto College, Texas
Wayne Stephens, Wharton County Junior College, Texas
Mark Stoltenberg, Brazosport Community College, Texas
Robert Walls, Del Mar College, Texas
Scott Wells, Brazosport Community College, Texas

CENTER FOR THE ADVANCEMENT OF PROCESS TECHNOLOGY STAFF

Bill Raley, Principal Investigator
Jerry Duncan, Director
Melissa Collins, Associate Director
Angelica Toupard, Instructional Designer
Scott Turnbough, Graphic Artist
Cindy Cobb, Program Assistant
Joanna Perkins, Outreach Coordinator
Chris Carpenter, Web Application Developer

This material is based upon work supported, in part, by the National Science Foundation under Grant No. DUE 0202400. Any opinions, findings, and conclusions or recommendations expressed in this material are those of the author(s) and do not necessarily reflect the views of the National Science Foundation.

CHAPTER 1

Introduction to Instrumentation

Objectives

After completing this chapter, you will be able to:

- Discuss the evolution and importance of process instrumentation to the petrochemical and refining industry.
- Describe the major process variables controlled in the process industries:

  process variables
  differential (delta, Δ)
  pressure
  temperature
  flow
  level
  analytical
  other
- Define terms associated with instrumentation:

  local
  remote
  indicating
  transmitting
  recording
  controlling
  control loop
  pneumatic
  electronic
  analog
  digital
- Explain the relationship between common process variables:

  What happens to the pressure in a closed container when temperature increases or decreases?

  What happens to the temperature in a closed container when pressure increases or decreases?

  Explain what happens to the temperature in a closed container when temperature increases or decreases.

  What happens to vessel bottom pressure when height of liquid increases or decreases?

What happens to boiling point of a material when pressure increases or decreases?
What happens to the volume of a material when temperature increases or decreases?
What happens to the density of a material when temperature increases or decreases?
What happens to the flow when the differential pressure increases or decreases?
How do variable changes affect accurate measurement?

Key Terms

Analog—analogous or similar to something else; in an analog system, process variable measurements change in a continuous manner mirroring the actual process variable.
Analytical—the use of a logical technique to perform an analysis; in instrumentation, a measurement of a physical or chemical property.
Control Loop—a group of instruments working together to control a single process variable such as pressure, temperature, level, or flow.
Controlling—keeping a variable at a specific quantity.
Differential pressure (Δp, d/p)—the difference between two related pressures; usually used in measurements of process variables (pressure, temperature, level and flow).
Digital—a transmission method that employs discrete electrical signals as opposed to continuous signals or *waves.*
Electronic—powered by electricity.
Flow—a fluid in motion.
Indicating—the showing of a current condition via a readout (e.g., digital or analog).
Level—the height of a liquid in reference to a zero point.
Local—located at or near the point of measurement.
Pneumatic—powered by a gas.
Pressure—a force applied to a unit of area (force ÷ area = pressure).
Process variable (PV)—measured property of a process.
Recording—the keeping of historical data by making a physical record.
Remote—located away from the point of measurement (usually in a control room).
Temperature—a specific degree of hotness or coldness as indicated on a calibrated scale; the measurement of the average kinetic energy of the molecules of the substance being measured.
Transmitting—communicating via a signal from one place to another.

Introduction

This chapter describes the evolution of instrumentation in the process industries from the earlier versions through the many technological advances and then on to the way they are used today. An introduction to various common terms is discussed with a preview of coming in-depth material on instrumentation. In addition, various relationships between process variables are introduced through the posing of questions and answers.

Evolution of Process Instrumentation

In prehistory, as well as in modern times, humankind has attempted to control machine processes. Possibly the first example of feedback control was the application of the flyball governor (Figure 1-1) in 1775. Although there had been previous attempts, James Watt was the first to successfully apply it to the steam engine as a speed control mechanism.

Rudimentary forms of automatic control (e.g., pneumatic) had begun to appear in the mid to late 1920s. However, the vast majority of processes were controlled manually. In manual control a process technician looks at a gauge or sight glass (Figure 1-2) and

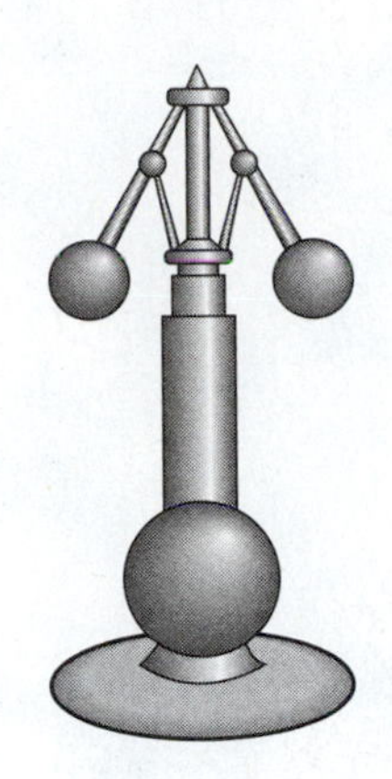

FIGURE 1-1 Flyball Governor

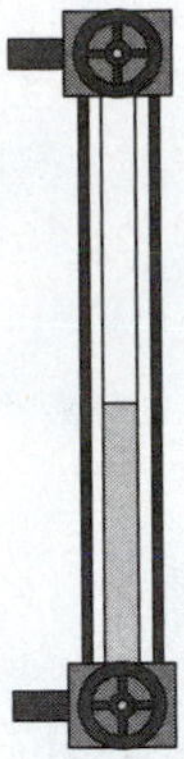

FIGURE 1-2 Manual Control (Tubular Sight Glass)

then makes an appropriate adjustment to a piece of equipment that directly affects a process variable. Throughout the workday, technicians had to control each variable in every piece of equipment using this time and energy-consuming manner.

Local control systems, using these newly found forms of control, began to appear in the 1930s. In 1932, Harry Nyquist (Figure 1-3) published the first theoretical paper on automatic process control.

The implementation of modern control theory was slow, yet steady, until the early 1940s when Japan attacked Pearl Harbor pressing the United States into war. In those days, the workforces in plants were comprised almost entirely of men who were leaving these jobs to join the war effort. This left a huge need for technicians to fill these positions. Concurrently, the war was generating an even greater need for the products that these plants were producing. Something had to be done. The answer to the problem was to hire and train women (Figure 1-4) and utilize automatic control.

In 1942, John Zeigler and Nathaniel Nichols described an innovative new way to tune PID control loops in another breakthrough technical paper on automatic process control.

Where new ideas and new hardware are concerned, there is always a development period followed by an implementation period. Although analog electronic instrumentation (Figure 1-5) was around in the 1950s (and before), it really came of age in the 1960s.

Even though the first use of computers in industry dates back to 1958, there were only about 3,000 computers in use a decade later. This began to change in 1971 when the Intel Corporation introduced the Model 4004 microprocessor, the world's first processor on a chip (Figure 1-6). The microprocessor paved the way for development of the modern digital control that is in use today. Digital control evolved from direct control through supervisory control to the most modern Distributed Control Systems today.

FIGURE 1-3 Nyquist

Courtesy of the American Institute of Physics

FIGURE 1-4 Rosie the Riveter

Credit: National Archives – WWII Public Domain Photographs

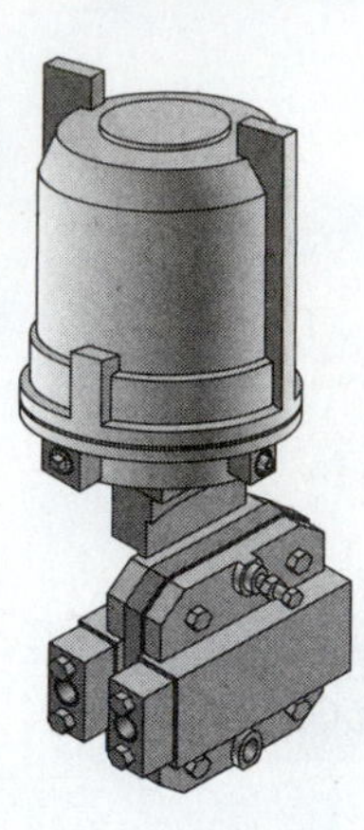

FIGURE 1-5 Analog Electronic Instrumentation

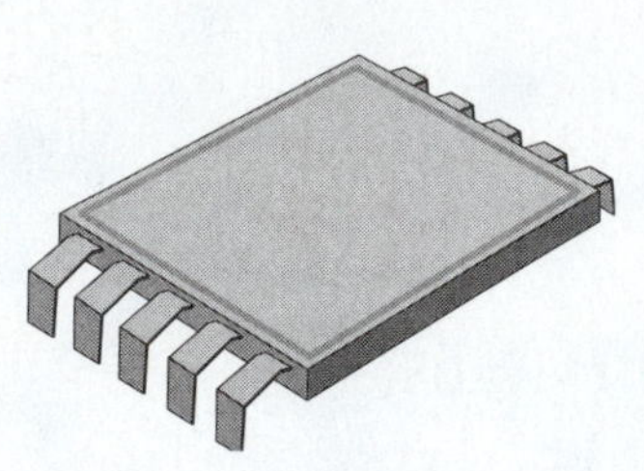

FIGURE 1-6 Computer Chip in the Early 1970s

Control theory and practice are continuing to change. As new technology emerges, such as fiber optics (Figure 1-7), industry continues to find ways for implementation.

Process Variables

The primary need for instrumentation in industry is to *measure* and *control* process variables such as pressure, temperature, level, flow, and analytical. Instrumentation is the eyes and ears as well as the decision-making element of a piece of equipment. In order to do the job, an instrument system must first measure a condition that is controllable. This controllable condition, known as a **process variable (PV)**, is any measured property of the process. The following information provides a brief description for each of these common process variables and also includes various terminologies used when describing process variables.

Process variables include the following:

- pressure
- temperature
- flow
- level
- differential pressure
- analytical

PRESSURE

Pressure can be defined as a force applied to a unit of area. Pressure is one of the most common measurements taken in industry. Nearly all industrial processes use or produce liquids, gases, or both. Gases and vapors apply force uniformly over all surfaces while liquids apply force in accordance to their depth and density. Various instruments can be used to measure pressure such as a pressure gauge (Figure 1-8).

TEMPERATURE

Temperature is defined as a measure of the average kinetic energy (hotness or coldness) of a substance as indicated on a reference scale (Figure 1-9). The two most common

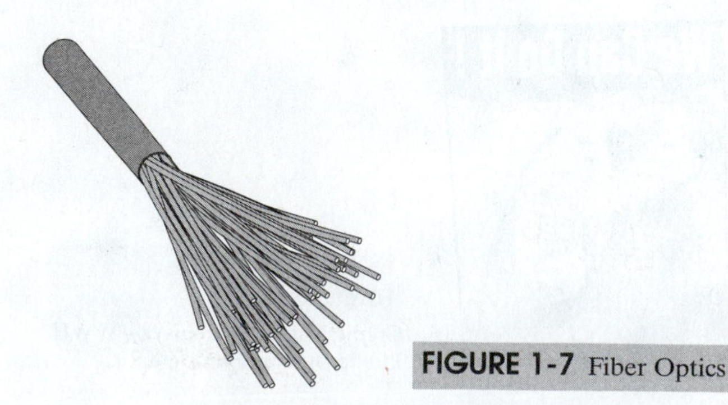

FIGURE 1-7 Fiber Optics

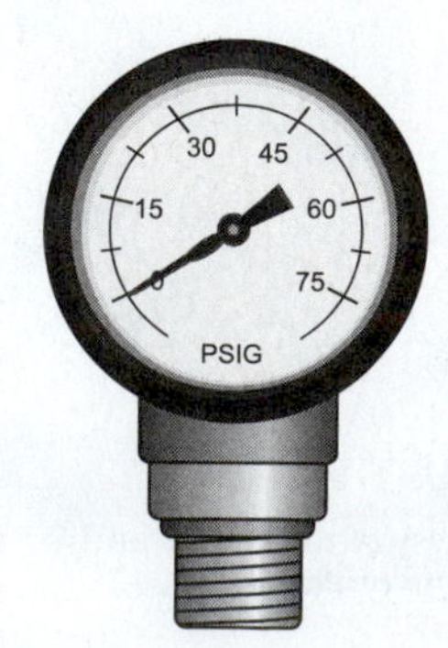

FIGURE 1-8 PSIG Scale

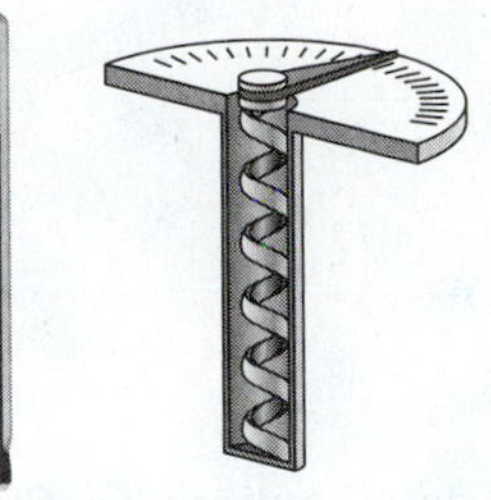

FIGURE 1-9 Thermometers

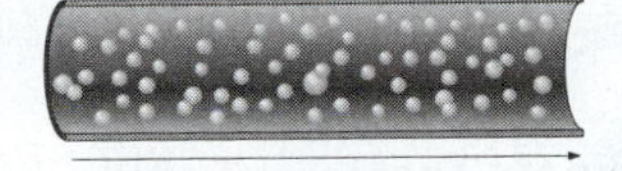

FIGURE 1-10 Flow

temperature measurement scales are Fahrenheit (degrees F) and Celsius or Centigrade (degrees C). Process plants control temperature in almost every major process vessel.

FLOW

In process industries, the word **flow** is used interchangeably with the term flow rate. Flow rate can be defined as the quantity of fluid that moves through a pipe (Figure 1-10) or channel within a given period of time. Flow rate is usually expressed in volume or mass units per unit of time, such as gallons per minute (gpm) or pounds per hour (pph).

LEVEL

Level is defined as the position of either height or depth (Figure 1-11) along a vertical axis. In industry, the term level specifically means the surface position of a material in a vessel. For example, checking levels is very important when controlling a liquid phase reactor where there is a need for a continuous flow of reactants into the vessel and a continuous flow of reacted product leaving the vessel.

DIFFERENTIAL (DELTA, Δ)

Simply put, differential means the difference between measurements taken from two separate points. Although differential is not a distinct process variable like the others mentioned here, the designation does deserve special attention due to the massive number of **differential pressure** (Figure 1-12) and temperature devices encountered in industry. Two commonly used ways to express differential are the Greek letter delta (Δ) and the letter combination d/p.

ANALYTICAL

Analytical instruments (Figure 1-13) are those instruments that measure the chemical and/or physical properties of a process stream. Some common analyzers include chromatographs, pH meters, and viscosity meters. Analytical instruments can be located within the process area as well as in a laboratory.

OTHER

There are many other process variables measured and controlled in the process industry besides the types listed above. Some examples are weight, speed, vibration, and

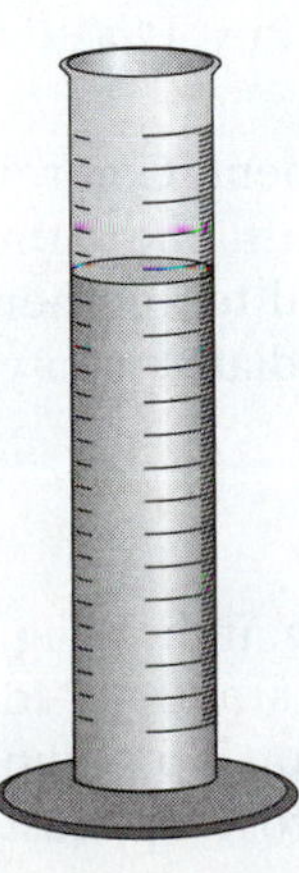

FIGURE 1-11 Level Measurement

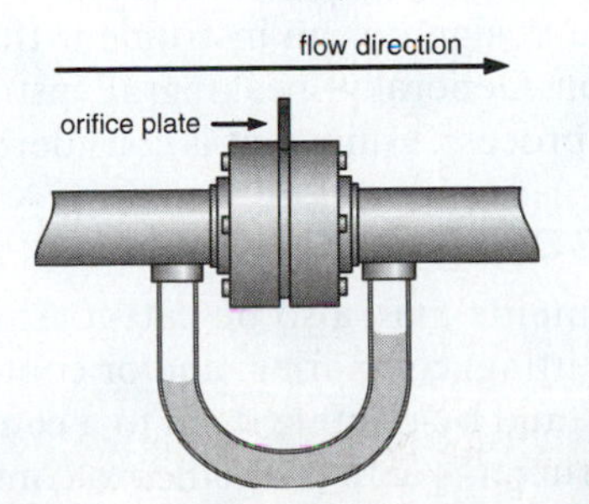

FIGURE 1-12 Process Variables–Differential Data

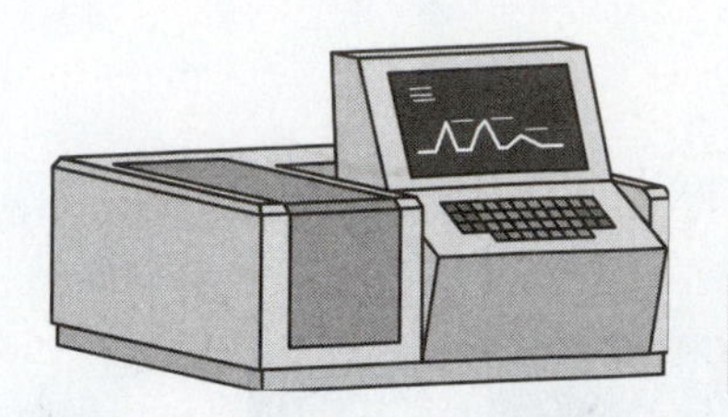

FIGURE 1-13 Analytical Instrument

acceleration. These are discussed later in this textbook in greater detail but are briefly introduced below:

- Weight: weighing materials in tanks, drums, or even on a conveyor belt
- Speed: measuring and controlling the rotational speeds of motors, etc.
- Vibration: detecting vibration of rotating machinery
- Acceleration: generally measuring the natural motion of rotating equipment to ensure that it stays within safe limits

INSTRUMENT CATEGORIES AND TYPES

Process instrumentation may be categorized by location, function, and power sources. Due to the natural overlapping of these categories, a single instrument may fall into more than one category at the same time. For example, a pneumatic pressure transmitter is also a local instrument because it is located in the processing area and it is a pressure-transmitting instrument powered by air. Use the following category designations to explore the differences in location, function, and power sources and to make sense of the predicament in attempting to categorize instruments.

CATEGORIZED BY LOCATION

An instrument may be categorized by location with respect to its proximity to the actual process it is measuring and/or controlling.

Local or Field

An instrument located at or near the process is called a **local** instrument (Figure 1-14). Another name for a local instrument is field instrument.

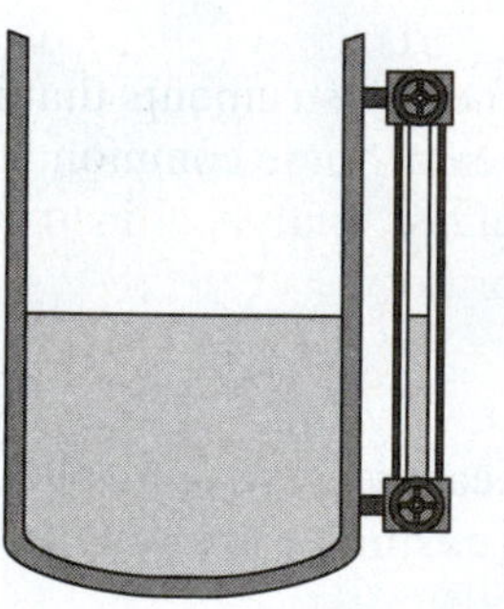

FIGURE 1-14 Local Level Instrument

Remote

An instrument located away from the process is called a **remote** instrument. Control room instruments are examples of the most common remote instruments. The term remote designates an instrument that receives and/or transmits a signal to another location. Generally speaking, all instrumentation not located in the immediate vicinity of the process equipment is considered as remote.

CATEGORIZED BY FUNCTION

Instruments may also be categorized by their function such as sensing, indicating, transmitting, comparing, and/or controlling. These function designations are easier to understand by relating them to a **control loop** (Figure 1-15). A control loop is a group of instruments acting together to control a single process variable. A control loop must

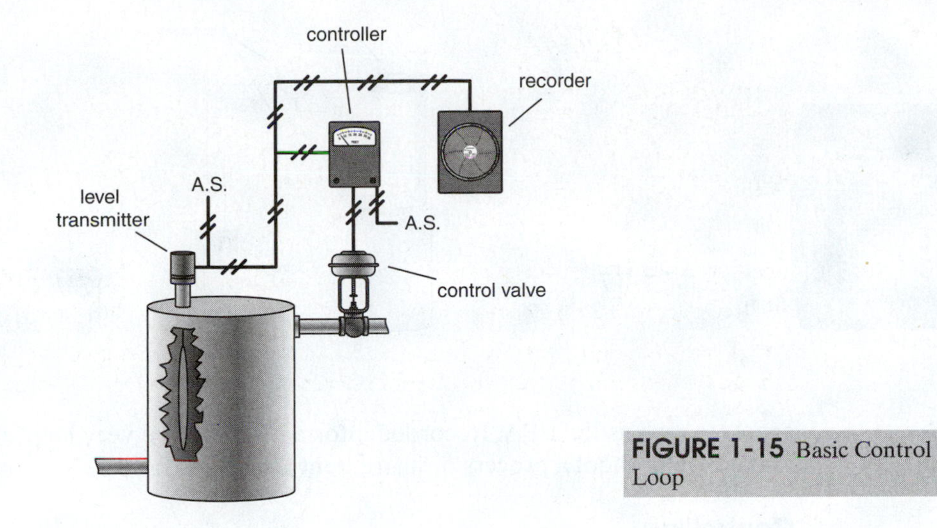

FIGURE 1-15 Basic Control Loop

consist of all the parts necessary to control a process variable. Typical control systems have a measurement sensor, a transmitter(s), a controller, and a final control element that is usually a valve.

The sensing or measuring or transmitting device is the first instrument in the loop. Once it measures the process variable, it then must communicate that value to the next instrument in the loop, typically the controller. The controller would have to interpret that incoming signal, compare it to a setpoint, process the difference, and then produce an output signal that indicates to the final controlling element, usually a control valve, to open more, close more or just stay the same.

Indicating

Instrumentation showing the current conditions, such as temperature, flow, etc., is categorized as **indicating**. Any instrument that has a digital readout or graduated scale with a pointing device is called an indicating instrument. Transmitters, recorders, and controllers could be indicating instruments.

Transmitting

Instrumentation that sends a signal (measurement represented in a standardized form such as 3-15 psi or 4-20 mA) (Figure 1-16) to a remote location is categorized as **transmitting**. Generally speaking, transmitters are both a measuring and a transmitting device built into a single unit. There may also be a transducer in the same unit. For example, a unit may perform a pressure sensing activity and convert this measurement to a pneumatic (3-15 psi) or electronic (4-20 mA) signal, and then transmit the signal to a controller.

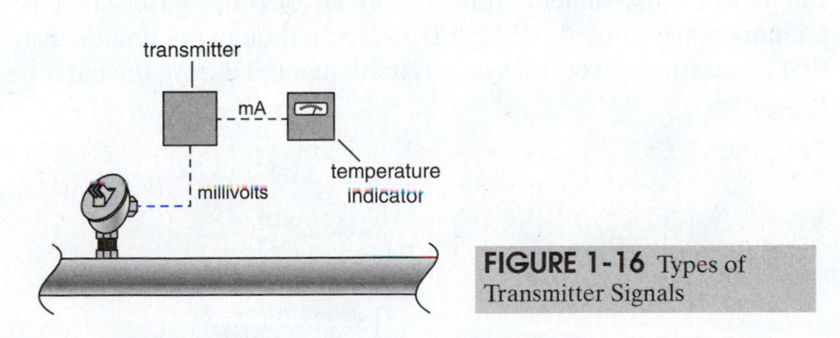

FIGURE 1-16 Types of Transmitter Signals

Recording

Instrumentation that records/registers (Figure 1-17) a process variable is categorized as **recording**. Recorders keep historical data on process variables. Sometimes these records are kept for extended periods of time due to strict directives from regulatory

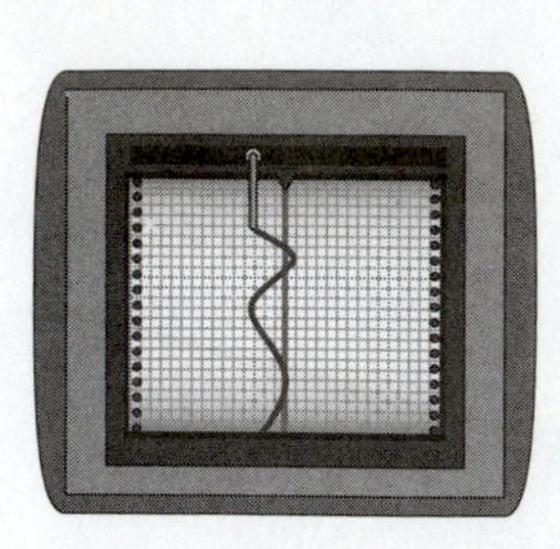

FIGURE 1-17 Pressure Recorder with Strip Chart

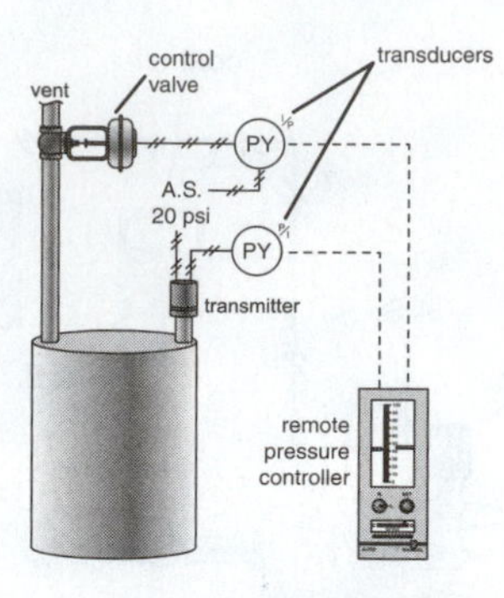

FIGURE 1-18 Remote Pressure Controller

agencies such as the EPA. Recorded information may be very helpful to someone trying to troubleshoot a process or instrument related problem.

Controlling

Instrumentation that maintains a process variable at a specific quantity is categorized as **controlling**. A controller is a device that performs the control function and may exist as either a physical entity (local controller) or an algorithm (subroutine or program) in a remote location (Figure 1-18) such as in a computer or control room rack or cabinet.

CATEGORIZED BY POWER SOURCE

Using the power source as a categorization for instrumentation would depend on how all the instruments function together inside a control loop or how the instrument functions by itself.

Entire Control Loop

As stated previously, an instrument control loop consists of all the parts necessary to control a process variable and usually includes a sensor, transmitter, controller, and a final control element. If the loop uses *air* as the primary power source to convey information between the instruments, then the loop (Figure 1-19) would be categorized as **pneumatic**. If a transducer were placed in the loop, then the pneumatic signal could be changed to another signal such as electrical.

Individual Instruments

Where combination power sources (Figure 1-20) are used within a loop, categorization cannot be by power source and the categorizations then fall to individual instruments within the loop. These categories (pneumatic and electronic) are the same as described previously for control loops.

Figure 1-21 shows an illustration of the various types of incoming signal power.

Pneumatic Instruments powered by air or other gases such as nitrogen are called pneumatic instruments (Figure 1-22). Even though pneumatic instrumentation was the first generation of control systems implemented during the early part of the 1900s, they

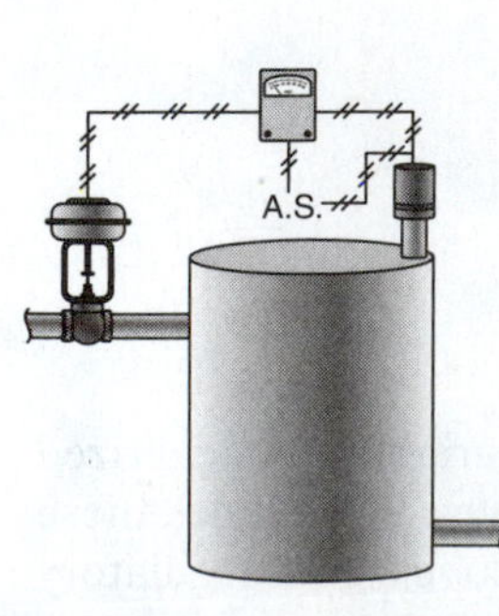

FIGURE 1-19 Pneumatic Pressure Control Loop

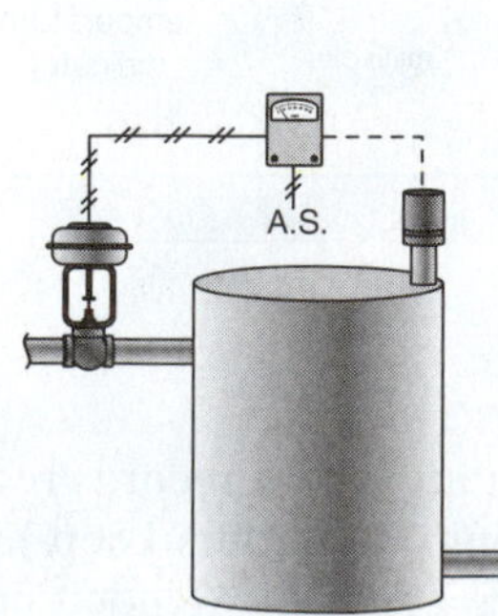

FIGURE 1-20 Combination Power Source Control Loop

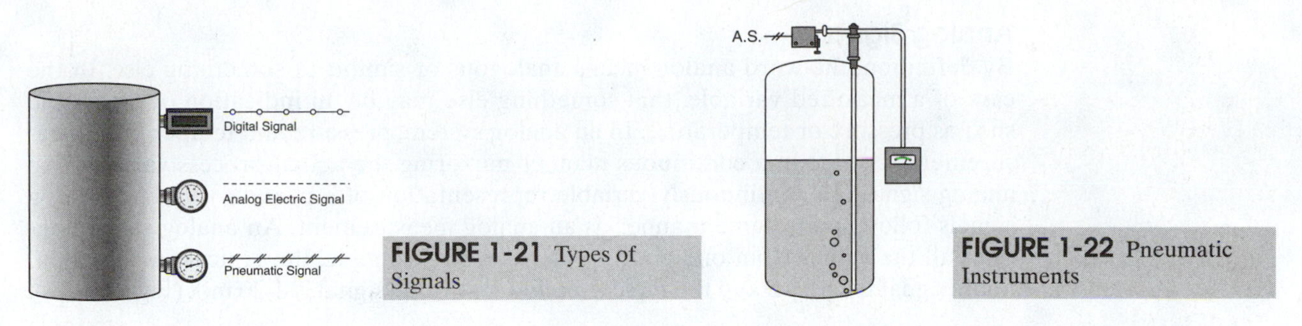

FIGURE 1-21 Types of Signals

FIGURE 1-22 Pneumatic Instruments

are still in use today. With the exception of control valves, the use of pneumatic instrumentation has been on a steady decline for over four decades. Pneumatic instruments are still a viable option in many industrial plants and should maintain a niche market in the process industry for a long time to come for several reasons. First, they are self-purging and nonelectrical, which makes them desirable in highly corrosive environments or where any electrical potential is unacceptable. Also, most control valves are still pneumatically actuated. The typical pneumatic signal is 3-15 pounds per square inch (psi).

Electronic Instruments powered by electricity are called **electronic** instruments (Figure 1-23).

Subcategories under the umbrella of electronic instrumentation are divided by the type of signal and include the following:

- Analog:
  Analog electronic instruments are those instruments that use continuously variable electrical quantities to measure, amplify, and/or produce a standard output signal such as 4-20 milliampere (mA).
- Digital:
  Digital instruments (Figure 1-24) are microprocessor based and can produce a digital as well as an analog output signal. A digital signal is presented in a coded form by packets of ones and zeros. Generally, they use a serial communication format similar to a computer network in an office or classroom.
  Example of a binary packet: 10101101
- Analog-digital hybrid:
  Analog-digital hybrid electronic instruments such as smart transmitters may be microprocessor based yet they produce an analog output such as a 4-20-mA signal. The microprocessor in these instruments can be communicated via a frequency signal that shares the same wires as the analog signal.

CATEGORIZED BY SIGNAL PRODUCED

When using the type of signal produced as the categorization, the two predominant types are analog and digital.

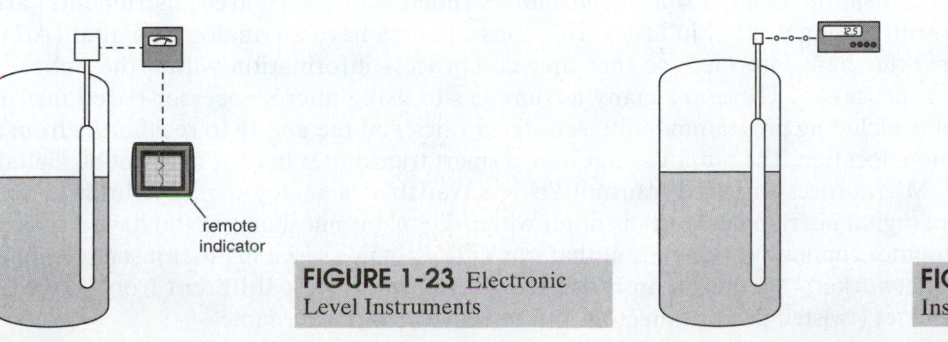

FIGURE 1-23 Electronic Level Instruments

FIGURE 1-24 Digital Level Instrument

Analog Signals

By definition, the word analog means analogous or similar to something else. In the case of a measured variable, that something else may be an indication of a variable such as pressure or temperature. In an analog system, pressure and temperature measurements change in a continuous manner mirroring the actual process variable. An analog signal is a continuously variable representation of a process variable. Analog signals follow in the same manner as an analog measurement. An analog signal contains all the values from one point to another. For example, the typical analog pneumatic signal is 3-15 psi and the typical *analog* electronic signal is 4-20 mA (Figure 1-25).

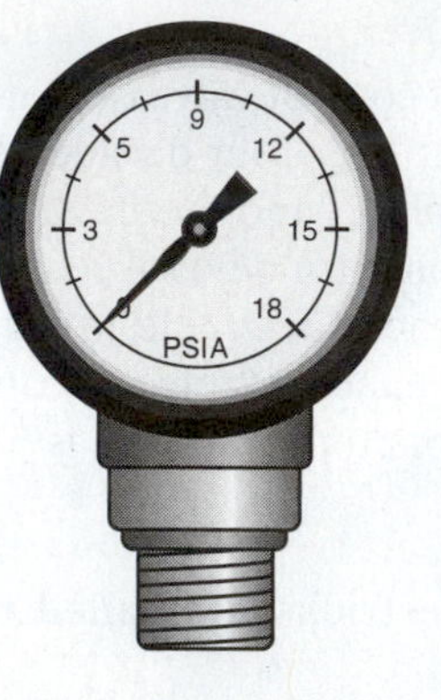

FIGURE 1-25 Analog Pressure Indicator

Digital Signals

Digital is a term applied to a device that uses binary numbers to represent continuous values or discrete (individual or distinct) states. Measurements taken from a process are continuously variable quantities (analog) that must be converted into binary numbers to enter into a computer for processing. A binary number representing an analog quantity can be produced by means of a device called an analog to digital (A/D) converter.

As stated above for analog signals, there are an infinite number of values between any two points. That is, the pressure in a vessel does not jump from 1 psi to 2 psi without going through all values between the two. A digital readout device (Figure 1-26) must jump from one numerical value to the next, whereas an analog pressure gauge with an indicating pen moves smoothly across all the values between 1 psi and 2 psi. When a digital indicator changes from 1 psi to 2 psi, it is only the readout device that is skipping numerical values rather than the actual pressure jumping from one value to another.

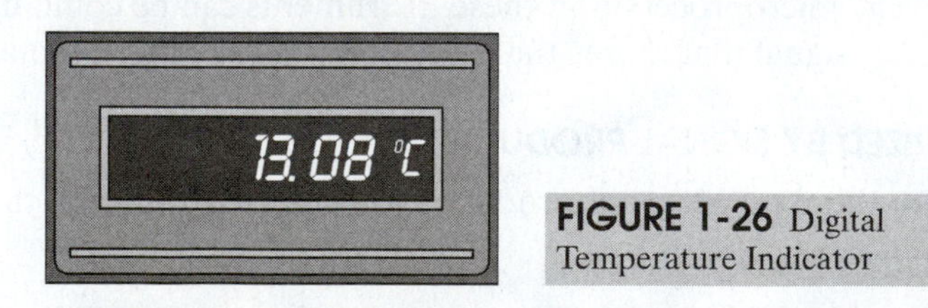

FIGURE 1-26 Digital Temperature Indicator

Transmitters called *smart transmitters* (microprocessor-based instruments) are currently being used in industry. These instruments have an analog to digital (A/D) converter built into them so that they can process information within the onboard microprocessor. There are many advantages to using microprocessor-based instruments including programmability, self-diagnostics, and the ability to recalibrate from a remote location. The output signal from a smart transmitter may or may not be digital.

Microprocessor-based transmitters are available as analog-digital hybrids and as pure digital instruments. A transmitter with a digital output signal would have a typical computer communication output that transmits data in a serial manner just as computers networked together in an office would. The protocol is different from a typical Ethernet (twisted pair) connection, but effectively works the same.

In short, industry uses both analog and digital systems because they both have qualities desirable in specific situations. The true power of a digital system lies in its ability to manage information far beyond the capabilities of an analog system. Digital instrumentation is certainly on the rise and analog systems are on the decline. However, as far as the near future is concerned, analog equipment continues to retain a significant role in the process industry.

Process Variable Relationships

Common process variables affect each other on a continuous basis in industry. Processes are designed to account for these reactions. To help illustrate these relationships, a discussion through the use of questions and answers is posed below.

WHAT HAPPENS TO THE PRESSURE IN A CLOSED CONTAINER WHEN TEMPERATURE INCREASES OR DECREASES?

Answer

A temperature increase causes a pressure increase; a temperature decrease causes a pressure decrease. Temperature and pressure are directly proportional to one another.

Example

The pressure in the radiator of a car rises as its temperature increases. You would never uncap a hot radiator because you know it is pressurized.

Relationship: Between Pressure, Volume, and Temperature in a Closed Container

Consider the measurement conditions inside of a typical tank. Process plants are like large chemistry labs with different types of vessels that are used to store, blend, or even react chemicals in a controlled environment. Individual processes have their own unique characteristics, but physical attributes such as pressure and temperature follow strict laws of physics. Pressure and volume are inversely proportional to one another.

The following equation shows the relationship between pressure (P), volume (V), and temperature (T) as applied to gases. The pressure, volume, and temperature on the left side of the equation represent condition (1), that is, the condition found prior to a change in conditions. The conditions on the right represent the changed condition (2). Pressure and temperature values must be expressed in absolute units.

$$\frac{P_1V_1}{T_1} = \frac{P_2V_2}{T_2}$$

In this equation, pressure and volume are directly proportional to each other and temperature is inversely proportional to both of them.

Relationship: Between Pressure and Temperature

The previous equation explains that there is a specific relationship between the volume, temperature, and pressure of gases. Common sense dictates that if the temperature in a closed container is increased, its pressure rises. A simple and very common example of this phenomenon is the pressure change that occurs in the tires on a car between the time it is at rest and after it has been driven a while.

Here is an industrial example: A vessel, somewhat like a tire, is a rigid structure and its volume therefore remains constant. By keeping the volume constant, it can be effectively removed from the previous equation leaving a more simplified relationship between temperature and pressure (Charles' Law).

$$\frac{P_1}{T_1} = \frac{P_2}{T_2}$$

Consider an actual example of how temperature affects pressure in a fixed volume. If a tank were blocked in (all inlets and outlets are closed) late in the evening with a relatively mild temperature of 68 degrees F (528 degrees R), a pressure gauge attached to it would indicate an ambient (surrounding) pressure (14.7 psia). The next day, as the sun begins to shine directly on the tank, its inside temperature would begin to rise and so would the pressure. If the temperature inside the tank rises to 158 degrees F (618 degrees R), how much pressure would now be in the tank?

NOTE: Pressure (psi, psia, psig, etc.) is covered in Chapter 2. Temperature scales and scale conversions are covered in Chapter 3.

$$P_2 = \frac{P_1 T_2}{T_1}$$

$$P_2 = \frac{(14.7 \text{ psia})(618°R)}{528°R}$$

$$P_2 = 17.2 \text{ psia}$$

or

$$\text{If converted to psig, } (17.2 - 14.7) = 2.5 \text{ psig}$$

Since tanks are designed to withstand pressure from the inside out, this increase in pressure may not create a problem. But, what would happen if the opposite were the case?

WHAT HAPPENS TO THE TEMPERATURE IN A CLOSED CONTAINER WHEN PRESSURE INCREASES OR DECREASES?

Answer

A pressure increase causes a temperature increase; a pressure decrease causes a temperature decrease.

Example

As in the previous example, heat causes gases to become more active, thus taking up more space. So, these expanded molecules are compressed and their total energy is concentrated resulting in a measurable rise in temperature. Notice in the following equation how the pressure (P_2) increases as the temperature (T_2) increases.

$$T_2 = \frac{P_2 T_1}{P_1}$$

$$T_2 = \frac{(17.2 \text{ psia})(528°R)}{14.7 \text{ psia}} = 618°R$$

$$T_2 = 618°R - 460°R = 158°F$$

Again, the inverse of this situation is that the pressure of a gas decreases as the temperature decreases.

Relationship: Between Temperature and Pressure Inside a Closed Container

Using a similar situation to the previous example, what would happen to the pressure in a tank if it were blocked in during the heat of the day (158 degrees F) and then a cold wet storm swept across the area later that evening (68 degrees F)?

$$P_1 = 14.7 \text{ psia}$$

$$T_1 = 158°F + 460°R = 618°R$$

$$T_2 = 68°F + 460°R = 528°R$$

$$P_2 = \frac{P_1 T_2}{T_1}$$

$$P_2 = \frac{(14.7\text{ psia})(528°\text{R})}{618°\text{R}} = 12.6\text{ psia}$$

This is 2.1 psia below atmospheric pressure.

Since vessels are usually built to withstand internal pressure, the vessel walls could collapse under vacuum conditions. Fortunately, most storage vessels are equipped with vacuum breakers that guard against too much negative pressure placed on a tank.

WHAT HAPPENS TO THE PRESSURE AT THE BOTTOM OF A VESSEL WHEN THE HEIGHT OF THE LIQUID INCREASES OR DECREASES?

Answer

A level increase causes the vessel bottom pressure to increase; a level decrease lowers the pressure.

Example

Consider a tank containing 10 feet of water that is vented to the atmosphere so problems of a pressurized tank are taken out of the scenario leaving head pressure as the only type of pressure considered. The pressure at the bottom of the tank is affected by the height of the liquid in the tank and by its density. The following equation expresses this concept:

Where:
Head = height of the fluid (h)
For conversion to psig:

As h increase, p increases

$h =$ liquid height

$p =$ pressure at the bottom

$$d = \text{density} = \frac{62.4\text{ lb}}{\text{Cubic Foot}} = \frac{62.4\text{ lb}}{\text{ft}^3}$$

$$p = \left(\frac{62.4\text{ lb}}{\text{ft}^3}\right)(10\text{ ft})\left(\frac{\text{ft}^2}{144\text{ in.}^2}\right) = \frac{(62.4\text{ lb})(10\text{ ft})}{144\text{ in.}^2 - \text{ft}}$$

$$= \left(\frac{0.433\text{ lb}}{\text{in.}^2 - \text{ft}}\right)(10\text{ ft}) = 4.3\text{ psi}$$

Relationship: Bottom Pressure and Height of Liquid

Pressure exerted by a liquid is called head pressure. Atmospheric head pressure is a result of the height of the liquid times its density. As the liquid height increases, the head pressure (Figure 1-27) also increases. Consider a tank containing a fluid with the equivalent of three-quarters of the density of water or a specific gravity (SG) of 0.75.

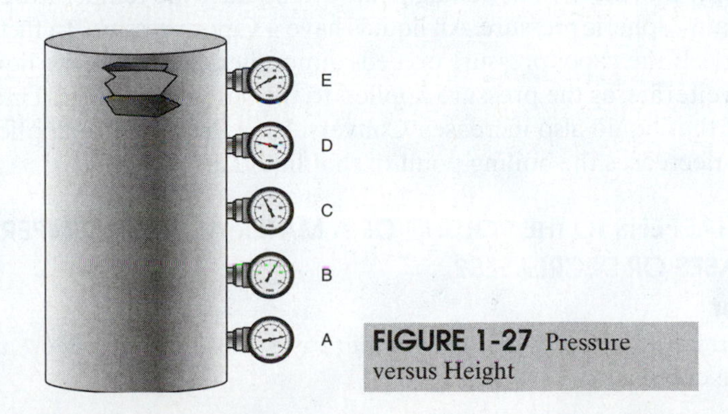

FIGURE 1-27 Pressure versus Height

Compare the following table to Figure1-27 to see the various pressure readings at each of the various levels within the tank.

Pressure Reading Level Points	*Pressure (psig)*	*Pressure (psia)*
E	0	14.7
D	0.3	15.0
C	0.6	15.3
B	0.9	15.6
A	1.2	15.9

The following equation expresses this concept:

$$P_B = (h)(SG)$$

Where:
P_B = Bottom pressure
h = Height of liquid (in feet)
SG = Specific gravity of liquid
D_w = Density of water (0.433 lb/in^2)
As h increases, P increases.

WHAT HAPPENS TO THE BOILING POINT OF A MATERIAL WHEN PRESSURE INCREASES OR DECREASES?

Answer

The boiling point increases when the material pressure increases; the boiling point decreases when pressure is lowered.

Examples

- The boiling point of a liquid is directly proportional to applied pressure. There is less atmospheric pressure at the top of the mountain than at sea level; atmospheric is the highest at the surface of the ocean (sea level). This is why water boils at a lower temperature on a mountain as compared to sea level. Pressure, such as atmospheric pressure, on the surface of a liquid holds the molecules in the liquid state. The greater the pressure, the less likely that the liquid molecules can vaporize and break the surface (boil).
- In a pressure cooker the pressure is allowed to increase: As the pressure above a liquid increases, the boiling point of the liquid also increases. This allows the food being cooked to heat up quicker and cook more rapidly and in a shorter time period.

Relationship: Boiling Point and Pressure

Pressure applied to the surface of a liquid directly affects its boiling point. To understand how the boiling point of a liquid is affected by pressure, an understanding of boiling point is necessary. An easy way of identifying when a liquid is boiling is by the bubbles being produced and through the release of vapor. This is a good observation, but what is really happening?

Simply put, the *normal* boiling point of a liquid is the temperature at which the liquid boils at atmospheric pressure. All liquids have a vapor pressure. In fact, that is how liquids evaporate. If the vapor pressure exceeds atmospheric pressure, the liquid evaporates.

To reiterate, as the pressure applied to the surface of a liquid increases, the boiling point of that liquid also increases. Conversely, as the pressure applied to the surface of a liquid decreases the boiling point of that liquid also decreases.

WHAT HAPPENS TO THE VOLUME OF A MATERIAL WHEN TEMPERATURE INCREASES OR DECREASES?

Answer

The volume increases when the temperature increases and decreases when the temperature decreases.

Example

The coolant system in a car is, once again, a good example of this concept. Look at the coolant level in the overflow/fill reservoir of a car when it is cool and then look at it again when it is hot. The reservoir should show a higher level when it is hotter. This is why there are level markings (look for a COOL line and a HOT line) to use when filling the reservoir. The extra space in the reservoir is needed for the expanding material. If the expansion allowance space were not calculated into the total filling volume, the expanding volume would drain out of the reservoir.

Relationship: Volume and Temperature

The volume of a substance is directly affected by temperature. Almost all materials expand when heated and contract when cooled. With any given temperature change, the change in volume experienced by gases is much greater than that experienced by liquids and solids.

The volume changes of liquids and solids are based more upon the characteristics of each individual material than to a general volume vs. temperature relationship that applies to gases. In the following formula, the volume (V) of a specific material responds directly (ΔV) to a temperature change (ΔT) by the factor (β), which is its coefficient of volume expansion.

$$\text{Liquids:}\Delta V = (\beta)(\Delta T)$$

Where:
ΔV = change in volume
β = coefficient of volume expansion
ΔT = change in temperature

To illustrate this concept, consider again the coolant level in the overflow/fill reservoir of a car. When the car is cool, there is less coolant volume in it as compared to it when it is at operating temperature (hot). In Figure 1-28, notice how the volume of coolant increased with a temperature increase.

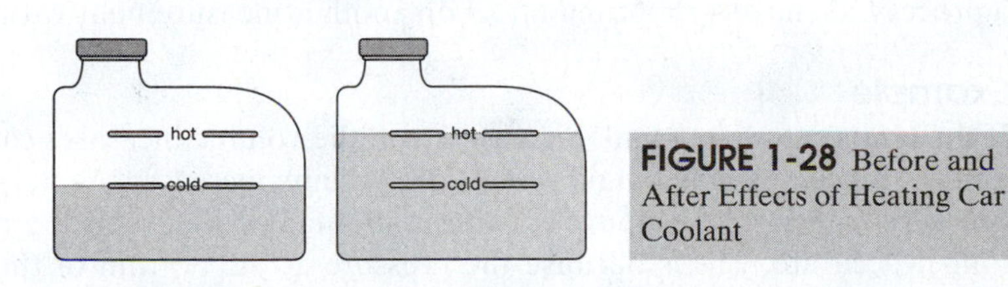

FIGURE 1-28 Before and After Effects of Heating Car Coolant

WHAT HAPPENS TO THE DENSITY OF A MATERIAL WHEN TEMPERATURE INCREASES OR DECREASES?

Answer

Density (mass or unit volume) decreases when the temperature increases; the density increases when the temperature decreases.

Example

In the radiator example above, the cooling system contains a certain amount of liquid (e.g., about 22 pounds of coolant). As the temperature of this system increases, its volume increases. If the volume expands and the weight (mass) remains the same, then the density of the fluid must have changed.

Where:

$$\text{Density} = \frac{\text{mass}}{\text{volume}}$$

Lower temperature:

$$\text{Density} = \frac{22\text{ lb}}{2.5\text{ gallons}}$$

$$\text{Density} = \frac{8.8 \text{ lb}}{\text{gallon}}$$

Higher temperature:

$$\text{Density} = \frac{22 \text{ lb}}{2.625 \text{ gallons}}$$

$$\text{Density} = \frac{8.38 \text{ lb}}{\text{gallons}}$$

If density is defined as mass per unit volume, then it stands to reason that if a material is expanding, its density (mass/volume) must be decreasing.

WHAT HAPPENS TO THE FLOW OF A MATERIAL WHEN THE DIFFERENTIAL PRESSURE ACROSS AN ORIFICE PLATE INCREASES OR DECREASES?

Answer

A flow rate change causes a change in differential pressure; reduced flow results in decreased differential pressure.

Relationship: Flow and Differential Pressure

In the 1700s a scientist by the name of Daniel Bernoulli described many principles of fluid flow. One of them was how the total energy of a material flowing through a pipe would react to a change in pipe diameter. His principle, known as the Bernoulli principle, states that as the speed of a moving fluid increases, the pressure within the fluid decreases. Through experimentation, scientists have found that the flow rate of a fluid is proportional to the square root of the differential pressure drop across a restriction such as an orifice plate. This concept is widely used in industry today.

HOW DO VARIABLE CHANGES AFFECT ACCURATE MEASUREMENT?

Answer

Because of the interrelationship of one variable to another, a change in one variable in a process can have a profound affect on another measurement variable.

Example

If the temperature in a tank increases, then the volume increases causing a decrease in density. Since a tank has rigid sides, extra volume increases the height of the level. If a pressure transmitter is being used to read a tank level, then the measurement is no longer accurate. This is because the pressure at the bottom of the tank remains the same, but the actual height of the liquid (level) has increased (Figure 1-29).

$$P = \left(\frac{0.433}{(\text{in.}^2)(\text{ft})}\right)(\text{Height})(\text{SG})$$

NOTE: Not all variables in a process are measured, but they may still have to be kept constant so that an accurate measurement of the controlled variable (the variable that is meant to be controlled) may be obtained.

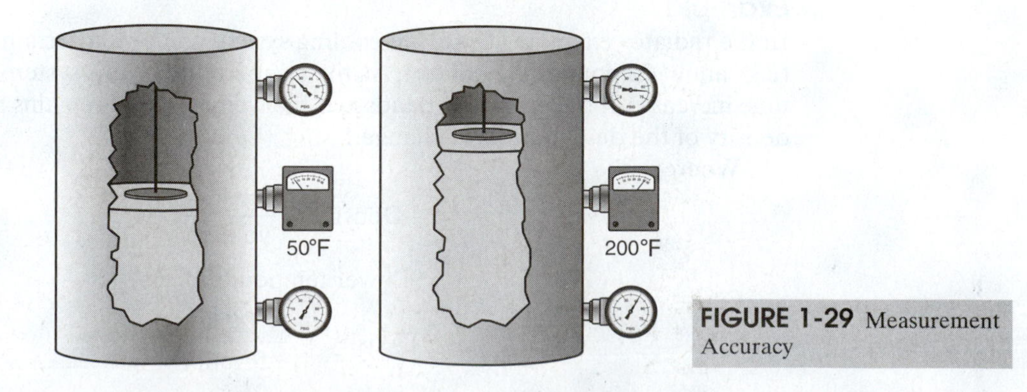

FIGURE 1-29 Measurement Accuracy

Relationship: Variable Changes and Accurate Measurement

Design engineers should ensure that all reasonable conditions are accounted for before the plant ever starts up. Even with all the thought and detail put into the design, chemical consistencies can change. In a plant, process technicians are part of the first line of defense against process problems and they need to be cognizant of the potential ramifications of these changes.

Summary

Instrumentation used to control processes date back to 1775 when a flyball governor was used to control the speed of a steam engine. Manual control has been around a longer time than automatic control that did not appear until the late 1920s. Modern control theory began with Harry Nyquist in the early 1930s and was slow in implementation until WWII. Zeigler and Nichols introduced PID (Proportional, Integral, Derivative) control in the early 1940s. Analog electronic devices were developed as early as the 1950s and are still common in many plants today. Computer control began to appear in the late 1950s and usage multiplied with the invention of the microprocessor (chip) in the early 1970s. From the development of microprocessors came Distributed Control Systems (DCS) used widely today.

Instrumentation is used to measure and control process variables such as pressure, temperature, level, flow, and various analytical properties.

Instruments have been categorized by various methods to include the following:

- *Location* [e.g., local (or field) or remote]: Local instruments are located close to the process equipment and remote are located away from the process.
- *Function* (e.g., sensing, indicating, transmitting, comparing, and/or controlling): These functions are usually related to a particular control loop that has all the parts necessary to control a process. These parts include a measurement sensor, a transmitter, a controller, and a final control element.
- *Power source* (e.g., pneumatic or electronic): These power sources depend on how all the instruments within a control loop function together or how each individual instrument functions as an entity. For an entire loop to be categorized as pneumatic, each instrument must use air (or other gases) as the power source. Instruments may also be electronic (analog, digital, and analog-digital hybrid) if their power source is electricity.
- *Type of signal produced* (e.g., analog or digital): This is a stronger designation than when used in conjunction with a power source. Analog signals are continuous as compared to digital signals that jump from one value to the next. Analog signals can be converted to digital by an analog to digital converter (A/D) converter or smart transmitters.

The many relationships between process variables include the following:

- Between pressure and volume when temperature is increased: In a closed vessel, when temperature increases, pressure increases. In an open vessel, when temperature increases, fluid volume increases. When temperature decreases in a closed vessel, the pressure decreases.
- Temperature in a closed container when pressure increases or decreases: A pressure increase causes a temperature increase and a pressure decrease causes a temperature decrease. Opening vents on closed containers prevents over-pressurization or negative pressure (vacuum) from occurring on temperature changes.
- Pressure at the bottom of a vessel when the height of the liquid increases or decreases: A level increase causes the vessel bottom pressure to increase whereas a level decrease lowers the pressure. The pressure at the bottom of the tank is affected by the height of the liquid in the tank and by its density. Pressure exerted by a liquid is called head pressure.
- Between the boiling point of a material when pressure increases or decreases: The boiling point increases when the material pressure increases and the boiling point decreases when pressure is lowered. Pressure applied to the surface of a liquid directly affects its boiling point.

- Volume of a material when temperature increases or decreases: The volume increases when the temperature increases and decreases when temperature decreases Most all materials expand when heated and contract when cooled, but the change in volume of gases is much greater than that experienced by liquids and solids.
- Density of a material when temperature increases or decreases: Density (mass or unit volume) decreases when the temperature increases and density increases when temperature decreases. If a material is expanding, then its density must be decreasing.
- Bernoulli's principle states that *as the speed of a moving fluid increases, the pressure within the fluid decreases*. Flow rate, then, is proportional to the square root of the differential pressure drop across an orifice plate.
- Variable changes and accurate measuring: A change in one variable in a process can have a profound affect on another measurement variable.

Checking Your Knowledge

1. Place the following historical events in chronological order from the earliest (1st) to the most current (5th).
   a. Analog electronic instrumentation becomes commonplace in the industry.
   b. Zeigler and Nichols publish a paper describing a break through method for tuning PID control loops.
   c. Fiber optic technology is integrated into process instrumentation.
   d. Digital process control emerges into the form of Distributed Control Systems.
   e. Nyquist proposes the theoretical basis for modern process control theory.
2. Match the following terms to their appropriate definition:

Term	*Definition*
Flow	a. Force applied to a unit area
Level	b. Quantity of fluid moving within a given period of time
Differential	c. Height of the surface of a material compared to a zero reference point
Temperature	d. Specific degree of hotness or coldness as indicated on a reference scale
Pressure	e. Difference between measurements taken from two separate points

3. Match the following terms to their appropriate definition or description:

Term	*Definition*
Local	a. An instrument located away from the process
Remote	b. An instrument located near the process
Pneumatic	c. An instrument powered by electricity
Electronic	d. An instrument powered by air or other gases

4. Fill in the blanks on the following illustration with the function of each instrument.

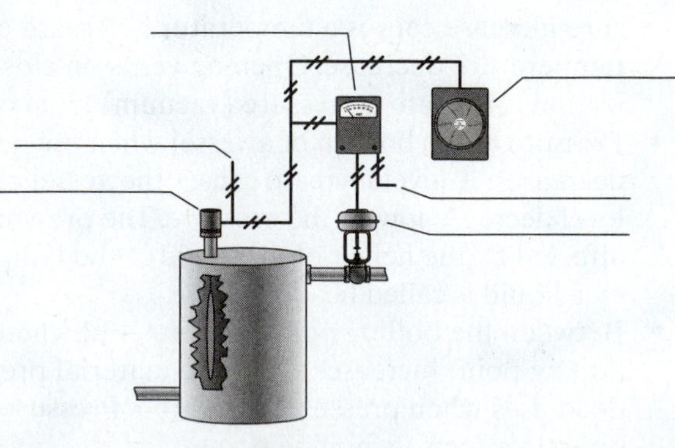

5. When pressure applied to a liquid increases, the boiling point _____.
   a. Increases
   b. Decreases
   c. Stays the same
   d. Fluctuates
6. *True or False* The Bernoulli principle states that as the speed of a moving fluid increases, the pressure within the fluid decreases.
7. *True or False* When the temperature of a liquid increases density decreases.
8. *True or False* When flow increases, pressure increases.

Student Activities

1. Given your earlier process training or experience with instrumentation think of at least five scenarios whereby you can create a discussion about the affects of various process variables in plant situations. Record your comments and share with other classmates to determine if you have discovered all the possibilities for each scenario.
2. Tour local plant facilities with an experienced process technician and have the technician point out various types of instrumentation within their working environment. If this type of tour is not available, then your instructor should perform for you a similar demonstration of available instrumentation within the classroom or lab environment.

CHAPTER 2

Process Variables, Elements, and Instruments: PRESSURE

Objectives

After completing this chapter, you will be able to:

- Define pressure.
- Discuss the formula used to calculate pressure: P = F/A.
- Identify the three components that affect the force exerted by molecules:
  - speed
  - mass (and weight)
  - number of molecules (density)
- Define terms associated with pressure and pressure instruments:
  - differential pressure (delta Δ)
  - pressure measurement scales
    - psig
    - psia
    - atmospheres
    - bars
    - inches Hg
    - inches H_2O
- Identify common types of pressure-sensing or pressure-measuring instruments used in the process industries:
  - manometers
  - pressure gauges (internals)
  - differential pressure (d/p) cells
  - strain gauge transducers
  - capacitance transducers
- Describe the purpose and operation of pressure-sensing or pressure-measuring instruments used in a lab or industrial setting.
- Given a standard calculator and conversion formulas, convert between the following pressure scales:
  - pounds per square inch gauge (psig) and pounds per square inch absolute (psia)
  - inches of mercury (in. Hg) and inches of water (in. H_2O)

Key Terms

Atmosphere(s)—the pressure at any point in the atmosphere due solely to the weight of the atmospheric gases above the point concerned; 14.7 psia is at sea level.

Atmospheric pressure—weight of the air comprising the atmosphere; 14.7 lb. per square inch is the basic reference point for pressure gauges.

Bars—measurement of pressure equal to 0.987 atmospheres.

Capacitance transducer—a device that contains a measurement diaphragm and capacitor plates; changes pressure measurement to an electronic signal.

Density—mass per unit volume.

Differential pressure (d/p cell)—an instrument that measures the difference between two pressure points and produces a corresponding output signal for tank level.

Force—a push or pull exerted on an object that causes the object to change direction.

Inches of mercury (H_g)—most common measurement scale for a manometer; measures the level of mercury in pounds per cubic inch.

Inches of water (H_2O)—a very small measurement of pressure equal to 0.036 pounds per square inch per inch at 4 degrees Celsius.

Local indicator—instruments that are placed on equipment in the field only to be read in the field; may be used for comparison with transmitted instrumentation readings or, used on non-critical processes to indicate pressure values.

Manometer—a gravity-balanced pressure-measuring device with two fluid chamber tube gauges connected by a U-shaped tube so fluid (a liquid or mercury) flows freely between the chambers.

Mass—the amount of matter in a body or object; mass has to do with the amount of matter in a molecule rather than its size.

Molecular speed—molecular speed changes as a result of numerous factors. Energy, normally in the form of heat energy, may change, which will increase or decrease molecular speed. Speed can also be affected by collisions with other molecules. The collisions may be with molecules from the material (for example, a product being agitated) or from other molecules (for example, the walls of a vessel).

Pressure gauge—the most common instrument for measuring pressure; the three primary types of pressure gauge measuring scales are absolute, gauge, and vacuum.

psia—pounds per square inch absolute; the absolute scale at zero would be no pressure at all.

psig—pounds per square inch gauge; pressure measurement that references 14.7 psia as its zero point.

Specific gravity—the density of a substance relative to the density of water that is defined as 1.0 at 39 degrees Fahrenheit (4 degrees Celsius).

Strain gauge transducer—a pressure-measuring device consisting of a group of wires that stretch when pressure is applied, creating resistance, thereby changing the process pressure into an electronic signal.

Vacuum—where the pressure measured is less than atmospheric pressure; usually measured in inches of mercury (Hg).

Introduction

This chapter reviews the concept and definition of pressure and, in general terms, how to calculate the amount of pressure for various applications. Also discussed are the various types of measurement scales used to determine the amount of pressure in a given situation and how to convert the pressure reading from one scale to another. Different types of pressure measuring devices are identified as to their purpose and operation including their respective locations where they may be used in a processing environment.

Pressure Defined

One of the most common process variables encountered is pressure. Pressure (P) is defined as the amount of force (F) per unit of area (A). If force is expressed in pounds and area in square inches, then pressure would be expressed in pounds per square inch (psi). Pressure can be defined mathematically in the following equation:

$$P = F/A$$

Where:

P = pressure (psi)
F = force (pounds)
A = area (square inches)

To illustrate how both force and area affect pressure, think of a solid object resting on a table (Figure 2-1): If the object weighs one pound and is resting on one square inch of surface, then it is applying 1 psi of pressure. If the same weight is instead resting on one-half square inch of area then it is now applying 2 psi of pressure.

Pressure directly affects the boiling point of a substance. The greater the amount of pressure on a substance, the greater the amount of heat required to bring the substance to its boiling point. Because of the effects of this principle on the properties of a substance, process industries must account for pressure in the mixing, creation, and/or separation of chemicals. A process technician is responsible for both monitoring and controlling pressure in pipes and equipment. Failure to adequately control pressure can have disastrous consequences.

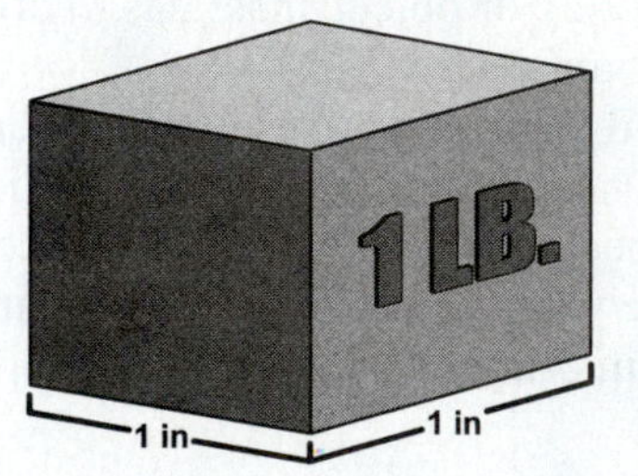

FIGURE 2-1 Pressure

Calculating Pressure

As stated earlier, pressure can be defined as a force applied to a unit of area. If the area is smaller, then the pressure is greater.

Force is a push or pull exerted on an object (solid) that causes the object to change direction. Liquids and gases, both considered to be fluids in processing, behave differently when force is exerted on them in a static (at rest), or stationary, environment. The total height of a liquid column in a given vessel (e.g., storage tank), in addition to the surface pressure (atmospheric pressure), and density (compactness) of the material, determines the amount of force at the bottom of the liquid column for each unit of area.

In the example in Figure 2-2, *h* represents the height of the liquid. As the column of liquid rises in the vessel, the amount of pressure on the bottom also increases. The

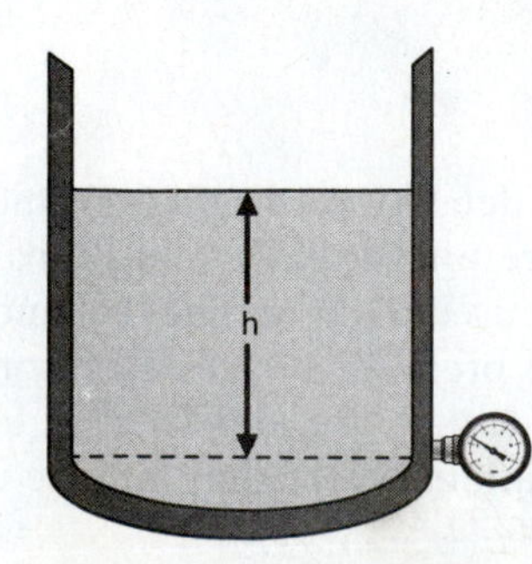

FIGURE 2-2 Force

height of a liquid column (e.g., product in a storage tank) and the amount of force (head pressure) it exerts at the bottom is often used to move (force) the liquid to another destination. The pressure is not released until a valve is opened allowing the liquid to flow out from the vessel. The vessel itself is designed to resist the internal forces caused by the liquid. All vessels and equipment are designed to handle a certain amount of total pressure.

Other factors affecting the force exerted by molecules on each other include the following:

- speed (temperature)
- mass (atomic weight)
- density (number of molecules per volume)

SPEED

Gases and vapors, in contrast to liquids and solids, apply pressure equally to all surfaces in a container. The pressure produced by this state of matter is primarily caused by the extraordinary energy of the individual molecules and, to a lesser degree, by gravity. Gases and vapors are different from liquids and solids because each molecule exists apart from the others. They are highly energized and react accordingly. Imagine billiard balls bouncing off one another as well as any surface they may encounter. The more energized they become, the faster they move, producing more pressure.

The temperature of a substance is usually the greatest factor affecting **molecular speed**. The greater the amounts of heat applied, the more the molecules become excited and move around (Figure 2-3). The more they move, the greater the distance the molecules try to put between themselves. The greater the amount of heat applied to a liquid, the more the molecules increase in energy and the greater their ability to overcome surface tension and become a vapor. If confined, the increase in speed also increases the amount of pressure exerted on the container (vessel) walls. The vessel itself confines the movement of the molecules and promotes pressure increase as the molecules increase in speed.

MASS

Mass is the amount of matter (atomic or molecular weight) in a body or object. Mass has to do with the amount of matter in a molecule rather than its size. The greater the collective weight of the individual atoms that make up a molecule, the more the molecule weighs. For example: An atom of iron (Fe) weighs more than either hydrogen (H) or oxygen (O). Water is the standard used to measure the mass of other liquids and solids.

The example of a solid object resting on a table (Figure 2-1) is only one way of expressing pressure. A liquid applies pressure that varies with its depth to that portion (walls) of the vessel that contains it. The force that produces the pressure created by solids and liquids is primarily derived from its mass being drawn towards a center of gravity. The more mass that is applied to a unit of area, the greater the pressure. Additional mass could be in the form of more molecules or just heavier ones.

DENSITY

Another factor affecting the pressure of gases and vapors is its **density**. Density is the total number of molecules per unit volume at a given temperature. With gases and

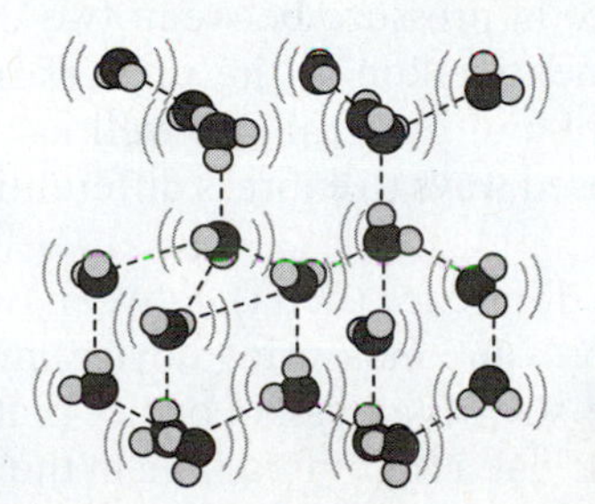

FIGURE 2-3 Molecular Motion

vapors, the number of molecules per unit of volume defines density. The greater the number of molecules, the greater the number of collisions that could occur between the molecules. What this implies is that there are a greater number of molecules colliding within the container walls of a vessel creating a higher pressure. These collisions create more force within the material.

Using a container of water (more molecules) versus the same type of container of paraffin (less molecules), and applying heat to both, the result would show that the container of water would begin to vaporize more readily than the paraffin.

The density of fluid gases is dependent on the allowable space that the molecules occupy. The greater the space, the less dense the gas would become since the molecules would spread out until they filled the entire space. If the container volume size were to be reduced for the same amount of gas molecules, then internal pressure would increase on the walls of the container. This is why substances that are gases at ambient temperatures are confined in specially designed pressure vessels.

SPECIFIC GRAVITY

Specific gravity is the density of a substance relative to the density of water that is defined as 1.0 at 39 degrees F (4 degrees C). Other liquids, or solids, have a specific gravity that is either greater or less than water. Specific gravity of a gas is the weight of a volume of gas divided by an equal weight of air at standard temperature and pressure. Specific gravity of a liquid is the weight of a volume of liquid divided by an equal volume of water.

Pressure Measurement

Pressure measurement is more often taken at unique points in a process operation. An example of this is a single local pressure gauge located on a pump suction or discharge line. When this measurement is vital to controlling the process, the measurement taken is converted to a pneumatic or electronic signal and transmitted to another location. This measurement signal may be used as a single numeric value in a control room monitoring process, or it may be used to signal a process change in a control loop arrangement.

Measuring and controlling pressure in the process industry allows for a safer and more productive plant. Controlling the pressure inside of a reactor helps optimize the chemical reaction. This results in higher yields and increased profits. From a safety standpoint, controlling pressure is a matter of common sense. Pipes or vessels exposed to pressures that exceed their pressure ratings can rupture, causing death, bodily injury, equipment destruction, and environmental consequences.

One problem encountered by new technicians is the misunderstanding of pressure units and their magnitudes. The consequences can be catastrophic. The difference between thinking psia and reacting with psig could rupture a tank. The difference between psi and inches water column can cause damage and pose a danger.

Knowing how pressure is measured and the appropriate scale to use for each application is critical. The following definitions and descriptions of pressure measurements are given to help process technicians make more informed decisions during daily operations.

DIFFERENTIAL PRESSURE (DELTA P, ΔP, D/P)

Differential pressure is the difference in pressure between two distinct points. This is one of the most common measurements taken in the process industry. Differential pressure is used to infer liquid level and flow rate as well as pressure differences between two points. Two commonly used ways to express differential is the Greek letter delta (Δ) and the abbreviation d/p.

Sometimes knowing the pressure difference (delta) between two points in a process is important. A d/p cell (differential pressure measuring device) may be located so that it measures the difference in pressure, such as across an orifice plate (Figure 2-4) that is installed in a pipeline. The orifice plate has a calibrated hole in the plate. The plate itself

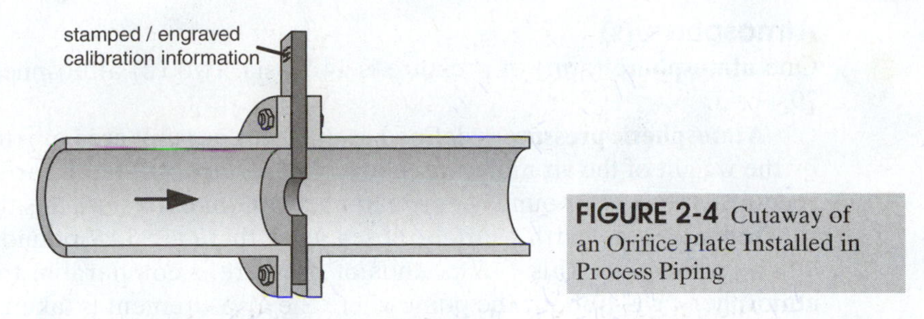

FIGURE 2-4 Cutaway of an Orifice Plate Installed in Process Piping

restricts the flow and pressure is lower on the downstream side of the plate. When the fluid passes through the orifice, the pressure energy is converted to velocity. Velocity is the measurement for how fast a particular amount (as determined by the pump capacity and pipeline size) of fluid can be pumped to a specified point downstream. When the pressure decreases, the velocity increases.

A simple example of this principle is when someone places a thumb over a water hose while watering the garden. The more restriction on the flow of water coming out of the hose, the more velocity is added to the stream causing it to shoot out further from where the person is standing.

PRESSURE MEASUREMENT SCALES

Whether from a single local pressure indicator (e.g., pressure gauge), or a more sophisticated d/p cell, pressure is still measured against a particular scale. Typical pressure measurement scales include the following:

- pounds per square inch (psi)
  - pounds per square inch absolute (psia)
  - pounds per square inch gauge (psig)
- atmospheres or bars
- inches of water (H_2O) or mercury (Hg)

PSI (Pounds per Square Inch)

Pounds per square inch is the basic formula for pressure: force (expressed in pounds) divided by the affected area (expressed in square inches or length times width). This psi scale originates at true zero pressure, that is, a total vacuum.

PSIA (Pounds per Square Inch Absolute)

Pounds per square inch absolute (**psia**) is the basic formula for pressure caused by the specific gravity of air (a combination of various gases). At sea level, 14.7 psi is the standardized (averaged) pressure measurement. This is the equivalent to one atmosphere (1 atm). Zero on an absolute scale would be the same as absolutely no pressure at all. A total vacuum is considered to be 14.7 psi below atmospheric pressure (0 psia). On an absolute scale, a total vacuum would be −14.7 psig. Vacuum is often measured in inches mercury (in. Hg). The higher the inches of mercury, the lower the absolute pressure.

If a person travels higher in the mountains, the atmospheric pressure is much less than at sea level and much greater if the same person were to travel to a place that was below sea level. This principle of atmospheric pressure (+ or −) must be taken into account if the processing facilities are significantly above or below sea level. For example, liquids boil at much lower temperatures when they are not affected by as much atmospheric pressure.

PSIG (Pounds per Square Inch Gauge)

This psi scale originates at atmospheric pressure. That means that regardless of where you are, the ambient (surrounding) pressure establishes zero on this scale.

The pounds per square inch gauge (**psig**) starts the scale at zero (0) with the 14.7 atmospheric pressure already taken into account in the scale. This is the most common pressure measuring scale used in modern processing facilities.

Atmosphere(s)

One **atmosphere** (atm) of pressure is 14.7 psia. Two (2) atmospheres of pressure are 29.4 psia.

Atmospheric pressure is defined as the force per unit area exerted against a surface by the weight of the air molecules above that surface. As stated earlier, one atmosphere is equivalent to 14.7 pounds per square inch absolute. To state this differently, for every square inch of the earth's surface at sea level, there are 14.7 pounds of force weighing down on the area. This 14.7 pounds of pressure is comparable to the weight of the atmospheric gases above the point where the measurement is taken.

Atmospheric pressure is measured with an instrument called a barometer. Atmospheric pressure is often referred to as barometric pressure. In process operations, atmospheric pressure may be factored into the calibration of a pressure measuring device.

Bar(s)

A **bar** is a metric unit primarily used to measure atmospheric pressure.

1 bar = 14.5 psia (See Tables 2-1 to 2-3)

Measurement of pressure is equal to 0.987 atmospheres. The term *bar* is the equivalent to 1,000 millibars. A millibar is the unit of measurement used on barometers to determine atmospheric pressure. The scale is based on inches of Hg. Most weather reporters use the phrase *atmospheric pressure* or *barometric pressure* rather than saying millibars or bars.

Inches H_2O

Inches of water (H_2O) is apressure unit equal to the pressure exerted by the height of 1 inch of water.

27.7 in. H_2O = 1 psi

TABLE 2-1 Gauge Pressure Measurement Comparisons

	Gauge (psig)	*Absolute (psia)*	*Vacuum (Inches Hg)*
Above Atmospheric Pressure	15	29.7	
Atmospheric Pressure	0	14.7	0
Below Atmospheric Pressure	−14.7	0	30

TABLE 2-2 Pressure Measurement Scale Comparisons

	psi/psia	*psig*	*Inches Hg*	*Feet H_2O*
1 atm	14.7	0	29.92 In. Hg at 0 °C	33.96' H_2O at 32 °F
1 bar	14.504	0	29.53 In. Hg at 20 °C	33.52' H_2O at 32 °F

TABLE 2-3 Pressure Unit Conversion Chart

	psi	*bar*	*mbar*	*In. Hg*	*In. H_2O*	*mmHg*	*mmH_2O*
psi	**1**	14.504	0.014504	0.49118	0.036127	0.019337	0.0014223
bar	0.068946	**1**	0.001	0.033865	0.0024908	0.0013332	9.8068×10^{-5}
mbar	68.946	1000	**1**	33.865	2.4908	1.3332	0.098068
In. Hg	2.0359	29.529	0.029529	**1**	0.073552	0.039368	0.0028959
In. H_2O	27.68	401.47	0.40147	13.596	**1**	0.53525	0.039372
mmHg	51.714	750.06	0.75006	25.401	1.8683	**1**	0.073558
mmH_2O	703.05	0.10197	10.197	345.32	25.339	13.595	**1**
atm	0.068045	0.98692	0.00098692	0.033422	0.0024583	0.0013158	9.6788×10^{-5}

Inches Hg

Inches of mercury (Hg) is a pressure unit equal to the pressure exerted by the height of 1 inch of mercury.

$$2.04 \text{ in. Hg} = 1 \text{ psi}$$

Unit Equivalency

$$1 \text{ atm} = 1.013 \text{ bars} = 14.7 \text{ psia} = 407 \text{ in. } H_2O = 29.92 \text{ in. Hg.}$$

In common usage, 1 bar is equated to 1 atmosphere.

Pressure-Sensing and Measurement Instruments

Many devices are used to indicate the pressure of the process, one of the most common being the **pressure gauge**. In process industries, three types of pressure are measured: absolute, gauge, and vacuum. All pressure is actually the weight exerted on a certain area. For example (Figure 2-5), if we were at sea level and examined a one-square-inch area at the surface, we would find that the pressure or force would be 14.7 lb. This means that a column of air one square inch and extending from the surface to the space beyond our atmosphere would weigh 14.7 lb. This is referred to as atmospheric pressure (weight of the air comprising the atmosphere) and is considered to be 14.7 lb. per square inch and is used as the basic reference point for pressure gauges. In other words, if the gauge reads zero, there is actually 14.7 psi on the gauge element. If the gauge reading increased to 50 psi, the pressure would now be 50 psi above 14.7 psi, which is referred to as gauge pressure.

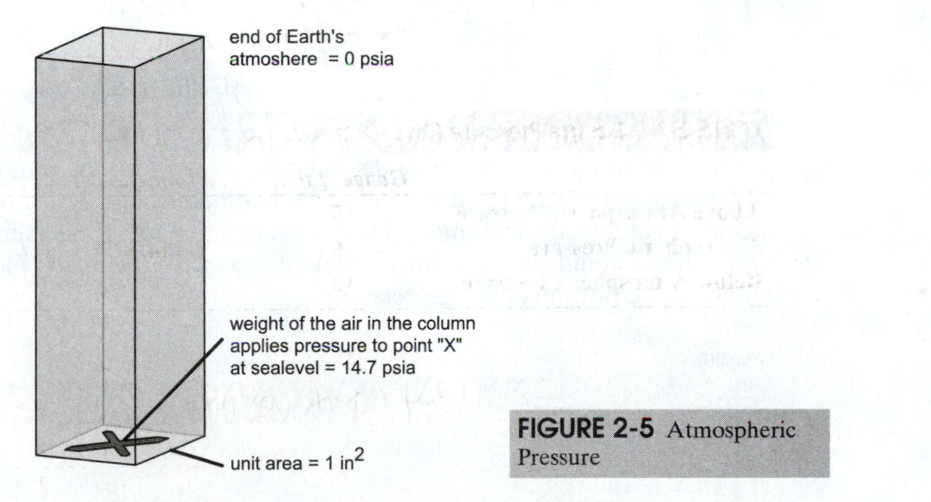

FIGURE 2-5 Atmospheric Pressure

If it were possible to remove all the pressure from a line or vessel so that it is actually 14.7 psi below atmospheric pressure or zero pressure, we would say our pressure is now *absolute zero* and any pressure measured above this absolute zero point is called absolute pressure. For example, if the gauge above were replaced with an absolute pressure gauge, the new gauge would read 64.7 psia instead of the 50 psi.

In industry, we are sometimes interested in maintaining pressures less than atmospheric; this is referred to as **vacuum**. Vacuum is measured from the 14.7 psi atmospheric pressure point in units of inches or millimeters of mercury vacuum.

Common types of pressure-sensing/measuring instruments used in the process industry include the following:

- manometers
- pressure gauges
- differential pressure (d/p) cells
- strain gauge transducers
- capacitance transducers

MANOMETERS

Visual-type liquid **manometers** have been used to measure pressure for hundreds of years. They are among the most reliable measuring devices in industry, provided they remain clean and dry. However, from a practical standpoint, manometers are indicating instruments only.

The following are our common manometer types (Figure 2-6):

- U-tube manometer
- well (reservoir) manometer
- inclined manometer
- barometer

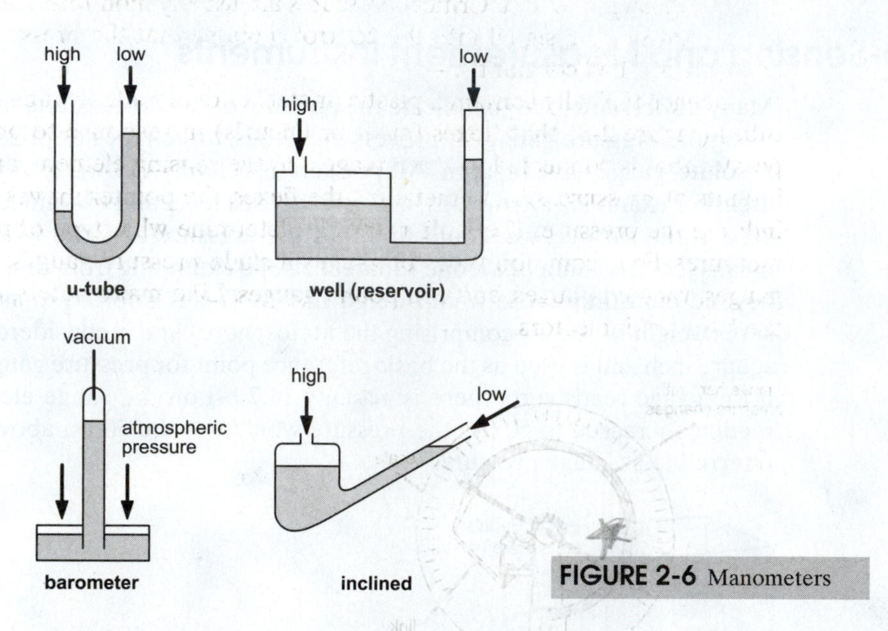

FIGURE 2-6 Manometers

The accuracy of manometers is limited by how well the reading is taken by the technician. For example, when reading a manometer filled with a liquid that produces a convex column, the reading should be taken on the top of the meniscus (the rounded dome of the liquid column). The reading of a concave meniscus (Figure 2-7) should be taken on the bottom of the depression.

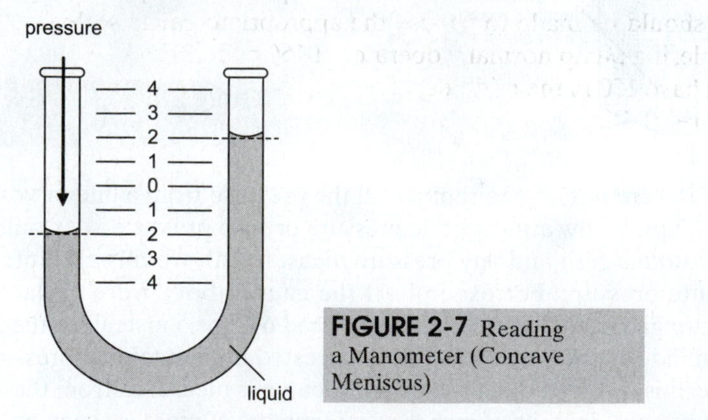

FIGURE 2-7 Reading a Manometer (Concave Meniscus)

The most common type of manometers are the U-shaped tubes that are filled with liquid or mercury and two fluid chambers that are connected by the U-shaped tube so that the fluid is free to move between them. When additional pressure is applied to one chamber, the fluid in that chamber flows through the connecting tube into the other chamber causing its level to rise until the head pressure created by the offset equals the pressure applied to the other chamber. Manometers are called gravity-balanced pressure devices because of this characteristic.

U-shaped tube gauges are filled with a colored liquid and two fluid chambers that are connected by a U-shaped tube so that fluid is free to move between them. This type of manometer is a visual type and is a reliable pressure-measuring device primarily used as an indicating instrument. On a U-tube manometer, read the meniscus on both sides of the tube and add the two readings together.

PRESSURE GAUGES

Pressure gauges, like manometers, are used primarily as **local indicators**. They can be found throughout the plant attached to process piping and vessels. Pressure gauges provide technicians with a convenient quick check of process conditions while they are walking through the unit. Critical pressures are usually monitored by means of a transmitter that sends a signal into the control room so that the pressure can be watched and/or controlled constantly.

Gauges typically contain a plastic or steel body, a metallic tube (called a Bourdon tube) (Figure 2-8) that flexes (curls or uncurls) in response to pressure changes, a pointer that is connected to the linkage and the sensing element, and a scale marked in units of pressure. As the metallic tube flexes, the pointer moves along the scale to indicate the pressure. The scale markings determine what type of pressure the gauge measures. Four common types of gauges include pressure gauges, absolute pressure gauges, vacuum gauges, and compound gauges. Like manometers, they are primarily used as local indicators.

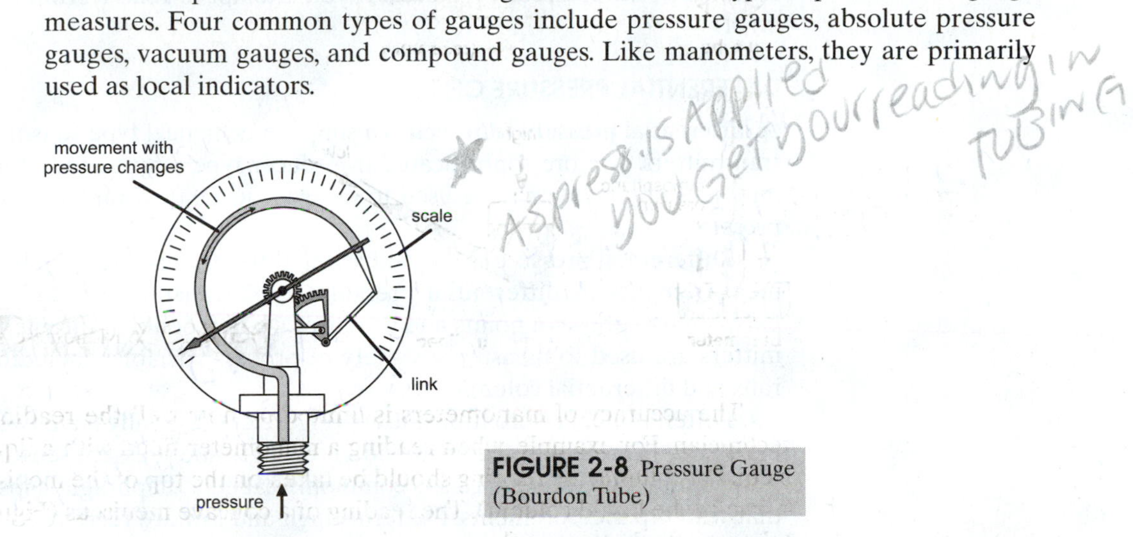

FIGURE 2-8 Pressure Gauge (Bourdon Tube)

Process technicians are allowed to replace some pressure gauges in kind. Every effort should be made to choose the appropriate gauge scale for the application. For example, if a pump normally operates at 150 psig, then the appropriate pressure gauge would have 150 in the middle of the scale. If a gauge were chosen that had a maximum pressure of 150, then the pump pressure would probably blow (overpressure) the gauge and render it inoperable.

For some applications, using a compound vacuum/pressure gauge (Figure 2-9) may be necessary when there is a need to measure pressure below one atmosphere, at atmospheric pressure, or above atmospheric pressure. The zero on a compound gauge is equivalent to 14.7 psia. From zero moving clockwise, the gauge is read in psig. From zero moving counterclockwise, the gauge is read in inches of mercury (in. Hg) vacuum.

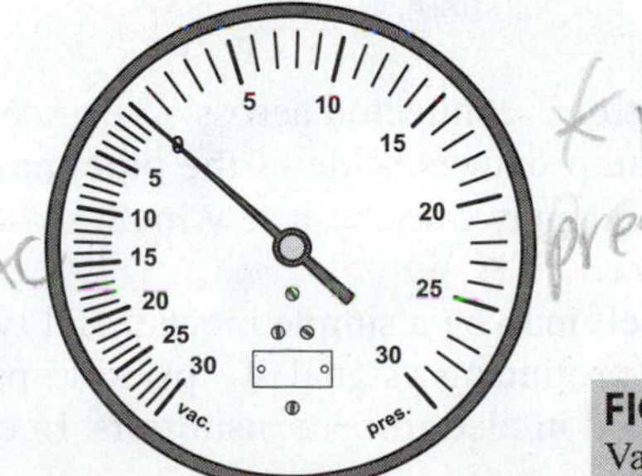

FIGURE 2-9 Compound Vacuum/Pressure Gauge

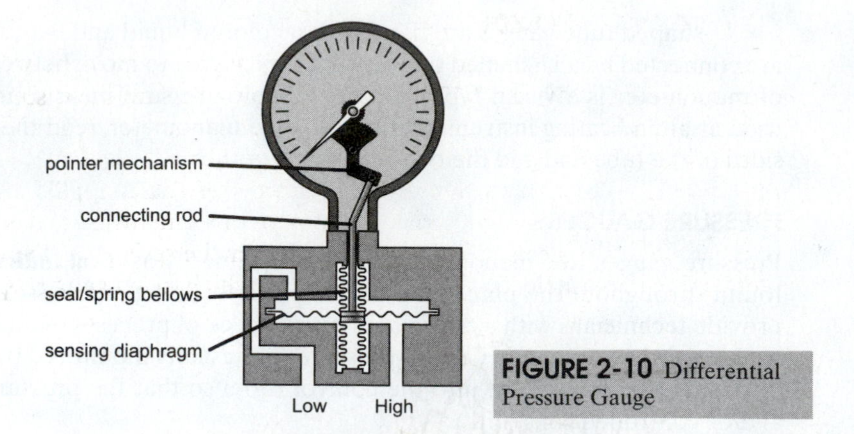

FIGURE 2-10 Differential Pressure Gauge

For other types of applications, a differential pressure reading may be desired. This type of reading may be obtained locally by a differential pressure gauge (Figure 2-10).

As with all instruments, there is a proper way to read gauges. First, look straight onto the faceplate making sure that your eyes are directly over the pointer tip. Then, read the position directly beneath. If you attempt to read it from the side, you induce error.

DIFFERENTIAL PRESSURE CELLS

A **differential pressure (d/p) cell** is a simple mechanical type sensor normally found in transmitters. A more sophisticated transducer type may be found in electronic transmitters. Either type may be used to measure level, flow rate, and differential column pressure.

Differential pressure is defined as the difference in pressure between two related measurements. A differential pressure (d/p) transmitter measures the difference between two pressure points and produces a corresponding output signal. These transmitters are used to measure a variety of process variables including tank level, flow rate, and differential column pressure.

Differential pressure transmitters (Figure 2-11) usually have two pressure measuring sensors. One sensor is considered as the high-side pressure sensor and the other the low-side pressure sensor. This is commonly called a diaphragm capsule. Keep in mind that with d/p measurements, the terms high and low are relative to each other and do not reflect the actual pressures sensed individually.

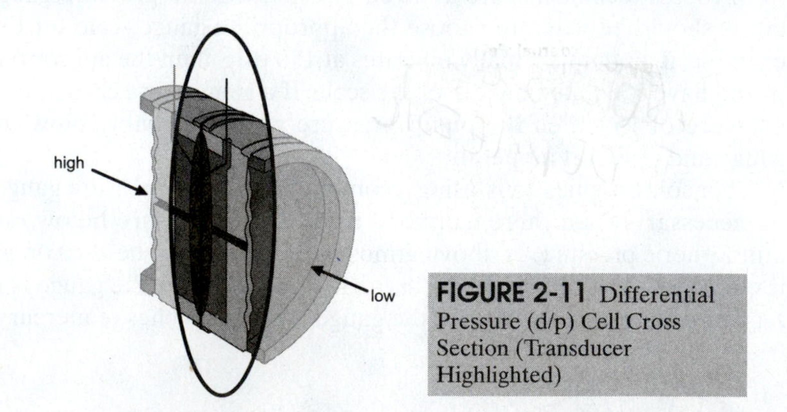

FIGURE 2-11 Differential Pressure (d/p) Cell Cross Section (Transducer Highlighted)

For example, if a d/p transmitter is connected across an orifice plate in a pipe that has a liquid flowing through it, the pressure inside of the pipe may be 200 psi, yet the differential pressure measurement may only be a few inches water column. This is called a proportional signal.

As illustrated, the d/p cell itself may be a simple mechanical type sensor normally found in pneumatic transmitters (producing a signal of 3 psi to 15 psi output) or a more sophisticated transducer type found in electronic transmitters. In either case, d/p is an important process variable.

STRAIN GAUGE TRANSDUCERS

A transducer is a device that converts one form of energy into another form. A **strain gauge transducer** (Figure 2-12) consists of a group of wires that stretch when pressure is applied and changes one form of energy (process measurement) into another (electronic signal) when the device responds to strain. As current flows through the wires, resistance changes proportionally to the stress or strain applied. Strain is the act of changing the dimensions of a solid. When the dimensions of a strain gauge are changed, its resistance also changes. The greater the resistance, the greater the change in the process variable. This is the premise on which strain gauges operate.

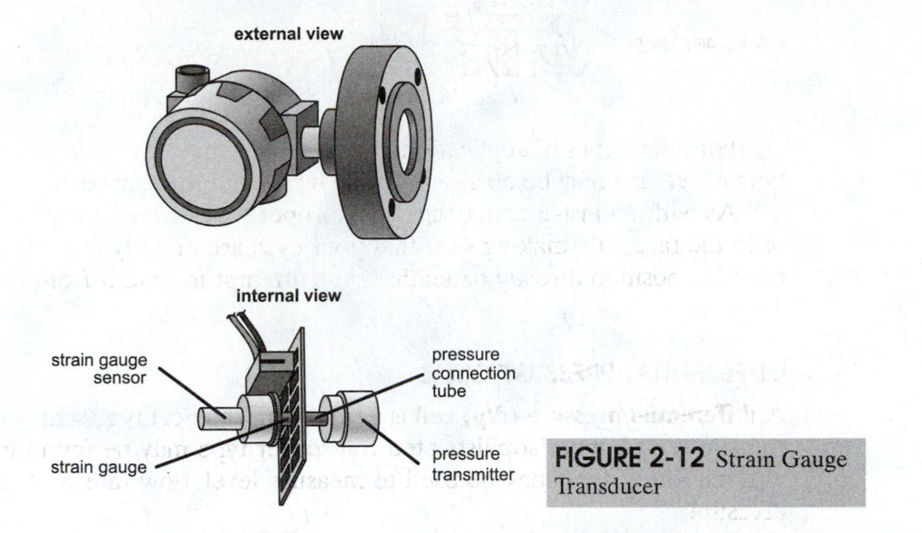

FIGURE 2-12 Strain Gauge Transducer

Generally speaking, transmitters are transducers. Technically, the measuring sensor manipulates the actual transducer, which in turn influences an electronic circuit in the transmitter that produces an output signal of between 4 mA and 20 mA.

Even though process technicians probably never come in contact with an actual measurement transducer (Figure 2-13), the concepts of how they operate and the technology involved in how they associate with other process instrumentation is important to understand.

At the present time, strain gauges are pressure transducers used in industry. Applying pressure to a strain gauge measuring sensor causes it to respond with a change in resistance. The resistance is converted in the transmitter to a standard output signal.

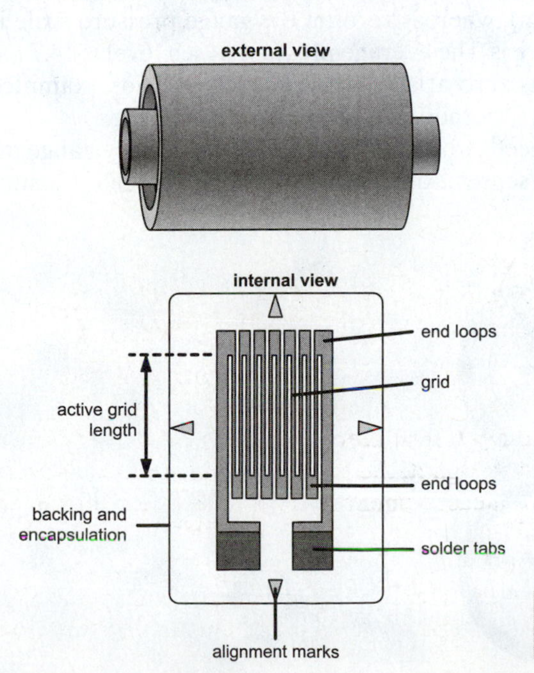

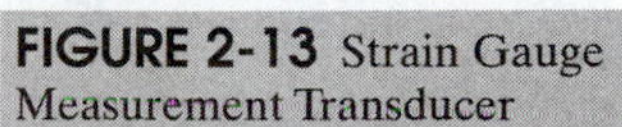

FIGURE 2-13 Strain Gauge Measurement Transducer

CAPACITANCE TRANSDUCERS

A **capacitance transducer** (Figure 2-14) is a device containing a measurement diaphragm and capacitor plates. They are second only to strain gauges in popularity as pressure measurement transducers. They are available in single or differential pressure designs.

When pressure is applied to the measurement diaphragm, the capacitor plates (Figure 2-15) are caused to move closer together. Since capacitance is inversely proportional to the distance between the plates, as the plates get closer together, capacitance increases.

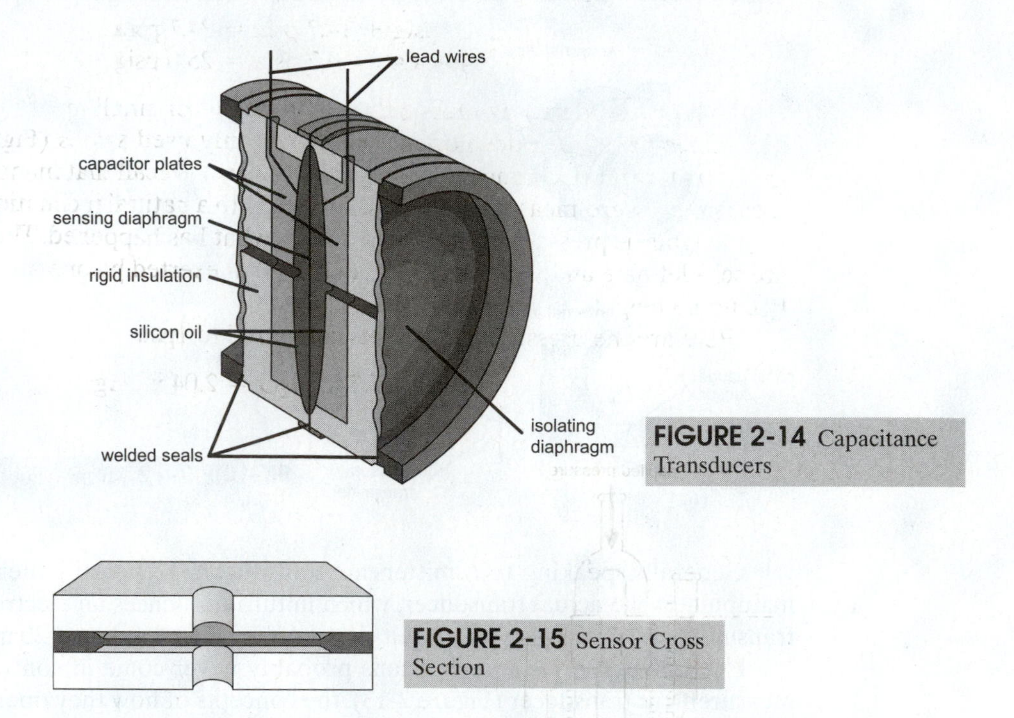

FIGURE 2-14 Capacitance Transducers

FIGURE 2-15 Sensor Cross Section

Pressure Conversions

Let us compare gauge pressure to absolute pressure (Figure 2-16). In the earlier definitions, a statement was made that the absolute pressure scale begins at true zero pressure (a total vacuum), whereas zero on the gauge pressure scale is equal to whatever atmospheric pressure is. The average pressure at sea level is 14.7 psia and 0 psig. Gauge pressure always sets zero at atmospheric pressure. For example, if the atmospheric pressure in Denver, Colorado, is 14.2 psia, it is still 0 psig.

So, why do we need two scales? The answer is that the gauge pressure scale was the only one that made sense at one point in time. How much pressure was inside a vessel

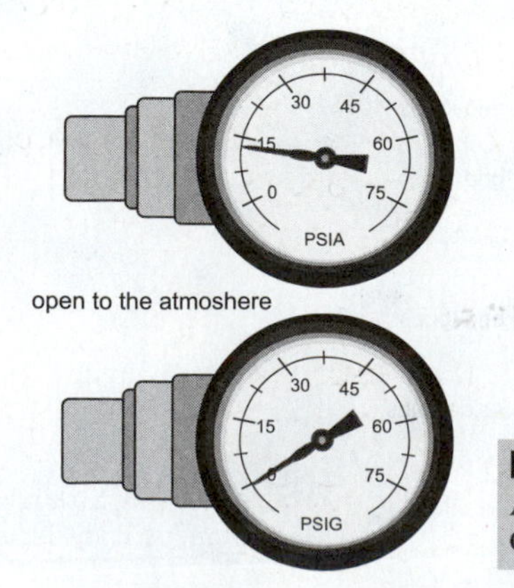

FIGURE 2-16 Gauge and Absolute Pressure Comparison

compared to the outside pressure was important to know. In time, knowing and controlling pressure at an exact point, independent from atmospheric pressure variations, also became important. Scientists must use an absolute scale to be accurate.

Process technicians need to know how to move easily between the two scales. Here are the mathematical relationships:

$$\text{psia} = \text{psig} + 14.7 \text{ psi}$$
$$\text{psig} = \text{psia} - 14.7 \text{ psi}$$

Example pressure measurement conversions:

$$20 \text{ psig} + 14.7 \text{ psia} = 34.7 \text{ psia}$$
$$40 \text{ psia} - 14.7 \text{ psia} = 25.3 \text{ psig}$$

To accommodate low-pressure measurements, or small measurement spans, several scales have been developed. Two commonly used scales (Figure 2-17) are the inches water (in. H_2O) and inches mercury (in. Hg). Recall that manometers are highly accurate pressure-measuring devices. This leads to a natural inclination to use their column heights as pressure units. In fact, that is what has happened. These column inches are self-defining and the pressure is equal to that exerted by one inch of their prospective liquid heights.

Here are the pressure equivalents compared to 1psi:

$$1 \text{ psi} = 27.7 \text{ in } H_2O = 2.04 \text{ in. Hg}$$

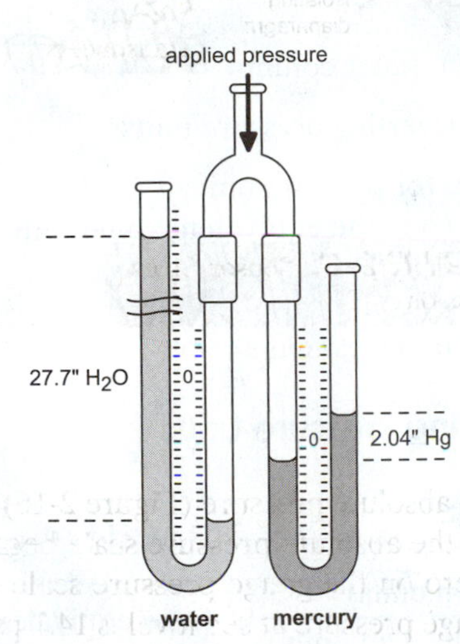

FIGURE 2-17 Inches Water and Inches Mercury Pressure

PRESSURE UNITS

In a simpler world there would be only one pressure unit, but this is no more reasonable than having one denomination of money. If the public had only one-dollar bills to use for currency, people would have to pack a suitcase to purchase a car and would have to spend an entire dollar just to buy a stick of gum. This does not make any more sense than having only one pressure unit to fit all situations. Using psi works fine for most pressure measurements, but sometimes it's just too large. Take the example of measuring the pressure of flue gases leaving a boiler. The pressure of flue gases is about 0.1 to 0.2 psi of pressure. Obviously, one pressure unit cannot possibly provide adequate resolution at both extremes.

PRESSURE CONVERSION CALCULATIONS

Process technicians are responsible for learning to convert between the following pressure units: psi, inches water column, and inches mercury. The following method takes a little longer to set up than simply multiplying with a factor. This method does allow the conversion of any pressure unit to another by simply knowing its value equivalent to 1 psi. This method is more efficient than trying to remember all of the factors for

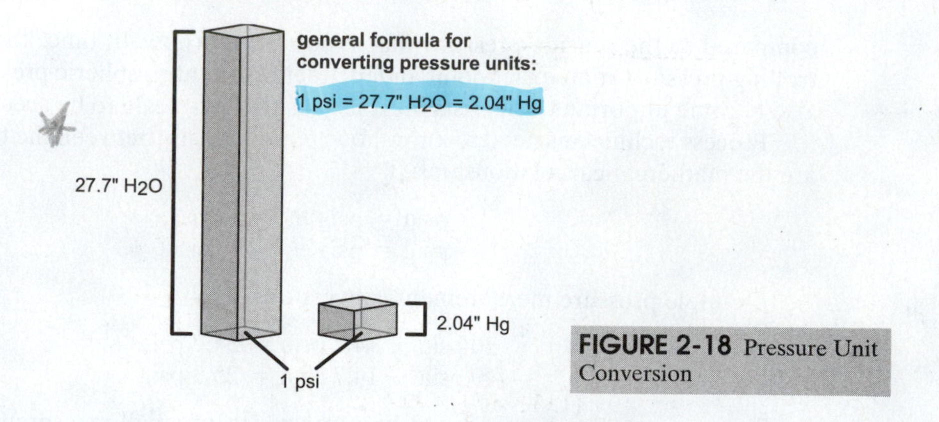

FIGURE 2-18 Pressure Unit Conversion

every conversion both to and from each unit. Process technicians need to memorize the general formula to always know how to perform these pressure conversions. Once memorized, this formula is a tool as real as a screwdriver or a pair of channel locks that can be used over and over again.

This method is referred to as the equivalent units method. This method produces a *factor* on the fly each and every time a conversion problem is set up. The following values are quantitatively equal.

Equivalent Units

$$1 \text{ psi} = 27.7 \text{ in. water column} = 2.04 \text{ in. Hg}$$

Use these steps to use when converting pressure units:

1. Find or recall equivalent values for pressure units.
2. Place the value to be converted and the equivalent values into their proper place in the formula.
3. Perform calculation.
4. Check your answer to see that it makes sense.

General Formula for Converting Pressure Units

Where:

$$1 \text{ psi} = 27.7 \text{ in. } H_2O = 2.04 \text{ in. Hg}$$

$$\text{New units: } \frac{\text{(known units)}}{\text{(known equivalent units)}} \times \text{(new equivalent units)}$$

Examples

- Convert 10 in. Hg to inches water.

$$\text{New units} = \frac{10 \text{ in. Hg} \times 27.7 \text{ in. } H_2O}{2.04 \text{ in. Hg}}$$

$$\text{New units} = 135.78 \text{ in. } H_2O$$

- Convert 3.6 psi to inches Hg.

$$\text{New units} = \frac{3.6 \text{ psi}}{1 \text{ psi}} \times 2.04 \text{ in. Hg}$$

$$\text{New units} = 7.34 \text{ in. Hg}$$

Example pressure-measurement conversion calculations:
Use the following formulas:

- psig to psia = psig + 14.7
- psia to psig = psia − 14.7

- psig to in. H_2O = psig × 27.7
- psig to in. of Hg = psig × 2.04

$$\text{inches of } H_2O \text{ to psig} = \frac{\text{height of liquid}}{27.7}$$

$$\text{inches Hg of psig} = \frac{\text{height of liquid}}{2.04}$$

Convert the following:

- 150 psig to psia (164.7 psia)
- 150 psia to psig (135.3 psig)
- 42 psig to in. H_2O (1163.40 in. H_2O)
- 55 psig to in. Hg (112.20 in. Hg)
- 300 in. of H_2O to psig (10.83 psig)
- 100 in. of Hg to psig (49.02 psig)

Summary

Pressure is a common process variable and is defined as the amount of force per unit of area. Mathematically, pressure is defined as Pressure = Force/Area or P = F/A. Mass applied to a unit of area depends on the density of the molecules within the mass and how heavy the molecules are. Gases and vapors apply pressure equally to all surfaces in a container while liquids and solids may vary according to the size or shape of the container.

When calculating pressure, factors that need to be considered include the following:

- liquids and gases are considered to be fluids
- liquids and gases behave differently when force is applied
- total height of the liquid column in a given vessel
- surface pressure
- density of the material

Speed, mass, and density of molecules exerted on each other affect force. Temperature is the greatest factor affecting molecular speed with increases in temperature directly proportional to increases in speed. If the increased speed is confined, then pressure increases as well. The greater the collective amount of mass, either atomic or molecular weight, applied to a particular unit of area, the greater the pressure applied. How dense a substance is determines the number of collisions occurring between the molecules. Where there are more molecules, there is more potential for a higher number of molecular collisions, creating more force. For gases, special pressurized containers must be used to contain the molecules since they fill up the entire space they are allowed to occupy. Specific gravity is the density of a substance relative to the density of water (expressed as 1.0). This makes the measurement of all other liquids or solids to have a specific gravity either greater than or less than water.

Pressure measurement make be taken at unique points within processing operations. In some cases, knowing the difference between two points is important. This is called differential pressure or delta p (d/p, Δp) and is accomplished by forcing the process stream through a calibrated orifice plate. Sensors on either side of the orifice plate sense the pressure and send the reading to a differential pressure transmitter where the difference in the two pressures is calculated and the result is transmitted appropriately.

Typical measurement scales include pounds per square inch (psi), pounds per square inch absolute (psia), pounds per square inch gauge (psig), atmospheres (atm) or bars, and inches of water (in. of H_2O) or mercury (in. of Hg). At sea level, 14.7 psi is the standardized pressure measurement that is equivalent to 1 atmosphere with zero on an absolute scale being equivalent to absolutely no pressure at all. Vacuum is most often measured in inches of mercury. Since the psi scale originates at atmospheric pressure, regardless of where you are, the ambient or surrounding pressure establishes zero on this scale with the 14.7 atmospheric pressure already taken into account. This is the most common measuring scale used in modern processing facilities.

Pressure-sensing and measuring instruments are commonly used to indicate absolute, gauge, and vacuum pressures in processing facilities. The oldest type of pressure gauge still in operation is the manometer and the most common of these is U-shaped. Manometer accuracy depends mainly on the reading taken from the meniscus. Pressure gauges, like manometers, are primarily used as local indicators. Pressure gauges are generally made of steel bodies and have an internal metallic tube that flexes in response to pressure changes. This response is linked to a pointer by a mechanism that causes the pointer to move along a predetermined scale (psig, psia, vacuum, or compound). Other process situations require measuring pressure differential. In these cases, a differential pressure (d/p) cell and transmitter configuration may be used to sense the high and low side of a process stream. In this case, the output would be read remotely. Another type of pressure measuring device is a strain gauge transducer that converts one form of energy (applied pressure to a group of wires that stretch) into another form of energy (an electronic signal) when the device responds to strain. Similarly, a capacitance type transducer may be used as a pressure measurement transducer and they are available in single or differential pressure designs. Capacitance is inversely proportional to the distance between the plates.

The two mathematical relationships between the two pressure measurement scales are:

$$\text{psia} = \text{psig} + 14.7 \text{ psi}$$
$$\text{psig} = \text{psia} - 14.7 \text{ psi}$$

Low-pressure measurement scales use inches of H_2O or inches of Hg. When using manometers, their liquid column height can also be factored into the equation or scale.

Checking Your Knowledge

1. Pressure is described as _____ per unit area.
   a. Flow
   b. Pounds
   c. Force
   d. Inches
2. Pressure is increased when:
   a. The number of molecules per unit area is decreased
   b. Heavier molecules per unit area are introduced
   c. Molecules begin to move faster
   d. The number of molecules are spread out over a larger unit area
3. Atmospheric pressure at sea level is _____ psia.
   a. 0
   b. 2
   c. 14.7
   d. 29.92
4. Match the instruments listed below with their method of operation.

Instruments	Method of Operation
Manometer	a. Uses wires that stretch when pressure is applied
Capacitance Transducer	b. Uses a measurement diaphragm to check pressure
d/p Cell	c. Uses two sensors to measure a change in pressure
Strain Gauge	d. Uses a liquid-filled tube to measure pressure

5. Convert 1 psi into inches of H_2O.
   a. 0
   b. 2
   c. 14.7
   d. 29.92
6. Convert 14.7 PSI into inches of H_2O.
   a. 109.6
   b. 247.17
   c. 398.76
   d. 407.19

Student Activities

1. Hold your hand on your chest while you take a VERY deep breath. What happened? Did you see or feel any expansion? Explain what happened when you did this.
2. Using an EMPTY plastic gallon milk jug with a screw-on top, fill it at least ¼ full with VERY hot water being careful not to burn yourself. Cap the jug quickly and let it stand for up to an hour. What happened?
3. Hold an empty plastic bottle upside down over a pan of boiling water. Ensure the bottle is positioned so as not to burn your hand by using tongs or a protective glove to hold the bottle in place for several minutes. The bottle will fill with steam. Remove bottle (still upside down) from boiling water area and immediately cap the bottle. As the bottle and the steam cool down, the bottle sides collapse. Explain the principles that cause this to occur.
4. Do you think atmospheric pressure is powerful enough to crush a can? Use the following materials and procedure to determine the answer:

   Materials
   - hot plate
   - empty soft drink can with one tablespoon of water inside
   - tongs (to pick up the hot can)
   - shallow pan filled with cool water

   Procedure
   a. Turn on the hot plate to a medium temperature.
   b. After ensuring a tablespoon of water has been inserted into the soft drink can, place it onto the hotplate.
   c. Observe how the water temperature rises and fills the can with water vapor allowing steam to escape from the top opening.
   d. Using the tongs, quickly pick up the soda can, turn it upside down and place the opening into the shallow pan of cool water.
      NOTE: The can should immediately crush and a dull popping sound should occur. If this does NOT happen then start over with the procedure ensuring the tablespoon of water inside the can has had enough time to vaporize.
   e. Explain what happened. Air pressure decreases as the water vapor condenses inside the can. The can crushed due to the air pressure inside the can becoming less than the atmospheric pressure outside the can.
5. Using a large mouth plastic bottle, insert an air-filled balloon taking care not to burst it. Cap the bottle. Using a small hole on the plastic container, force air into the bottle by using an air pump/injection needle apparatus (tire air pump). Describe what happens to the balloon.
6. Identify pressure-sensing or measurement devices in the lab.
7. In a pilot plant or on a table-top model, have students correctly read pressure gauges.
8. Use a vacuum pump with a pressure gauge hooked to a one-gallon tin can with tubing. Start pump and note the pressure when the can implodes due to the pressure of the atmosphere.
9. Make a U-tube manometer from tubing and fill with colored water to measure very low pressures.
10. **PRESSURE CONVERSION WORKSHEET**
    Instructions
    Using the Conversion Chart, perform the calculations on a clean sheet of paper. Turn in your completed calculations as per your Instructor's guidelines. Ensure problem numbers and corresonding answers are clearly written.

Conversion Chart	
Convert psig to psia	psig + 14.7
Convert psia to psig	psia − 14.7
Convert psig to inches of H_2O	psig = 27.7
Convert psig to inches of Hg	psig = 2.04
Convert inches of H_2O to psig	$\frac{\text{height of liquid}}{27.7}$
Convert inches of Hg to psig	$\frac{\text{height of liquid}}{2.04}$

Use this conversion chart to complete the following conversion exercises.

Pressure Conversions:

1. 3.00 psi = ________	in. w.c.	6. 1.00 psi = ________	in. Hg
2. 3.61 psi = ________	in. w.c.	7. 10.2 psi = ________	in. Hg
3. 9.00 psi = ________	in. w.c.	8. 15.8 psi = ________	in. Hg
4. 11.0 psi = ________	in. w.c.	9. 100 psi = ________	in. Hg
5. 15.0 psi = ________	in. w.c.	10. 500 psi = ________	in. Hg
11. 100 in. w.c. = ________	psi	16. 74.7 in. w.c. = ________	in. Hg
12. 74.7 in. w.c. = ________	psi	17. 100 in. w.c. = ________	in. Hg
13. 23.2 in. w.c. = ________	psi	18. 145 in. w.c. = ________	in. Hg
14. 50.0 in. w.c. = ________	psi	19. 235 in. w.c. = ________	in. Hg
15. 300 in. w.c. = ________	psi	20. 5.00 in. w.c. = ________	in. Hg
21. 4.47 in. Hg = ________	psi	26. 2.00 in. Hg = ________	in. w.c.
22. 9.35 in. Hg = ________	psi	27. 16.0 in. Hg = ________	in. w.c.
23. 15.7 in. Hg = ________	psi	28. 24.2 in. Hg = ________	in. w.c.
24. 44.7 in. Hg = ________	psi	29. 29.9 in. Hg = ________	in. w.c.
25. 29.9 in. Hg = ________	psi	30. 100 in. Hg = ________	in. w.c.

psia to psig Conversions at Standard Atmospheric Pressure:

31. 4.00 psia = ________	psig	36. 0.00 psig = ________	psia
32. 14.7 psia = ________	psig	37. 21.0 psig = ________	psia
33. 22.2 psia = ________	psig	38. 35.4 psig = ________	psia
34. 37.7 psia = ________	psig	39. 45.0 psig = ________	psia
35. 50.0 psia = ________	psig	40. 50.0 psig = ________	psia

11. VACUUM PRESSURE MEASUREMENT Laboratory Procedure

Background Information

Pressure is defined as a force per unit area. A common unit of pressure is pounds per square inch or simply psi. At sea level, atmospheric pressure averages 14.7 psi on the absolute pressure scale and zero (0) psi on the gauge pressure scale. Atmospheric pressure is dynamic, meaning that it changes according to weather conditions. If a vessel is blocked in, the internal pressure drops below atmosphere and the pressure across its walls exceeds its rated capacity, then extensive damage to the vessel can occur in the form of crushing. Most vessels are not designed to withstand significant pressure applied from the outside inward.

Consider the following example. A storage tank is prepared for maintenance by removing all process materials from the vessel and then steaming out the vessel. All connections on the tank are closed. Then, cold wet rain drenches the tank for several hours during the evening time causing the steam in the tank to condense. When the day shift process technicians make their first rounds on the following morning, they find the tank has been crushed on one side causing extensive damage.

Materials Needed

- vacuum source
- copper tubing
- vacuum gauge or a psig gauge that reads negative pressure or a psia gauge
- a vessel to vacuum source connecting apparatus

  NOTE: The materials needed for this apparatus can be found at most hardware stores.
- a gallon tin can or plastic liter bottle

Safety Requirements

Safety glasses are required in the lab due to the potential dangers associated with imploding or exploding vessels.

Procedure

a. Connect the container to be tested to a vacuum source. If the vacuum source does not have a pressure indicator mounted to it, then you will have to connect one to the apparatus with a tubing tee.
b. Make sure the gauge is indicating zero before starting the experiment.
c. Slowly apply vacuum to the plastic container.
d. Record the reading at the initial wall collapse and then again when it is completely collapsed.
e. Apply the same amount of positive pressure to the plastic bottle.
Safety Tip: There is a danger of explosion.
f. Record your observations.
g. Do this to several containers if time allows.
h. Complete the final report for this lab.

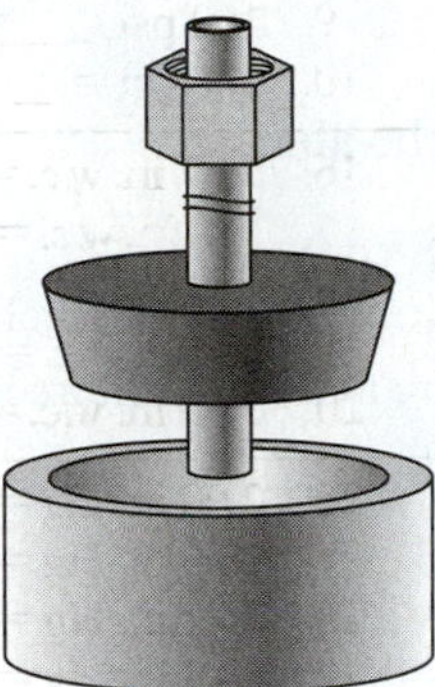
connect this end to a vacuum pump
drill a hole through the rubber stopper and insert the copper tubing
insert the rubber stopper into the plastic bottle making a tight fit

Findings

Compile your observations and readings into a report that includes your recorded data, an explanation of the results of the experiment, and a conclusion. Include a paragraph explaining how a process technician could guard against the scenario provided in the background information.

* tear gauges measures
PSIA=14.7

CHAPTER 3

Process Variables, Elements, and Instruments: TEMPERATURE

Objectives

After completing this chapter, you will be able to:

- Define temperature.
- Describe the affect heat energy has on the movement of molecules.
- Define terms associated with temperature and temperature instruments:
  - differential temperature
  - Fahrenheit
  - Celsius/centigrade
  - Rankine
  - Kelvin
- Identify the following for Celsius and Fahrenheit:
  - boiling/freezing points for H_2O
  - minimum/maximum water temperature at sea level
- Identify common types of temperature-sensing and measurement devices used in the process industries:
  - thermowell
  - thermometer
  - bimetallic strip
  - resistance temperature detector (RTD)
  - thermocouple
  - temperature gauge
- Describe the purpose and operation of various temperature-sensing andmeasurement devices used in the process industries.
- Given a standard calculator and conversion formulas, complete Fahrenheit and Celsius conversions.

Key Terms

Bimetallic strip—two dissimilar strips of metal bonded together that expand and contract at different rates when exposed to temperature change, causing a blending or rotating effect; used as the primary element in a temperature gauge or bimetallic thermometer.

British thermal unit (BTU)—the amount of heat required to raise the temperature of a pound of water 1 degree Fahrenheit.

Celsius/centigrade—scale of measurement to determine temperature; freezing point of water is 0 degrees C and boiling point of water is at 100 degrees C.

Conduction—the transfer of heat through solid matter by moving from one molecule to the next.

Convection—the transfer of heat through the circulatory motion occurring in a fluid when there is a difference in temperature from one region to another; density differences between the hot and cold molecules causes convection motion in fluids.

Fahrenheit—scale of measurement to determine temperature; freezing point of water is defined as 32 degrees F and the boiling point of water is at 212 degrees F.

Filled thermal system—a temperature-sensing bulb filled with a liquid, vapor, or gas and connected by means of a capillary tube to a pressure-measuring element.

Glass-stem thermometer—a temperature-measuring device constructed of a glass bulb and a capillary (small) tube made of glass with a numbered scale; the bulb and the capillary tube contain a liquid such as mercury or colored alcohol that expands whereby a reading can be made from the etched scale on the length of the tube.

Heat—the energy that flows between bodies of differing temperatures.

Kelvin—scale of measurement that is sometimes called an *absolute scale* because 0 K is the point at which no heat exists; freezing point of water is 273 K and boiling point of water is 373 K.

Radiation—the transfer of heat through emitting radiant energy in the form of waves or particles.

Rankine—an absolute scale of measurement to determine temperature; freezing point of water is 492 degrees R and boiling point of water is 672 degrees R.

Resistance temperature device (RTD)—primary element that measures temperature changes in terms of electrical resistance.

Temperature differential (delta)—difference between two temperatures, normally across two points in a control loop.

Temperature gauge—an independent device with a sensing element such as a bimetallic strip, bourdon tube or bellows that is linked to a pointer displaying the temperature on a calibrated face.

Thermistor—a type of resistor used to measure temperature changes, relying on the change in its resistance with changing temperature.

Thermocouple—primary element consisting of two wires of dissimilar metals connected at one end.

Thermowell—a thick-walled, typically stainless steel device shaped like a tube that is inserted into a hole in piping or equipment specifically prepared to house a temperature sensing and measuring element.

Introduction

This chapter describes temperature and heat energy as they relate to molecular movement. Boiling and freeze points on commonly used temperature measuring scales are discussed along with the various temperature measuring devices to include their specific purpose and operation. Conversions between the various temperature measuring scales are also covered.

Temperature Defined

Temperature is defined as the average kinetic energy of a material. Temperature is also the indication of heat energy available to flow between bodies of differing temperatures. The sun gives off heat energy (Figure 3-1) to its surrounding area and this heat radiates outward flowing from extremely hot to what we feel here on earth on a daily basis.

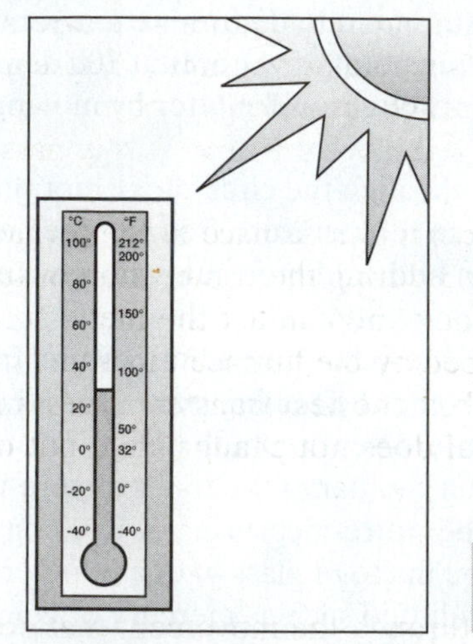

FIGURE 3-1 Temperature and Heat

As the temperature increases in a substance, its physical characteristics may also change. Water, with the increase of heat energy, becomes steam as the molecules increase in speed and overcome surface tension to vaporize (boiling point). Conversely, when taking away heat from water, the molecules slow down considerably and if enough heat is lost, the water turns to ice (freeze point).

This principle of physical characteristic change is used inside temperature-measuring devices, such as the common thermometer, to determine the amount of temperature available in an external substance (e.g., performed by placing the thermometer into a substance such as a liquid filled container). Inside a glass thermometer is a liquid that is usually mercury (Hg). When mercury is heated, it immediately expands away from the bulb and into the connected capillary tube. The expansion of the mercury has been calibrated in the markings on the outside of the thermometer to indicate, or measure, the amount of heat energy (temperature) in the substance being measured. When the thermometer is taken out of the substance being measured, the measurement begins to move readily to the surrounding temperature. Temperature is determined by observing the markings etched on the glass and to where the mercury level has expanded.

Effects of Heat Energy on Molecular Movement

A temperature measurement is one that is taken at a *given point in time* with a specific measuring device that has been calibrated to a particular temperature scale. By comparison, heat is the amount of energy that flows *over time* between bodies of differing temperature. **Heat** always flows from a higher temperature to a lower temperature until temperatures are equal. Heat is also a measurable quantity.

Since heat is the energy that causes molecules to become more active, they move faster when heat is added and their temperature also increases. Because a temperature increase or decrease is dependent on the amount of heat energy moving from one substance to another, there is a direct relationship between the two. Units of heat measurement are based on this relationship. For example, a **British thermal unit (BTU)** is defined as the amount of heat required to raise the temperature of a pound of water 1 degree Fahrenheit. In the metric system, 1 calorie is defined as the amount of heat required to raise the temperature of 1 gram of water 1 degree Celsius.

HEAT TRANSFER

Heat moves in one of three distinct ways:

- conduction
- convection
- radiation

Conduction

Conduction is the transfer of heat through solid matter as it moves from one fixed molecule to the next. An example of this would be a common household iron. Heat generated by electrical energy moves through the soleplate of the iron by conduction. The soleplate does not change form or state, but the heat energy passes through the solid quite easily.

In process units, conduction occurs in a furnace in the furnace tubes (solid walls of the tubes). Hot flue gases, formed during the combustion of fuel gas in the burner area, flow upward in the furnace box and contact the metal furnace tubes carrying material. The heat energy is absorbed by the furnace tubes and is then transferred to the material passing through the tubes. The heat transfer process in the furnace tubes is conduction since the solid material does not change state but only passes the heat through itself.

Convection

Convection is the transfer of heat through the movement that occurs in a fluid when there is a difference in temperature from one region to another. The density differences between the hot and cold molecules cause this convection motion in fluids.

Using the same example of material moving through the tubes in a furnace, when the heat is transferred through the tubes (conduction) to the fluid contents of the tubes, the fluid transfers heat by convection and conduction throughout its volume.

Radiation

Radiation is the transfer of heat by emitted radiant energy in the form of waves or particles. The sun is an example of a radiant energy source. The sun's heat energy travels through the vacuum of space.

Again using the furnace example, radiation occurs from the heat given off by the combustion process at the burners. Radiant heat flows from the burners to the outside walls of the piping and the furnace structure. The further upward the piping is from the burners, the less heat energy is transferred. This difference in the amount of heat energy is used to an advantage.

HEAT TRANSFER: PHASE CHANGES

Molecules within solids have little movement. The molecules have a force that tries to push them apart, and another force holding them together. When heat energy is applied, it increases the force pushing the molecules apart. As the temperature increases, the distance between the molecules in a solid also increases. The increased distance between molecules causes the volume within the solid to increase. The increased distance between molecules allows more room for the molecules to move. As the solid begins to melt, the molecules become active enough to lose much of their order and no longer have the strength between them to hold their shape against the force of gravity. When the substance is melting, adding heat energy does not raise the temperature. The temperature remains the same until all the substance melts.

When the solid has completely melted, the temperature of the liquid substance again increases. The increase in temperature increases the energy level of the molecules. They then begin to overcome the forces holding them together and the liquid converts to a gas. At this point, all the thermal energy goes to increasing the energy of the molecules, causing no additional increase in temperature. At this point, the liquid is boiling. Once the boiling liquid has completely changed phase to a gas and no liquid remains, then the continued addition of heat causes the temperature of the gas to rise. If temperature is increased beyond this point, the gas is called superheated. Superheated steam at high pressures is common in process plants.

Temperature Measurement

The concern for measuring temperature has been in the minds of humans for thousands of years. As early as 170 A.D., a Greek physician and writer by the name of Galen proposed a standard for temperature measurement. Galileo is credited with inventing the thermometer in 1592, but an accurate and repeatable mercury thermometer for fixed points on the low end of the temperature scale was not available until the early 1700s and was invented by Gabriel Fahrenheit (freezing point 32 degrees F and boiling point 212 degrees F). Then about 1742, Anders Celsius proposed that 0 should be used as the freezing point and 100 as the boiling point of water for his scale; therefore, the freezing point and boiling point of water became the benchmarks for both temperature scales.

In modern-day industry, temperature measurement is extremely important to the operation and safety of the plants due to its inherent relationship with pressure and volume. Remember the perfect gas law in physics ($pv = RT$). Since "R" is a constant, if temperature is rising and volume is constant, then pressure must be increasing. Of course, liquids also expand when heated. Since process systems are contained in finite volume piping and vessels, process technicians have to pay close attention to the temperature, pressure, flow, and level indications that are displayed on their instruments. Each part of the process has specific operating parameters for these process variables (e.g., degrees Celsius for temperature).

Technicians should be concerned that the process indications are maintained at those parameter values or especially not beyond the acceptable limit (usually upper) for the particular process. If the temperature is rising in a gas process, then the pressure is rising somewhere else as well. If that other place has a restricted volume (e.g., tank or blocked pipe), then that pressure is likely to cause a catastrophic rupture or a safety relief valve release. To prevent these occurrences, temperatures must be lowered to acceptable levels by some form of reasonable operational maneuver until all indicators can be returned to normal. If the temperature is rising and the pressure is not rising, there may be a path open to another volume. In the case of gas, this could be a place such as the atmosphere where it should not go. Technicians should be aware of any changes, but temperature readings should always stay on or close to desired values.

Temperature can be measured in the following ways:

- single points
- multiple averaging points
- temperature differential (delta)

Single point measurement (Figure 3-2) is the most common. When the temperature of a fluid traveling through a line is needed, temperature can be measured. When the average temperature in a reactor zone is needed, a single temperature point may

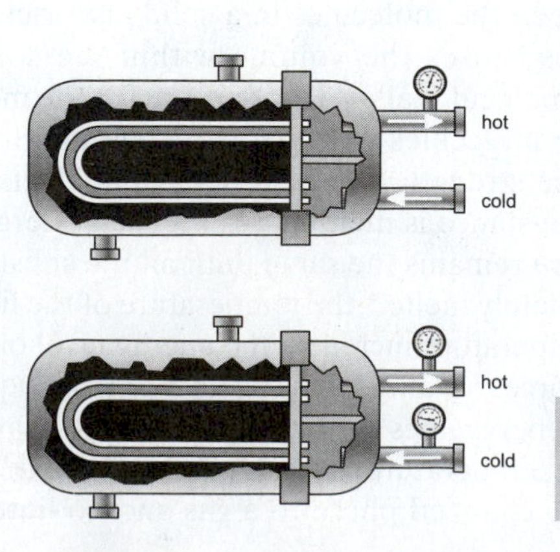

FIGURE 3-2 Single Point and Differential (delta) Temperature Indicators

be installed at various points in the reactor. Single points can be observed independently or they can all be added together and an average calculated. When observing two different temperature points in a process, a **temperature differential (delta)** measurement may be calculated. An example of a differential temperature measurement is the inlet versus the outlet temperature of a heat exchanger (Figure 3-2).

TEMPERATURE MEASUREMENT SCALES

Each temperature measurement may be represented, or determined, on a different type of scale (Figure 3-3). These scales include the following:

- Fahrenheit
- Celsius/centigrade
- Rankine
- Kelvin

Each of these four temperature scales are in use in the United States and this is the only community using Fahrenheit while the rest of the world uses centigrade or its equivalent Celsius. Kelvin is used worldwide. Absolute values of temperature (e.g., Rankine and Kelvin) must be used for Gas Law calculations.

Fahrenheit

The **Fahrenheit** temperature scale is divided into 180 degrees between the freezing and boiling point of water. The freezing point of pure water is defined as 32 degrees F and the boiling point of pure water is at 212 degrees F at standard pressure. As a matter of interest, the temperature of 0 degrees F was the lowest attainable temperature under laboratory conditions in 1714, the year Fahrenheit introduced his thermometer.

Celsius/Centigrade

The **Celsius/centigrade** scale is divided into 100 degrees between the freezing point and boiling point of pure water. On this scale, pure water freezes at 0 degrees C and boils at 100 degrees C at standard pressure.

Kelvin

The **Kelvin** scale is called an absolute scale because 0 K is the point at which no heat exists. Theoretically there is no molecular motion and kinetic energy ($KE = \frac{1}{2}mv^2$) is zero. The Kelvin scale uses the size of the Celsius degree unit. Water freezes at 273 K and boils at 373 K. Notice that a degree mark is not used in front of the K. That is by standard practice.

Rankine

The **Rankine** scale (degrees R), an absolute scale, is the Fahrenheit equivalent of the Kelvin scale. On this scale, the freezing point of pure water is 492 degrees R and its boiling point is 672 degrees R.

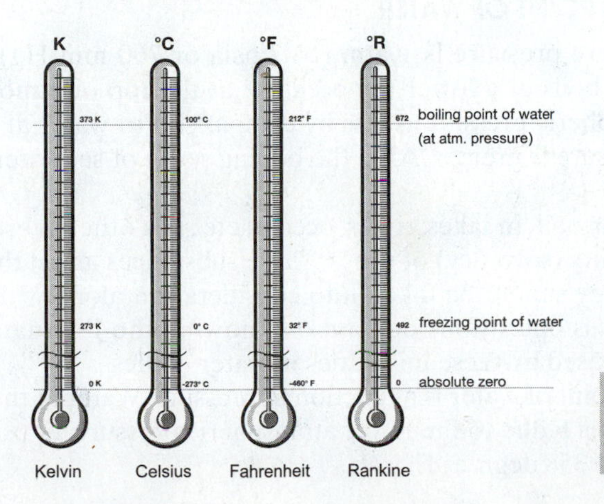

FIGURE 3-3 Temperature Measurement Scale Comparisons

BOILING AND FREEZING POINTS OF WATER

All matter has a freezing point and a boiling point. Even though certain matter is thought of as eternally solid, that doesn't mean that the solid can't exist as a liquid or even as a gas. Rock is a solid, but rock exists in the earth's core as magma and flows from volcanoes as lava. Solid steel melts away at the tip of a cutting torch. In comparison, water exists in all three states: solid (ice), liquid (water), and gas (water vapor or steam).

The most important reason that water is used as the reference point for boiling and freezing is that pure water can be produced in almost any laboratory anywhere in the world. Historically, water was distilled to make it pure. This was easily accomplished hundreds of years ago. In addition, water has a relatively high thermal capacity that adds to its stability as a calibrating standard.

An interesting phenomenon occurs at the melting point of water and then again at its boiling point. At both of these points, as long as two states (ice and liquid or liquid and vapor) are present, a thermal equilibrium exists stabilizing the temperature. The reason for this phenomenon is that as water or other materials change state, the latent (hidden) heat requirements have to be met before the sensible (measurable) heat can cause the temperature to change again. For example, if a thermometer is placed into a pot of water on a stovetop before turning the heat on, the thermometer should register ambient temperature. As the water heats up, its temperature changes steadily until it reaches the boiling point. At that point, the temperature becomes stable (doesn't change) until all the water boils off (vaporizes). The temperature reading would then begin to climb higher on the thermometer. In this illustration, where is all of the heat going during the state change of the water? The heat is going into the molecular structure of the substance (water) itself. This latent heat requirement to change a material's state makes the boiling and freezing points of a substance highly stable for calibration points.

A basic comparison of the temperature measuring scales is found in the boiling and freezing points (Table 3-1).

TABLE 3-1 Temperature Measurement Scale Comparisons

Units	*Absolute Zero*	*Melting/Freezing Point*	*Normal Boiling Point*	*Normal Room Temp.*	*Normal Body Temp.*
°F	−460	32	212	77	98.6*
°C	−273	0	100	25	37
K	0	273	373	298	310
°R	0	492	672	569	591

*Average touch tolerance temperature for humans is 120 °F.

EFFECTS OF PRESSURE AND CONTAMINANTS ON THE BOILING POINT OF WATER

At sea level, where pressure is 1 atm (14.7 psia or 760 mm Hg) water boils at 212 degrees F. Water boils at a lower temperature at the top of a mountain (Figure 3-4) where the atmospheric pressure is less by comparison to water at sea level where the atmospheric pressure is greater. Also, the boiling point of seawater is higher than that of pure water.

Water, fresh or salt, in lakes, rivers, oceans, etc., has other substances mixed in due to the solvent ability (solvency) of water. These substances affect the boiling and freezing points, and they should be taken into consideration along with ambient temperatures when process operations depend on knowing the fluctuations in boiling and freezing points caused by these impurities in water bodies.

The boiling point of water is a function of pressure. Water at the top of Pike's Peak boils at 195 degrees F due to the lower atmospheric pressure 12 psia. Water in a boiler at 150 psia boils at 358 degrees F.

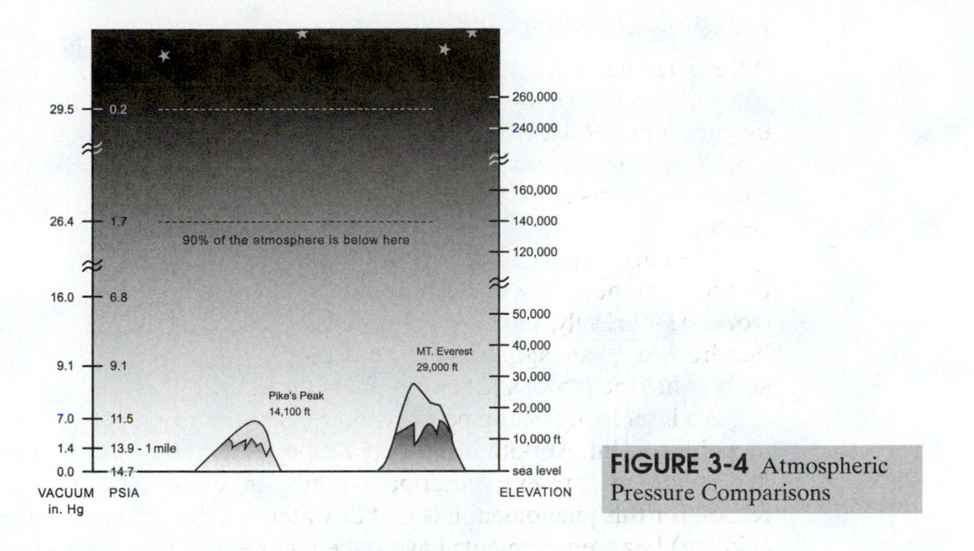

FIGURE 3-4 Atmospheric Pressure Comparisons

Temperature-Sensing and Measurement Instruments

While there are many places in a processing facility to measure temperature, there are a few common temperature-sensing and measurement instruments. These devices include the following:

- thermowell
- thermometer
- bimetallic strip
- resistance temperature device (RTD)
- thermocouple
- thermistor
- temperature gauge

THERMOWELL

Thermowells, while not actual temperature-sensing and measuring devices, are protective devices that are used to isolate temperature-sensing elements from the adverse conditions of the process. The design of the thermowell (Figure 3-5) also allows the actual sensing and measuring element to be removed from the process without having to shut down and isolate the equipment first. This is a very important advantage in continuous processing units.

A typical **thermowell** (Figures 3-5 and 3-6) is a thick-walled (e.g., stainless steel) device shaped like a tube with threaded top. The device may be constructed of another type of material and may be attached by a flange or welded into place depending on the service where the thermowell is to be used.

Each thermowell, or protection well as it is called in some facilities, is constructed from a solid piece of bar stock that has been cut down on a lathe to its specified dimensions. A hole is then drilled from the top so that a temperature-sensing element can be inserted. Where the thermowell is to be installed, a hole is drilled and a weld-o-let

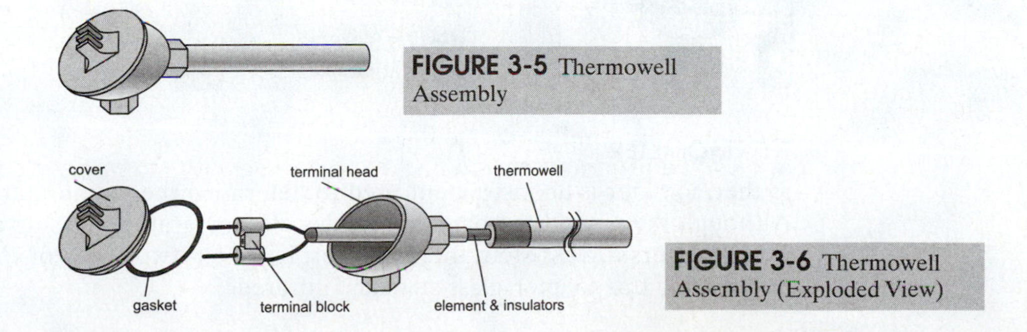

FIGURE 3-5 Thermowell Assembly

FIGURE 3-6 Thermowell Assembly (Exploded View)

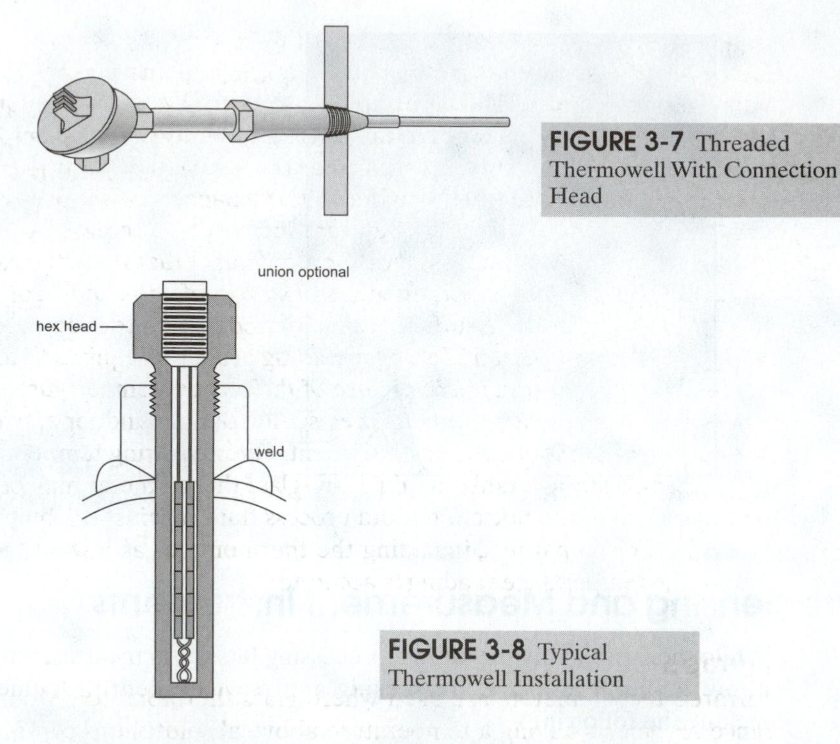

FIGURE 3-7 Threaded Thermowell With Connection Head

FIGURE 3-8 Typical Thermowell Installation

taped with pipe threads is welded to the line or vessel. The thermowell is screwed into place (Figure 3-7) or attached by a flange on a nozzle or welded (Figure 3-8) into place depending on the service. The temperature sensing/measurement element is inserted into the thermowell and held in place by an attaching gland. A wiring termination head may be screwed to the top of the thermowell to allow connection of the thermocouple wiring to the signal wiring so as to provide indication or control input.

Thermowells must be robust and corrosive resistant since they come directly into contact with the process. Good heat conducting characteristics through the proper selection of construction material (e.g., stainless steel) is required. If the process temperature is varying, there is a conductive time lag associated with the time taken for process heat to transfer through the thickness of the thermowell wall and to the actual element sensing the temperature. The advantages derived by having a thermowell usually outweigh any disadvantages or negative effects of the added response time. Several variations of the probe end (Figure 3-9) of the thermowell have been designed to expedite the transfer of heat through the thermowell walls.

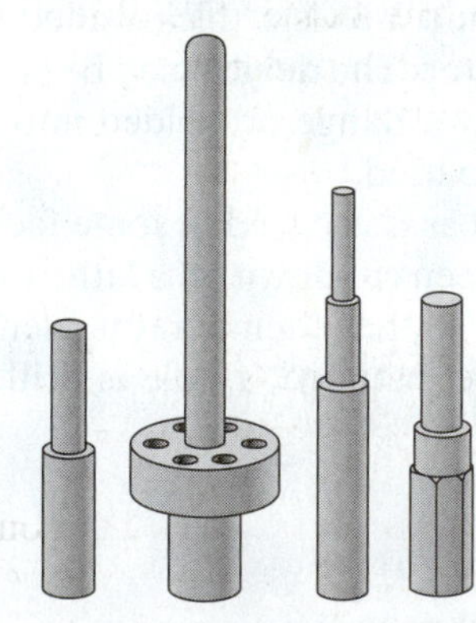

FIGURE 3-9 Thermowell Probe Variations

THERMOMETER

A thermometer is an instrument used to determine the temperature of a substance. Although the term thermometer includes all temperature-sensing devices, the term usually infers a glass-stem thermometer; however, two types of thermometers are discussed in this chapter: glass-stem and infrared.

Glass-stem

Glass-stem thermometers are constructed of a bulb and a capillary tube (Figure 3-10) with a numbered scale. The bulb and the capillary tube contain a liquid such as mercury or colored alcohol. The liquid in the sensing bulb responds to temperature changes by expanding. Since the bulb is connected to the capillary tubing, the additional liquid moves upward into the tube until the volume change has stopped. Volume expansion is directly proportional to the temperature increase or decrease. A temperature reading is taken directly from the etched calibration marks that flank the capillary tube.

To read a thermometer, view it so the top of the liquid column is at eye level (Figure 3-10). Allowing a sufficient time period for the thermometer to reach a steady condition is also important. Taking a reading properly requires that process circulation is adequate to represent a true picture of the process temperature.

Liquid-in-glass thermometers are usually found in indoor applications such as laboratories or on indoor process equipment. For measuring temperatures while gauging or sampling a storage tank, the liquid-in-glass thermometer may be placed in a holder that has a cup at the bottom to hold process fluids against the bulb until a reading can be taken. This apparatus, including the thermometer as lowered into the tank, helps ensure the temperature reading is accurate.

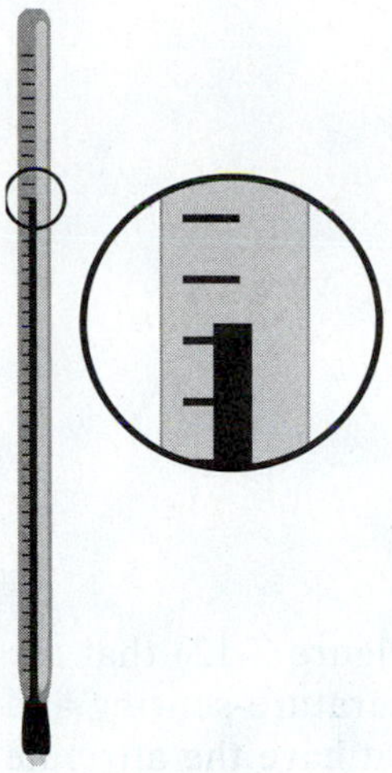

FIGURE 3-10
Thermometer

Infrared

Infrared thermometers are used where glass thermometers would not be practical. Since all objects having a temperature above absolute zero continuously emit energy by radiation in the form of electromagnetic waves or by conduction or convection, then this energy can be measured and temperature determined. Infrared (IR) radiation means thermal energy and it is the only part of the electromagnetic spectrum that runs from high-energy gamma and X rays to long wavelength radio waves. IR wavelengths are usually 1 μm (micrometer) to 1 mm (Figure 3-11).

FIGURE 3-11 Electromagnetic Spectra

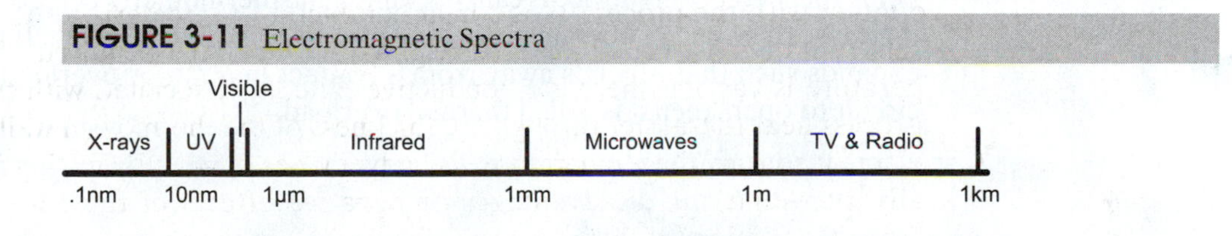

When measuring the IR radiation, the hotter the object, the more the peak energy shifts to shorter wavelengths. The area under the radiation curve is the total energy emitted. IR radiation behaves similarly to light and can travel at or near the speed of light. IR radiation can also be reflected, refracted, absorbed, and emitted. Generally, the amount of IR energy an object radiates is more convenient to measure; then one can infer the temperature of the object. An IR thermometer can calculate the temperature of the object if the emissivity is known. Emissivity is a measure of the blackness of an object or how much it resembles a blackbody. A blackbody is an ideal object that absorbs all radiation that hits it, thus the blackbody is also a perfect emitter.

A spot IR thermometer is noncontact and measures a spot on a target. The measurement spot should be smaller than the object. This device can be used to measure as low as –50 degrees F to as high as 6500 degrees F. This type of IR thermometer can be used to control oven, furnace, and heater temperatures. Line scanners are also available that measure a process line by breaking it up into several spots. Both spot and line thermometers can be used on moving targets while thermal imagers are generally used for stationary targets. Thermal imagers provide information on an entire target area and newer versions act much like night vision glasses by providing scanning capability of an outfall pond or pipe rack.

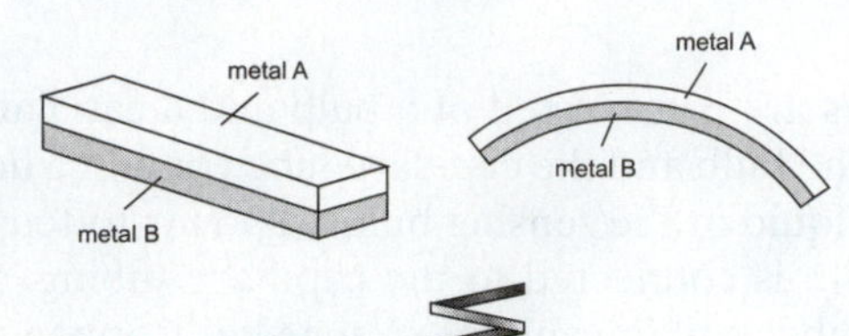

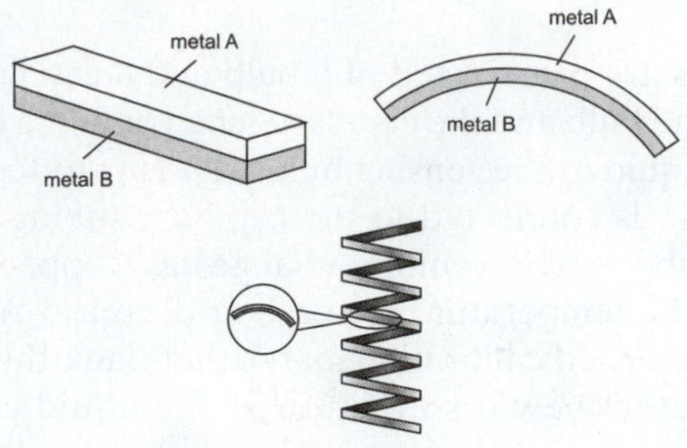

FIGURE 3-12 Bimetallic Element

BIMETALLIC STRIP

A **bimetallic strip** consists of two dissimilar strips of metal (Figure 3-12) that are bonded together. The operating principle of this type of temperature-sensing and measuring device is based on choosing two dissimilar metals that have the attribute of easily expanding and contracting at different rates with the addition or subtraction of heat. The expression used to describe this effect is referred to as the coefficient of thermal expansion. The bimetallic element is often shaped into a spiral or helix for compactness.

Since all materials react to temperature changes at differing rates, each metal in the bimetallic strip must have a significantly different rate of expansion and contraction when exposed to temperature changes and cause a bend or rotation to occur on the strip. When heated, the thermal expansion of the metal with the highest coefficient actually bends the bar in the direction of the other lower coefficient metal.

If a short length of a compound bar (bimetallic strip) can bend enough to be measured, then a longer length of the same bonded metals could bend much more (Figure 3-13). If a long bimetallic strip is coiled into a tight spiral and placed within a metal tube where one end of the coil is attached to an indicator pen, then the expansion/contraction of the strip would move an indicator to represent how much temperature is being measured. This device is called a bimetallic thermometer (Figure 3-14). If the bending causes a switch to operate, then it is called a temperature switch. If the bimetallic strip expands such that it bends away from a contact in a motor overload relay causing the circuit to open, then it is called thermal overload.

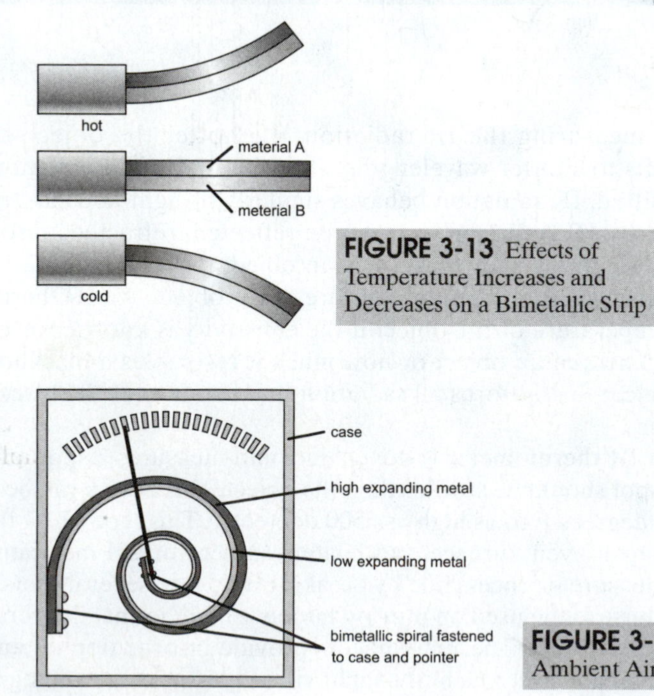

FIGURE 3-13 Effects of Temperature Increases and Decreases on a Bimetallic Strip

FIGURE 3-14 Bimetallic Ambient Air Thermometer

FIGURE 3-15 Dial-Type Thermometer (Bimetallic Coil)

Dial thermometers (Figure 3-15) are the most common thermometers in industrial use and they are frequently of the bimetallic type with circular dials. Filled systems are also available. The dials come in a wide range of temperature scales and styles.

Thermometers placed in thermowells and used with process equipment located in outdoor facilities are generally bimetallic in nature due to their material structure. Bimetallic strips are commonly used in industry for lots of reasons.

RESISTANCE TEMPERATURE DEVICE (RTD)

A **resistance temperature device (RTD)** is a primary element (calibrated resistor) that measures temperature changes in terms of electrical resistance. RTDs are mounted or inserted in thermowells (Figure 3-16) just like any other thermometers and they are second only to thermocouples as the most popular electronic temperature-sensing elements in industry. They are more linear than thermocouples and generally more accurate, but they lack the operating range of a thermocouple. The standard temperature range of an RTD is from –200 to 900 degrees F. Special ranges go as high as 1475 degrees F.

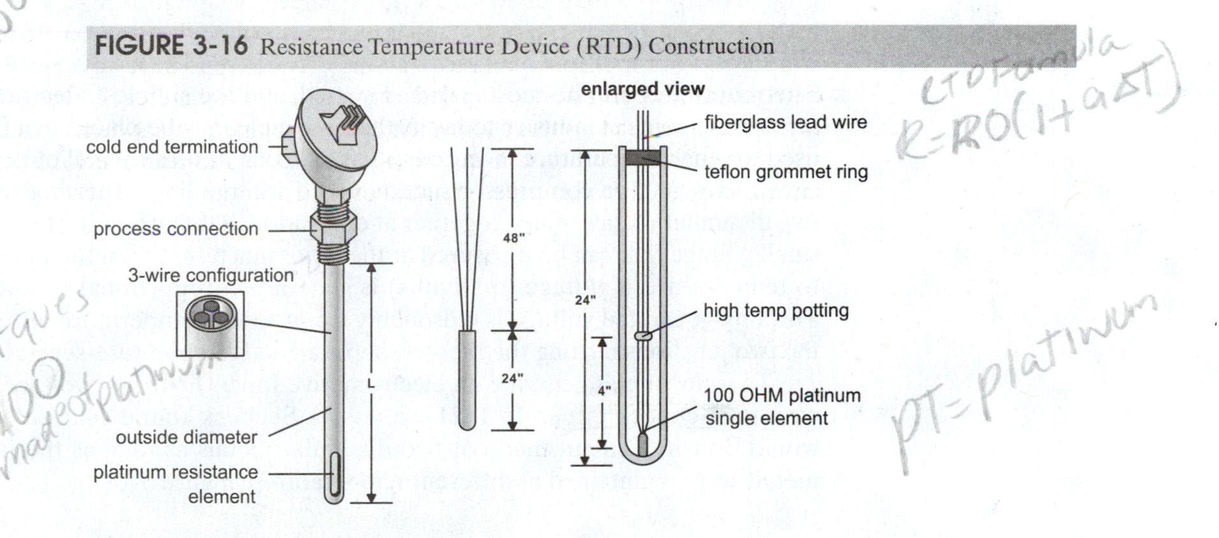

FIGURE 3-16 Resistance Temperature Device (RTD) Construction

The fundamental operating principle of an RTD is that as the temperature of a conductor changes, so does its resistance. The most common RTD is no more than a fine platinum wire wound around a mandrel of nonconductive material. The mass of the wound wire is small and allows quick response to temperature changes.

All conductors respond to changes in temperature. For example, the resistance of slightly doped (where a minute amount of an impurity is added to a pure substance to alter its properties) industrial standard platinum changes by 0.00385 ohms/degrees C. What this means is that a platinum wire with a resistance of 1.0 ohms would change by 0.00385 ohms for every 1 degree Celsius change in temperature. If a platinum wire were long enough to have a resistance of 100 ohms, the wire would change by 0.385 ohms per degree C. This resistance-to-temperature relationship is known as its temperature coefficient of resistance. Other metals such as nickel, iron, and copper are also used as RTDs.

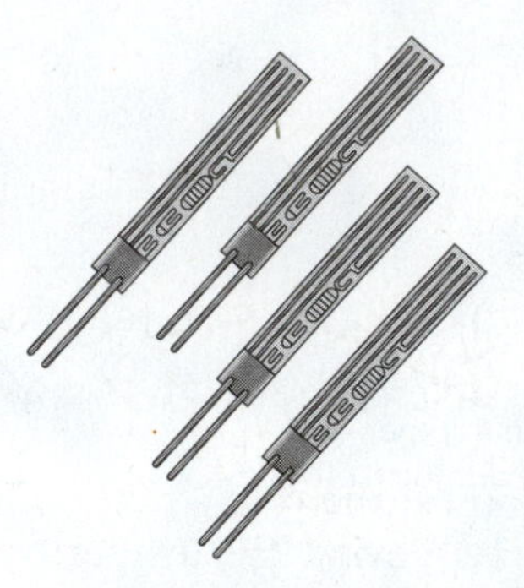

FIGURE 3-17 RTD Probes

The most commonly used RTD in industry is the 100-ohm platinum type (Figure 3-17). This type is called a 100-ohm RTD because its resistance at 0 degrees C is 100 ohms. Since this RTD has a resistance change of 0.385 ohms per 1 degree C, then at the boiling point of water (100 degrees C) the resistance would be 138.5 ohms (100 ohms + 38.5 ohms).

$$\text{RTD resistance} = \text{base} + \text{delta } ^\circ\text{C}$$

In processing industries, many RTDs are connected to transmitters. The transmitter responds to a change in resistance from the RTD and produces a corresponding instrument signal. Since the transmitter cannot distinguish resistance produced by the RTD from the resistance added to it by the lead wires connecting them together, lead wire compensation must be provided. The transmitter normally does this compensation for us by providing input from a third or even fourth wire. These additional wires provide a solution to the lead wire problem by allowing the resistance of the lead wires to be measured and subtracted from the total sensed resistance. By eliminating the lead wire resistance, only the resistance of the RTD is available for processing.

THERMOCOUPLES

Thermocouples are the most commonly used, and the simplest electrical temperature sensing elements in industry today. A thermocouple may be placed in a thermowell and used to sense temperature in various locations such as furnace coil outlets, boiler tubes, steam lines, and on compressor suction and discharge lines. Thermocouples consist of two dissimilar metals joined together at one end such that when heated, they generate a small voltage that can be measured at the other junction. When the junction is exposed to temperature, a voltage (millivolts) is generated proportional to that temperature. This tiny generated voltage is reasonably linear to the temperature difference between the two junctions making the thermocouple a good temperature-sensing element.

Thermocouples generate an electromotive force (EMF), or voltage, in accordance with the Seebeck effect. In 1821, Thomas J. Seebeck found that an electric current would flow in a circuit made of two dissimilar metals as long as the junctions of the metals were maintained at different temperatures (Figure 3-18).

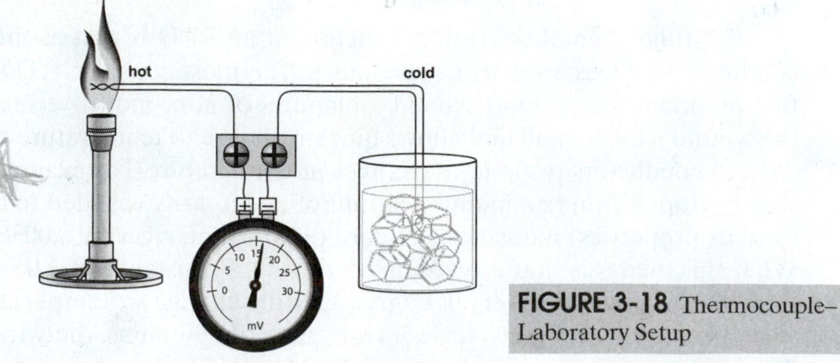

FIGURE 3-18 Thermocouple–Laboratory Setup

Identifying the two junctions (Figure 3-19) is important. The first one, or the *measured junction,* is generally the one that is connected to the process. The second

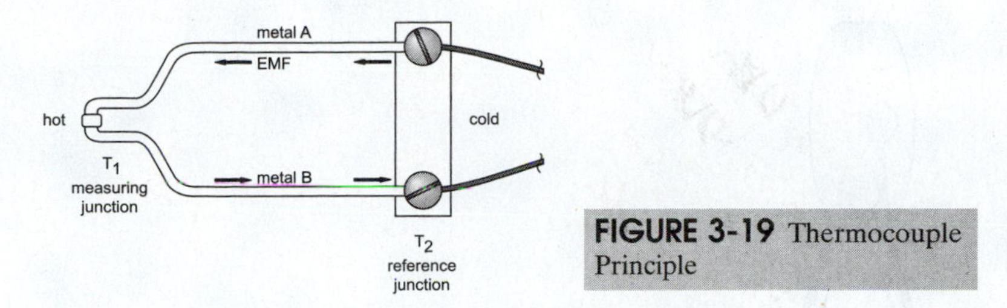

FIGURE 3-19 Thermocouple Principle

junction is called the reference junction (or the cold junction). The second junction is connected to the readout or transmitting device. The temperature difference between the two junctions is all that a thermocouple is capable of measuring. This measurement is no more than a delta voltage measurement. The cold end junction must connect to the readout instrument. If other wires are connected to the cold junction, these wires cause a false reading. In these cases compensations must be made. This is called cold junction compensation.

Thermocouple junctions, wires, and welds are protected by the primary and secondary connection tubes of the thermowell (Figure 3-20).

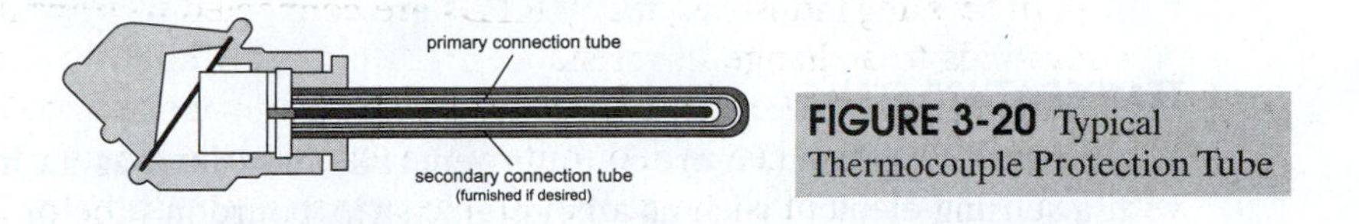

FIGURE 3-20 Typical Thermocouple Protection Tube

Most thermocouples used today are prefabricated and sold or purchased in large quantities. Prefabricated thermocouples are butt-welded. Butt-welding (Figure 3-21) is popular because it provides a joint of minimum mass and yet provides good physical strength.

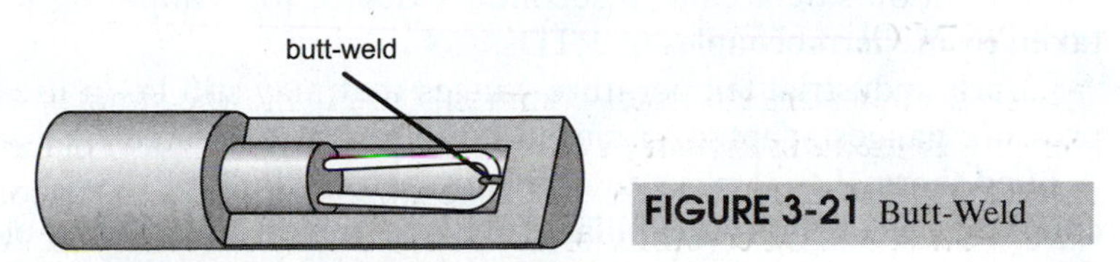

FIGURE 3-21 Butt-Weld

The specifics for providing a good method for cold compensation is beyond the scope of this text, but all the methods available must determine the temperature at the cold junction so that the temperature differences between the junctions can reflect an actual temperature measurement rather than a delta temperature. There may be other temperature measurement devices available that would be easier to use, but thermocouples are usually the best option since they have the widest temperature measuring ranges of all the temperature measuring elements. This fact alone makes selecting them more desirable.

Thermocouples usually fail at the hot junction. The two dissimilar metals are likely to fail at the bonded joint. This failure mode produces an open circuit with the voltage going to zero. If a thermocouple suddenly reads zero or low scale, it is wise to check for thermocouple failure.

THERMISTORS

Thermistors are very small ceramic resistors with a high temperature coefficient of resistance. Thermistors are usually made of a sintered mixture of metallic oxides. For a given change of temperature, the resistance of a thermistor changes approximately 10 times as much as the resistance of a platinum RTD. Thermistors are sensitive to very small changes in temperature. When mounted in small thermowells, thermistors respond very quickly because of their small thermal mass.

Thermistors are usually made in the form of a tiny bead that is encased in glass. Disc thermistors (Figure 3-22) are also available. Since thermistors are not interchangeable, temperature instruments must be calibrated to match the specific thermistor design.

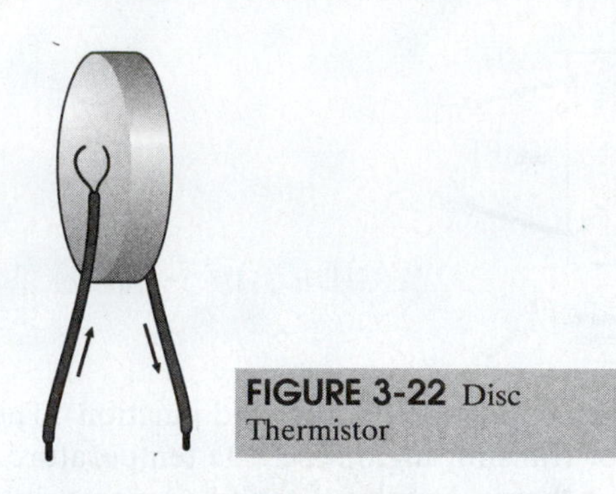

FIGURE 3-22 Disc Thermistor

They are also noted for their long-term drift due to aging. Accuracy and ambient temperature compensation are usually less than conventional temperature sensors.

A thermistor is a semiconductor that exhibits a large (usually negative) temperature coefficient of resistance. Thermistors are very sensitive and permit full-scale operating ranges of less than 1 degree F. The upper operating temperature is determined by physical changes in the semiconductor material and is typically 200-750 degrees F (93-398 degrees C).

TEMPERATURE GAUGE

A **temperature gauge** (Figure 3-23) is generally described as an independent device with a sensing element such as a bimetallic strip, bourdon tube or bellows linked to a pointer that subsequently displays changes in temperature on a calibrated face. Temperature gauges, like pressure gauges, are used in the processing area for local indication and can usually be seen from a distance. The gauge face and pointer provide an easy to read temperature indication from a distance. Temperature gauges may be used in an outside area as a secondary device for comparing temperature readings taken from thermocouples or RTDs.

Early industrial temperature gauges that may still be in use today were actually pressure gauges adapted to measure the pressure created by a *filled thermal system*. A **filled thermal system** is a temperature-sensing bulb filled with a liquid, vapor, or gas connected by means of a capillary tube to a pressure-measuring element. This is where the term temperature gauge began and the faceplate was changed to reflect temperature units.

The temperature bulb (Figure 3-24) of a filled thermal system is usually inserted into a thermowell connected to the process. Filled thermal systems are closed. This means that changes in the process temperatures change the pressure in the volume of the fluid or gas within the temperature-sensing bulb and directly affect the pressure-measuring sensor in the gauge housing.

Temperature-sensing devices such as bimetallic thermometers and any other temperature measuring system (Figure 3-25) that can provide dial-type readouts are loosely

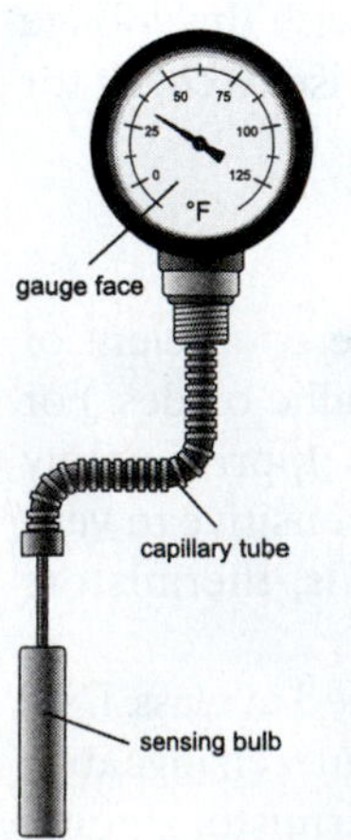

FIGURE 3-23 Temperature Gauge

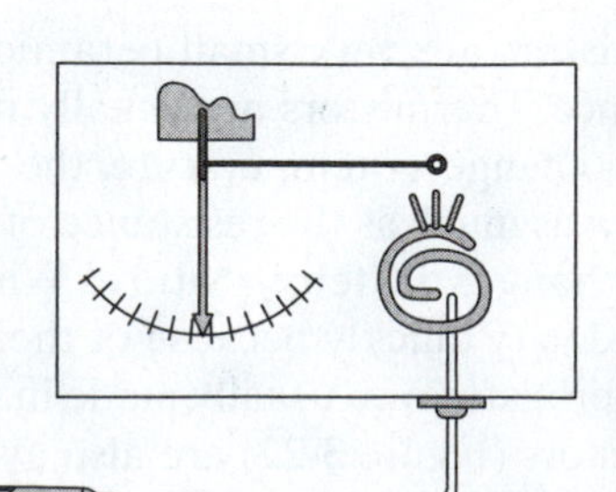

FIGURE 3-24 Filled Thermal System

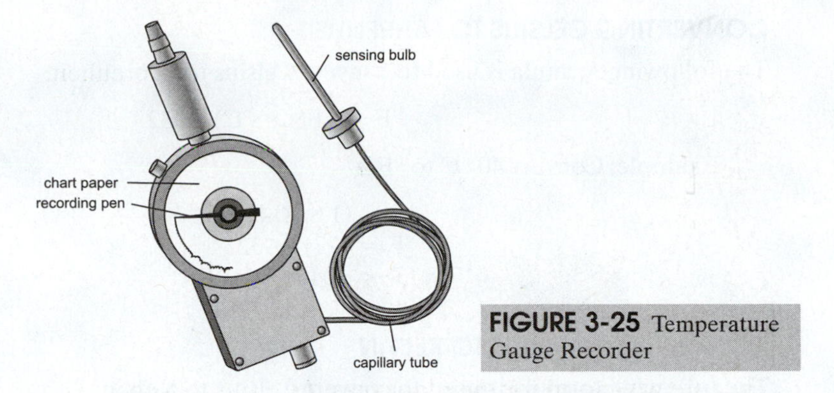

FIGURE 3-25 Temperature Gauge Recorder

referred to as temperature gauges. Common pocket-type thermometers with a dial indicator are usually bimetallic. The measuring element in a bimetallic thermometer is a spiral-shaped bimetallic strip contained in a metallic sheathing (long slender tube capped at one end). As the temperature changes, the bimetallic coil twists in response. One end of the coil is attached to a shaft that is threaded down through the center of the bimetallic coil. The indicator pen is attached to the tip of the shaft and moves accordingly.

Temperature Conversion

Both the Fahrenheit and Celsius temperature scales are important since the Celsius scale is the preferred scale abroad and the Fahrenheit scale dominates the United States. The Fahrenheit scale has 180 units between the freezing point (32 degrees F) and the boiling point (212 degrees F) of water whereas the Celsius scale has 100 units (100 degrees -0 degrees C). Both scales are linear making them easier to convert from one scale to another, but since the zero value is different, another step is added.

When viewing the scales side by side and comparing the unit size, a difference can be seen. In actuality, the Celsius scale unit is 1.8 times larger than the Fahrenheit unit. The two scales also differ by 32 Fahrenheit units. This relationship is shown mathematically below.

$$1.8\,°C = °F - 32$$

The Kelvin scale is an absolute temperature scale that uses the Celsius degree unit. The freezing point of water is 273 K with the zero point on the scale equaling –73 degrees C. When writing Kelvin, the degree mark is not associated with the numerical value.

$$K = °C + 273\,K$$

Conversion formulas are necessary in order to find the corresponding temperature on a different scale. In the following formulas, the scale to convert to is on the left side of the equation. This means that the known value is placed into the right side of the equation. To convert, place the known value into the equation and then use a calculator to do the math.

CONVERTING FAHRENHEIT TO CELSIUS

The following formula is used to convert Fahrenheit to Celsius:

$$°C = \frac{°F - 32}{1.8}$$

Example: Convert 68 °F to °C.

$$°C = \frac{68\,°F - 32}{1.8}$$

$$°C = \frac{36}{1.8}$$

$$°C = 20$$

CONVERTING CELSIUS TO FAHRENHEIT

The following formula is used to convert Celsius to Fahrenheit:

$$°F = (1.8 \times °C) + 32$$

Example: Convert 40 °C to °F.

$$°F = (1.8 \times 40\ °C) + 32$$
$$°F = (72) + 32$$
$$°F = 104$$

CONVERTING CELSIUS TO KELVIN

The following formula is used to convert Celsius to Kelvin:

$$K = °C + 273$$

Example: Convert 100 °C to K.

$$K = 100\ °C + 273$$
$$K = 373$$

CONVERTING KELVIN TO CELSIUS

The following formula is used to convert Kelvin to Celsius:

$$°C = K - 273$$

Example: Convert 500 K to °C.

$$°C = 500\ K - 273$$
$$°C = 227$$

Summary

In this chapter, temperature is described as the degree of hotness or coldness of a substance as indicated on a temperature measuring scale. The amount of heat (temperature) measured is the quantity of potential heat that could be given off from the substance. The giving off of heat equalizes with the surrounding substances until both are at the same temperature. When heat is added to a substance it may change state (solid → liquid → gas). The amount of heat required to change the state of a substance is called latent heat.

The effects of heat energy on molecular movement cause the molecules to become more active and to move faster. This increase in molecular movement also causes a corresponding temperature increase. Heat can be transferred from substance to substance by conduction, convection, and radiation.

Temperature measurements may be taken at single points, multiple averaging points, and/or places where differential (delta) temperatures are required. These measurements are read on temperature measurement scales with the most common ones being Fahrenheit (used almost exclusively in the United States), Celsius (used exclusively in other countries), and their corresponding absolute scales (Rankine and Kelvin). All four scales are based on the boiling and freezing points of pure water.

The boiling point for pure water is 100 degrees on a Celsius scale and 212 degrees on a Fahrenheit scale. The freezing point on a Celsius scale is 0 degrees while the corresponding freeze point on a Fahrenheit scale is 32 degrees. When measuring temperatures at sea level, other things need to be considered when determining the appropriate temperature-sensing device so that the appropriate scale may be applied, understood, and reported correctly for the environment.

The following are common types of temperature-sensing, measurement, and supporting devices used in the process industries:

- *Thermowell*: a protective device usually constructed of stainless steel and used to isolate temperature-sensing elements from adverse process conditions

- *Thermometer*: usually a glass-stem or liquid-in-glass temperature measuring device where the liquid in the bulb has a much greater volume than the volume in the indicator and is inserted into the substance to be measured; the volume increases as the material in the bulb expands and moves upward in the indicator tube until it reaches a stable point where the value is read
- *Bimetallic strip*: two dissimilar metals bonded together so that when heat is applied one of the metals expands or contracts more readily than the other causing a bend or rotation of the strip whereby the amount of movement is calibrated on a dial type readout of the measured temperature
- *Resistance temperature device (RTD)*: a primary element connected to transmitters that measures temperature changes in terms of electrical resistance of a conducting material such as platinum
- *Thermocouple*: the most commonly used temperature-sensing element in industry; it consists of two dissimilar metals joined together at one end so when they are heated, they generate a small voltage that can be measured at the other junction
- *Thermistor*: a small ceramic resistor with a high temperature coefficient of resistance that is sensitive to very small changes in temperature
- *Temperature gauge*: an independent device with a sensing element such as a bimetallic strip, Bourdon tube or bellows linked to a pointer that displace temperature on a calibrated face; a filled thermal system may also be used as a temperature gauge in a closed system where internal pressure does not affect the volume of fluid or gas within the temperature-sensing bulb

The basic formulas for converting between the different temperature scales are as follows:

- Fahrenheit to Celsius:

$$°C = \frac{°F - 32}{1.8}$$

- Celsius to Fahrenheit:

$$°F = (1.8 \times °C) + 32$$

- Celsius to Kelvin:

$$K = °C + 273$$

- Kelvin to Celsius:

$$°C = K - 273$$

Checking Your Knowledge

1. Temperature is defined as a specific _____ of hotness or coldness as indicated on a definite scale.
   a. Range
   b. Unit
   c. Type
   d. Degree
2. _____ is the manner in which heat moves through solid matter.
   a. Conduction
   b. Convection
   c. Radiation
   d. All of the above
3. Which temperature scale(s) contain(s) 100 degrees between the freezing point of water and the boiling point of water?
   a. Centigrade/Celsius
   b. Fahrenheit
   c. Rankine
   d. Kelvin

4. Explain how and why temperatures are converted from Fahrenheit to Celsius or vice versa.
5. List the formulas used for temperature conversions.
6. What is the result after converting 212 degrees F into degrees C.
   a. 0
   b. 32
   c. 100
   d. 373
7. Make the following temperature scale conversions:
   a. Convert 27 C to Fahrenheit.
   b. Convert 95 °F to Celsius.
   c. Convert 32 °F to Celsius.
   d. Convert 100 °C to Fahrenheit.

Student Activities

1. Find a thermometer with a typical glass-with-liquid configuration that is normally used outside. Take it outside and observe the temperature changes while in direct sunlight, in the shade, on a hot day, on a cold day, etc. Be sure to allow the thermometer to adjust to the temperature differences by allowing it to remain in a place for several minutes before taking a reading. Take your readings and find the relationships between them such as delta, averaging, and/or identify the differences that may occur on a hot day versus a cold day. Write down your observations as they would apply to a process technician working outside.
2. Using a dial readout (with a pointer) thermometer commonly used in outdoor gardens, perform the same activities as the liquid-in-glass thermometer. Compare your results. Look inside the dial readout device and see how it was made. Be careful not to destroy the device.
3. Cut a circle out of common copy paper and then starting at an angle, cut a spiral around and around until the center is reached. Punch a small hole in the center. Tie a knot in the end of a piece of string and thread the other end through the hole in the spiral. Light a small candle and hold the spiral by the string over the candle taking care NOT to put the paper too close to the flame. Observe what happens to the spiral. Is this heat transfer radiation, convection or conduction? How would this experiment apply to a plane flying over a refining facility?
4. Identify temperature-sensing and measurement devices in the lab.
5. Use a multimeter and hook up leads to a thermocouple. Heat the thermocouple with a heat gun (the type used on carpet because it gets much hotter). Have students read the mV ouput on the meter. After the mV output is reached, have students use conversion tables to get from millivolts to temperature for the type of thermocouple used (J or A).
6. Use sling psychrometer to measure wet and dry bulb temperature to determine dew points and percentage of relative humidity.
7. **TEMPERATURE CONVERSION WORKSHEET**
   Instructions
   - Using the Temperature Conversion chart, perform the calculations on a clean sheet of paper.
   - Turn in your completed calculations as per your instructor's guidelines.
   - Ensure problem numbers and corresponding answers are clearly written.
   - Use the following conversion chart to complete the following conversion exercises:

Temperature Conversion

What is to be converted:	*Formula to be Used:*
Fahrenheit to Celsius	$°C = \frac{°F - 32}{1.8}$
Celsius to Fahrenheit	$°F = (1.8 \times °C) + 32$
Celsius to Kelvin	$K = °C + 273$
Kelvin to Celsius	$°C = K - 273$

   - Perform the following temperature conversions on a sheet of paper and then write an e-mail listing the problem number and your answer. Send it to your instructor.

Temperature Conversion Problems	
1. 0 °F =	°C
2. 32 °F =	°C
3. 68 °F =	°C
4. 100 °F =	°C
5. 212 °F =	°C
6. 0 °C =	°F
7. 25 °C =	°F
8. 37 °C =	°F
9. 50 °C =	°F
10. 200 °C =	°F
11. 0 °C =	K
12. –50 °C =	K
13. 100 °C =	K
14. 400 K =	°C
15. 50 K =	°C

8. **TEMPERATURE MEASUREMENT Laboratory Procedure**

Background Information

The student must be capable of connecting a digital multimeter to a thermocouple and then reading the millivolt scale appropriately.

A thermocouple is one of the most common temperature-sensing elements in the process industry. Thermocouples are nothing more than two dissimilar wires joined (usually fused) at one end and connected to a measuring device at the other. A thermocouple operates according to the Seebeck Effect, which states that a voltage is produced when there is a temperature difference between two junctions of dissimilar metals in a circuit. The measuring junction is also called the *hot* junction and the reference junction is called the *cold* junction. Since the voltage (actually millivolt) output from the thermocouple is based upon the difference in temperature between the two junctions, the thermocouple actually measures differential temperature. If a thermocouple is to provide an actual temperature on a scale such as the Fahrenheit or Celsius, the reference junction temperature must be known. In fact, the conversion tables that are used to convert the raw mV signal generated from the thermocouple into a temperature is referenced to the freezing point of water (also known as ice-point). Modern electronic transmitters have the ability to compensate for the temperature difference between the real reference junction temperature and true ice point, whereas the original thermocouple work was based upon dipping the *cold* junction into ice water. In this lab, there is no luxury of ice-point compensation so ice water is used instead.

Another useful bit of information is that thermocouple wires are color coded so that a technician can readily identify them according to their type. For example, a type J thermocouple has a white positive conductor and a red negative conductor. A type K thermocouple has a yellow positive conductor and a red negative conductor. The industry standard for thermocouple wire is to color the negative lead either red or a shade of red.

Materials Needed

- type J thermocouple or type J extension wire
- ice
- 250-ml beaker (minimum size)
- hair dryer or other regulated heat source
- digital multimeter with a millivolt selectable position or scale
- millivolt-to-temperature table for type J thermocouples
- glass thermometer

Safety Requirements

Students should always wear safety glasses in the lab.

Simplified Procedure

a. Identify the sensor or instrument.
b. Describe the function of the sensor or instrument.
c. Calculate an expected result of the sensor or instrument.
d. Conduct an experiment to verify the calculated results.

Detailed Procedure

To perform this lab, the following steps should be performed:

a. Connect a multimeter to the lead wires of a type J thermocouple as illustrated.
   Record the mV reading: _____mV
   NOTE: Connect the ampmeter and the positive and negative thermocouple leads (see Additional Information below) according to their proper polarity.
b. Place the thermocouple into a well-stirred beaker of mostly ice in water.
   Record the mV reading: _____mV
c. Determine the ice water temperature according to the conversion table.
   Temperature derived from the conversion table: _____°C
d. Compare this temperature to a glass thermometer reading ambient temperature.
   Temp: _____°C
e. Using a heat source such as a hair dryer, heat the thermocouple again. Wait until the reading has had time to stabilize, then record the mV reading.
   Record the mV reading _____mV
f. Determine the temperature of the hot air generated by the hair dryer.
   Hot air temperature: _____°C

Additional Information

If manufactured thermocouples are not available, twisting the two conductors of thermocouple lead wire together works.

Findings

Write an observational conclusion to this lab. In this conclusion, be sure to explain what happens to a thermocouple when the hot and cold junctions are at equal temperatures and when they are not. Include an explanation about ice point and how it affects the raw millivolt reading taken by the digital multimeter.

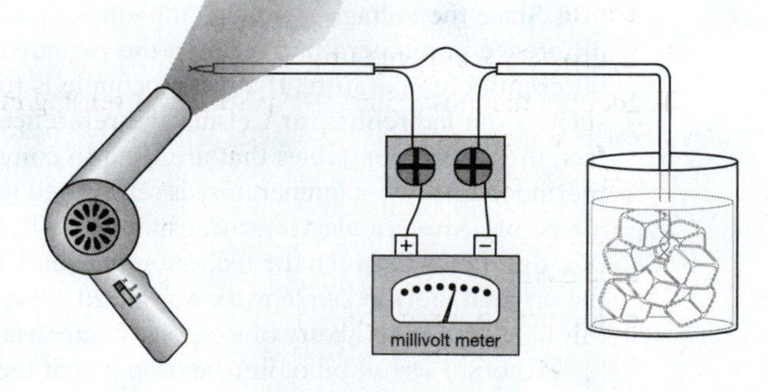

CHAPTER 4

Process Variables, Elements, and Instruments: LEVEL

Objectives

After completing this chapter, you will be able to:

- Define terms associated with level and level instruments:
  - level
  - innage
  - ullage (outage)
  - direct and indirect measurement
  - interface level
  - meniscus
  - density
  - hydrostatic head pressure
- Identify the most common types of level-sensing/measuring devices used in the process industries:
  - gauge or sight glass
  - float
  - tape gauge
  - differential pressure cell
  - bubbler
  - displacer
  - ultrasonic device
  - nuclear device
- Describe the purpose and operation of various types of level-sensing and measuring devices.
- Describe the relationship between temperature and level control as it relates to the density of liquid.
- Describe the relationship between temperature and level control as it relates to the volume of a liquid.

Key Terms

Bubbler—a special kind of head pressure measuring method allowing the measurement of head pressure in a liquid without the pressure sensor coming in contact with the process fluid.

Continuous level measurement—monitors all level points in the tank from the zero percent level (bottom of the measuring device) to 100 percent full.

Depth—a distance from the surface of a liquid that extends downward.

Direct level measurement—measures the process variable directly in terms of itself (e.g., a sight glass or dipstick).

Displacer—a sealed cylindrically shaped tube used to measure buoyancy.

Flat-glass level gauge—there are three types of flat-glass level gauges: reflex, transparent, and welded pad. The heavy, thick-bodied glass promotes safety in high-temperature and high-pressure hydrocarbon service.

Float gauge—an instrument that uses cables, pulleys, levers or any other mechanism to convey the position of a float to a liquid level.

Height—a distance from a zero reference point to the surface.

Inches of water column—liquid head pressure measurement expression; also expressed as pounds per square inch (psi).

Indirect level measurement—measures another process variable (e.g., head pressure or weight) in order to infer level.

Innage—the measurement from the bottom of a tank to the surface of the product.

Interface level—a plane where two materials meet; can be between two liquids, a liquid and a slurry, or a liquid and foam.

Level measurement—the act of establishing the height of a liquid surface in reference to a zero point.

Liquid (hydrostatic) head pressure—the pressure exerted by the height of a column of liquid; the most common indirect level measurement.

Load cell—a device comprised of a strain gauge bonded to a robust support column called a force beam; together these are capable of supporting the entire weight of a vessel and its contents; a transducer that measures force or weight.

Meniscus—the curved upper surface of a column of liquid having either a convex or concave shape.

Nuclear device—uses a tightly controlled gamma radiation source with a detector to infer a level in a tank and used when other technologies are unsuccessful; located on the outside of a tank and impervious to the effects of adverse process conditions; common gamma radiation sources are the radioisotopes cobalt 60 and cesium 137.

Percent level measurement—a measurement of level based upon percentage with zero percent level at the bottom of the measuring device and 100 percent level being at the top.

Point level measurement—level is measured at one distinct point in a tank.

Pounds per square inch (psi)—liquid head pressure measurement expression; also expressed as inches of water column.

Reflex level gauge—has a single flat-glass panel capable of refracting light off a prism-like backside creating a silvery contrast above the liquid level that allows it to be seen from a distance.

Tape—a narrow strip of calibrated ribbon (steel) used to measure length.

Tape gauge—a level-measuring device consisting of a metal tape that has one end attached to an indicator and the other end attached to a float.

Transparent level gauge—has two transparent flat-glass panels, front and back, that form a vertical chamber in conjunction with the metal sides where the process level can be seen and measured using this vertical chamber.

Tubular-type sight glass—a reflex or clear glass tube open into a vessel on top and bottom.

Ullage (outage)—the measurement from the surface of the product to the top of the tank.

Ultrasonic and radar device—an accurate distance-measuring instrument usually inserted into the top of a vessel; emits a pulse of energy that is reflected off the surface of the material back to the receiver; may be either a single unit (transponder) or two separate units (transmitter and receiver).

Welded pad gauge—generally made of flat glass and integrally mounted to the vessel by either a welded or flanged connection with threaded pipe; extremely rugged to prevent vessel drainout.

Introduction to Level and Terms

This chapter defines the process variable identified as level as well as the most common terms associated with level measurement. After this, the most common types of level-measuring devices are discussed as to their respective purpose and operation. Then, a discussion of the relationships between level, temperature, density, and volume brings everything together to show the effects of each one on the other.

WHAT IS LEVEL?

Gravity causes the surface of an undisturbed liquid to be a flat level plane. In the process industry, the word *level* describes this plane in terms of its height above a reference point. Accordingly, **level measurement** can be defined as the act of establishing the height of a liquid surface in reference to a zero point.

Level measurements are usually concerned with quantifying the position of the surface of the liquid as compared to a reference point. The most common way to express level is in liquid height. **Height** is the distance (Figure 4-1) from a zero reference point in a container to the surface of the material. **Depth** is a distance measurement starting at the surface (Figure 4-1) of a liquid and then extends downward. Depth measurements are uncommon in industry, but the process technician should be able to differentiate between the depth of a liquid and the more common height measurement. In the United States, the actual height of a liquid in a tank is expressed in feet and/or inches (Example: 10 feet 3 inches). Level measurements may also be expressed in feet with decimal subunits (Example: 10.25 feet).

The most common way to indicate the level of a tank is by **percent level measurement**. This measurement of level is based upon percentage with a zero percent level indicating empty, a 50 percent level indicating half full, and a 100 percent level indicating that the tank is full.

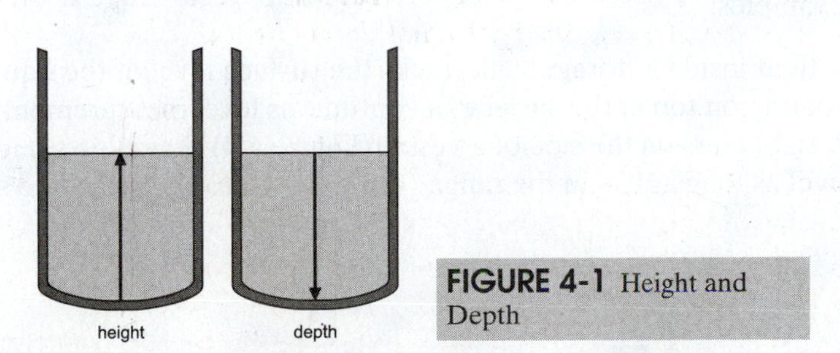

FIGURE 4-1 Height and Depth

INNAGE AND ULLAGE (OUTAGE)

Process technicians should always be cognizant of tank levels. Tanks can be overfilled or under filled. Tanks are often charged (filled) with several different materials and should be monitored accordingly. Depending upon the specific needs of a process, one tank may require a larger vapor space than another. Concerns such as these require our knowledge of two terms that are used in and around industry that express this very

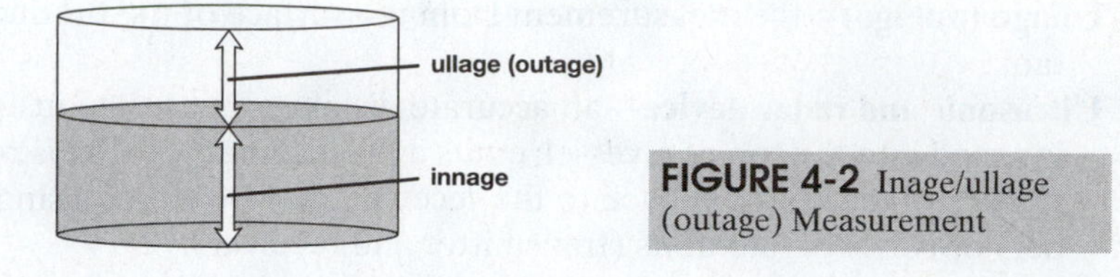

FIGURE 4-2 Inage/ullage (outage) Measurement

concept. They are innage and ullage. **Innage** is the measurement from the bottom of the tank to the surface of the product and **ullage**, also known as **outage**, is the measurement from the surface of the product to the top of the vessel (Figure 4-2).

LEVEL MEASUREMENT CATEGORIES AND METHODS

Level measurements are divided into two distinct categories: point level measurement and continuous level measurement (Figure 4-3).

- **Point level measurement:** Where level is measured at one distinct point in a tank. The most common reason for providing point measurements is to establish high and low level alarm points.
- **Continuous level measurement:** Used to monitor all level points in the tank from zero percent at the bottom of the measuring device to 100 percent full.

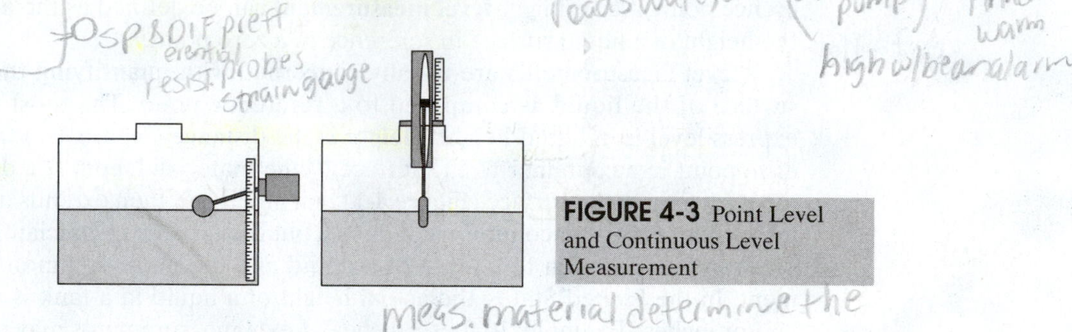

FIGURE 4-3 Point Level and Continuous Level Measurement

The measurement terms direct and indirect are differentiated by how they actually measure a variable such as level. A **direct level measurement** (Figure 4-4) measures the process variable in terms of itself.

The following are examples of how the direct level method may be applied to measure level:

Examples:

- A float inside a storage tank tracks the surface level of the liquid contents by floating on top of the surface (a continuous level measurement).
- A sight glass on the side of a vessel (Figure 4-4) that allows one to see the actual level as long as it is in the range (top to bottom) shown by the sight glass.

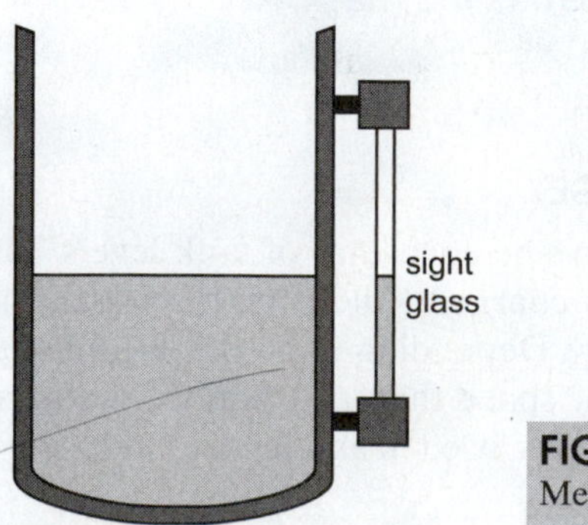

FIGURE 4-4 Direct Level Measurement

- A dipstick, when inserted into a tank vertically until the end of the dipstick rests on the bottom (a point level measurement), is wetted from the bottom end of the dipstick to the top of the liquid surface. When the dipstick is raised, the actual measurement can be read by observing the place where the liquid stopped wetting the stick.

Indirect level measurements are characterized by how they measure one property of the contained material such as head pressure or weight to infer another measurement, which in this case is level. As shown in Figure 4-5, the pressure gauge at the bottom of the vessel indicates the total amount of head pressure. The greater the amount of pressure observed, the higher the level is in the tank. Load cells and displacers are examples where indirect measurements are translated to levels. These devices are described in detail within the area of discussion on level measuring devices.

The term *datum* or *datum line* is used to describe the zero reference point for level measurement. For a dipstick, this is the bottom of the stick where the measurement starts at zero. For a sight glass, the zero reference point is the lowest point at which the scale can be read. For a transmitter, the zero reference point, or datum line, is the calibrated point at which the zero point is established.

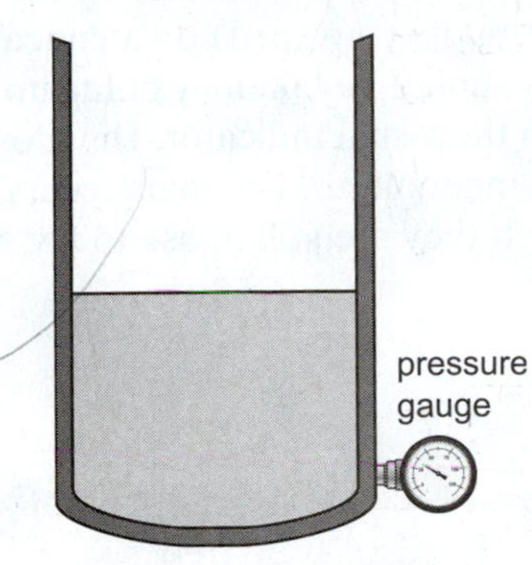

FIGURE 4-5 Indirect Level Measurement

INTERFACE LEVEL

The common boundary, or plane, between two immiscible (i.e., incapable of mixing) liquids is called an **interface level**. This interface (Figure 4-6) can be between two liquids, a liquid and slurry, or a liquid and foam. If gasoline and water are added in a clear jar and then shaken, the two liquids almost immediately separate. The water (heaviest liquid) naturally flows to the bottom of the jar and the gasoline to the top. Once the two liquids have settled, there are two surfaces in the jar. One surface is the water surface at the water and gasoline interface and the other is the customary surface, the air and gasoline interface.

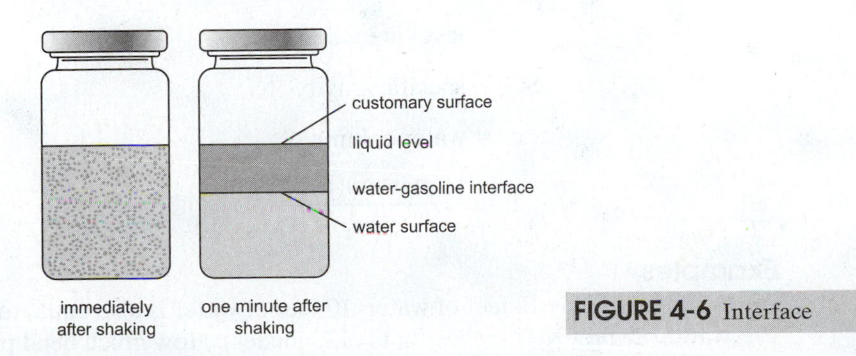

FIGURE 4-6 Interface

In industry, if separating two immiscible liquids were necessary, the mixture would be transferred into a separating tank (Figure 4-7) where a decanting action could take place. The lighter liquid spills over a weir (which is like a dam or holding plate) into a separate compartment for removal while the heavier liquid is removed in front of the

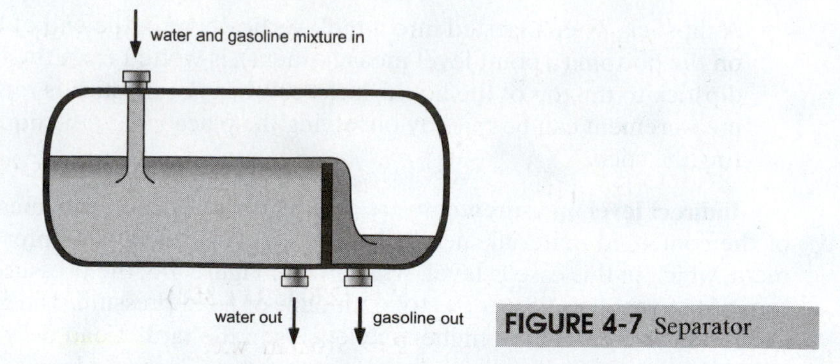

FIGURE 4-7 Separator

weir through a pipe connected to the bottom of the tank. Keeping track of the liquids' interface ensures that the outflows, both top and bottom, can remain free of the other material.

MENISCUS

When taking direct measurements, understanding the concept of a meniscus is important. A **meniscus** (Figure 4-8) is the curved upper surface of a column of liquid. A meniscus can have a convex shape (bulging upward) or a concave shape (swaying downward). In either case, the reading should be taken by lining up the highest or lowest portion of the meniscus curve with the visual indicator. The most common place to observe a meniscus is in a glass tube manometer. For many years, manometers were associated with storage tanks. Although they are still in use today, their numbers have diminished significantly.

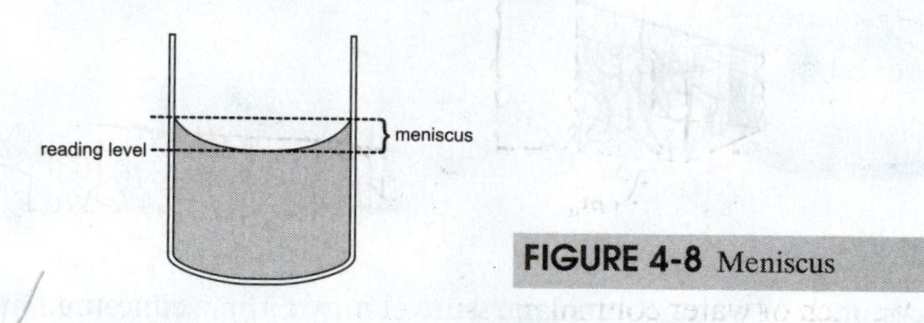

FIGURE 4-8 Meniscus

DENSITY AND HYDROSTATIC HEAD PRESSURE

Density is mass per unit volume. A convenient way of expressing the density of a material is through specific gravity. The specific gravity of a liquid is the ratio of the density of a liquid to the density of water.

Formula for liquid (hydrostatic) head pressure:

P = hydrostatic head pressure (inches w.c.)

h = level in inches

SG = specific gravity

w.c. = water column

$$P = \frac{(1 \text{ in. w.c.})(h)(SG)}{\text{in.}}$$

Examples

- A tank contains 10 feet of water; 10 feet of water is also equal to 120 inches of water (10 feet X 12 inches/foot = 120 inches). How much head pressure is exerted by the full height of the liquid?

$$P = \frac{(1 \text{ in. w.c.})(120 \text{ in.})(1.0)}{\text{in.}}$$

$$P = 120 \text{ in. w.c.}$$

This example incorporates the one-inch water column pressure per one-inch height factor. This is a constant that equates to "1" and can now be assumed for all future examples and word problems.

- An open top tank holds a liquid with a specific gravity of 1.350. The liquid has a vertical height of 230 in. How much head pressure (in inches water column) does this liquid exert?

$$P = (h)\ (SG)$$

$$P = (230 \text{ in.})\ (1.350)$$

$$P = 310.5 \text{ in. w.c.}$$

In the process industries, most liquid level measurements are typically provided through indirect methods. The most common indirect level measurement method is **liquid (hydrostatic) head pressure**. Head pressure is the pressure exerted by a column height of liquid. Liquid head pressure (Figure 4-9) is usually expressed in **inches of water column** (in. w.c.) or in pounds per square inch (psi).

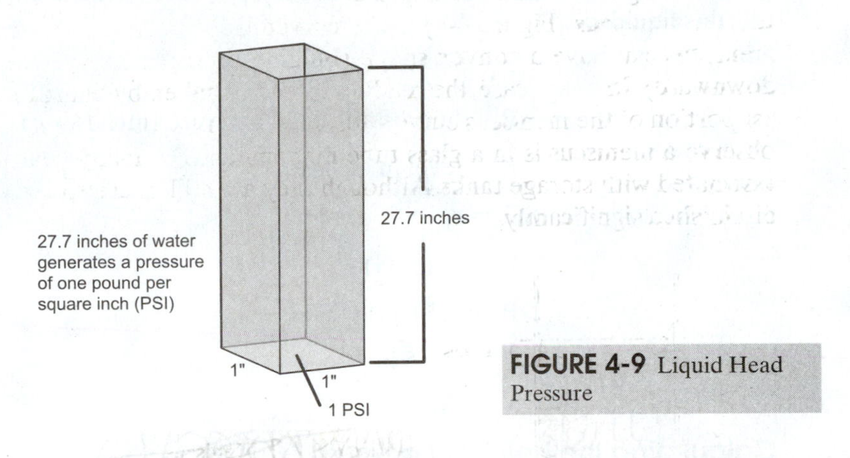
27.7 inches
27.7 inches of water generates a pressure of one pound per square inch (PSI)
1"
1"
1 PSI

FIGURE 4-9 Liquid Head Pressure

One inch of water column pressure (1 in. w.c.) is the amount of pressure exerted by the height of 1 in. of water. So, 2 in. of water exerts 2 in. water column pressure and 100 in. of water exerts 100 in. of water column pressure. As you can see, this pressure unit is self-defining and very easy to use when calculating head pressure.

As stated above, liquid head pressure is also expressed in the form of **pounds per square inch (psi)**. If the height of the liquid and its density are known, one can calculate the equivalent head pressure in pounds per square inch. One foot of water exerts 0.433 psi of head pressure. Hence, ten feet of water exerts 4.33 psi.

Calculating Head Pressure

Head pressure can be calculated if the height of the liquid and its density are known. A convenient way of expressing the density of a material is through specific gravity. The specific gravity (SG) of a liquid is the ratio of its density to the density of water. This makes calculating the head pressure of any liquid easy. As mentioned earlier, one inch of water exerts a one-inch water column (w.c.) pressure. So, to calculate head pressure, take the height of a liquid in inches and multiply it by its specific gravity.

Examples

- A tank contains 10 feet of water; 10 feet of water is also equal to 120 inches of water, expressed mathematically as:

$$10 \text{ ft} \times \frac{12 \text{ in.}}{1 \text{ ft}} = 120 \text{ in.}$$

How much head pressure is exerted by the full height of the liquid?

Where:

$$P = \text{hydrostatic head pressure (inches w.c.)}$$

$$h = \text{level in inches}$$

$$SG = \text{specific gravity}$$

$$P = \frac{(1 \text{ in. w.c.})\,(h)\,(SG)}{\text{in.}}$$

$$P = \frac{(1 \text{ in. w.c.})(120 \not{\text{in.}})(1.0)}{\not{\text{in.}}} = 120 \text{ in. w.c.}$$

This example incorporates the one-inch water column pressure per one-inch height factor. This constant can now be assumed for all future examples and word problems.

- An open top tank holds a liquid with a specific gravity of 1.350. The liquid has a vertical height of 230 in. How much head pressure (in inches water column) does this liquid exert?

  Where:

$$P = \text{hydrostatic head pressure (inches w.c)}$$

$$h = \text{level in inches}$$

$$SG = \text{specific gravity}$$

$$P = (h)\,(SG)$$

$$P = (230 \text{ in.})\,(1.350)$$

$$P = 310.5 \text{ in. w.c.}$$

Calculating the Height of a Liquid

Using this same simple approach to solve for head pressure, solve for liquid height instead. To do this, the total head pressure and the density of the liquid must be known. Liquid height is equal to head pressure divided by its density ($h = P/SG$).

Examples

- A pressure gauge attached to the bottom of a tank (vented to atmosphere) is indicating 200 in. w.c. The specific gravity of the liquid in the tank is 1.25.

  Where:

$$P = \text{hydrostatic head pressure (inches w.c.)}$$

$$h = \text{level in inches}$$

$$SG = \text{specific gravity}$$

$$h = \frac{P}{SG}$$

$$h = \frac{(\text{head pressure})}{(\text{specific gravity})}$$

$$h = \frac{(200 \text{ in. w.c.})}{(1.25)}$$

$$h = 160 \text{ in.} = 13 \text{ ft } 4 \text{ in.}$$

Formula for height of a liquid:

$$P = \text{hydrostatic head pressure (inches w.c.)}$$

$$h = \text{level in inches}$$

$$SG = \text{specific gravity}$$

$$h = \frac{P}{SG}$$

- A pressure gauge attached to the bottom of a tank (vented to atmosphere) is indicating 200 inches w.c. The specific gravity of the liquid in the tank is 1.25. What is the height of the liquid?

$$h = \frac{200 \text{ inches w.c.}}{1.25}$$

$$h = 160 \text{ inches w.c.}$$

Level-Sensing and Measurement Instruments

The most common types of level-sensing and measuring devices used in the process industry include the following:

- gauge or sight glass
- float
- tape gauge
- differential pressure cell
- bubbler
- displacer
- ultrasonic device
- nuclear device
- load cell

GAUGE OR SIGHT GLASS

A common sense law of physics states that a liquid seeks its own level. Gauge glasses operate just like U-tube manometers. If they have an equal pressure applied to both sides, their levels are the same. Since gravity applies equally to both sides, if the height of the liquid is greater on one side, the height naturally adjusts by moving liquid between the two chambers. For the most part, gauge glasses are very accurate. However, phenomena exist that can cause a slight offset in column height between the two chambers. One such phenomenon is capillary action and the other is an extreme temperature difference that causes the density of the material contained in the gauge glass (which is on the *outside* of the vessel) to be heavier.

Gauge glasses are among the oldest direct level measurement devices still in use today. The fundamental purpose of a gauge glass is to indicate level. Gauge glasses serve as a *window* into the process. This can be very helpful when troubleshooting the system. There are two basic types: the tubular glass type and the flat-glass type. Also, as described below, there are several variations of the flat-glass type.

The externally mounted tubular, reflex and transparent gauge glass types should have block valves between their top and bottom connection and the tank. In fact, most of these block valves are equipped with an internal ball check valve. The ball check valve seats off (i.e., closes) if there is an unusually large amount of flow through rate. These valves are added as a safety feature so that if the glass breaks, only a small amount of fluid is able to get out.

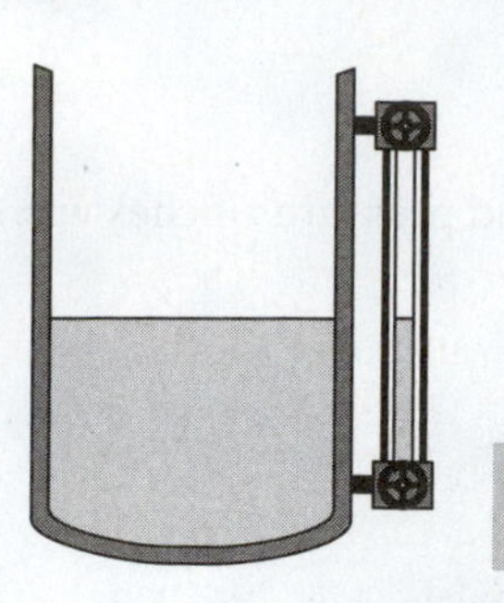

FIGURE 4-10 Tubular-Type Sight Glass

Tubular-type Sight Glass

The **tubular-type sight glass** (Figure 4-10) is the oldest type of gauge glass still in use today and serves as a window into the process. The tubular type sight glass has a reflex or clear glass tube and is open into a vessel on the top and the bottom. They are relatively inexpensive but fragile, which is why they are used less frequently. Tubular type sight glasses operate just like U-tube manometers; the sight glass level matches the liquid in the vessel; if they have equal pressure applied to both sides, their levels will be the same.

Flat-glass Level Gauges

Flat-glass level gauges (Figure 4-11) are designed with heavy bodies and thick glass to promote safety in high temperature and high-pressure hydrocarbon service. Special valves are commonly used at the top and bottom of these gauges so the gauge glass can be cleaned and/or removed while material is still in the tank.

There are three types of flat-glass level gauges: reflex, transparent, and welded pad.

- Reflex level gauges:

  Reflex level gauges have a single flat-glass panel capable of refracting light off of a prism-like backside creating a silvery contrast above the liquid level. The level in a reflex gauge glass can be seen from a distance, which is helpful when making rounds in a darkened process area. Reflex gauges are not capable of distinguishing between immiscible liquids.

- Transparent level gauges:

  Transparent level gauges have two transparent flat-glass panels, front and back, that form a vertical chamber in conjunction with the metal sides. The process level can be seen and measured using this vertical chamber. However, transparent level gauges are not as easy to read as the reflex level gauge even though one can see right through them. Transparent gauges are required to measure an interface level.

- Welded pad gauges:

  Welded pad gauges are generally made of flat glass, but they are integrally mounted to the vessel. They can be welded, flanged or connected with threaded pipe. This type must be extremely rugged because unlike the others, if the weld is compromised, all the material in the tank can drain out.

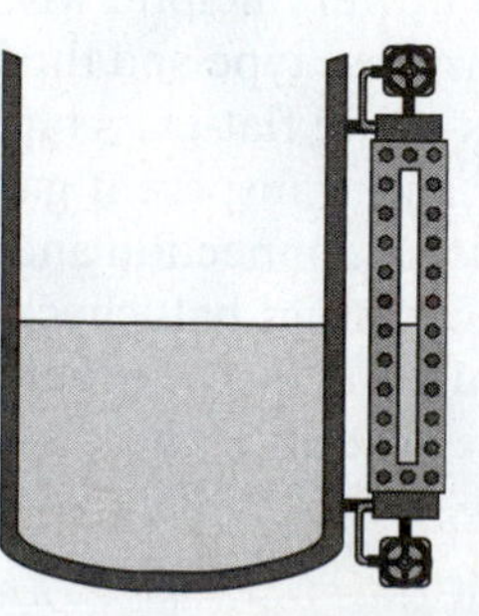

FIGURE 4-11 Flat-Glass Level Gauge

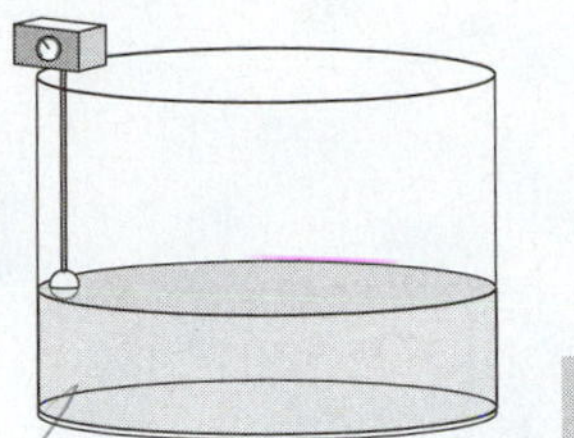

FIGURE 4-12 Float Gauge

FLOATS AND FLOAT GAUGES

A *float* is a level-sensing element that tracks the surface of a liquid by floating on it. The float must be buoyant enough to float on the surface and heavy enough to provide the force necessary to actuate the level tracking mechanism.

A **float gauge** (Figure 4-12) is an instrument that uses cables, pulleys, levers or any other mechanism to convey the position of a float to a liquid level. Instruments that are called float gauges are usually point level devices. Most point level floats are used to actuate high/low level alarms and shutdown circuits.

Float level gauges operate on two simple principles: buoyancy and mechanical action. The float senses the surface of the liquid and then a mechanism or transducer equates its position into a level measurement.

Floats can also directly actuate a control valve. Imagine a simple toilet bowl float and valve mechanism. If you connect the industrial float to one end of a metal rod that pivots on a fulcrum point, then the other end can actually open and close a control valve.

Another type of float system employs a magnetic coupling (Figure 4-13). This type is primarily used as a point level device although it can also be used for continuous level measurements. When the magnet in the float is pushed by the float or comes close enough to the magnetic follower, a switch snaps open and/or snaps closed providing an input into a high or low-level alarm circuit.

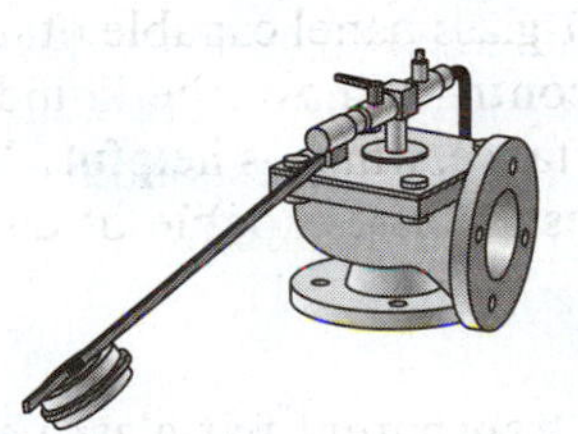

FIGURE 4-13 Float System–Magnetic

In a continuous level-measuring instrument, the float and the following device are also equipped with a magnet so that they stay magnetically coupled with each other. As the float rises and falls with the surface of the liquid, the magnetic follower moves with it. The following mechanism is isolated from the process fluid by a nonmagnetic housing. A cable to a readout device is connected to the magnetic follower that can then provide a level indication.

Although there are applications where a direct float-operated continuous measuring device is found, most of them fall under the category of tape gauges and are discussed below.

TAPES AND TAPE GAUGES

A **tape** is usually defined as a narrow calibrated strip or ribbon (steel) used to measure length. A **tape gauge** is a level-measuring device consisting of a metal tape that has one end attached to an indicator and the other end attached to a float.

Metal tape gauges are generally used to measure continuous level changes in a tank. Specifically, metal tapes are used because they do not stretch under normal conditions. A taut tape provides an absolute distance measurement between the surface of the level and the indicator mechanism. Consider this, a storage tank with a diameter of 50 feet has 1,224 gallons of material per inch of level. If the material in this tank were to sell for several dollars per gallon, the accuracy of this level gauge would be critical.

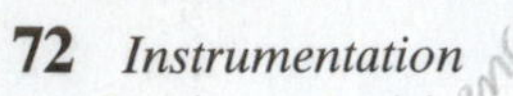

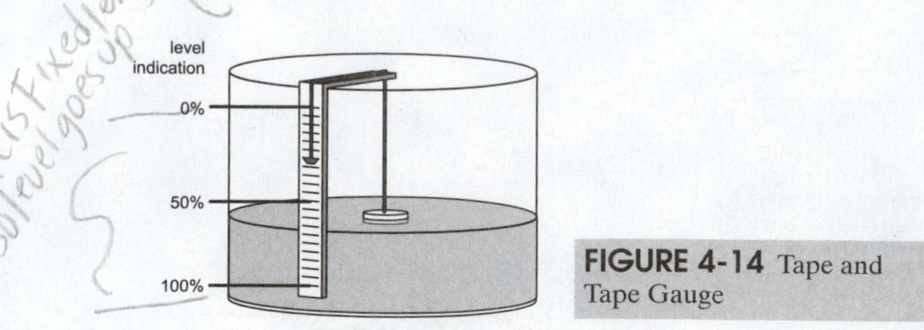

FIGURE 4-14 Tape and Tape Gauge

Float and gauge board devices (Figure 4-14) are simple in construction and function. This gauge is comprised of a float that rides on the surface of the liquid, a connecting metal tape (or cable) and an indicator board. Notice that the indicator board is numerically inverted because the indicator is attached to the float by means of a specific length of tape. As the level in the tank decreases, the float rides down with the level pulling the indicator upward on the board. Conversely, as the level increases, the indicator slides downward towards the 100% mark on the indicator board.

As the level increases, the more sophisticated tape gauges rely on a spring assisted reeling action to take up the slack in the tape instead of a counterweight. The indicators in these mechanisms most probably possess a digital readout instead of an indicator board. In either case, the tape gauge works the same. As the level changes, the length of the tape is carefully measured providing an accurate level measurement.

Tape gauges can also measure solids in a vessel. Instead of the float constantly sitting on the surface, a sensing bob or sounder can be lowered from a drum and sensor mechanism located on the top of the tank to the surface of the solid. The drum is connected to a reversible servo-motor (a motor specially designed for high dynamics) and torque sensor. When the sensing bob hits the surface, the torque sensor notes the spot and the reversible motor rewinds the cable and sensing bob back to the top of the vessel.

DIFFERENTIAL PRESSURE (D/P) CELLS

In a previous chapter, a differential pressure cell (Figure 4-15) was described as a special type of pressure sensor that simultaneously measures two different pressure points in reference to one another and then produces a corresponding output signal (e.g., for tank level). One of the two inputs to the d/p cell is considered as the high-side pressure input and the other, the low-side pressure input.

FIGURE 4-15 Differential Pressure (d/p) Cells

When a d/p transmitter is used to measure level (Figure 4-16), the high-pressure side of the d/p cell is connected to the bottom of the tank and the low-pressure side is connected to the top of the tank. The high side measures total pressure that is a combined head pressure of the liquid plus any pressure applied to its surface. The low side measures only vapor space pressure. Since a d/p cell measures the difference between two pressures, any pressure applied to both sides of the cell cancels out. This effectively eliminates the vapor space pressure from the measurement. With the vapor space pressure eliminated, the only pressure remaining is the head pressure exerted by the liquid.

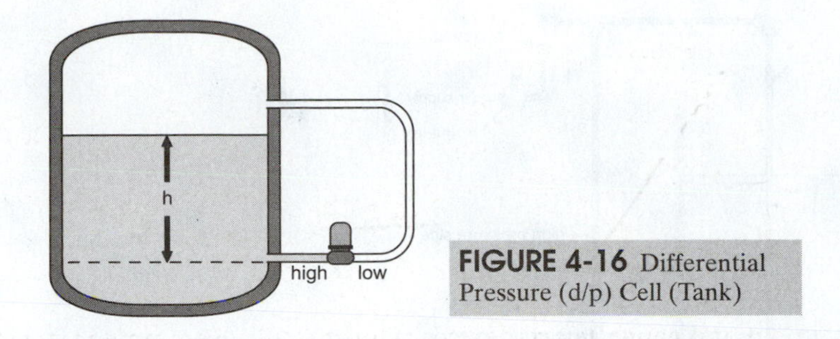

FIGURE 4-16 Differential Pressure (d/p) Cell (Tank)

The upper tap-sensing line is occasionally filled with a reference fluid to the upper tap location. This arrangement causes the former low side to transmit a constant high pressure that reverses the usual operation of a d/p cell level transmitter. This is sometimes referred to as a wet leg (Figure 4-17) level transmitter.

FIGURE 4-17 D/P Cell Level Transmitter–Dry Leg vs. Wet Leg

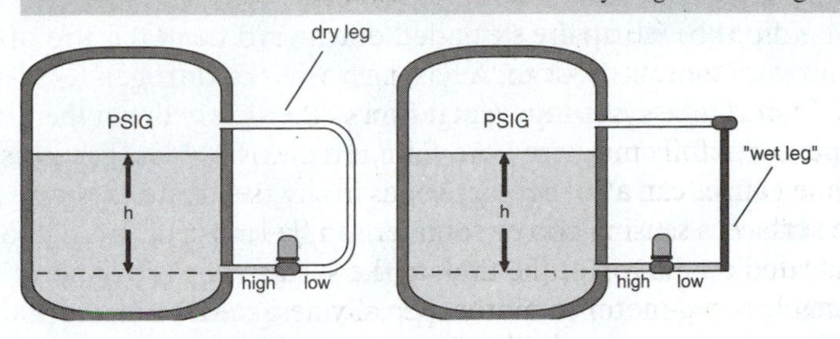

Note that a vessel indicates empty if the liquid level is below the lower tap and below the measuring range of the level instrument. The zero range (Figure 4-18) suppression is required to indicate the correct value.

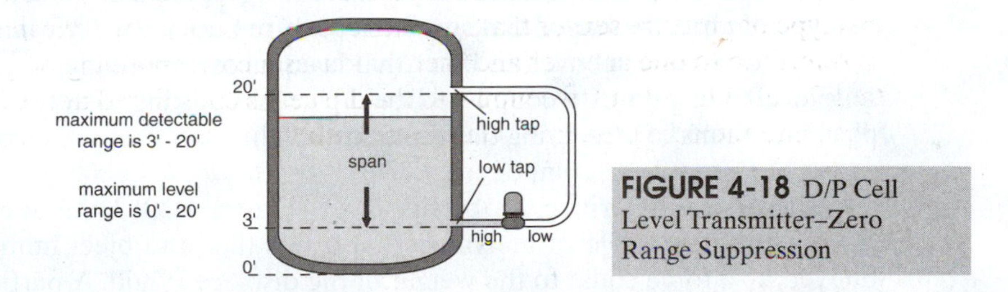

FIGURE 4-18 D/P Cell Level Transmitter–Zero Range Suppression

Span is the difference between the upper and lower detectable values (e.g., 20' – 3' = 17'). If the instrument range is in inches of water, correction must be made for the fluid specific gravity if different from water. This is especially important if a wet leg transmitter is employed with a reference fluid.

Head pressure is the most common variable used to infer level in industry. Recall that head pressure is the pressure exerted by a column height of liquid. If all tanks were open to atmosphere, then measuring head pressure could be done with a simple pressure-sensing instrument. But in reality, almost every tank in industry is a closed environment where a certain amount of pressure is trapped above the liquid in the vapor space. As the temperature (or level) changes, the vapor space pressure also changes. The d/p cell (transmitter) works best in this type of situation because of its ability to measure the head pressure produced by the liquid in reference to the varying pressure in the vapor space.

BUBBLERS

A **bubbler** is a special kind of head pressure measuring method that allows us to measure the head pressure of a liquid without the pressure sensor coming in contact with the process fluid. Corrosive properties or other potentially problematic physical properties

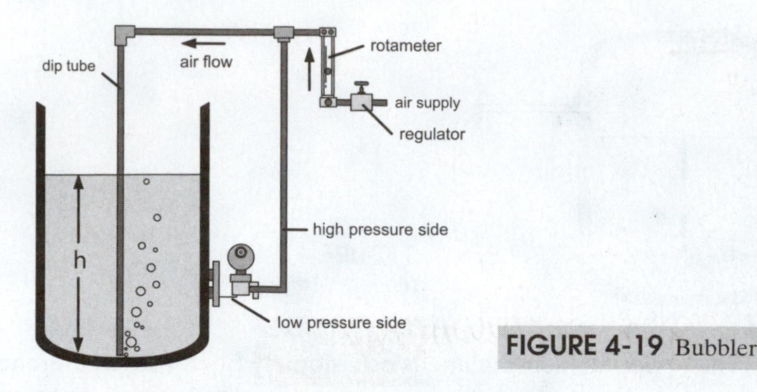

FIGURE 4-19 Bubbler

of the process fluid can be lessened or eliminated by using a bubbler system instead of a directly connected sensing device. A bubbler is comprised of a purging gas source, pressure and flow regulating device(s), a pressure-sensing device and an open-ended tube called a dip tube (Figure 4-19).

The dip tube is usually extended downward from the top of the tank to about an inch or two above the bottom. A gas, such as air or nitrogen (N_2), is supplied to this otherwise closed tube system so that it flows very slowly down the tube and escapes out of the open end of the tube into the fluid creating bubbles. The pressure of the gas in the tube must equal the total pressure exerted by the liquid at the end of the tube if it is to bubble out. Also, the diameter of the dip tube must be large enough so that the pressure throughout the tube is the same. A d/p transmitter or other pressure-sensing device is connected to the dip tube providing a measurement of the back pressure. The transmitter is calibrated so that its output represents the measured pressure as a level measurement.

DISPLACERS

Displacers are among the oldest level-measuring devices in industry. A **displacer** is a sealed cylindrically shaped tube used to measure buoyancy. *Buoyancy* is defined as an upward force on a submerged body that is equal to the weight of the displaced fluid. Displacers are part of a complete liquid level measuring system comprised of a displacer, a means of measuring the apparent weight change of the displacer, and an indicating or transmitting component.

The operating principle of a displacer (Figure 4-20) level device is based upon Archimedes' principle of buoyancy that states that an object immersed in a fluid is buoyed by a force equal to the weight of the displaced fluid. A partially or completely covered displacer exerts a buoyant force equal to the weight of the displaced fluid. As the liquid rises over the length of the displacer, its effective weight changes due to the volume of the fluid displaced. This resultant weight change is directly proportional to the change in level.

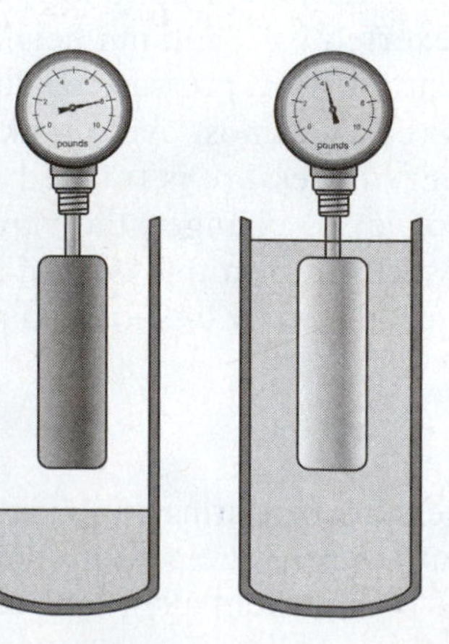

FIGURE 4-20 Displacer Principle of Operation

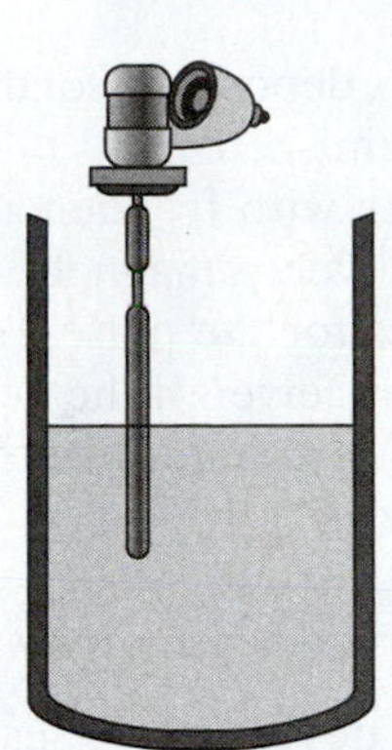

FIGURE 4-21 Displacer Torque Tube

Displacers are also used to measure level indirectly by measuring the change in the buoyancy experienced by the displacer as the liquid rises across its length. They are used in refineries and chemical plants throughout the world to provide point and/or continuous level measurement. Displacers can provide an interface level measurement as long as there is a sufficient difference in densities between the two liquids.

The upper level maximum for the gauge is when the displacer is completely submerged. To measure tall tanks several displacers are added together. The displacer is hung from an extension arm connected to a device called a torque tube (Figure 4-21), which acts in the same manner as a weight measuring spring. Another mechanism or transducer converts the change in rotation of the torque tube (in response to the weight change) into a level indication or signal output.

The cylinder shape body of the displacer provides a consistent buoyancy change per incremental change in level. Also note that the displacer must weigh more than the fluid it displaces so that it always remains in a vertical position. The basic premise of a displacer is that it displaces the fluid and does not float on it. The other factor to remember is that displacers do not measure the true bottom level or the true top most level, but rather the low or high level as determined by the displacer placement within the vessel.

DISTANCE-MEASURING DEVICES

Ultrasonic and *radar* electromagnetic distance-measuring devices (Figure 4-22) are distance-measuring instruments that are usually inserted into the top of a vessel, although ultrasonic types can be located on the bottom. **Ultrasonic and radar devices** are very accurate. Both instruments emit a pulse of energy that is reflected off of the surface of the material back to the receiver. They can be two separate units, a transmitter and a receiver, or they may be a single unit called a transponder.

Ultrasonic and radar devices make excellent noncontact level measuring instruments. Since neither of these devices come in direct contact with the process material, they can be used when many other level-sensing types cannot. Some process fluids demand a noncontact sensing device because of their chemical or physical properties. These fluids may be highly corrosive, which would reduce the life of the sensor, or they

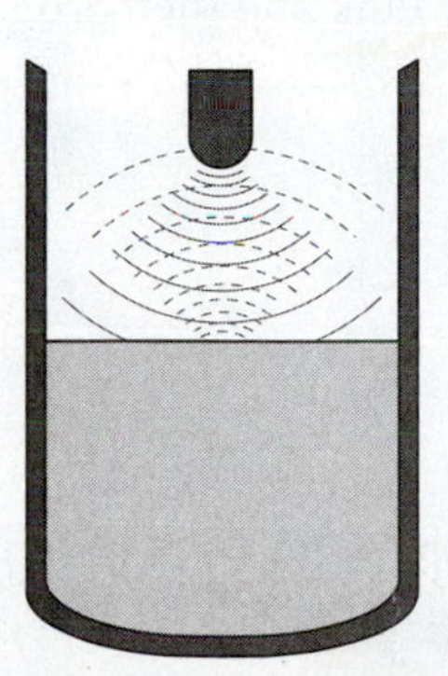

FIGURE 4-22 Ultrasonic or Radar

may simply have a tendency towards caking (i.e., depositing solids onto the sensor). In either case, the effectiveness of the sensor is diminished.

The word ultrasonic identifies sound waves with frequencies higher than those detectable by the human ear. Radar pulses are electromagnetic radio waves. In both cases, the instrument calculates the time it takes for the pulse to reach the surface of the material, bounce off, and return back to the receiver. If the density of the vapors in the space above the liquid (or solid) is known, then a simple distance calculation can be provided resulting in an inferred level measurement.

NUCLEAR DEVICES

A nuclear level instrument (Figure 4-23) is a device that uses a tightly controlled gamma radiation source with a detector to infer a level in a tank. The two most commonly used gamma radiation sources are the radioisotopes cobalt 60 and cesium 137.

CAUTION:* *Nuclear devices are VERY dangerous and should be handled properly at all times.

Nuclear devices can be used for point and continuous level measurements. When other technologies are unsuccessful, nuclear devices may be the only choice. Nuclear devices are located on the outside of the tank making it impossible for adverse conditions within the tank to affect them.

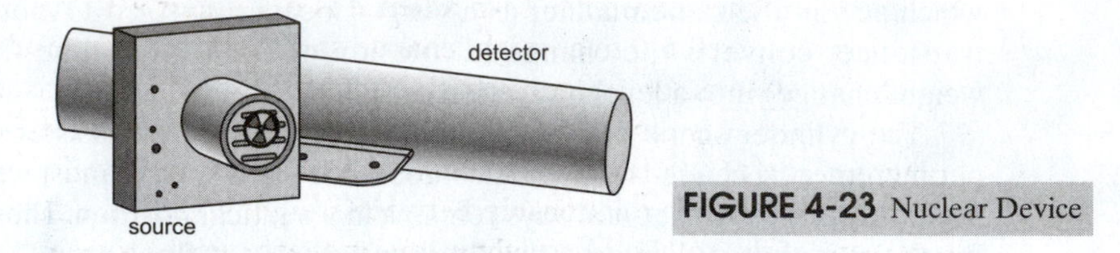

FIGURE 4-23 Nuclear Device

The most common nuclear radiation is cobalt 60 (Co^{60}) and cesium 137 (Cs^{137}) gamma rays. The gamma rays emitted from a nuclear source penetrate the vessel walls and the process material inside and are detected on the other side of the vessel. If level is being measured, the measurement can be continuous level or point measurement. Level is measured by how many gamma rays get absorbed. For example, an empty vessel may be equivalent to 2,000-counts/sec. and a full vessel equivalent to 50 counts/sec. The transmitter linearizes the counts for percent level in between these counts, or a linear source may be matched to the vessel.

Some of the properties of gamma radiation are similar to visible light. However, unlike visible light, the short wavelength and higher-energy gamma rays can penetrate the vessel walls and the process material inside. Imagine standing in a dark room with a bright light shinning through a translucent plastic tank containing a liquid. Your eyes would probably be able to see a diminished, yet detectable amount of light from the other side of the tank. This is similar to how a gamma source and detector operates.

The typical installation (Figure 4-24) of a nuclear level system is to place the source (a point source or a strip source) on one side of the vessel and the strip detector on the other side. The strip detector can discern the point where the gamma radiation is diminished (shadowed) by the liquid in the tank and then produces an output corresponding to a level.

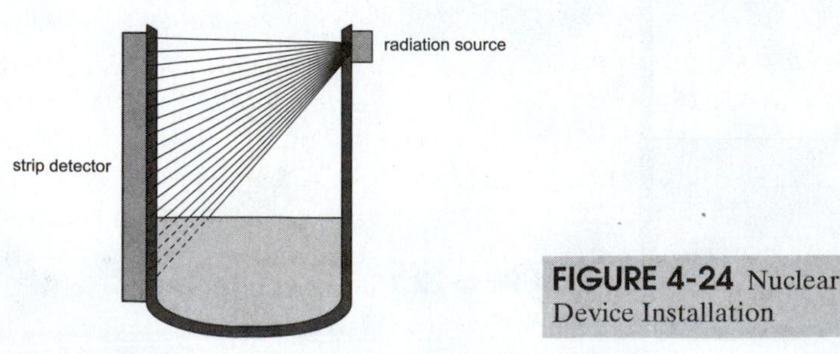

FIGURE 4-24 Nuclear Device Installation

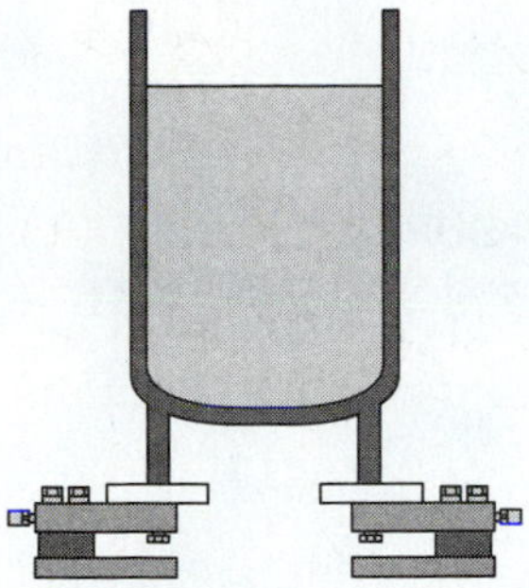

FIGURE 4-25 Load Cell

LOAD CELLS

A **load cell**, an indirect level measuring device, is a transducer used to measure force or weight. A typical load cell (Figure 4-25) is comprised of a strain gauge bonded to a robust support column called a force beam. This force beam is usually a short piece of solid steel capable of supporting the entire weight of a vessel and its contents. As weight is applied to the load cell, a readout and/or transmitting device provides a corresponding output measurement. As weight is applied to the load cell, the associated strain gauges measure the deformity of the force beam. This produces a change in resistance that can be measured by an electronic circuit. Recall that a strain gauge under strain has a higher resistance than when it is not under a strain. This change in resistance can be correlated to a weight change and therefore, a level change (Figure 4-25).

The most common process variable inferred by a load cell is level. If the vessel is weighed when it is empty and then again when it is full, a relationship between its weight and level can be established. If the tank is cylindrically shaped and standing upright, then a linear relationship between weight and level exists. For example, when 20 percent of the total tank weight minus the tare weight is measured, then the level in the tank would also be 20 percent. Load cells are capable of measuring either tension (hanging weight) or compression (supporting weight). Hoppers (a special type of vessel that enhances the movement of solids) are filled with solids and usually hang from a tension (pulling or stretching the load cell) type load cell while tanks containing liquids usually sit on a compression (compressing the load cell) type load cell.

Electronic load cells (Figure 4-26) are the most common providers of direct weight measurement in a plant. Weighing tank trucks and rail cars transporting materials into and out of a plant are two of the many direct weight-measuring applications found in the process industry. Weight can also be used to infer the flow rate. A solid material moving across a weigh point (load cell) as it speeds down a conveyor belt can provide a weight per unit of time measurement.

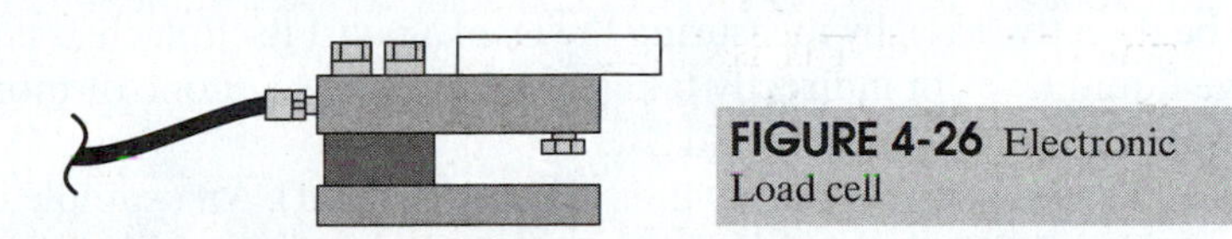

FIGURE 4-26 Electronic Load cell

Level, Temperature, Density, and Volume

Level measurement was defined earlier as the act of establishing the height of a liquid surface in reference to a zero point. Recall also that when the mercury in a sealed glass bulb thermometer is heated, it expands up into the capillary tube. When the water in the radiator of a car heats up (Figure 4-27), the water volume increases as indicated by the HOT versus COLD line marks on the cooling system overflow reservoir. Fluids characteristically act in this manner.

Temperature (Figure 4-28) affects the density of all matter. The physical manifestation of this phenomenon is that a fixed amount of matter experiencing an increase in

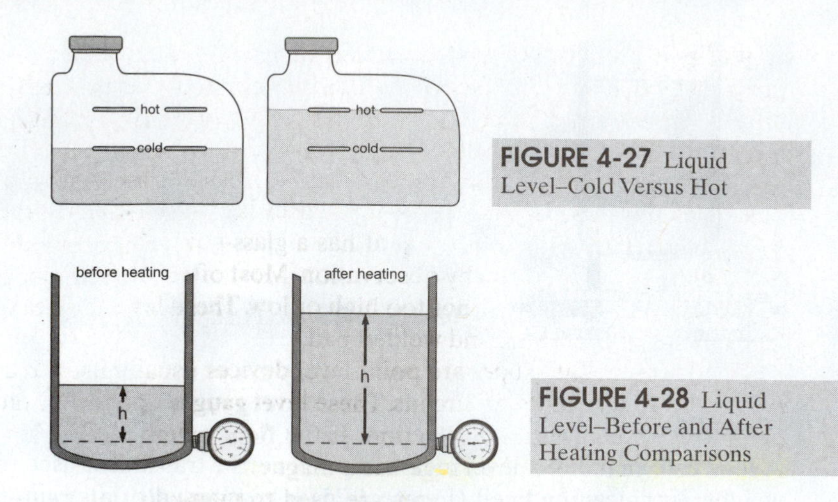

FIGURE 4-27 Liquid Level–Cold Versus Hot

FIGURE 4-28 Liquid Level–Before and After Heating Comparisons

temperature requires more room because its volume increases proportionally. If its volume has increased and it still has the same mass, then its density must have changed to compensate for the new relationship. Recall that density is defined as mass per unit of volume. Again, if the fixed amount of mass occupies a larger volume, then its density has to decrease.

There are several ramifications to this physical phenomenon. First, if a tank is filled to its capacity and then heated, its contents may expand to a problematic level. Under normal conditions, standard operating procedures would generally take this into account; however, a new process technician may not know to do this.

A second problem is an instrument problem. Level measuring instruments, such as differential pressure instruments, rely on certain characteristics of the process remaining constant. Recall that head pressure is equal to the height of the liquid times its density. If the density of the liquid in the tank decreases, which means it occupies more volume, then the liquid has to respond by expanding upward in the tank increasing its height. Interestingly though, the head pressure stays the same. Bear in mind that head pressure is a function of the height of the liquid times its new density.

Summary

The process industry defines level as the act of establishing the height, a flat level plane, of an undisturbed liquid surface in reference to a zero point. The measurement of the liquid height may be made in several methods to include measuring what is in (i.e., innage) the container (e.g., storage tank) or what is not (i.e. ullage/outage) in the container. Height and depth may also be described as a percentage, such as 50 percent full. Liquid level measurements may be taken at only one point (point level) in a vessel or at all level points (continuous level) from zero to 100 percent. Measurements may also be directly taken by measuring the level against itself, such as a float on the surface of the liquid level, or indirectly taken through the use of one or more properties (head pressure or weight) of the liquid.

Liquids may stratify due to their density (weight). An example is a waste tank containing both oil and water. Water is generally heavier and it falls out to the bottom while oil is lighter and floats on top. The point at which the two materials join is called the interface. Both liquids combine for a total level in the container, but each liquid has its own percentage level. Process technicians must ensure that the water draw takes only the water portion while sending the oil portion to reclamation.

When observing a liquid level in a jar or other small container such as a glass tube manometer, understanding how the liquid tends to have either a convex or concave appearance is very important when taking a reading. Level readings should be taken on the highest or lowest part of the surface.

In the process industries, most liquid level measurements are typically provided through indirect methods. The most common indirect level measurement method is head pressure. Head pressure is the pressure exerted by a column height of liquid.

Liquid head pressure is usually expressed in inches water column (in. w.c.) or in pounds per square inch (psi). To determine head pressure, the density, mass per unit volume, must also be known. Density may be expressed as specific gravity, which is a ratio of the density of the liquid to the density of water.

There are many common level-sensing and measurement instruments. One of the most common direct-level-measuring devices is a level gauge commonly known in the field as a sight glass. This instrument has a glass-covered portion allowing the process technician to read the level by observation. Most often the technician is only interested in knowing that the level is not too high or low. These level gauges come in three basic types: reflex, transparent, and welded pad.

Floats and float gauges are point level devices usually used to actuate high or low level alarms and shutdown circuits. These level gauges operate on the simple principles of buoyancy and mechanical action. If the float gauge is used in a continuous level measuring application, the gauge has a magnet.

Simple calibrated metal tapes are used to manually determine level. These tapes may also be connected to an indicator on one end and a float on the other. This apparatus measures continuous level changes in the associated tank or vessel. Accuracy on this type of level gauge is very important. When used to measure solids in a vessel, the float is replaced with a sensing bob and sounder mechanism.

When a d/p cell is installed to measure level, the high-pressure side is connected to the bottom of the tank and the low-pressure side is connected to the top of the tank. The high side measures total pressure (combined head pressure plus surface pressure) while the low side measures only the vapor space pressure above the liquid level. The vapor pressure measurement is cancelled out from both measurements leaving only the pressure exerted by the head pressure of the liquid itself.

Bubblers are special head pressure measuring devices where the pressure sensor does not come into contact with the process fluid. This level measuring system prevents, lessens, or eliminates problems from corrosive or other problematic physical properties of the substance being measured. A dip tube is positioned inside the vessel to about one or two inches above the bottom surface. A d/p transmitter or other pressure-sensing device is connected to the end of the tube to measure back pressure.

Displacers measure buoyancy by using a sealed cylindrically shaped tube (equal to the weight of the fluid it displaces) submerged into the liquid. Displacers measure the level indirectly by measuring the change in buoyancy experienced by the displacer as the liquid rises across the length of the displacer. The resultant weight change is directly proportional to the change in level.

Ultrasonic and/or radar devices are distance-measuring instruments usually inserted into the top of a vessel. They emit a pulse of energy that is reflected off the surface of the material back to the receiver. Since this device type is noncontact, it can be used where many other types cannot be used due to the chemical or physical properties of the substance. Ultrasonic devices use a sound wave while radar pulses are electromagnetic radio waves.

Nuclear level measurement devices use a tightly controlled gamma radiation source with a detector to infer tank levels. The most common gamma radiation sources are the cobalt 60 and cesium 137 radioisotopes. These devices can be used in either point or continuous level measurements. Typical installations have a strip on the inside and outside where detection is determined by the shadowing of the light when comparing the liquid covering the strip to where it is not covering the strip.

Load cells are indirect-level-measuring devices that are a transducer used to measure force or weight. A strain gauge connected to a force beam measures the resistance to strain that correlates to a weight change and therefore a level change.

There is a great correlation between level, temperature, density, and volume. When one changes, the others also change. Temperature affects the density of all matter causing an increase or decrease in volume. Volume increases caused by density fluctuations do not increase head pressure since the total amount of head pressure is based on the specific gravity of the substance being measured. Knowing these characteristics helps to understand why the levels may change in process vessels.

Checking Your Knowledge

1. Level can be defined as the _____ of a material's surface distance from a zero reference.
   a. Location
   b. Depth
   c. Point
   d. Height
2. Match the terms below with their proper definitions or descriptions:

Term	Definition
Ullage	a. The measurement of another property of a contained material to determine level
Innage	b. The measurement from the bottom of a tank to the surface of a material
Indirect measurement	c. The measurement from the surface of a material to the top of a tank
Direct measurement	d. The measurement of the point where two immiscible materials meet
Interface level	e. The measurement of the process variable (level)

3. The hydrostatic head pressure exerted by 36 in. of water is equal to _____.
   a. 36 in. w.c. ÷ 1.25
   b. 36 in. w.c.
   c. 36 in. w.c. ÷ 1.350
   d. 12 in. w.c.
4. Match the terms below with their proper definitions or descriptions:

Term	Definition
Displacer	a. Measures with a spring-assisted reeling action
Float gauge	b. Measures the difference between total pressure and vapor space pressure
D/P transmitter	c. Measures its own buoyancy force
Nuclear device	d. Detects the point where gamma radiation is shadowed by the tank level
Tape gauge	e. Senses the surface of a material

5. If the mass remains constant and the density of a liquid in a tank decreases, the level of the liquid in the tank _____.
   a. Increases
   b. Decreases
   c. Remains constant
   d. Does nothing because density and level are not related
6. When the volume of a liquid in a tank decreases, the level of the liquid in the tank _____.
   a. Increases
   b. Decreases
   c. Remains constant
   d. Does nothing because density and level are not related.

Student Activities

1. Take a large container and fill it with water. Determine the height of the container where the highest level the water could obtain would be. Pour out differing amounts of water (¾ full, ½ full, ¼ full) for each set of level measurements. Using a metal calibrated measuring tape, extend the tape into the water until it reaches the bottom. Ensure the tape is as straight vertically as possible and read the level of the water. Also, place the end of the tape at surface level and read the measurement at the top of the tape where the highest liquid level could be obtained. Which measurement would be considered innage and which one ullage (outage)? Do the two measurements add up to the total measurement (top to bottom of the container) that was taken earlier?

   NOTE: To show how a process technician uses a tape on a reel to measure the level of the contents of a storage tank, perform the same type of measurements on a tall, thin container filled with lube oil or any other *safe* oil that may be available. After allowing the metal tape to hit the bottom of the container, pull the tape out slowly and read the level of oil on the tape since the oil will leave a distinct wetted line on the tape. Be sure to clean the tape before reeling it back into its housing.

2. Using the conversion chart, perform the calculations on a clean sheet of paper. Turn in your completed calculations as per your instructor's guidelines. Ensure problem numbers and corresonding answers are clearly written.

Use this conversion chart to complete the following conversion exercises.

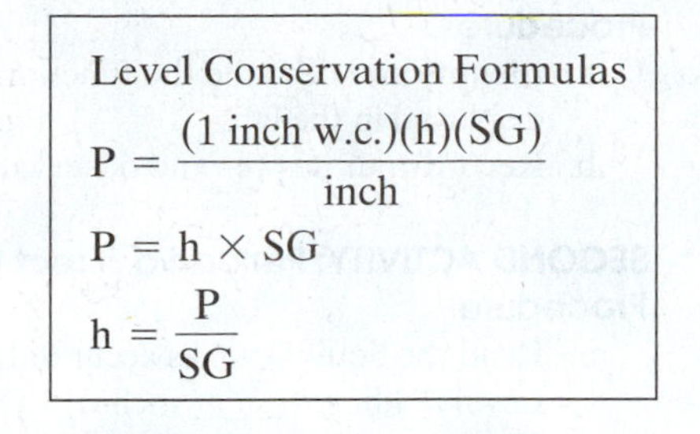

Perform the following level conversions on a sheet of paper, and then write an email listing the problem number and your answer. Send it to your instructor.

Head Pressure Problems

a. An open tank contains a liquid with a specific gravity of 1.735. If the height of the liquid is 10 ft, how much head pressure, in inches w.c., will it exert?
b. An open top tank has a liquid with a specific gravity of 0.95. If the height of the liquid is 200 in., how much head pressure, in inches w.c., will it exert?
c. An open top tank is filled with a liquid that has a specific gravity of 1.735. The liquid exerts a head pressure of 450 in. w.c. What is the level in inches?
d. An open top tank is filled with a liquid that has a specific gravity of 0.873. The liquid exerts a head pressure of 200 in. w.c. What is the level in inches?
e. A sealed tank has a liquid with a specific gravity of 1.00 in it. The pressure in its vapor space is 1 psi. If the height of the liquid is 10.0 ft, how much head pressure, in inches w.c., will it exert?

3. Identify pressure-sensing and measurement devices in the lab.
4. Take apart a level instrument to locate the displacer float. Hook up air to see how the transmitted signal varies with displacer position.
5. Use cutaways of gauge glass valve(s) to show internals (especially ball check).

6. **LEVEL MEASUREMENT Laboratory Procedure**

Background Information

Direct Level Measurement

A device called a gauge glass, or sight glass, can be used to read the level of a tank directly. Transparent glass allows the technician to actually *see* the level and read it in a direct manner. The principle of operation is simple; gravity causes the level in two interconnected containers to be the same. Therefore, the liquid level seen in the sight glass should be the same as the level in the tank.

Indirect Level Measurement

A *differential pressure (d/p) transmitter* can be used to infer level by measuring the hydrostatic head pressure exerted by the liquid above the transmitter's datum (zero) line. When using a differential pressure transmitter to measure the level in a tank, a correlation between the head pressure produced by the liquid and the output of the transmitter must be established. The act of establishing this correlation is called calibration. A d/p transmitter is an indirect method as opposed to the direct reading provided by the sight-glass.

When a transmitter is properly calibrated, its output signal will have a linear relationship to the level in the tank. For example, the output of a typical 4-20 mA analog transmitter at 0% level should be 4 mA and at 50% level should be 12 mA.

Materials Needed

- tall cylindrical tank (2-4 feet tall)
- pressure or differential pressure transmitter with an appropriate calibrated range matching the height of the cylinder
- gauge glass instrument attached to a tank containing a liquid

- 24 VDC power supply to supply power to the transmitter
- digital multimeter with a milliampere test scale

Safety Requirements
Safety glasses are required in the lab.

Procedure
Perform the following lab activities:

FIRST ACTIVITY: Identify and describe level instruments
Procedure

a. Identify and describe the function of each of the level measuring instruments displayed in the lab.
b. Record your answer and observations.

SECOND ACTIVITY: Perform a Direct Level Measurement
Procedure

a. Read the liquid level as accurately as possible and record your results.
Level reading = _____ inches
b. Convert this reading to hydrostatic head press for use in Activity #3.
Head pressure = _____ inches w.c.
c. Explain how you lined up the liquid in the gauge glass with the associated scale.

THIRD ACTIVITY: Perform an Indirect Level Measurement
Procedure

a. Attach or make sure that the high-pressure side of the d/p transmitter is connected to the lowest point in the test cylinder.
b. With no liquid in the cylinder and power applied to the transmitter, confirm that its output is reading 4 mA (3.0 psi for a pneumatic transmitter). Output reading: _____ mA (psi)
c. Fill the cylinder to a point above the datum line and open the bleed valve on the high side of the d/p transmitter to bleed air from the tubing and transmitter chamber. Once all the air is bled and water is flowing steadily, close the bleed valve.
d. Fill the cylinder with water until the output of the transmitter is equal to 8 mA, then stop and record the results. Then move to 12 mA, 16 mA, and 20 mA stopping and recording each time in the same manner as with 8 mA. (Pneumatic signal: 6 psi, 9 psi, 12 psi, and 15 psi)

Percent Level	*Transmitter Output*	*Record Height in Inches*
0%		
25%	8 mA (6 psi)	Inches
50%	12 mA (6 psi)	Inches
75%	16 mA (12 psi)	Inches
100%	20 mA (15 psi)	Inches

Findings
For Activity 1, write a description of each type of level transmitter studied in this lab. Explain the basic concept behind each instrument as well as any additional observations you may have made while examining it. Look in your book and/or handout for additional information about these transmitters.
For Activities 2 and 3, use all readings taken in this lab to further explain the basic concept behind the specific measurement techniques used. For example, explain how the liquid in the gauge glass directly represents the liquid level of the tank or how the head pressure equates to liquid level. Finally, explain the difference between direct and indirect level measurement using the information gathered in this lab.

Attach the following information to the back of the written lab report:

- All measurements taken during the lab experiment.
- All pages associated with this lab including the lab handout itself.
- Answers to the following questions. Use classroom notes to help answer these questions.
  - What can cause the level of the gauge glass (sight glass) to be different from the level in the tank?
  - If the liquid in the gauge glass were concave in appearance, would you take your reading with the liquid next to the tube wall or with the bottom of the meniscus?

CHAPTER 5

Process Variables, Elements, and Instruments: FLOW

Objectives

After completing this chapter, you will be able to:

- Define flow and terms associated with flow:
  - fluids (gases and liquids)
  - laminar
  - turbulent
- Define terms associated with flow measurement:
  - direct and indirect flow measurement
  - positive displacement flow measurement
  - percent flow rate
  - volumetric flow units
  - mass flow units
- Identify the most common types of flow-sensing and measuring devices used in the process industries:
  - orifice plate
  - venturi tube
  - flow nozzle
  - pitot tube
  - annubar® tube
  - rotameter
  - electromagnetic meter (Magmeter)
  - turbine meter
  - mass flow meter
- Describe the purpose and operation of flow-sensing/measurement devices used in process industry.
- Explain the difference between total volume flow, flow rate, and volumetric flow.

Key Terms

Annubar® tube—a trademark name for one manufacturer's multiport tube having four impact points spaced across the pipe and facing the flow with another tube sensing static pressure; measures the average pressure produced by the four impact points.

Electromagnetic flow meter (Magmeter)—a magnetic flow meter designed to determine volumetric flow of electrically conductive liquids, slurries, and corrosive and/or abrasive materials.

Flow measurement—flow rate (an instantaneous flow measurement) and total flow (a summation of instantaneous flow rates over a time interval or an accumulation of counts provided by a positive displacement device); measured in volume or mass units without respect to time.

Flow nozzle—similar to a venture tube but has an extended tapered inlet commonly installed in a short piece of pipe called a spool piece.

Flow rate—a measure of how much fluid moves through a pipe or channel within a given period of time.

Fluid—flowing liquids and gases that mix easily because of the continual movement between molecules.

Indirect flow measurement—measuring one variable of a process to infer another.

Laminar flow—smooth-flowing fluid.

Mass flow meter coriolis—the most common type of true mass flow meter; a meter that eliminates the need to compensate for typical process variations such as temperature, pressure, density, and even viscosity.

Mass flow units—a unit of measure for the weight being passed through a certain location per unit of time; usually expressed in pounds per unit of time as in pounds per minute.

Orifice flange—a flange with holes drilled in the flange through to the pipe to allow pressures upstream and downstream of an orifice plate to be measured.

Orifice plate—a piece of 1/8-inch to 1⁄12-inch-thick metal with a calibrated hole drilled (or cut) through; types of orifice plates include the concentric plate, eccentric bore, and segmental plate.

Percentage flow rate—a common way to indicate a flowing process with 100 percent equating to an actual quantity such as gallons per minute (gpm).

Pitot tube—an L-shaped tube that is inserted into a pipe with its open end facing the flow and another tube sensing static pressure.

Positive displacement flow measurement—when flow is measured in absolute volumes where the flowing materials is admitted into a chamber of known volume and then transferred to a discharge point; a counter registers the number of times the chamber fills and discharges.

Positive displacement meter—piston, oval-gear, nutating-disk, and rotary-vane types of positive displacement flow meters.

Reynolds number—a mathematical computation describing the flow of fluids numerically.

Rotameter—a direct-read variable area (tapered) flow tube where the fluid enters through the bottom, then flows upward, lifting a free-floating indicator plummet (float); the position of the float references to the calibrated marks on the glass tube to indicate flow rate.

Turbine meter—a flow tube containing a free-spinning turbine (fan) wheel where the revolutions per minute (rpm) are proportional to flow rate; the faster the flow, the faster the turbine spins; the output is fed directly into an instrument to indicate either flow rate or total flow.

Turbulent flow—fluid flowing with turbulence.

Venturi tube—a primary element used in pipelines to create a differential pressure such that when converted to flow units (e.g., gpm), the flow in the line is measured.

Volumetric flow units—a unit of measure for the volume being passed through a certain location in a pipe or channel per unit of time; usually measured in gallons per minute and cubic feet per minute.

Introduction to Flow

Flow is a fluid in motion. Gases, like liquids, flow from one point to another depending on pressure and temperature. When **fluids** or gases flow, they mix easily because of the continual movement between molecules. Material always flows from a high pressure area to a low pressure area. The rate that a fluid flows is a measure of how much fluid is flowing or moving through a pipe or channel within a given period of time. The term *flow* is often used interchangeably with **flow rate** in industry.

Molecules that are flowing continually change how they move among themselves. This is a basic characteristic of fluid flow whether it is a liquid or gas.

Fluid movement may be either laminar or turbulent depending on the path taken. When the path is smooth and without obstruction, the flow is called laminar since the molecules line out to a smooth flowing pattern. This type of movement is characteristic of long pipelines where the molecules have a chance to settle down and move in a more orderly fashion. **Laminar flow** (Figure 5-1) is also called streamlined flow. The velocity of the flowing fluid changes smoothly and equally from the pipe wall inward to the center of the pipe where the velocity is at its highest. Imagine the annular rings of a tree defining the velocity zones as they increase towards the center.

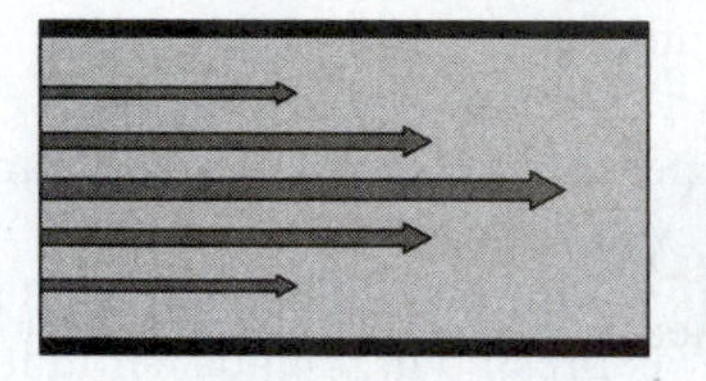

FIGURE 5-1 Laminar Flow

Piping bends, corrosion, valves, or any other obstruction to the flow inside a line causes turbulence or **turbulent flow** (Figure 5-2). If a stream of water passes over rock formations, the water looks like it is boiling and may even cause bubbles to form as air is entrained into the water from the turbulence. If an airplane goes through a section of turbulent air flowing by the plane, then the plane moves violently in response to the turbulence.

At first this may seem undesirable, but in most cases, the opposite is true. A turbulent flow is consistent and therefore manageable.

FIGURE 5-2 Turbulent Flow

REYNOLDS NUMBER

The **Reynolds number** can be used to identify whether a flow is laminar or turbulent. The Reynolds number is a mathematical computation known as the fluid velocity profile (Figure 5-3) that describes flowing fluids numerically. Flowing fluids fall into one of two categories (laminar and turbulent) as described previously, although there is a transitional zone where characteristics of both overlap. To calculate the Reynolds number, the following must be known:

- velocity
- density
- viscosity of the fluid
- inside diameter of the pipe

The following equation describes how the Reynolds number is obtained:

$$\text{Reynolds number} = \frac{(\text{velocity of fluid})\ (\text{inside diameter of pipe})\ (\text{density of fluid})}{(\text{absolute viscosity of fluid})}$$

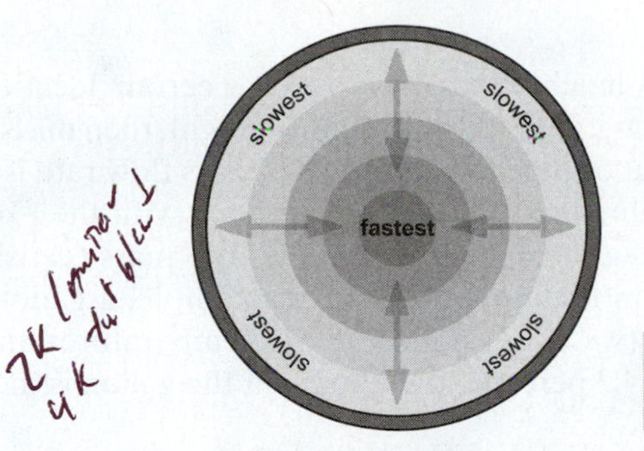

FIGURE 5-3 Fluid Velocity Profile

In this equation, an increase in fluid velocity, the inner diameter of the pipe or the fluid density, makes the Reynolds number larger, whereas an increase in absolute viscosity makes the Reynolds number smaller.

When a flowing process material has a Reynolds number of over 4,000, the flow is turbulent. To be fully turbulent, the Reynolds number should be greater than 10,000. When the Reynolds number is less than 2,000 the flow is laminar. The region above 2,000 and below 4,000 is said to be transient, which means having characteristics of both turbulent and laminar types.

TYPES OF FLOW MEASUREMENT

Direct and Indirect Flow Measurement

Total flow is a direct form of flow measurement as determined by totalizers. Flow indicators such as flappers and paddlewheels seem to be direct but indicate the result of an impinging fluid force, categorizing them as indirect flow measurement devices.

Flow rate is measured indirectly. **Indirect flow measurement**, as described in previous chapters, measures one variable of a process to infer another. For example, differential pressure (d/p) can be measured and flow rate can be inferred. Although there are devices known as direct-read flow meters (such as rotameters, weirs, etc.), these are still indirect methods when examined closely.

Flow measurements for liquids or gases may be made in exact or absolute quantities, percentages, volumetric equivalents, and also in mass (weight) units for solids.

Positive Displacement Flow Measurement

Positive displacement flow measurement is used within industry where the measurements of absolute volumes are required. Flow meters are used to determine exact quantities of material going into a tank (e.g., additives) and/or to measure the transfer of products between adjacent plants.

Positive displacement meters operate by admitting a flowing material into a chamber with a known volume capacity and then transferring (discharging) all the contents to another point (e.g., gasoline blending tank). A counter registers the number of times the chamber(s) is filled and discharged. A total amount of metered material can be read directly from the counter. Positive displacement meter types include piston, oval-gear, nutating-disk, and rotary-vane.

Percentage Flow Rate

Percentage flow rate is a common way to indicate a flowing process. A 100 percent flow rate is equated to an actual quantity such as gallons per minute (gpm). If 100 percent flow rate is equal to 200 gpm, then a 50 percent flow rate equals 100 gpm, and so on.

Volumetric Flow Units

Volumetric flow units are instantaneous flow rate measurements for how much volume passes through a certain location in a pipe or channel per unit of time. Volumetric flow rates are measured in units such as gallons per minute (gpm) and cubic feet per minute (cfpm).

Mass Flow Units

Mass flow rate is a measure of how much actual mass passes a certain location per unit of time. Since mass in a constant gravity field has a definite weight, then **mass flow units** can be expressed in pounds per unit of time (e.g., lb./ min.). Mass flow rate is an instantaneous measurement just like volumetric flow measurements with the exception of quantifying the total mass occupying the measured volume. The mass (density) of any given volume of liquid depends on its temperature. For example, a gallon of water weighs 8.345 lb. at 39.2 degrees F (water at its most dense temperature) and 7.998 at 212 degrees F. The change is about 4.3 percent; the volume of the gallon is the same but its mass has changed.

Flow-Sensing and Measurement Instruments

A primary flow element is a device that creates a measurable variable that is proportionally equal to flow rate. The most common primary flow elements measure differential pressure. All the following primary flow elements operate according to the flow principles established by Daniel Bernoulli in the 1700s. In his work he found that as velocity of a flowing fluid in a pipe increases, the static pressure of the fluid decreases. This differential pressure can be equated to a flow rate.

The following are the most common types of primary flow-sensing and measuring devices used in the process industry:

- orifice plates
- venturi tubes
- flow nozzles
- pitot tubes
- Annubar® tubes (multiport pitot tubes)
- rotameters
- electromagnetic meters (Magmeters)
- turbine meters
- mass flow meters

Orifice Plates

The orifice meter is a commonly used flow meter. An **orifice plate** (Figure 5-4) is a piece of ⅛-inch to ½-inch-thick metal with a precise hole drilled (or cut) through it. They are the simplest and most practical way to create a pressure drop in a flowing process.

There are several types of orifice plates. One type, the concentric plate, which is the most common, has the hole drilled directly in the center, whereas the eccentric bore is off center. Another type is called a segmental plate, which has a half-moon-shaped opening cut into it. Both the eccentric and segmental plates are designed to allow solids to pass freely through them rather than accumulating in front of them. Lastly, there are the quadrant-edged and the conical plates. These differ from the others because of their upstream face. The quadrant-edged plate has a smooth, rounded inlet and the conical has a beveled or cone-shaped inlet. The quadrant and conical are used with more streamlined and viscous flows.

All these plates have a sharp upstream face and most have a beveled downstream side. Manufacturers stamp or stencil the word INLET on the upstream side. Other information such as line and bore size is usually imprinted as well.

An orifice plate often has a small hole drilled next to the metering hole called a *weep hole*. This small hole allows vapors or liquids that are likely to be stopped by the plate to pass through keeping trapped materials on the upstream side of the plate from reducing the accuracy of the meter.

Orifice plates are generally mounted in a special set of flanges called **orifice flanges**. These flanges have pressure-measuring taps drilled through the flange into the pipe allowing the pressure upstream and downstream of the orifice plate to be read by a differential pressure transmitter.

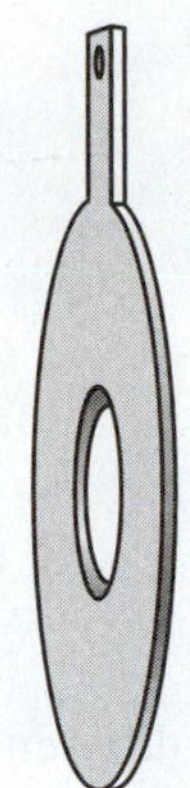

FIGURE 5-4 Orifice Plate

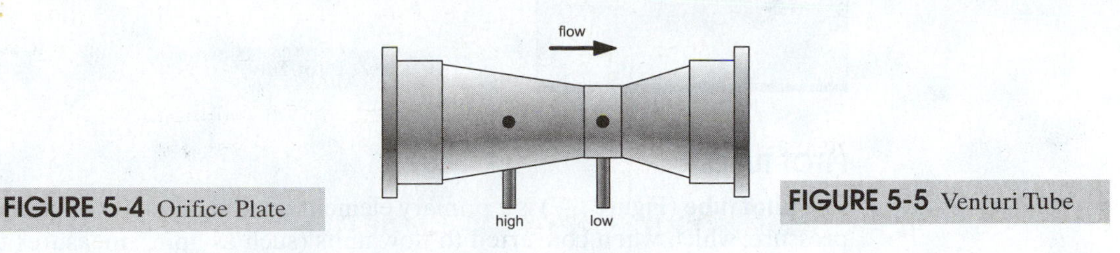

FIGURE 5-5 Venturi Tube

VENTURI TUBES

A **venturi tube** (Figure 5-5) is a primary element used in piping to create a differential pressure; when converted to flow units (such as gpm), the flow in the line can be measured. Venturi tubes cost more than orifice plates, but their day-to-day operating cost can be significantly less because their design provides less permanent pressure loss. Pressure loss in any restrictive primary device is expected. However, in some larger applications, permanent pressure loss is a significant factor in determining which primary device should be used.

The shape of a venturi tube is characterized by its smooth, cone-shaped inlet and outlet components. As the fluid speeds up in the smaller-throat-diameter section, the pressure is reduced according to Bernoulli's principle (if a fluid's static pressure decreases, its velocity increases). This lower pressure is compared to the high pressure measured at the tap located in front of the inlet cone. A d/p transmitter can be connected to the two taps to measure the difference in pressure.

FLOW NOZZLES

A **flow nozzle** (Figure 5-6) is similar to a venturi tube in that a flow nozzle has an extended tapered inlet. The differences between a venturi tube and a flow nozzle are that a flow nozzle can be inserted into a pipe at a flange connection and the flow nozzle does not have a recovery (outlet) cone. A flow nozzle may also be installed in a short piece of pipe called a spool piece.

Flow nozzles perform the same function as all other restriction devices; that is, they provide a differential pressure that can be used to infer a flow rate. Because of their design, they are capable of allowing more flow to pass through them per unit of pressure drop than an orifice plate. They are also a good choice for use with slurries due to their sloped inlets allowing solids to pass freely through them.

The operation of a flow nozzle is the same as a venturi tube on the front side and like an orifice plate on the downstream side. The smooth conical shape of the inlet is the primary reason why the flow nozzle is able to handle more than twice the flow of an orifice plate with the same pressure drop. The cost of a flow nozzle is more than an orifice plate and less than a venturi tube.

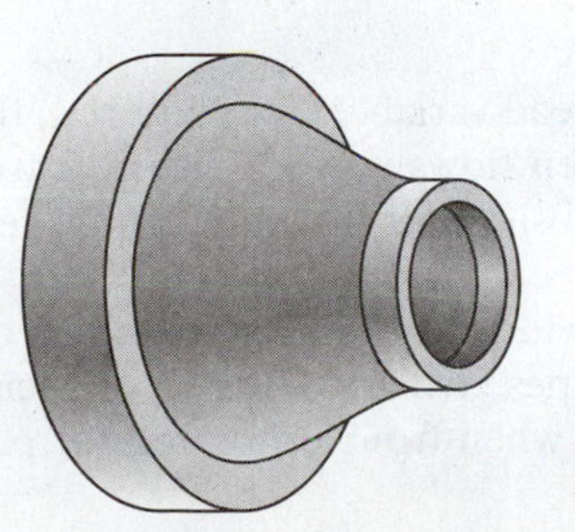

FIGURE 5-6 Flow Nozzle

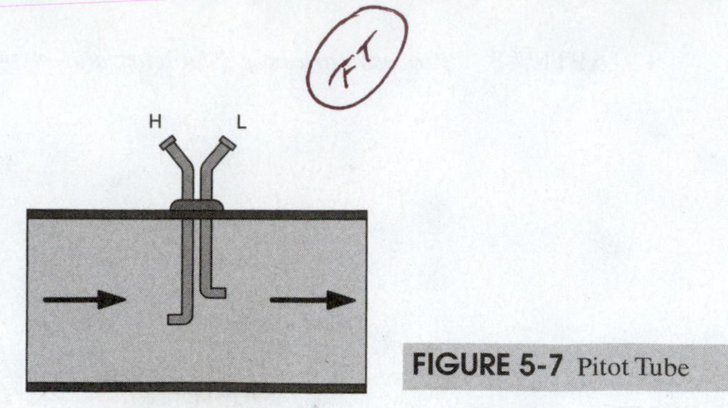

FIGURE 5-7 Pitot Tube

PITOT TUBES

The **pitot tube** (Figure 5-7) is a primary element used in pipelines to create a differential pressure, which when converted to flow units (such as gpm), measures the flow in the line. A pitot tube is shaped like an "L" and is inserted into a pipe with the open end facing directly into the flow and another tube measuring or sensing static pressure in the same vicinity. By comparing the impinging pressure produced by the flowing fluid to the static line pressure, flow rate can be inferred.

A disadvantage of the pitot tube is that it detects the velocity of a flowing fluid at only one point in the pipe. If the Reynolds number of the flowing material is high enough, then any point away from the wall represents the average flow rate. However, if the Reynolds number is too low (suggesting a laminar flow characteristic), then a pitot tube cannot provide an accurate average velocity measurement.

ANNUBAR® OR MULTIPORT PITOT TUBES

Annubar® (Figure 5-8) is a trade name one manufacturer has given to its *multiport tube*. This tube has four impact points spaced across the pipe facing the flow and another tube sensing static pressure. The multiport tube has a smaller tube inside of the tube that is designed to measure the average pressure measurement across a pipe produced by four impact points.

The multiport pitot tube measures velocity like the pitot tube except that it has multiple impact points rather than one. The multiport tube takes an average of all the impact points to produce a single measurement. Like the pitot tube, a static pressure measurement is compared to the average impact pressure producing a d/p that can be measured and converted into a flow rate.

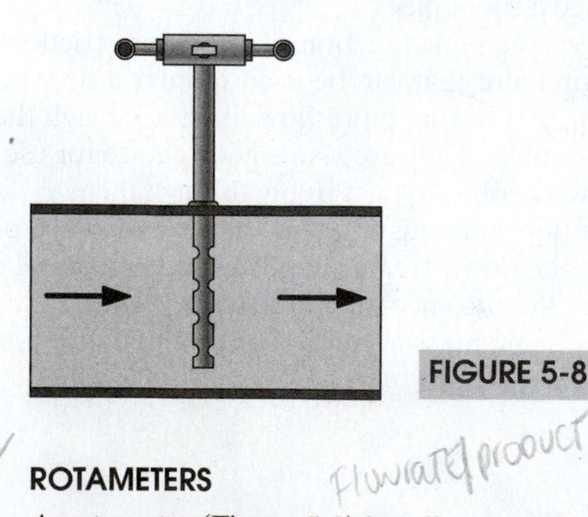

FIGURE 5-8 Annubar Tube

ROTAMETERS

A **rotameter** (Figure 5-9) is a direct-read variable area (tapered) flow tube in which the fluid enters through the bottom, then flows upward, lifting a free-floating indicator plummet (also called a float). The position of the *float* can be referenced to the calibrated marks on the glass tube to indicate flow rate.

Rotameters are designed to provide a flow indication. They can be made of glass, plastic, or metal. Glass and plastic types give the process technician an opportunity to see the process fluid. This is helpful when troubleshooting for process problems. The

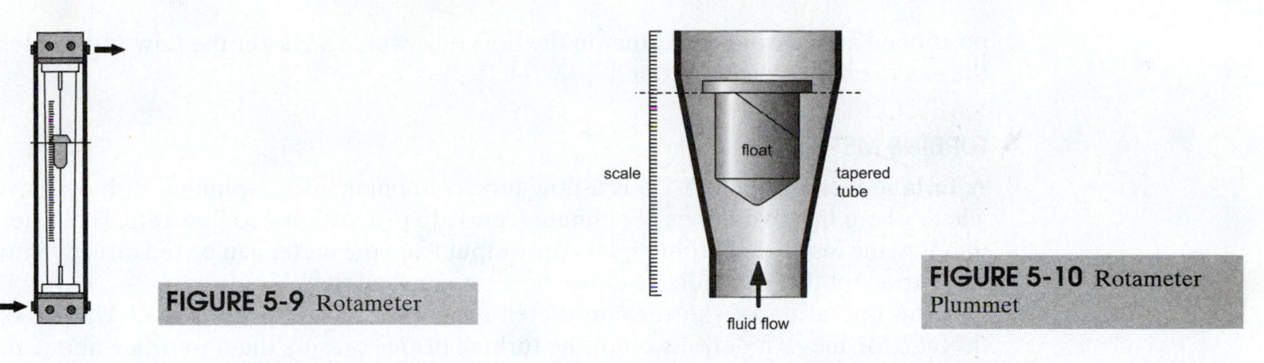

FIGURE 5-9 Rotameter

FIGURE 5-10 Rotameter Plummet

downside of using glass and plastic is that the materials are fragile. Glass is inherently prone to breakage. The armored or metal tube type removes this problem, but hides the fluid from inspection. Generally speaking, rotameters are used for indication, but the metal tube types can be equipped with a magnetic follower to both indicate and transmit a signal.

The operating principle of the rotameter is based on gravity and impinging flow pressure. Physically, the smallest diameter of the tube is on the bottom with the diameter increasing steadily as it goes up. The plummet (Figure 5-10), or float, requires that a minimal amount of impinging force be applied to the float for it to rise (float) in the flowing fluid. As the fluid velocity increases, it pushes the plummet upward to a new position in the tube. This new position is directly related to the impinging force applied to the plummet. At any point in time, the velocity at the smaller end of the tube is the highest, requiring less flow to *float* the plummet. As the diameter of the tube increases, however, the velocity slows down requiring increasingly more fluid flow through the tube to push the plummet higher. The tube is marked in flow units providing a flow indication.

The shape and weight of the plummet along with the size and shape of the flow tube provide an accurate flow rate. Each rotameter is sized to indicate a specific type of liquid or gas. If the rotameter is improperly sized or replaced with a different tube and/or plummet, accuracy suffers.

ELECTROMAGNETIC METERS (MAGMETERS)

A magnetic flow meter is designed to determine volumetric flow of electrically conductive liquids, slurries, corrosive and/or abrasive materials. **Electromagnetic flow meters (Magmeters)** (Figure 5-11) are used to measure flowing liquids having at least some ability to conduct electricity. Generally, this means aqueous (water-based) materials.

Electromagnetic meters are important for several reasons. One is that they are designed with a completely obstruction free flow path. Another is that they are noninvasive, meaning that there are not any sensing parts protruding into the process. The noninvasive characteristic is particularly important to the food and drug industry where residual materials deposited in and around sensors cannot be tolerated.

The electromagnetic flow meter operates according to Faraday's law of induction, which states that an electrical potential is produced when a conductor moves at a right angle through a magnetic field. The Magmeter produces a magnetic field that penetrates the nonferrous (nonmagnetic) flow tube. The liquid flowing through the tube is the conductor that Faraday describes as moving at a right angle to the magnetic field. This action creates an electrical potential, which is sensed by two electrodes that are

FIGURE 5-11 Electromagnetic Meter (Magmeter)

positioned across from each other in the flow tube wall. The faster the flow, the greater the electrical potential generated.

TURBINE METER

A **turbine meter** (Figure 5-12) is a flow tube containing a free spinning turbine (fan) wheel where the revolutions per minute (rpm) are proportional to flow rate. The faster the flow, the faster the turbine spins. The output from the meter can be fed directly into an instrument that can indicate either flow rate or total flow.

The operation of the turbine meter is simple. As the process material flows through the meter it deflects off of the turbine blades causing them to rotate about an axis like a toy pinwheel. With the induction-type sensor, one of the turbine blades is magnetic. When the blade rotates past an induction pickup coil located on the outside of the tube, the break in the magnetic field produces a pulse. Each pulse represents a specific volume of material. Turbine meters must be calibrated occasionally so that an adjusted K-factor can be established. A K-factor is the number of pulses generated per gallon of flowing material. Properties like viscosity and specific gravity can affect flow meter calibration.

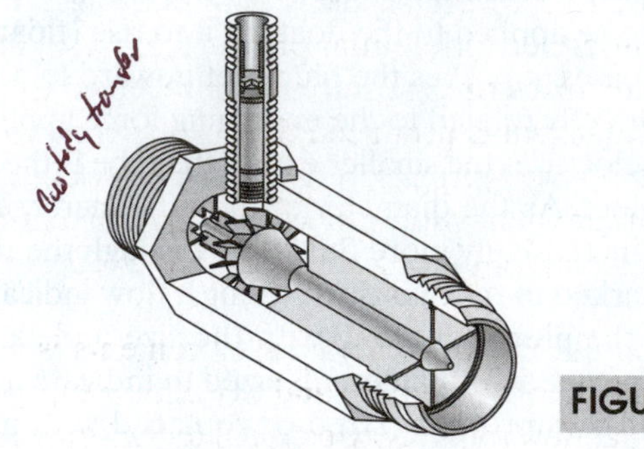

FIGURE 5-12 Turbine Meter

MASS FLOW METERS

Although *volumetric flow meters* can be combined with a *density sensor* (which may be a pressure and temperature measurement) to infer mass flow, there are some meters that read mass flow directly. The most common type of true **mass flow meter** (Figure 5-13) is the **coriolis** meter. True mass flow meters eliminate the need to compensate for typical process variations such as temperature, pressure, density, and even viscosity.

The drive coil causes the tube to vibrate. Two velocity sensors detect the natural frequency of the filled tube to establish a baseline measurement. As material starts flowing through the tube, the vertical motion produced by the drive mechanism causes the fluid entering the tube to push against the tube wall in one direction and push in the opposite direction leaving the tube. These resultant forces cause the tube to twist producing a displacement and velocity difference that can be measured by the two sensors. Mass flow is determined by the difference between the two velocity detector signals. Most mass flow meters read flow, temperature, and density. This type of flow meter is generally not affected by changes in fluid temperature, pressure, or density.

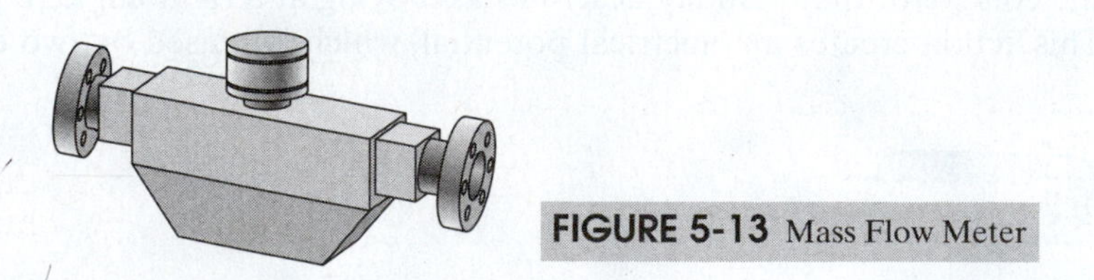

FIGURE 5-13 Mass Flow Meter

DIFFERENTIAL (D/P) TRANSMITTERS

D/P transmitters (Figure 5-14) are the single most commonly used flow measuring devices in the processing industry. The majority of fluid flow rates are measured by differential pressure. A process technician needs to know how to read and interpret the

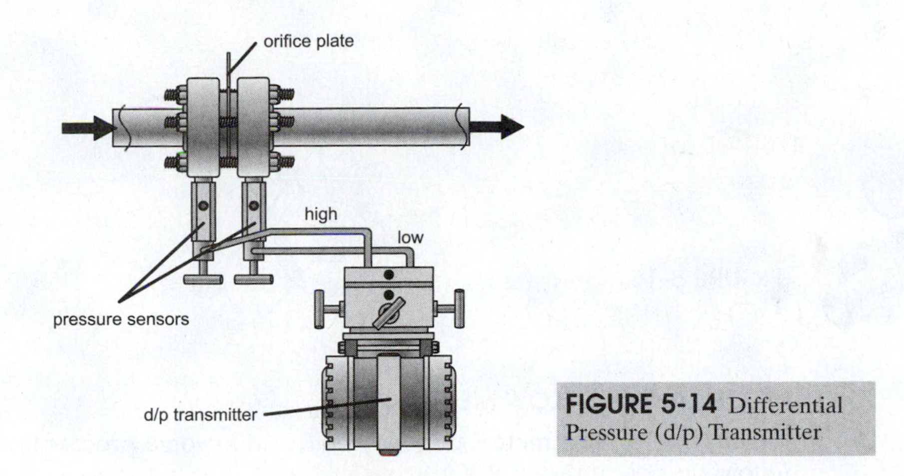

FIGURE 5-14 Differential Pressure (d/p) Transmitter

output signals from all flow transmitters while making their rounds. The knowledge of d/p versus flow rate is particularly helpful when troubleshooting a process problem.

The d/p transmitter responds to the pressure created by the primary device (orifice plate, venturi, flow tube, etc.). The pressure signal is then converted into a standard output signal that can represent either differential pressure or flow rate. When a d/p cell responds to the high and low-pressure measurements received from the primary element, a corresponding output signal is produced that is linear and proportional to the differential pressure. The electronic circuits in the transmitter are capable of producing both an output that is either linear with respect to differential pressure or linear with respect to flow rate. The standard analog d/p transmitter output is linear with respect to differential pressure.

An important concept inherent in all d/p flow measurements is the square root relationship that exists between the measured d/p and the inferred flow rate. The concept in its simplest form states that flow rate is proportional to the square root of the differential pressure. In other words, if the flow rate doubles, the pressure drop quadruples ($2^2 = 4$). If the flow rate triples, the pressure drop increases by 9 times ($3^2 = 9$). For example, if 25 percent of the calibrated d/p measurement is sensed across the orifice plate, then 50 percent of the flow rate is moving through the d/p cell.

To change percent d/p into percent flow rate, perform the following steps:

Step	***Calculation***
1. Insert the % d/p into the equation:	$\text{Percent Flow Rate} = \frac{\sqrt{\%\ d/p}}{100} \times 100$
2. Change percent into a decimal form:	$\text{Percent Flow Rate} = \frac{\sqrt{25\%}}{100} \times 100$
3. Take the square root:	$\text{Percent Flow Rate} = \sqrt{.25} \times 100$
4. Make the decimal a percent again:	$\text{Percent Flow Rate} = 0.50 \times 100$
5. Record your answer:	$\text{Percent Flow Rate} = 50\%$

Or,

Step	***Calculation***
1. Insert the % d/p into the equation:	$\text{Percent Flow Rate} = \frac{\%d/p}{100} \times 100$
2. Change percent into a decimal form:	$\frac{25\%}{100} \times 100$
3. Take the square root:	0.25×100
4. Make the decimal a percent again:	0.50×100
5. Record your answer:	50%

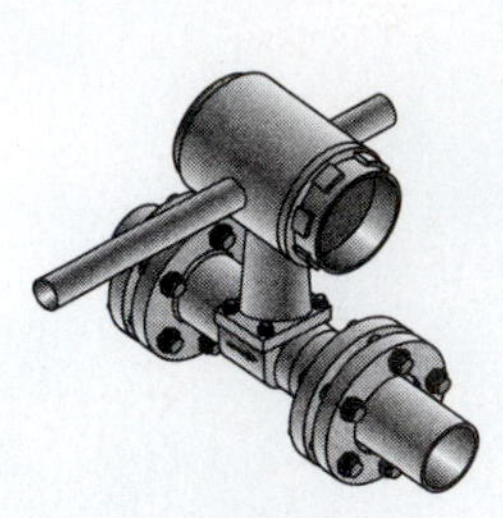

FIGURE 5-15 Vortex Flow Meter

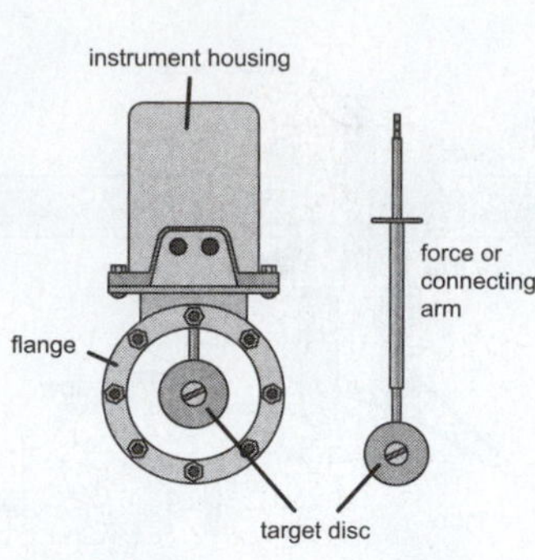

FIGURE 5-16 Target Flow Meter–End View

MISCELLANEOUS FLOW METERS

Other types of flow meters that may be found in some process facilities include the following:

- vortex
- target
- integral

Vortex Flow Meter

A vortex flow meter (Figure 5-15) measures most process fluids and causes only a minor pressure loss. Vortex flow metering does rely on the formation of vortices in the fluid flow and is not suitable for high-viscosity fluids or for most low-flow applications. A vortex flow meter uses vortex shedding to measure flow. Vortex shedding occurs when a bluff body (a flat fronted bar tapered toward the back) is placed in the flow path. This object causes the fluid to separate and form small eddies, called vortices that are shed on the sides of the bluff body. These alternating vortices cause a fluctuation in pressure from side to side. A sensor inside or at the top of the bluff body senses these pressure fluctuations and sends a signal that is then converted by a transmitter section to a standard signal proportional to the volumetric flow rate.

Target Flow Meter

A target flow meter (Figure 5-16) has a circular disc placed in the center of the flow path. Flow through the annular space between the target disc and the inside diameter of the meter causes a force on the disc that is proportional to the velocity head pressure. This force causes a mechanical motion that is transferred to the top of the instrument through a force or connecting arm. The force is measured by an element that responds to the amount of motion produced by the arm.

Integral Orifice Flow Meter

An integral orifice flow meter is a modified differential pressure transmitter (Figure 5-17) used to measure small fluid flows. The fluid flows into the body of the transmitter through a calibrated orifice that develops a differential pressure.

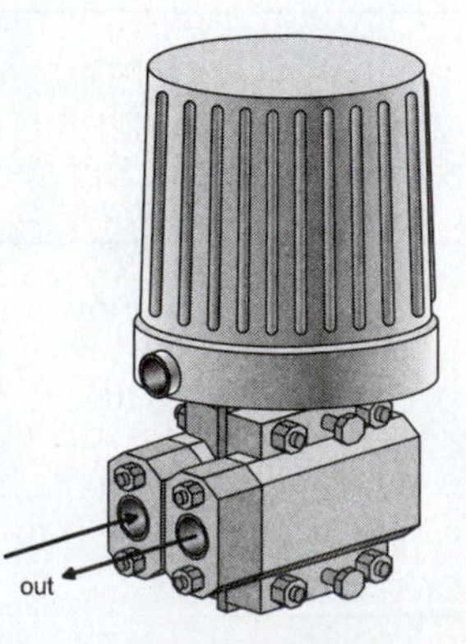

FIGURE 5-17 Integral Orifice Flowmeter–Modified d/p Transmitter

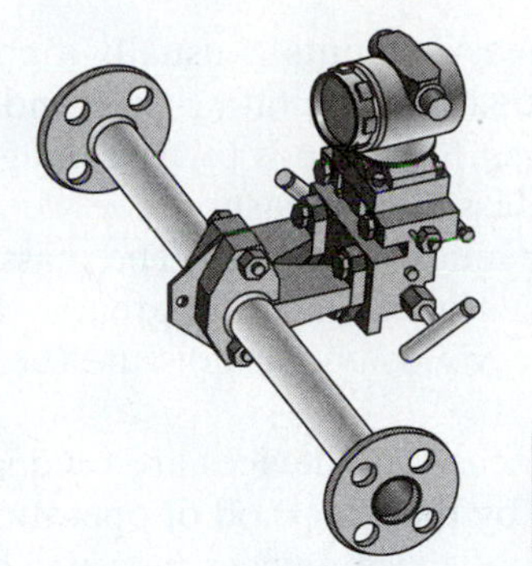

FIGURE 5-18 Integral Orifice Flow Meter–Meter Run

Some types of integral orifice flow meters are permanently installed in a meter run (Figure 5-18) that is installed as a whole in the appropriate-size process line. The integral orifice plate is built into the meter as a welded-in-place precalibrated orifice installation. Pressure sensor taps are positioned vertically on either side of the integral orifice. The pressure measurement (d/p) is sent to the transmitter that converts the signal to either a pneumatic or electronic signal proportional to the d/p.

Total Volume Flow, Flow Rate, and Volumetric Flow

There are two kinds of flow measurement: flow rate (Figure 5-19) and total flow. Flow rate is an instantaneous flow measurement and total flow is a summation of instantaneous flow rates over a time interval or an accumulation of counts provided by a positive displacement device. In either case, total flow is measured in volume or mass units over a specified time interval (e.g., gallons per day, etc.).

The units associated with each of these types of flow include the following:

- Volumetric flow: gallons per minute
- Mass flow: pounds per minute
- Total flow: gallons, pounds, cubic feet, etc. over a specified time interval

Liquid actively flowing into a drum is metered in volumetric flow units (gpm), while the total flow (the accumulated instantaneous flow value) is measured in gallons.

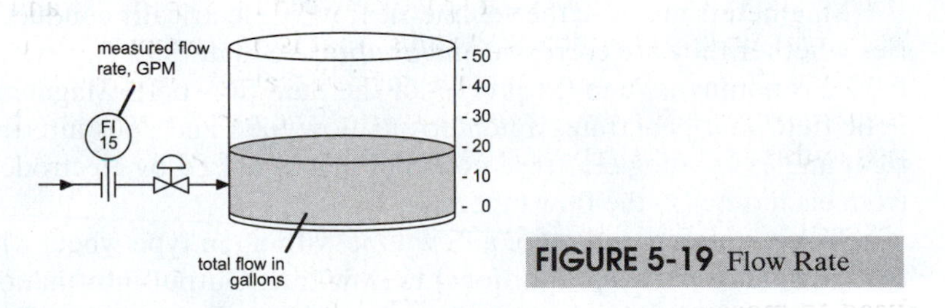

FIGURE 5-19 Flow Rate

Summary

Flow is either a gas or a liquid in fluid motion rather than static or standing still. Flowing fluids mix easily since the molecules are also moving within themselves. How fast the flowing fluid moves is the flow rate and this flow rate is a measure of how much by volume of the fluid is moving over a given time period such as a minute or hour. Fluid movement is either laminar or turbulent depending on the characteristics of the path in which the fluid moves. The Reynolds number is a mathematical computation describing how laminar or turbulent a fluid movement is.

There are several different types of flow measurement devices that include direct and indirect, positive displacement, percentage, volumetric, and mass flow units. The rate of flow in a fluid is always indirectly measured in each of these types of devices, even when the name of the instrument sounds like it could be of the direct measurement type.

Metering the flow for specific measurements is usually a positive displacement method such as when using a piston, oval-gear, nutating-disk, and/or rotary-vane type meter. A common method of measuring flow rate is by percentages such as 50 or 100 percent gallons per minute (gpm). Mass flow measures the weight (lb./min.) of the material flowing through a particular measuring point. The mass weight depends on the temperature of the fluid being measured since the greater the temperature, the lower the weight. Density, temperature, viscosity, and pressure (head) are other factors that affect flow.

Many types of flow-sensing and measuring devices are used in the process industry, but they fall into a few categories by their method of operation and how they are used in a process. The more common of these devices measure differential pressure (d/p). Bernoulli's principle, which states if a fluid's static pressure decreases, its velocity increases, is used to determine if differential pressure in a flowing fluid equates to a flow rate.

The first of these types of d/p devices is the orifice plate. There are several types of orifice plates but they operate on the principle of restricting the flow of a fluid and allowing it to pass through a calibrated hole with a specific inlet and outlet side. Similar to orifice plates in operation, a venturi tube costs more but is used in installations when pressure loss across the apparatus is unacceptable. The venturi tube is cone-shaped on either end with a smaller throat in the middle. Another device, a flow nozzle, is also similar to the orifice plate but it has an extended tapered inlet. A flow nozzle can be inserted into a pipe at a flange since the nozzle does not have a recovery outlet cone like a venturi tube.

Two other d/p flow measuring devices are the L-shaped pitot tube and the Annubar® tube. The open end of a pitot tube is inserted into the flow and another tube measures static pressure in the same vicinity. The difference in pitot and Annubar® is that the pitot measures only one place in the flow while the multiport Annubar® tube has several ports and takes an average of all the d/p points. The latter is more accurate.

Rotameters, which can be made of glass or plastic, allow process technicians the ability to see the movement of the process fluid since it pushes the calibrated plummet further upward as pressure increases in the fluid movement. This basic type is for an indication of flow only, but a metal tube may be equipped with a magnet follower so that the flow rate can be both indicated locally and transmitted via a signal to a remote location such as a control room.

Magmeters measure the volumetric flow of electrically conductive liquids and slurries whether they are corrosive and/or abrasive materials. This type of flow-measuring device is noninvasive to the process or the fluid flow path. Magmeters produce a magnetic field that penetrates a nonferrous flow tube and goes into the flowing liquid at right angles creating electrical potential that is sensed by electrodes positioned across from each other in the flow tube wall.

Turbine meters consist of a flow tube with a fan type wheel. The number of times the fan spins (rpm) is proportional to flow rate. Output information from the turbine meter is fed directly to another instrument that indicates either flow rate or total flow (percentage). Mass flow meters eliminate the need to compensate for process variations such as temperature, pressure, density and even viscosity. Two sensors measure the forces causing the tube to twist producing a displacement and velocity difference. Mass flow is the difference between the two signals.

D/P cells and transmitters are secondary flow meter devices that respond to the signals created by the primary devices described previously. The signals are converted to an output signal that may represent either the differential pressure or flow rate. Flow rate is proportional to the square root of the differential pressure.

When comparing flow measurements, there are two kinds: flow rate and total flow. Flow rate is an instantaneous flow measurement and total flow is a summation of instantaneous flow rates over a time interval. Flow rates may also be an accumulation of counts provided by a positive displacement device. Total flow is measured in volume or mass units without respect to time.

Checking Your Knowledge

1. To calculate the Reynolds number, which of the following variables is/are needed to perform the calculation? *Select all that apply.*
   a. Inside diameter of the pipe
   b. Viscosity of the fluid
   c. Density of the fluid
   d. Velocity of the fluid
   e. Laminar flow number
2. *True or False* When a flowing fluid possesses a Reynolds number that is greater than 4,000, it is considered to be laminar.
3. The temperature of the flowing fluid must be considered when measuring _____ flow rates.
   a. Positive displacement
   b. Percent
   c. Volumetric
   d. Mass
   e. Volumetric and mass
   f. Primary flow element
4. Properly identify each instrument by choosing the appropriate image for each flow measuring device.
   a. Orifice plate
   b. Flow nozzle
   c. Venturi tube
   d. Pitot tube
   e. Annubar®

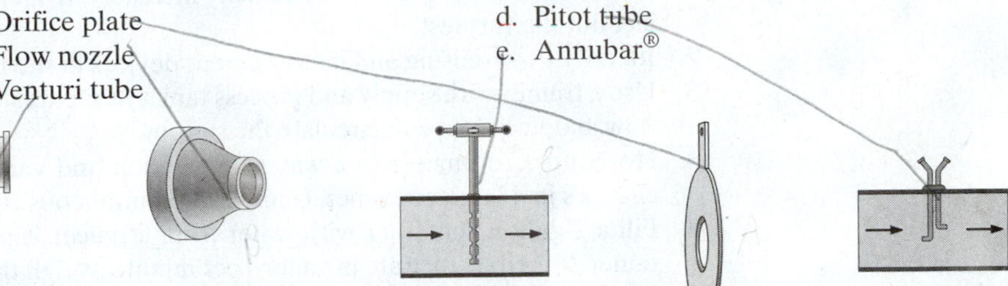

5. Match the terms below with their proper definitions or descriptions:

Term	*Definition*
1. Orifice plate	a. Measures flow using a metal disc containing a drilled opening.
2. Flow nozzle	b. Measures flow using a cone-shaped device with inlet and outlet components
3. Venturi tube	c. Measures flow using a tapered inlet device inserted into a flange connection (or spool piece)
4. Pitot tube	d. Measures flow using an L-shaped tube and another tube that compares the impinging pressure with static pressure
5. Annubar®	e. Measures flow using a tube with several openings and then averaging all flow measurements.

6. Properly identify each instrument by choosing the appropriate image for each metering device.
   a. Mass flow meter
   b. Rotameter
   c. Magmeter
   d. Turbine meter
   e. D/P transmitter

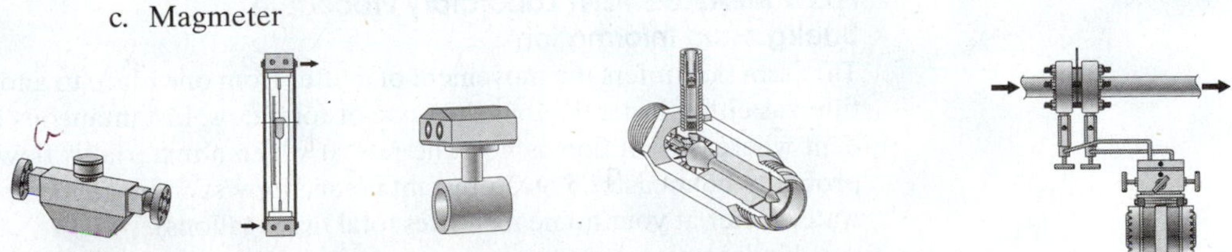

7. Match the terms below with their proper definitions or descriptions:

Term	*Definition*
1. Mass flow meter	a. Measures flow using a flow tube containing a spinning fan
2. Rotameter	b. Measures flow by sensing electrical potential
3. Magmeter	c. Measures flow by using a tapered flow tube and a float
4. Turbine meter	d. Measures flow using a U tube and frequency vibrations
5. D/P transmitter	e. Measures flow by responding to the signal created by the primary device and converting it into an output signal

8. If the flow through an orifice doubles, the differential pressure _____.
   a. Doubles
   b. Triples
   c. Quadruples
   d. Stays the same
9. Special flanges to mount orifice plates are called _____ flanges.
   a. Pipe
   b. Orifice
   c. Concentric
   d. Annubar®

Student Activities

1. Adjust water flow on a garden hose by opening the valve wide. When the flow is at maximum, place your thumb over varying portions of the nozzle opening. What happens when you do this? Does pressure or velocity increase? What thumb position makes the water project out the furthest?
2. Identify flow-sensing and measurement devices in the lab.
3. Use a trainer with supply and process tanks. Students should start up the trainer and control flow and level loops to circulate the system.
4. Hook up a rotometer to a water connection and vary the rate to observe the inventory changes in a large container. Compare instantaneous rate to volume.
5. Fill a 1-gallon container with water from a faucet. Time how long it takes to fill the container. Calculate the rate in gallons per minute. Weigh the filled container. Calculate the rate in pounds per minute.
6. Perform the calculations on a clean sheet of paper. Turn in your completed calculations as per your instructor's guidelines. Ensure problem numbers and corresponding answers are clearly written. Convert % differential pressure into % flow rate.

% Differential Pressure	*% Flow Rate*
50% d/p	_____ % flow rate
75% d/p	_____ % flow rate
80% d/p	_____ % flow rate
90% d/p	_____ % flow rate
60% d/p	_____ % flow rate
32% d/p	_____ % flow rate
66% d/p	_____ % flow rate
45% d/p	_____ % flow rate
83% d/p	_____ % flow rate
92% d/p	_____ % flow rate

7. **FLOW MEASUREMENT Laboratory Procedure**
   Background Information
   The term flow infers the movement of matter from one place to another. Flow can be quantified as either instantaneous flow rate or total flow. Instantaneous flow rate is time dependent whereas total flow is not. The rate at which a material is flowing through a pipe will probably be measured as an instantaneous flow such as gallons per minute whereas the water meter at your home measures total flow (gallons).

   Under normal automatic process control, a process technician should expect that the control instruments both measure and then stop a flowing material that is filling a tank. There are, in fact, metering instruments with preset shutoff capabilities that can mechanically accomplish this task as well as more sophisticated control schemes that are designed to do the same. In either case, the process technician should know how long it takes for the tank to fill so that he can be available to monitor the shutoff. For example, if the tank is to be charged with 1,000 gallons of a material and that material is flowing into the tank at 80 gallons per minute, then it will take 12 minutes and 30 seconds to charge. The process technician should check to make sure that the charge is completed accurately.

Materials Needed

- low-flow-rate rotameter with a throttling valve
- containment vessel with graduated volume marks on it (the marked units should match the rotameter units if possible).
- water supply and wet sink or drain apparatus
- timkeeping device that indicates minutes and seconds
- fittings and tubing for connecting the apparatus

Safety Requirements

Safety glasses are required in the lab.

Procedure

a. Set up the flow rate apparatus as illustrated below.

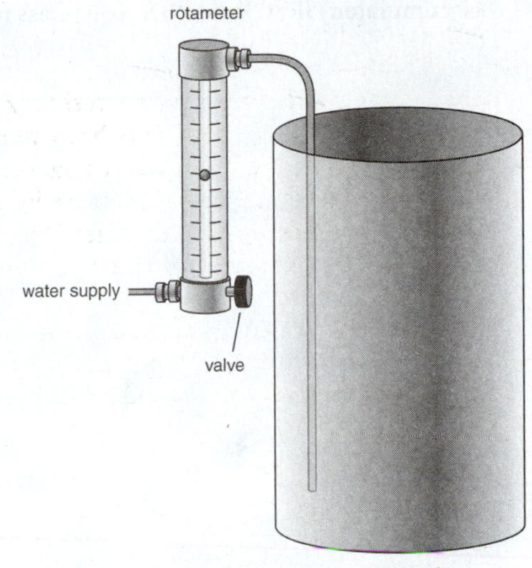

b. Place marks at 25%, 50%, 75%, and 100% volume on the vessel wall; 100% should equal a known volume.

c. Connect a water supply to the input of the rotameter and a piece of tubing to its outlet side.

d. With the open end of the tube placed in a basin or drain, fill the supply line, rotameter, and discharge tubing with water and adjust the flow rate to some point about midrange. Let it stabilize at that rate.

e. At the same time, place the discharge tubing from the rotameter into the vessel and start the timing device.

f. Fill the vessel until it reaches the 25%, 50%, 75%, and 100% volume marks. Record the time at each interval. Stop the timing device and remove the tubing from the rotameter.

Percent Level	*Elapsed Time*
25	
50	
75	
100	

g. Calculate the average flow rate.

$$\text{Flow Rate} = \frac{\text{Volume}}{\text{Time}}$$

Where: Volume units are the same as the rotameter units.
EXAMPLE:

$$\text{Flow Rate} = \frac{2.0 \text{ gallon}}{6.5 \text{ minutes}}$$

$$\text{Flow Rate} = 0.3077 \text{ gpm}$$

h. Compare the calculated flow rate to the indicated flow rate on the rotameter.
i. Complete the final report for this lab.

Findings

Compile your observations and readings into a report that includes a sketch of the equipment, the recorded data, an explanation of the results of the experiment, and a conclusion. Include a paragraph explaining the difference between instantaneous flow verses total (accumulated) flow. Reference your class notes if necessary.

CHAPTER 6

Process Variables, Elements, and Instruments: ANALYTICAL

Objectives

After completing this chapter, you will be able to:

- Define terms associated with analytical instruments:
  - pH (acid or base) and ORP
  - conductivity
  - optical measurements
  - chromatography
  - combustion (O_2 and CO)
- Identify the most common types of analytical devices used in the process industries:
  - optical analyzer (turbidity analyzer or meter and/or opacity analyzer or meter)
  - color analyzer
  - conductivity meter
  - pH and ORP meters
  - gas chromatograph (GC) and GC mass spectrometer
  - spectrophotometer (UV/VIS and/or infrared)
  - total carbon analyzer
- Describe the purpose of analytical devices used in the process industries.
- Discuss how analytical instrumentation affects the role of the process technician and how the process technician affects the operation of the analytical instrumentation.

Key Terms

Analyzer measurements—qualitative and quantitative.

Chromatogram—a graphic record of the separated components.

Chromatography—a process where molecular components of a liquid or gas are separated and identified by means of a tube called a column.

Color/optical analyzer—a photometer that operates in the visible light spectrum (400–800 mμ); two types include either the visual color analyzer and the photometer or spectral color analyzer.

Combustion—the rapid consumption of fuel resulting in its conversion to heat, light, and gases.

Conductivity—a measurement of the ability of a material to conduct an electrical current; the measure of the ability of a liquid (or any solution) to conduct electricity.

Conductivity analyzer—a device that measures all ions in an aqueous solution.

Conductivity meter—a device that measures the conductivity of a process sample by comparing it to a known value or standard cell.

Electrolyte—a nonmetallic substance that is an ionic conductor; the greater a substance conducts an electrical charge, the more ionic the substance becomes.

Gas chromatograph analyzer—an analyzer that provides the necessary means to accomplish the chromatographic separation and analysis.

Gas chromatography (GC)—an analytical method that can provide a molecular separation of one or more individual components in a sample.

Inline analyzer—an analyzer (either continuous or contiguous) installed directly in a process line that may be either a sample system or of the newer probe type.

Mass spectrometer (MS)—a device capable of separating a gaseous stream into a spectrum according to mass and charge.

Opacity analyzer—an optical analyzer used to determine how much particulate matter is in a gas sample.

Optical measurement—a measurement that uses reflection, refraction, or absorption properties of light to measure the chemical or physical properties of a sample.

Optical meters—any or all of the following: turbidity meters, opacity meters, and color meters.

ORP analyzer—an electrochemical analyzer that measures a small electromotive force (EMF) across a hydrogen-ion-sensitive glass bulb; an ORP (oxidation reduction potential) meter also measures a small voltage (EMF) across its electrode and has the capability of detecting all oxidizing and reducing ions in the solution.

Oxidation reduction potential (ORP)—a measure of redox potential created by the ratio of reducing agents to oxidizing agents present in the sample.

pH—a measurement of the hydrogen ion concentration of a solution indicating how acidic (below 7.0) or basic (above 7.0) a substance is from neutral (7.0).

pH meter—an electrochemical instrument that measures the acidity and alkalinity of a solution; an analyzer that measures a specific ion (e.g., hydrogen) concentration in a solution.

Photometer—a color analyzer that operates in the visible light spectrum (400–800 μm).

Photometry system—an automated system used to determine the color of a sample.

Process analyzer—an unattended analytical instrument that is capable of continuously monitoring a process stream.

Properties—characteristics of a substance such as pH, ORP, conductivity, optical measurement, composition, and/or combustion capability.

Qualitative—a measurement of the properties of a substance.

Quantitative—a measurement of the amount of properties within a substance.

Representative sample—a sample that contains portions of the process stream combined together over time or distance.

Sample system—the various components of a sampling apparatus that obtains, transports, and returns the sample with the analyzer itself conditioning and analyzing the sample.

Spectrogram—a graph where transmittance is plotted with respect to wavelength.

Spectrometer analyzer—an analyzer used to detect chemical components in a process sample by measuring variations in transmittance (or absorption) of a spectrum of light passed through the sample; may use visible light (VIS), ultraviolet light (UV), or infrared light (IR and NIR) as a source and detection measurement.

Total carbon analyzer—an analyzer used to determine how much carbon is in a sample; used to detect carbon-based contaminants in steam condensate and wastewater.

Turbidity analyzer—an optical analyzer used to determine the cloudiness of a liquid.

Turbidity and opacity meters—optical meters; measure the transmittance or absorption of light passing through a sample.

Visual color analyzer—compares a physical sample with a standard by shining visible light through the sample and the color comparison is made by the human eye/brain.

Introduction to Analytical Instruments and Terms

There are many types of instruments that analyze various streams within a process facility. Analytical instruments are designed to monitor the chemical and physical properties of the process stream. **Process analyzers** (meters) are laboratory instruments redesigned to withstand the rigorous environmental conditions presented by the processing area. A process analyzer can be defined as an unattended analytical instrument that is capable of continuously monitoring a process stream.

A properly designed **inline analyzer** (continuous and contiguous) system is capable of doing the following steps for each sample taken:

1. Obtaining a representative sample from the process stream
2. Transporting the sample to the analyzer in a way that maintains the physical and chemical integrity of the sample
3. Conditioning the sample so that it can be introduced into the analyzer
4. Analyzing the sample
5. Returning the sample to the process or discards the sample in another suitable manner

Of the five important design components listed above, a **sample system** needs to do three of them: Steps 1, 2, and 5. A sample system accounts for most of the problems encountered in an analyzer loop (Figure 6-1) because of the job it must do. The sample system can get plugged, contaminated, bent, broken, or compromised in any number of other ways. Many new inline, or probe-type, analyzers eliminate the need for the problematic and expensive sample system. However, most analyzers still require a well-designed sample system.

Analyzers are capable of measuring both **quantitative** and **qualitative** properties of a process. A qualitative analysis is a determination of the composition of a substance. For example, determining whether methane, ethane, propane, and butane are present in the sample is a qualitative test. A quantitative analysis is a determination of the amount or proportions of a substance. A typical quantitative measurement would be reported in terms of percentage or in parts per million (ppm) units.

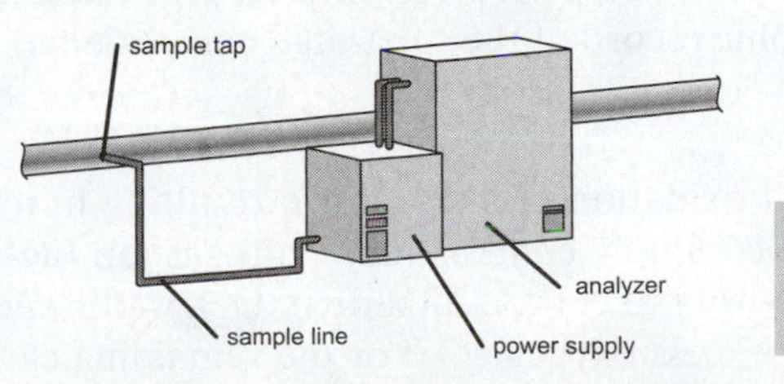

FIGURE 6-1 Analytical Instruments (Typical Outdoor Mounted Unit)

There are hundreds of different kinds of analytical instruments designed to analyze many different chemical and physical properties, but the **properties** that are of greatest interest within the process industry where analytical instruments are used to measure or monitor include the following:

- pH
- ORP
- conductivity
- optical measurement
- chromatography (concentration of components in process streams)
- combustion (concentration of components in flue gas)

pH

pH is a measurement of the hydrogen ion concentration $[H^+]$ of a solution that indicates how acidic or basic a substance is. Solutions having a pH below 7.0 are acidic. Solutions with a pH above 7.0 are basic. A pH of 7.0 is neutral—neither acidic nor basic.

$$pH = -\log_{10}(H^+)$$

Oxidation Reduction Potential (ORP)

ORP, or redox, is a measurement of potential created by the ratio of reducing agents to oxidizing agents present in the sample. Where pH measurement is specific to hydrogen ion concentration, ORP measurement can be thought of as being specific to free electron concentration.

Conductivity

Conductivity is a measurement of the ability of a material to conduct an electrical current. Industry uses conductivity as a means of determining the level of dissolved solids in a liquid, which in turn tells them when they need to blow down or treat a process fluid. Otherwise, harm to equipment or processes may occur.

$$C = \frac{I}{R}$$

Conductivity is the inverse of resistance and is measured in siemens/cm^2 and micromhos.

Optical Measurement

An **optical measurement** uses reflection, refraction or absorption properties of light (electromagnetic waves) to measure the chemical or physical properties of a sample.

Chromatography

Chromatography is the process where molecular components of a liquid or gas are separated and identified by means of a tube called a column. A special packing material, located in the column, selectively retains molecules of components of the **representative sample** both momentarily and sequentially as the sample propagates through the column when a carrier gas pushes the sample. The sample components move through the packing material according to component affinity for the packing and individual vapor pressures. A detector located downstream of the column is then able to sense these separated components as they emerge from the column producing a **chromatogram** (a graphic record of the separated components).

Combustion

Combustion is the rapid oxidation of a substance resulting in its conversion to heat, light, and gases. During complete combustion, hydrocarbon fuels such as natural gas (CH_4) converts in its entirety to carbon dioxide (CO_2), water and heat. In a boiler, or furnace, monitoring the excess oxygen (O_2) or the remaining carbon monoxide (CO) in the flue gases can be used to enhance the combustion process by providing input into a control system that in turn limits the amount of excess air entering the firebox, the combustion compartment of the boiler or furnace. Since oxygen is the only component in air that affects the combustion reaction and accounts for only about one-fifth (20.9 percent) of its total volume, the rest (primarily nitrogen at 78.1 percent) occupies space and steals heat energy from the system by carrying it out the flue. Being able to measure the concentration of components in the flue gas allows technicians to optimize the combustion process and comply with environmental permits.

Analytical-Sensing or Measurement Instruments

PURPOSE OF ANALYTICAL DEVICES

Analyzers are used within industry for several reasons to include the following:

- environmental monitoring and reporting
- mechanical integrity of fixed equipment
- economics
- product quality assurance

Environmental Monitoring and Reporting

Analyzers, commonly referred to as Continuous Environmental Monitoring Systems (CEMS), may be used to report emissions directly to the Environmental Protection Agency (EPA). EPA guidelines are becoming increasingly confining. To be sure that emissions coming from each unit fall within a given spectrum, **analyzer measurements** must be accurate and reliable so that the analysis prevents potentially debilitating EPA fines as well as criminal charges. Ensuring the safety of both the environment and its citizens within the surrounding community is also critical since dangerous levels of contaminants emitted into the atmosphere could pose an emergency or life-threatening condition.

Mechanical Integrity of Fixed Equipment

Permanent damage could occur to process equipment when the process is cloudy, contaminated, too acidic, or too basic. Fixed equipment such as exchangers, pumps, valves, vessels, pipes, etc., can become corroded, eroded, or plugged if the process goes unchecked. Analyzing equipment can check for these problems or the potential for these problems to occur, thus preventing equipment failure.

Economics

Online product analysis gives technicians an "as it's happening" type of view of the product. This allows adjustments while the product is being made and helps to eliminate off specification product. Online analysis is more expensive to install, but less expensive over the long haul in less off-spec product.

Product Quality Assurance

Online analyzers are the first step in ensuring product quality, but analyzers in control laboratories are also used to run finished product samples to ensure all specifications are met before a product is sold to a customer. These analyzers are located in controlled environments and are calibrated under strict conditions. The measurements they make should be highly reliable, accurate, and repeatable.

COMMON TYPES OF ANALYZERS

The more common types of analyzers used in the process industry include the following:

- pH and ORP meters
- conductivity meter
- optical analyzer (color, turbidity, and/or opacity)
- gas chromatograph (GC) and GC mass spectrometer
- spectrometer (UV/VIS and/or infrared)
- total carbon analyzer

pH and ORP Meters

pH and ORP meters are among a group of instruments known as electrochemical analyzers. The pH analyzer, for example, is a specific ion analyzer that is capable only of measuring hydrogen ion concentration in a solution. A pH meter does this by measuring a small electromotive force (EMF) across a hydrogen-ion-sensitive glass bulb. The ORP meter also measures a small voltage (EMF) across the electrode within the meter, but it *sees* all oxidizing or reducing ions in the solution. Both of these analyzers play an important role in controlling the chemistry of industrial processes.

pH Meter The **pH meter** measures the acidity and alkalinity of a solution by measuring the hydrogen ion concentration. pH is defined mathematically as the negative logarithm of the hydrogen ion concentration: $pH = -\log[H^+]$. When an acid or base is mixed into water, the acid/base disassociates either partially or fully into positive and negative ions. The pH analyzer then measures the concentration of hydrogen ions (H^+) in solution by comparing the hydrogen ions to a buffered solution (exactly 7 pH) located inside the measuring electrode. The millivolt potential developed across the glass membrane is proportional to the pH of the solution. pH meters (Figure 6-2) are calibrated with standard pH solutions at several pH levels (typically 4.0, 7.0, and 10.0).

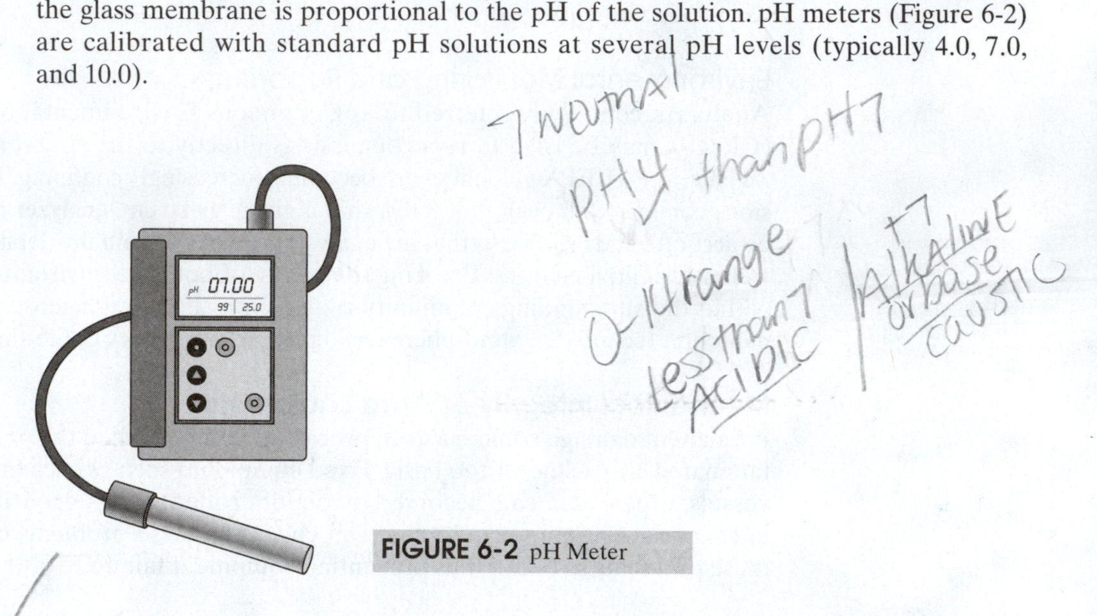

FIGURE 6-2 pH Meter

The pH scale ranges from 0 to 14 where 7 is considered to be neutral. A pH of less than seven is considered to be acidic and greater than seven to be basic (caustic). Every technician needs to understand that each pH number constitutes a tenfold change in concentration. For example, a pH of 4 is 1,000 times more concentrated with hydrogen ions than a pH of 7. See Table 6-1.

TABLE 6-1 pH Relationship to H Ion Concentration

	pH Range	*H Ion Concentration*
acid	pH 0–6	10^{-0}–10^{-6}
neutral	pH7	10^{-7}
base	pH8–pH14	10^{-8}–10^{-14}

Oxidation Reduction Potential (ORP) Meter Another type of electrochemical analyzer is the **ORP analyzer**. The **Oxidation Reduction Potential (ORP)** is also known as redox (short for reduction-oxidation). An ORP measurement indicates how much *free electron* potential exists in the sample. In the chemical world, oxidation means the loss of electrons whereas reduction means the gain of electrons. This potential is similar to pH in that they both are measuring ion potential based on concentration. ORP analyzers are used to monitor the electrochemical properties of a reaction as it is taking place. ORP analyzers let a process technician see into the chemical reaction of a process in a way unlike any other analytical method. ORP analyzers are used in control loops in the same manner as pH meters. ORP meters (Figure 6-4) are calibrated in millivolts. For the ORP scale, positive millivolts indicates oxidation and negative millivolts indicates reduction.

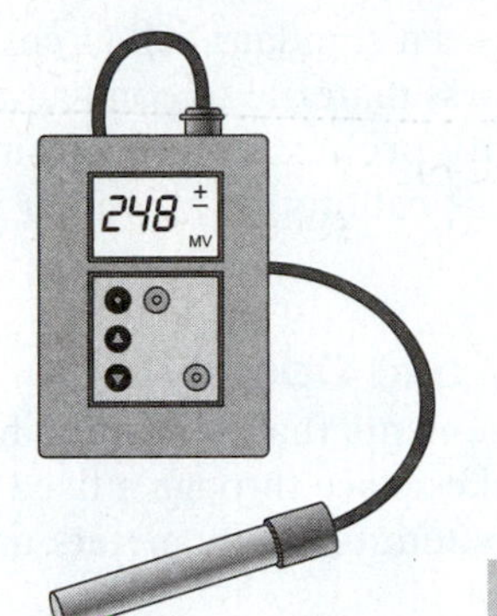

FIGURE 6-4 ORP Meter

Conductivity Meters

Conductivity is the reciprocal of resistance. Conductivity is an electrochemical property measured in micromhos and in the metric unit called siemens (per sq. cm.).

When a salt such as table salt, sodium chloride (NaCl), or another strong electrolyte, is dissolved in water (Figure 6-5), the salt disassociates into cations carrying a positive (+) charge, and anions carrying a negative (−) charge. Once this happens, the solution acts as an **electrolyte** (a nonmetallic substance that becomes an ionic conductor). Table salt is just one example of an electrolyte that acts in this manner. Other electrolytes can be either metallic or nonmetallic and act in a similar manner. The more ionic the solution becomes, the greater its ability to conduct an electrical charge. A **conductivity analyzer** measures all ions in an aqueous solution.

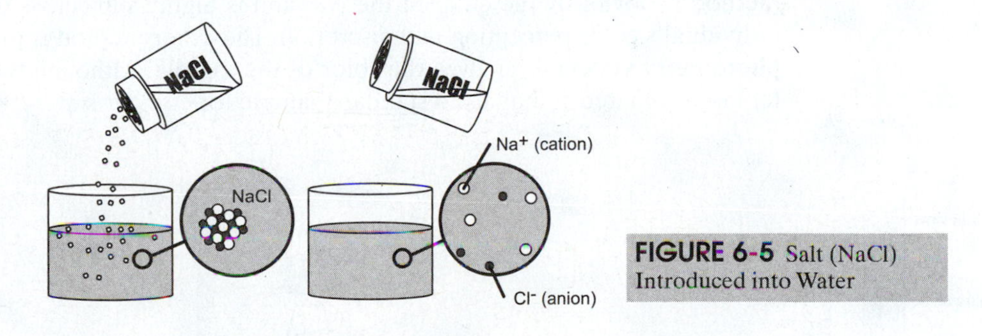

FIGURE 6-5 Salt (NaCl) Introduced into Water

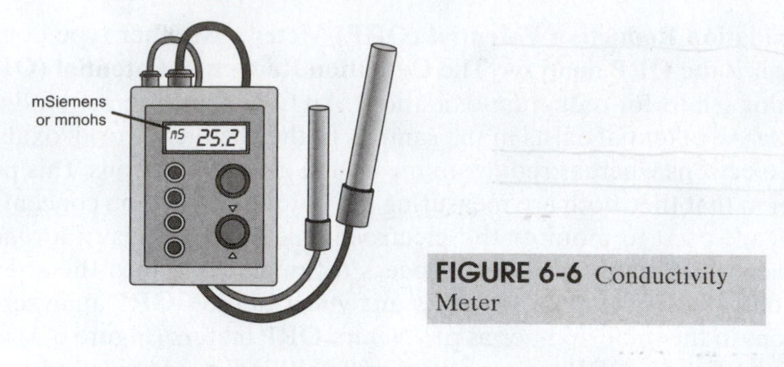

FIGURE 6-6 Conductivity Meter

One important industrial conductivity application is in measuring for dissolved solids in water. Pure water is nonconductive. There are various levels of conductivity in boiler water, condensate, and cooling tower water.

A **conductivity meter** (Figure 6-6) uses a standard cell (1 cm^2 in size) to quantify the measurement of the ability of a process material to conduct an electrical current. Numerous styles or types of conductivity probes are used to insert in the process. Conductivity meters use standards to check calibration at regular intervals and a reference solution is used for that purpose.

Optical Analyzers (Color, Turbidity, and Opacity)

Optical measurements are those measurements that determine how much reflection, refraction, or absorption of visible light takes place through a fixed distance in a sample. Generally, **optical meters** include color meters, turbidity meters, and opacity meters.

Color

Color measurement implies the measurement of visible light. **Color/optical analyzers** are **photometers** that operate in the visible light spectrum (400–800 μm) (Figure 6-7). A color analyzer is used to determine the color of a process sample. Some off-color products are impossible to sell. Take for example, bottled water. Most consumers would not buy or drink water if it had a tinge of yellow coloring. The same is true for other mass-produced products.

Color can also indicate process problems. If a chemical reaction goes too far or is heated beyond a certain point, a color change may be the result. Some products that are off spec with color can still be sold if they can be reworked. Color analyzers (Figure 6-8) are found in the process areas of the plant as well as in the lab.

There are two types of color analyzers: the **visual color analyzers** and the photometers or spectral color analyzers. In the visual type, a process technician or lab analyst visually compares the sample to a standard. A visible light is shined through the sample and is detected by the human eye and analyzed by the human brain. This method is obviously the older of the two and is highly subjective and depends on an individual's color perception and discretion. The other method is photometric using a **photometry system** to analyze the color of the sample. Although the photometry system is automated, it may use a standard sample for comparison.

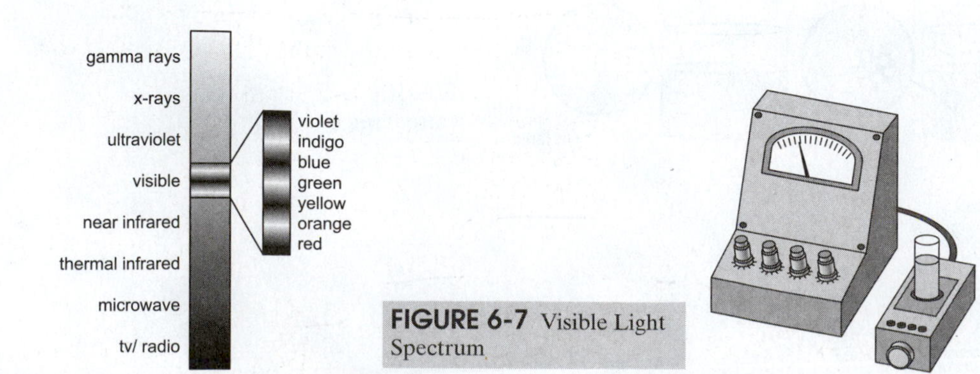

FIGURE 6-7 Visible Light Spectrum

FIGURE 6-8 Optical Analyzer–Color

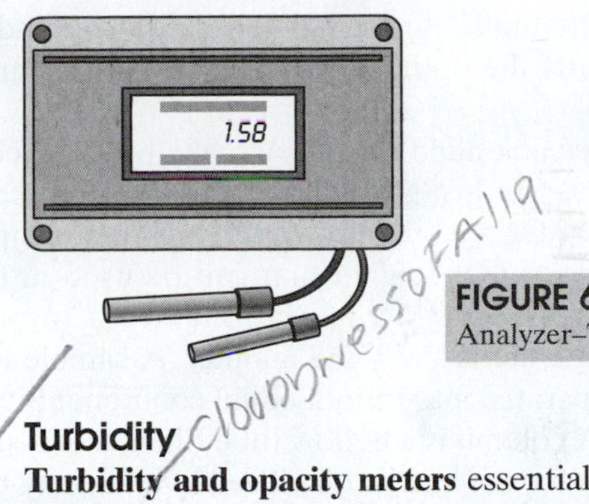

FIGURE 6-9 Optical Analyzer–Turbidity

Turbidity

Turbidity and opacity meters essentially do the same thing; they measure the transmittance or absorption of light passing through a sample. The **turbidity analyzer** (Figure 6-9) is used to determine the cloudiness of a liquid. Cloudiness is due to suspended particles. To detect the particles, the photometer is located across from a light source so that it detects the result of the scattering or absorption of the beam by the particulate matter in the sample.

Types of electromagnetic waves used for the light source within these meters include the following:

- gamma rays
- X rays
- ultraviolet
- visible
- near infrared
- thermal infrared
- microwave
- TV/radio

Opacity

Opacity analyzers (Figure 6-10) are used to determine how much particulate matter is in a *gas* sample. Opacity analyzers work essentially the same as turbidity analyzers. Again, a beam of light projecting from a source located on one side of a furnace or boiler stack penetrates the gases rising through the stack. A photo-detector across from the source provides an optical measurement indicative of the absorption and scattering of the light beam as it strikes particulate matter in the stack.

Chromatographs (GC)

Gas chromatography (GC) is an analytical method that can provide a molecular separation of one or more individual components in a sample. This method identifies the chemical composition of the molecules within the separated sample. A **gas chromatograph analyzer** provides the necessary means to accomplish the chromatographic separation and analysis. A GC takes a mixture and separates the mixture into its individual components with each being identified and quantified.

The GC (Figure 6-11) consists of an injection port that vaporizes the sample, a column (a hollow tube filled with a special material called packing), a detector that senses the separated gases, and an output device that generates a chromatogram (an illustration depicting the separated components as peaks on a graph). The GC may also

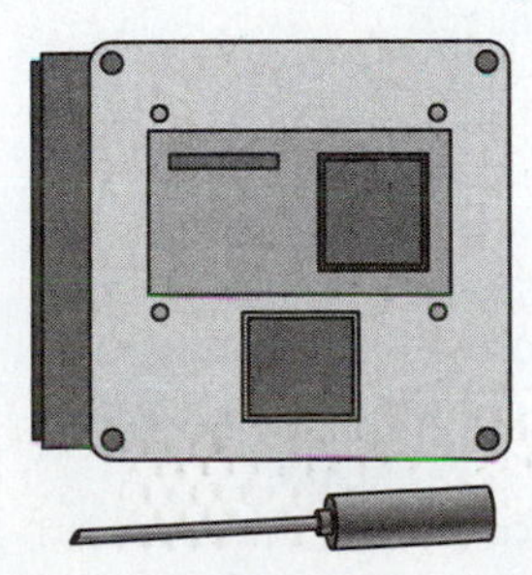

FIGURE 6-10 Optical Analyzer–Opacity

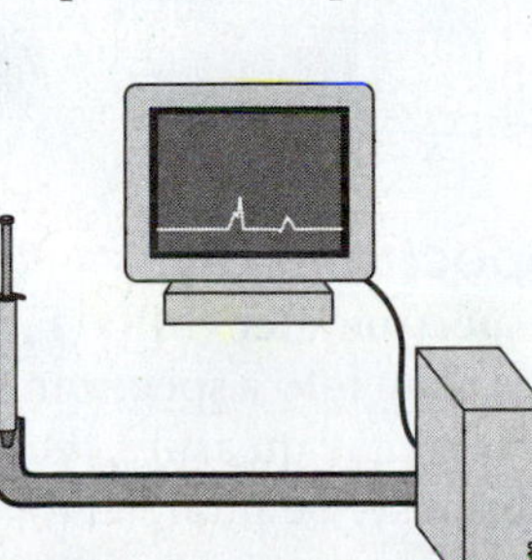

FIGURE 6-11 Gas Chromatograph

include a microprocessor that takes the signal generated by the detector and converts it into a chart and/or reports the output as a weight percent, parts per million (ppm) volume, etc.

In continuous processes, an online GC can actually provide feedback (closed-loop) to an instrument control loop or just give information that can be used in an informative (open-loop) manner. A GC can monitor as few as one component in the sample and produce a standard instrument signal that can be used in the same way as any other transmitter in a control loop.

All chromatographs operate in the same manner. A sample is introduced into the GC. This sample is then separated into its individual components while flowing through a device called a column. A column is a hollow tube filled with a special material called *packing* (capillary columns are the exception). The columns may be configured in coils (Figure 6-12) to conserve space within the GC.

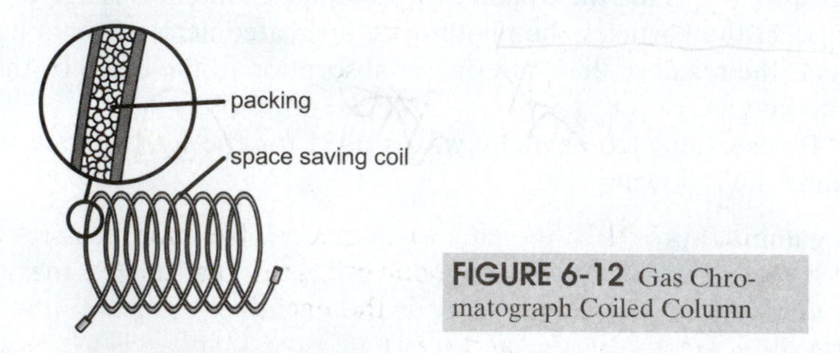

FIGURE 6-12 Gas Chromatograph Coiled Column

In a packed column, the packing material is comprised of a finely ground, inert porous material saturated with another material that is dry to the touch at room temperature. However, under normal operating temperatures in the column, the packing acts like a liquid providing a fixed surface where soluble gases can be easily absorbed and released. The packing is called the stationary phase and the carrier gas with vaporized sample gases is called the mobile phase. The carrier gas pushes the vaporized sample gases through the column into the detector for analyzing. Some chromatographs can be liquid.

All components of the sample enter the column simultaneously and move through the length of the column at different rates according to the coefficient of adsorption, mole weight, vapor pressure, and molecular size. There are several reasons why they move at different rates, but the solubility differences between each component and the stationary phase is one of the most important. Molecules that are not soluble in the stationary phase flow through the column as fast as the carrier gas while the other more soluble molecules take longer.

The detector is placed at the end of the column to sense the separated gases as they emerge from the column. A graphic representation called a chromatogram (Figure 6-13) illustrates the separated components as peaks on a graph, similar to fingerprints from an individual, for the identification of specific chemical molecules in the sample.

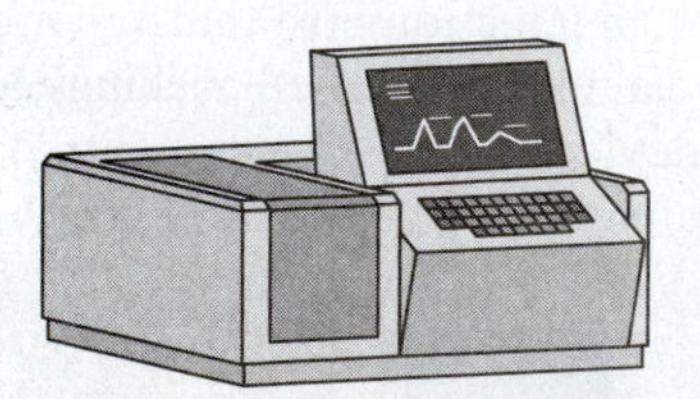

FIGURE 6-13 Gas Chromatogram

Mass Spectrometers

A **mass spectrometer (MS)** (Figures 6-14 and 6-15) is a device capable of separating a gaseous stream into a spectrum according to mass and charge. There are chromatograph applications that measure the composition of liquids without converting to a vapor state. These may be referred to as liquid chromatographs.

In one type of mass spectrometer analyzer, a sample of process gas is converted into an accelerated stream of highly charged ions. The sample is then sent through a

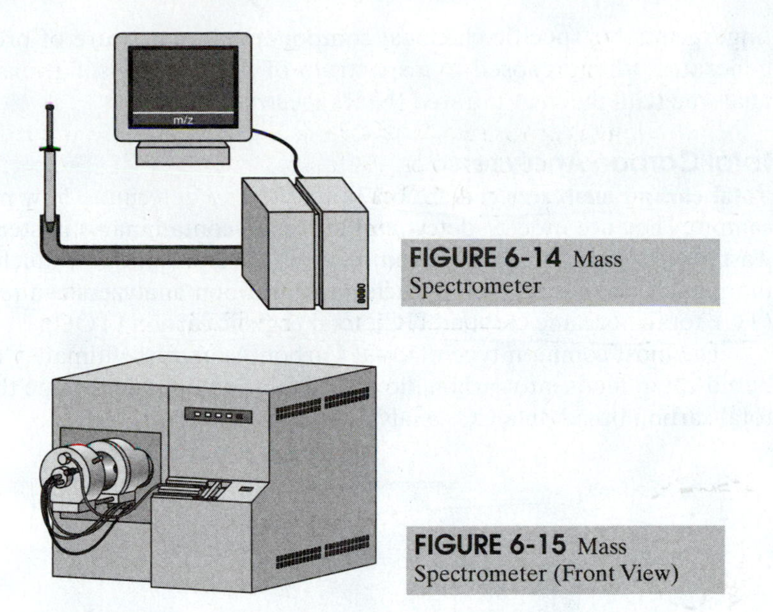

FIGURE 6-14 Mass Spectrometer

FIGURE 6-15 Mass Spectrometer (Front View)

magnetic filter (field) and then into a detector. The more massive particles tend to go in a straight line through the magnetic prism whereas the less massive particles tend to move away from the center towards the negatively or positively charged rods in the prism. The detector located at the exit of the magnetic prism senses both the mass and charge of each particle at their position of impact on the detector. Different molecules tend to have unique identifying patterns represented by these impact positions registered on the detector.

Mass spectrometers may be used in many of the same applications as gas chromatographs; however, GC output is on a volume percent data while mass spectrometers provide weight percent data. This can be useful in identification of components. Mass spectrometers are faster than a GC, but they are also a lot more expensive. Their numbers are significantly less in online applications as compared to the GC. The most common place to find a mass spectrometer is in a quality control lab.

A mass spectrometer analyzer is sometimes used in combination with a gas chromatograph. Effectively, the mass spectrometer analyzer becomes the detector for the chromatograph. There are several reasons why these two analyzers are used in combination. One is that the chromatograph separates the sample gases into partitioned components (or groups of components) so that they can have a more definitive analysis when they enter the MS. Another is that the gas chromatograph can provide a separation then backflush the unwanted components in the sample leaving only the components of interest to be analyzed. Of course, when the GC and MS are used in parallel with each other, the analysis time takes longer.

Spectrometers

A **spectrometer analyzer** (Figure 6-16) is used to detect and/or quantify chemical components in a process sample by measuring variations in transmittance (or absorption) of a spectrum of light passed through the sample. A spectrometer may use visible light (VIS), ultraviolet light (UV), or infrared light (IR and NIR) as a source and detection measurement.

The word spectrum implies a continuum of wavelength bands of light. When a beam of white light is passed through a prism, a spectrum of colors can be seen as a result. A spectrophotometer is capable of detecting a spectrum of wavelengths rather than a single-frequency band. Scientists have determined that unique and reliable

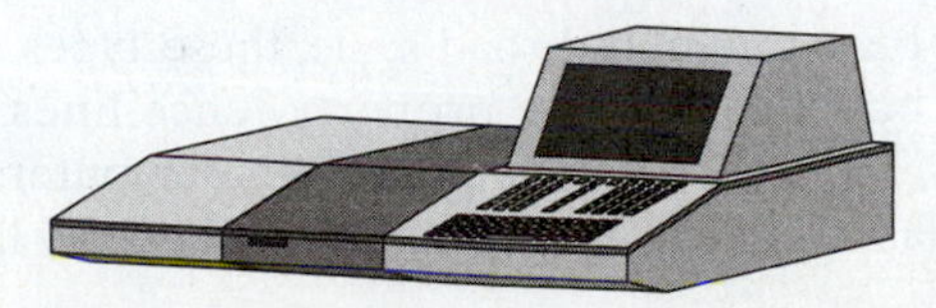

FIGURE 6-16 Spectrometer

fingerprints for specific chemical components in a mixture of process fluids can be generated when exposed to a spectrum of light analysis. For example, CO_2 is best analyzed with the near infrared (NIR) spectrum.

Total Carbon Analyzers

Total carbon analyzers (Figure 6-17) are used to determine how much carbon is in a sample. They are used to detect carbon-based contaminates in steam condensate and wastewater. Carbon analyzers can be used to determine how much organic as well as inorganic carbon is present in the sample. Carbon analyzers can test for: total carbon (TC), total inorganic carbon (TIC), total organic carbon (TOC).

The most common type of Total Carbon Analyzers ultimately convert all carbon-based compounds into carbon dioxide (CO_2) and then determine the concentration of total carbon based on a CO_2 analysis with an NIR analyzer.

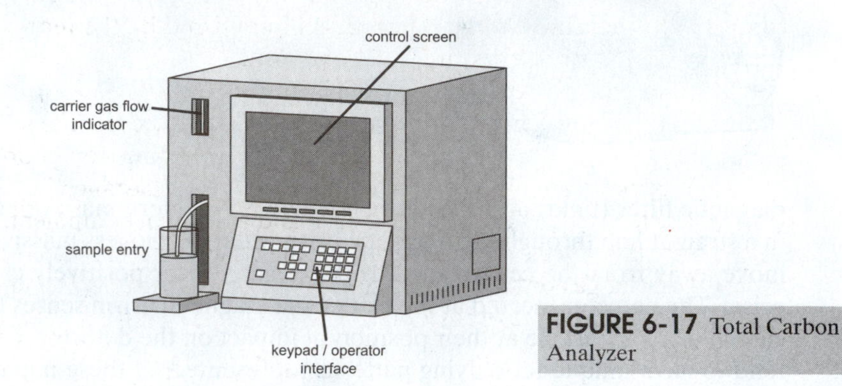

FIGURE 6-17 Total Carbon Analyzer

Miscellaneous Monitors

Process technicians encounter many obvious types of process analyzers as described previously in this section, but there are still many other types worthy of mentioning that are in fact analyzers and play a very important role in the day-to-day operations. These monitors are sensitive instruments, but they are usually only maintained by instrument personnel; however, process technicians should observe these monitors on their daily rounds in their work areas to ensure appropriate housekeeping is performed so that airflow around the monitors is maintained. In some cases, process technicians may come in contact with these types of monitors only in certain work sites or on special occasions.

Personnel Monitors

Plant workers sometimes wear this type of monitor to detect potential exposure to harmful substances. A ring badge or a pocket dosimeter may be used that is capable of indicating exposure to nuclear radiation and either type responds to X rays or gamma rays. Chemical badges respond by changing colors as a representation of the exposure levels in ppm while electronic pocket devices detect exposure and respond by various means. Process technicians need to ensure that these devices are kept in good operating condition and exposed badges are discarded appropriately.

Area Monitors

Specific types of monitors (e.g., Cl_2, CO, H_2S, etc.) are placed around loading and unloading stations, compressor stations, specific vessels, or piping manifolds where dangerous chemicals could potentially leak. These monitors work in conjunction with wind direction and speed or weather stations.

Unit or Plant Boundary Monitors

Similar to area monitors, but on a more broad scale, these types of monitors may be installed at or near unit boundaries or on property fence lines. Newer technology devices employ laser beams or passive IR coupled with computers to analyze for several types of gases. These laser beams may be shot down property lines using mirrors at various points with the final return beam going through a computer to be analyzed for escaped gases.

Other Monitors

There are other monitors to analyze pipe welds, pipe corrosion, scrubber and column tray placement, etc. These types of monitors may use radiation or ultrasonic types of detection methods. There are also portable types of nuclear analyzers that give proper chemical composition of pipes or valve parts that are capable of giving a printout listing the percent composition by common name.

Analytical Instruments and the Role of the Process Technician

Analytical instrumentation provides process technicians with a unique means of *looking* into the process. Analysis gives them an insight into the chemical composition and/or physical attributes of the process fluid missed by the more common measurements of temperature, pressure, level, and flow.

The new role of the process technician has changed significantly from expectations in the past. This modern role includes operator-assisted maintenance, where process technicians initiate the first steps in troubleshooting for problems in the unit. Operations personnel are also expected to provide assistance by doing routine tasks and checks that relate to instrumentation and analytical equipment. For example, they are expected to check for analytical sample flows, temperatures, pressures, and/or to monitor carrier gas bottles. Process technicians must have a basic understanding of the theories and operations of instruments, analytical devices, and their sample systems.

Summary

Process facilities have many types of analytical instruments that are designed to monitor various chemical and physical properties of process streams. Analyzers, sometimes called meters in the field, are specially designed instruments that are used in more harsh environments than their counterparts in controlled atmospheres such as in laboratories. These analytical instruments are also designed to be left unattended and continuously monitor process streams.

Sample systems are responsible for obtaining and transporting samples before they go into some analyzers. After the sample is analyzed, the sample system either returns the sample to the process or suitably discards the sample. Many analyzers are installed in line (inline) or have a probe that eliminates the need for a sample system. These innovations help to reduce problems and expenses associated with sampling systems.

Analyzers measure quantitative (proportion) and qualitative (composition) properties of process streams. With hundreds of types of properties that can be analyzed, there are a few that are most common within the industry.

The more common types of analyzers include the following:

- *pH or ORP meter*:
  Used in the control of the chemistry of industrial processes; electrochemical; pH meters only measure hydrogen ion concentrations in solutions; ORP meters measure a small voltage across an electrode and also see all the oxidizing or reducing ions in a solution
- *Conductivity meter*:
  Measures all ions in an aqueous solution with the more ions in the solution giving the solution a greater ability to conduct an electrical charge; commonly used to detect contaminants in condensate or spent steam
- *Optical analyzer (color, turbidity and/or opacity)*:
  Optics such as reflection, refraction, or absorption of visible light observed through a fixed distance in a sample; color meters (visual and photometer or spectral color) operate in the visible light spectrum with results in a simple on spec or off spec (off color) result; turbidity analyzers determine the cloudiness of a liquid using various forms of light with a photometer for detection of either the scattering or absorption of the beam; opacity analyzers are used to determine how much particulate matter

is in a gas by using a beam of light to penetrate rising gases and detect by a photo-detector like the turbidity meter

- *Gas chromatograph (GC) and GC mass spectrometer*:
  GCs provide a molecular separation of one or more individual components in a sample to identify the chemical composition (proportionally); GCs can be stand-alone units or included into a loop by transmitting the measured data to control an operation; GC mass spectrometers separate gas streams into spectrums according to mass and charge; GC mass spectrometers are more expensive and mostly found only in laboratory environments
- *Spectrophotometer (UV/VIS and/or infrared)*:
  Used to detect chemical components in a process sample by measuring variations in transmittance (or absorption) of a spectrum of light passed through a sample; each specific chemical component has its own fingerprint when exposed to a spectrum of light analysis with the transmittance plotted **(spectrogram)** with respect to wavelength
- *Total carbon analyzer*:
  Used to determine how much carbon (organic and inorganic) is in a sample in steam condensate and wastewater

Through the use of analytical instrumentation, process technicians have a way to look into various processes that may be missed by the more common measurements of pressure, temperature, level, and flow. Process technicians are expected to understand the theories and operations of these instruments so they can troubleshoot process problems.

Checking Your Knowledge

1. Match the terms below with their correct definitions.

Term	*Definition*
1. pH	a. Measurement of the consumption of fuel
2. ORP	b. Measurement of the ratio of reducing agents to oxidizing agents
3. Conductivity	c. Measurement of hydrogen ion concentration
4. Chromatography	d. Measurement of the molecular components of a liquid or gas
5. Combustion	e. Measurement of the ability of a solution to conduct electricity

2. Which of the following instrument capabilities would you use to measure pH?
   a. Measuring reflection, refraction or absorption of light
   b. Measuring the level of ionic concentration
   c. Measuring variations in the transmittance of light through a sample
   d. Measuring the molecular composition
   e. Measuring free electron potential
3. *True or False* Process technicians are often responsible for monitoring process analyzers and sampling systems.
4. Which type of analyzer would be used to find out what components and their respective percentage are in a sample stream?
   a. pH meter
   b. Gas chromatograph
   c. ORP meter
   d. Optical analyzer
5. Which type of analyzer is best suited to measure cloudiness in process water?
   a. pH meter
   b. ORP meter
   c. Turbidity meter
   d. Opacity analyzer
6. What is important about measuring the percentage of O_2 in the flue gas from fired equipment?
   a. Too much O_2 can lead to fouling.
   b. Too little O_2 can lead to contamination.
   c. Too much or too little O_2 can lead to dangerous conditions and inefficient combustion.
   d. Measuring % O_2 is not important.

Student Activities

1. Draw a horizontal linear scale from 0 to 14 using each increasing number (2, 3, 4, etc.) as a point on the scale. Label the diagram for pH. Perform research (library, Internet, etc.) on common household substances to determine their pH levels. List the verified items (at least 20) by their scale number.
   NOTE: Many commercial substances (e.g., shampoo) list pH levels on the label.
2. Construct a simple conductivity tester as per the diagram below. Determine the conductivity of various metal and non-metal objects by performing the following steps:
   a. Put a flashlight bulb in a socket and mount them onto a piece of wood.
   b. Connect one wire from the socket to one terminal of the battery.
   c. Connect another wire to the other side of the socket (NOT to the battery).
   d. Connect a third wire to the other battery terminal (NOT to the socket).
   e. Wrap (at least six times) each of the loose wire ends to clean head thumbtacks.
   f. Place the tacks (after wire wrapping) firmly into the erasers of two pencils that are used as the conductivity probes.
   g. Now that you have constructed a conductivity tester (a simple conductivity analyzer), test a variety of objects and record the results in chart form.

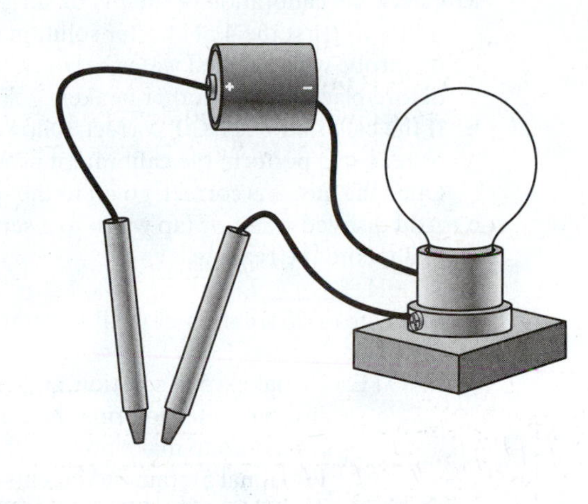

3. Identify analytical devices in the lab.
4. Use a pH meter to measure pH. Make a 5%, 10%, and 15% solution using Peter's fertilizer and distilled water. Use the solutions to define clarity and concentration. Determines the concentration mathematically, then uses a meter to determine the pH of each solution.
5. Use conductivity meter to show the concentration of dissolved solids in cooling towers and boilers.
6. Use oxygen analyzers to check oxygen concentration in air and other gas streams.
7. Use gas test meters to show flammability of gasoline and diesel.
8. **ANALYTICAL (pH) MEASUREMENT Laboratory Procedure**
   Background Information
   pH is the hydrogen ion concentration of an aqueous (water-based) solution. Pure water ionizes to a very small extent to produce a few hydrogen and hydroxyl ions in equilibrium with the water molecules. Pure water always contains equal amounts of each of these ions and is considered neutral. An acid is a substance that yields hydrogen ions when dissolved in water. Bases are substances that yield hydroxyl ions when dissolved in water. The relative strength of an acid or base is found by comparing the concentration of hydrogen ions in solution with that of pure water. Any solution that contains equal concentrations of hydrogen and hydroxyl ions is neutral. Any solution that contains an excess of hydrogen ions is acidic, and any solution that contains an excess of hydroxyl ions is basic.

   The pH probe is designed to selectively measure a hydrogen ion concentration difference between the inside of the pH sensitive glass probe and the solution it is immersed in. The small voltage (millivolt) produced by this probe is usually indicated on a meter marked 0-14 with 7 at midscale. Seven is considered to be neutral having equal amounts of hydrogen ions and hydroxyl ions.

   Finally, each whole number on the pH scale represents a ten-fold change in concentration. That means that a pH of 5 is ten times more acidic than a pH of 6.

Materials Needed

- a pH meter and pH probe(s)
  NOTE: The pH probe(s) may be a combination probe (singular) or it may be two separate probes. In this lab, assume that the probe is a singular combination probe.
- five 250-ml beakers
- buffer solutions: 4 pH and 10 pH (Alternatively, rather than a solution of one material in different concentrations, use different liquids. Try fruit juices, milk, Coca-Cola®, and liquid detergent.)
- squirt bottle filled with distilled water for rinsing the probe
- distilled water or tap water to make up fertilizer solutions
- fertilizers such as Peters or Miracle-Gro®

Safety Requirements

Safety glasses are required in the lab.

Precautions

Glass electrodes are very fragile. Electrodes should be left in liquid.

Procedure

a. Check the calibration of the pH meter by immersing the probe in the two buffer solutions (first the 4 pH buffer solution and then the 10 pH buffer solution). Rinse the probe with distilled water between separate immersions. Always rinse the probe before placing it in another beaker.

b. If the calibration is NOT correct, follow the instructions provided by the instructor or read and perform the calibration procedure found in the instruction manual. Once the meter is correct, go on to the next step.

c. Add distilled water or tap water to a separate beaker. Place the probe into the water and record the reading.

   pH = _____

d. Make up a 5% solution of fertilizer in the beaker. Record the new pH.

   pH = _____

   NOTE: To make a 5% solution, add 5 grams of fertilizer to the beaker and enough distilled water to bring the solution level to the 100-ml mark. Use this procedure to make up the other solutions. Another option is to add the additional 5-gram increments to the previous solution.

e. Make up a 10% solution of common fertilizer in the same or another beaker. Record the new pH.

   pH = _____

f. Finally, make up a 20% solution of common fertilizer in the same or another beaker. Record the new pH.

   pH = _____

g. Correlate the percent solution concentration to the pH recorded.

	Percent Solution	*pH*
1	5%	
2	10%	
3	20%	

h. Draw a graph representing the percent versus the pH.

Additional Information

The actual solution concentration (chemical to water weight concentration) will probably not have a linear relationship with the resulting pH. Specific chemicals will have different effects on the water and its resulting pH. If you were careful while conducting the lab procedure, your data should support a reasonable conclusion.

Findings

Compile your observations and readings into a report that includes your recorded data, an explanation of the results of the experiment, and a conclusion. Include a paragraph explaining correlation of the observed pH for each concentration level.

CHAPTER 7

Miscellaneous Measuring Devices

Objectives

After completing this chapter, you will be able to:

- Define terms associated with miscellaneous measuring devices:
  - vibration
  - rotational speed
- Identify common types of miscellaneous measuring devices:
  - vibration measurement
  - speed measurement

Key Terms

Accelerometer—a vibration measuring device (e.g., piezoelectric type).

Acceptable limits—the operating range within which a piece of rotating equipment can operate without causing excessive wear to the bearings, or other types of catastrophic failure.

Mass flow rate—where a solid material is weighed as it is conveyed on a moving belt (conveyor) and an instantaneous weight measurement is taken and the rate of motion of the belt is known.

Overspeed—a dangerous condition that can occur in a turbine or other type of equipment that moves too fast.

Rectilinear speed—linear speed expressed in distance per unit of time (e.g., feet per second).

Rotational speed—number of revolutions per unit of time (e.g., revolutions per minute or rpms).

Speed—the distance traveled per unit of time irrespective of direction (e.g., feet per second).

Speed monitor—a device that measures speed; comprised of a speed sensor and a readout/receiving device.

Velocity—speed with a specific direction.

Vibration—the periodic motion of an object.

Vibration meter—a device used to measure displacement, velocity, or acceleration due to vibration; consists of a pickup device, an electronic amplification circuit, and an output meter.

Vibration sensors or monitors—a device used to sense the effects of vibration by sending a signal to a meter or monitor, or to shut down a device if operating limits are exceeded.

Introduction

There are devices that do not fall in the main categories or types of instrumentation used in process facilities as described in Chapters 2–6, but in fact are forms of commonly used instrumentation. The two types that are discussed in this chapter include devices that measure vibration and speed. Also discussed are the various applications and applicable-type variations for each of these miscellaneous measuring devices.

Types of Miscellaneous Devices: Vibration

Vibration (Figure 7-1) in an object, device, or system is the random or periodic change in velocity, acceleration, or displacement from a predetermined point. A **vibration meter** (Figure 7-2) is a device used to measure velocity, acceleration, or displacement due to vibration. Vibration meters consist of a pickup device, an electronic amplification circuit, and an output meter.

There are **acceptable limits** within which rotating equipment can operate without causing excessive wear to the bearings or otherwise causing the equipment to be subjected to high stress forces. If operating limits are exceeded, severe and potentially

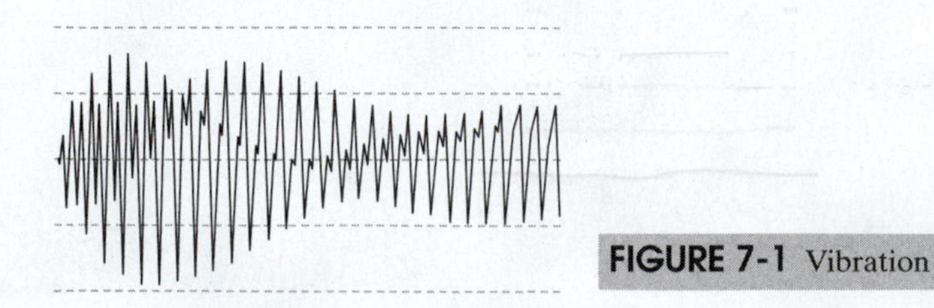

FIGURE 7-1 Vibration

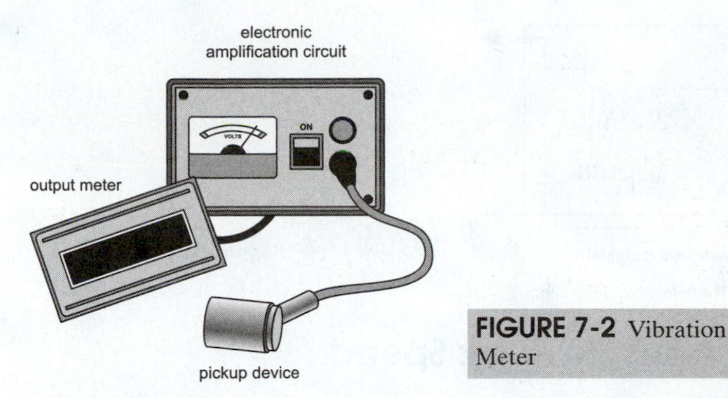

FIGURE 7-2 Vibration Meter

catastrophic damage can occur very quickly. To alleviate this problem, the vibration in large and/or high-speed rotating equipment must be monitored and controlled.

For example, most compressors have high-vibration alarms as well as shutdowns associated with them. The two types of vibration encountered are either axial or radial thrust on the shaft. If a vibration monitor works properly, the compressor shuts down before major damage is caused, hence avoiding a safety problem or excessive downtime in the process. Having **vibration sensors or monitors** on high-energy rotating equipment makes good sense from a safety standpoint as well as from an economic standpoint.

Vibration can be measured with an **accelerometer** (Figure 7-3). One type of accelerometer sensor is the piezoelectric type. The piezoelectric sensor is self-generating. It generates a small voltage when strained or pushed on by the operating equipment. This characteristic makes this sensor a good choice for measuring vibration. The piezoelectric crystal (a type of transducer) is attached to a sensing mass (Figure 7-4). The sensing mass is a small but relatively heavy piece of metal that provides enough mass so that when the rotating equipment vibrates, the transducer can be strained and subsequently creates an output signal. As the sensing mass changes directions, vibrating back and forth with the rotating equipment, the mass presses or pulls on the crystal that in turn responds with a voltage output. This voltage can be measured and transduced into an instrument signal. The signal drives an electronic circuit (Figure 7-5) providing a readout, high-vibration alarm, and/or shutdown point.

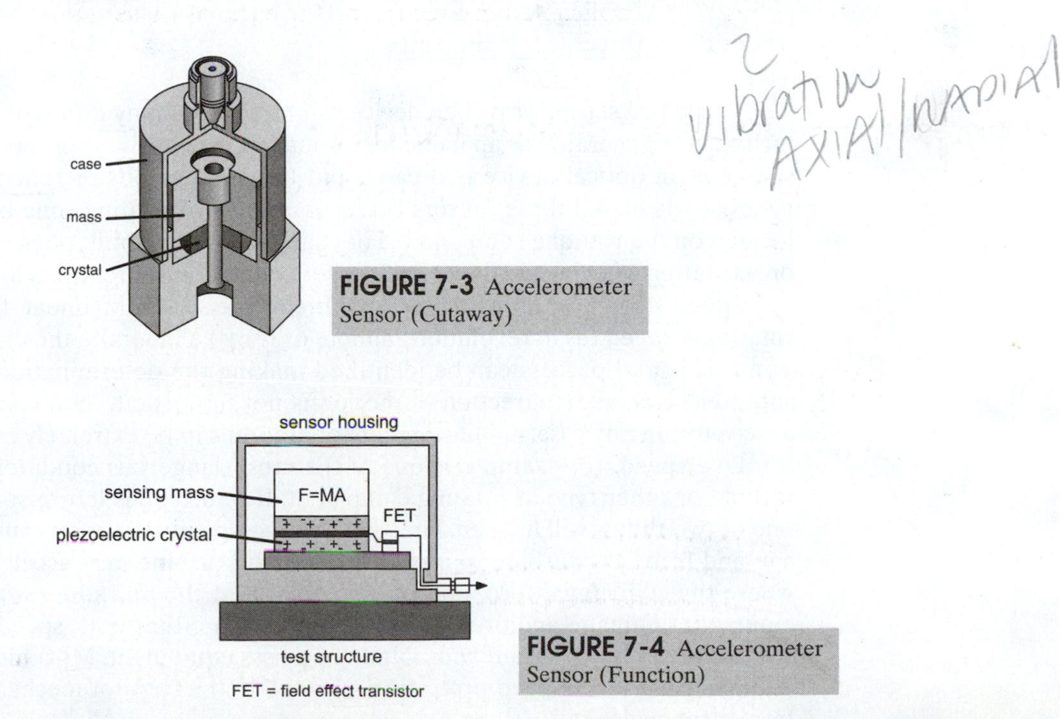

FIGURE 7-3 Accelerometer Sensor (Cutaway)

FIGURE 7-4 Accelerometer Sensor (Function)

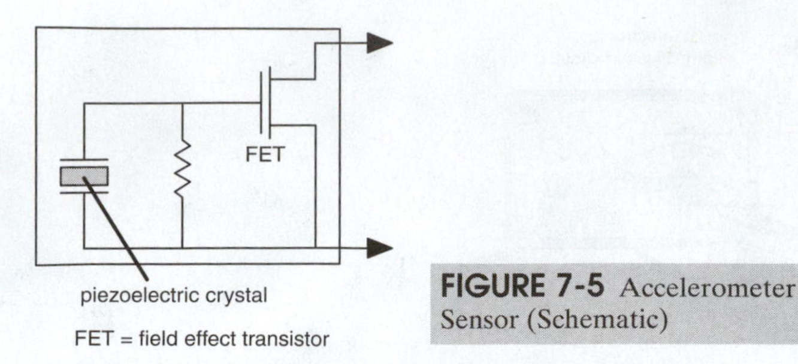

FIGURE 7-5 Accelerometer Sensor (Schematic)

Types of Miscellaneous Devices: Speed

Speed is the distance traveled per unit of time irrespective of direction. **Velocity** is speed with direction (Figure 7-6). A **speed monitor** (Figure 7-7) is a device that measures speed. As with other monitors, a speed monitor is comprised of a speed sensor and a readout/receiving device.

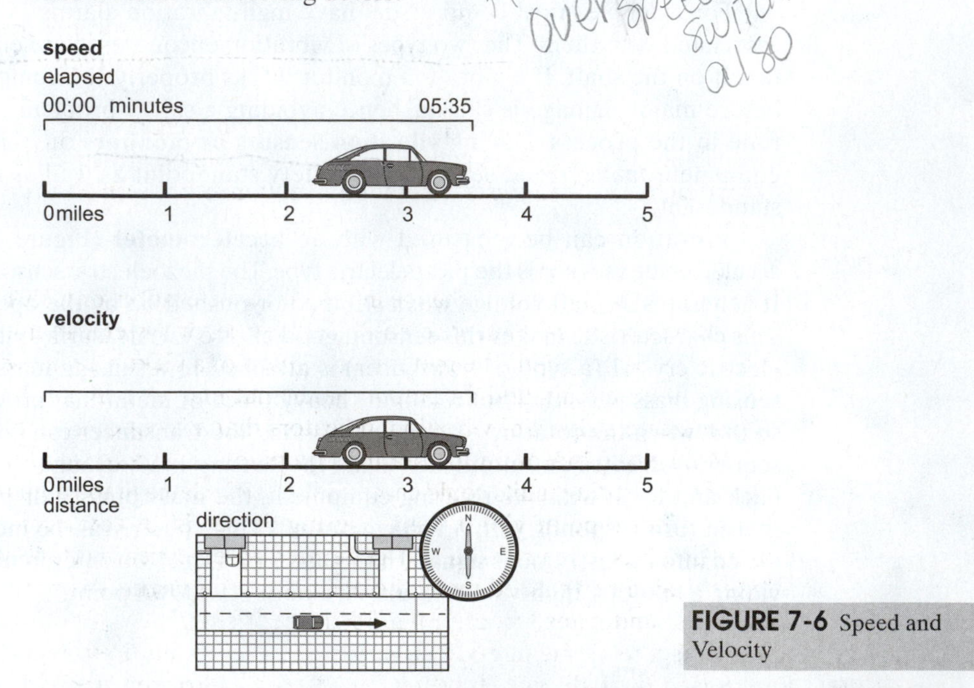

FIGURE 7-6 Speed and Velocity

The speed sensor or pickup device can be one of many different designs such as a tachometer generator or an induction or magnetic proximity sensors. Another type of sensor is an optical device that can count transparent slits or reflective patches in a rotating wheel. All these sensors detect rotation by counting (one or more) markers located on the rotating component. For example, the proximity sensors count the teeth on a rotating wheel while the optical sensor counts transparent slits in a rotating wheel.

Speed can be monitored as **rectilinear speed** (as in linear feet/second) and **rotational speed** (as in revolutions/minute or rpms). Generally, the direction of motion of a monitored process can be identified making the determination a velocity measurement instead. If direction of motion is not important, then speed is a sufficient descriptor. In any case, monitoring rate of motion can be extremely important.

Overspeed, for example, is one of the most dangerous conditions occurring in a turbine or other type of rotating equipment. If a turbine is left to spin out of control, one of two things will happen. First, the bearings eventually seize causing a lot of damage and bring the turbine to a stop. Second, the turbine may accelerate to the point where the centrifugal forces become so great that the machine explodes resulting in equipment damage and/or injury to personnel. In either case, speed monitoring and control is vitally important with rotating process equipment. Most high-speed rotating equipment has overspeed protection installed in the form of mechanical and/or electronic trip devices.

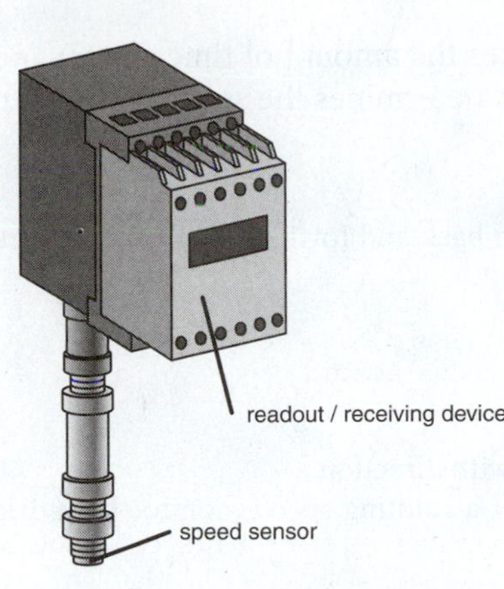

FIGURE 7-7 Rotating Speed Monitor

Another application for speed control is with rectilinear speed (or just linear speed). One way to measure the flow of solid materials is to observe the material on a moving belt. If the conveyor belt moves over a weigh scale, then a **mass flow rate** can be calculated as well as a total flow. To accurately do this, the instantaneous weight measurement must be accurate and the rate of motion must be known. The rectilinear speed of a conveyor belt, or the speedometer in a car, can be measured by converting the linear (straight-line) motion into rotational motion and then measuring it accordingly.

Summary

Vibration is the amount of change in velocity, acceleration, or displacement from a predetermined point as applied to an object, device, or system. To monitor this change a meter may be installed. The meter is composed of a pickup device sensor, an electronic amplification circuit, and an output meter that may or may not have a transducer to communicate a signal of vibration indicators that are out of an operating limit.

Monitoring vibration in rotating equipment is vitally important since vibration amounts exceeding limits could cause catastrophic damage to equipment and personnel.

Another type of vibration-measuring device is an accelerometer (piezoelectric type) that has a sensor that self-generates an electrical voltage when the sensor is either strained or pushed on by the operating equipment. The resulting voltage can be measured and transduced into an instrument signal that drives an electronic circuit providing a readout, high-vibration alarm, and/or a shutdown point.

Speed is the distance traveled per unit of time irrespective of direction while velocity is speed with direction. Speed monitors measure speed and are comprised of a speed sensor and a readout/receiving device. Speed sensors (pickup devices) can be made in several different designs such as a tachometer generator, induction or magnetic proximity sensor, or an optical-type device that counts transparent slits or reflective patches in a rotating wheel.

Speed can be measured rectilinearly (feet per second) or in rotational speed (revolutions per minute). The direction of motion of the velocity may or may not be important, but monitoring the rate-of-motion can be extremely important.

Machines have an overspeed trip to prevent the centrifugal forces from tearing the rotating assembly apart. Excessive overspeed for a prolonged period of time could also cause the bearings to seize or malfunction. This could cause major damage to the machine or even an explosion if the machine contained flammable material and this was allowed to escape through damaged seals.

The speed of solids across a conveyor path can be measured rectilinearly (linear speed versus rotating speed). A mass flow rate is established through an instantaneous weighing scale measurement and the rate of motion of the conveyor belt itself. Rotational motion can be converted to linear motion and vice versa depending on what type of speed needs to be measured. A car odometer is an example of converting the linear motion of driving a certain number of miles and measuring them by how many times they rotate fully (number of rotations times the circumference of the tire). The speedometer, by comparison, takes the amount of time (hour) and the amount of miles covered (as in the odometer) and determines the speed (miles per hour) of the car.

Checking Your Knowledge

1. _____ is the periodic or random back and forth motion of an object as it sweeps across the same path or predetermined point.
   a. Voltage
   b. Variance
   c. Vibration
   d. None of the above
2. *True or False* Speed is velocity with direction.
3. Identify the key components of a rotating speed monitor by indicating the name of each component on the graphic.
   a. Speed sensor
   b. Readout/receiving device

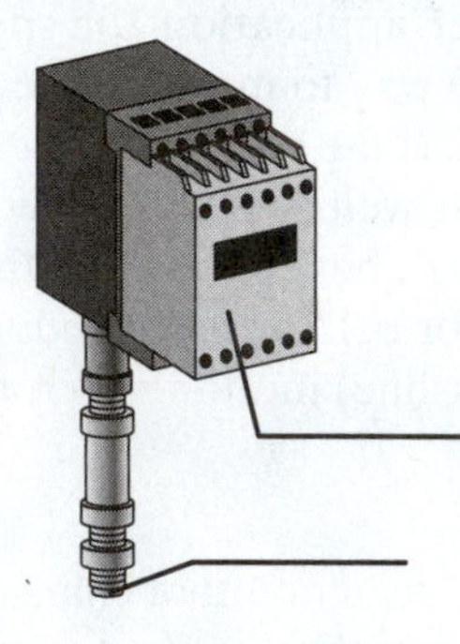

4. What are the two directions typically measured in vibration?
   a. Up and down
   b. Speed and velocity
   c. Axial and radial
   d. In and out
5. _____ is one of the most dangerous conditions occurring in a turbine.
   a. Rectilinear speed
   b. Overspeed
   c. Rotational speed
   d. An accelerometer
6. A piezoelectric sensor is used in what type of vibration device?
   a. Magnetic proximity
   b. Optical
   c. Tachometer generator
   d. Accelerometer

Student Activities

1. Draw a scale and label it from 0 to 100 with 10 (0, 10, 20, 30, etc.) evenly spaced and marked increments along the line. Look for things that produce vibrations in your home, at work, at school, or any other place you happen to be going. Make sure to have a full range of vibration intensities. On the scale you have drawn, try to visually measure how each vibration that you feel corresponds to the other vibrations. The less strong ones should be closer to the smaller numbers and the very strong ones should be charted closer to the higher numbers. Write a 300+ word description of your observations in relation to vibration.
   NOTE: Electromagnetic vibrations go from zero (0) to $+\infty$.
2. Research vibration via the internet of other library sources to find normal vibration measurements for various things (e.g., guitar strings, fan motors, etc.) and plot them on a similar scale as drawn in the activity above. Increase the scale increments to match the lowest and highest vibration measurements. Be sure to ONLY use the same measurement scales on one graph—if using more than one type of measurement scale be sure to plot the *like* measurements on different graphs.

3. Identify rotational vibration analytical sensing and measurement devices in the lab.
4. Use bench units to show flow, level, pressure, and temperature loops.
5. Purge a tank with nitrogen and measure the reduction in oxygen content.
6. Use cutaways of common primary sensors placed in their respective positions in the control loop. Use a show and tell type of demonstration.

CHAPTER 8

Introduction to Control Loops: Simple Loop Theory

Objectives

After completing this chapter, you will be able to:

- Describe process control:
  - variables
  - measuring means (primary element/transmitter)
  - controller (setpoint)
  - final control element (valve or louvers)
- Explain the function of a control loop.
- Describe the differences between *open* and *closed* control loops.
- Identify the components of a control loop:
  - sensing
  - measuring
  - comparing
  - controlling
  - manipulating
  - transducer (converter)
- Explain signal transmission:
  - pneumatic
  - electronic
  - digital
  - mechanical

Key Terms

Accuracy—how close a measurement corresponds to its true value.

Algorithms—preset mathematical functions calculated in a controller that can be mechanical, analog, or digital. The three most common output functions deal with proportional (P), integral (I), derivative, (D) tuning.

Basic control functions—sensing, measuring, comparing, calculating, correcting, and manipulating.

Closed control loop—when a control loop has feedback (e.g., controller in automatic mode).

Comparator—a component of a controller that compares the measurement to a predetermined setpoint.

Comparing, calculating, and correcting) element—the control loop component that receives the appropriate signal from the transmitter and compares the signal to a desired value (setpoint); if there is a difference, then the output of the comparison causes a calculation to be performed to cause a corrective response by the controller output signal to the final control element.

Controlled variable—a process variable that is sensed to initiate the control signal.

Controller—an instrument that receives a signal from the transmitter and compares it to a setpoint, and produces an output to a final control element.

Converting and transmitting element—the control loop component that converts the sensed process variable and transmits the measured signal.

Converting device—a device that receives information in one form of an instrument signal and changes it into another form of an instrument signal.

Device error—the accuracy of the instrument (+/–) full scale.

Digital signal—characterized by data that is represented as coded information in the form of binary numbers; used to transmit data to and from field transmitters on a twisted pair of wires; may also be between computers and computer components.

Electronic signal—either an analog or digital signal; current or voltage signal.

Feedback loop—the most common type of control loop where the change caused by the output of the controller is fed back to the process providing a self-regulating action.

Final control element—the last active device in an instrument control loop; directly controls the manipulated variable; usually a control valve, louver, or an electric motor.

Live zero—a standard bias has been added to the instrument signal (e.g., pneumatic 3–15 psig or electronic 4–20 mA); instead of reading zero the reading is 3 psig or 4 mA.

Loop error—the accumulated error of each device in the loop; calculated as the square root of the sum of the sum of the squares of individual device accuracy.

Manipulating element—the final control element (e.g., control valve) is manipulated by the corrective response of the controller output so that the process variable is maintained at the appropriate setpoint value.

Measured variable—a process variable that is measured.

Mechanical link—a way of mechanically transmitting the motion of a primary sensor to a controlling mechanism; conveys linear or rotary motion by using a pivoting crank.

Open control loop—when a control loop does NOT have feedback (e.g., controller in manual mode).

Parallel data communication—one wire per bit or 64+ wires for a 64-bit binary word; used primarily in short distances (a few feet).

Pneumatic signal—an instrument communication with a range of 3–15 psig; must have an air supply; has a lag time associated with the signal; relatively short transmission distances.

Precision—how close repeated measurements are versus the action; reproducibility; the closeness of repeated measurements of the same quantity; the agreement between the numerical values of two or more measurements made in the same way and expressed in terms of deviation.

Process control—the act of regulating one or more process variables so that a product of a desired quality can be produced.
Process error—the difference between setpoint and process variable (SP − PV).
Sensing—the act of detecting.
Sensing element—the control loop component that detects, or senses, the process variable.
Serial data communication—two data wires; the most common means of communication used between plant equipment.
Setpoint—the desired process value.
Transmitter—a device that transmits a signal from one device to another.

Introduction

The information found in this chapter introduces the concept of process control by combining various instruments to function together in a control loop. These instruments control the process by measuring, controlling, and manipulating it by communicating with each other. Loops may be either open or closed depending on the type of feedback communicated within the loop. The various components found in a control loop may sense, measure, compare, control or convert signals between the various components. Signal transmission between the instruments may be pneumatic, electronic, digital, or mechanical. Each of these characteristics of a control loop is discussed below.

Process Control and Control Loops

A process variable (PV) is usually a quantity that is measured or controlled such as pressure, temperature, level, flow, pH, etc. **Process control** (Figure 8-1) is the act of regulating one or more process variables so that a product of a desired quality can be produced. To control any process, a variable must first be sensed and measured. Then the process variable (PV) must be compared to a desired value or **setpoint** (SP–the target value); next, a calculation (some algorithm) is performed, and then a counter response (corrective output) must be produced that enables the measurement to move towards setpoint.

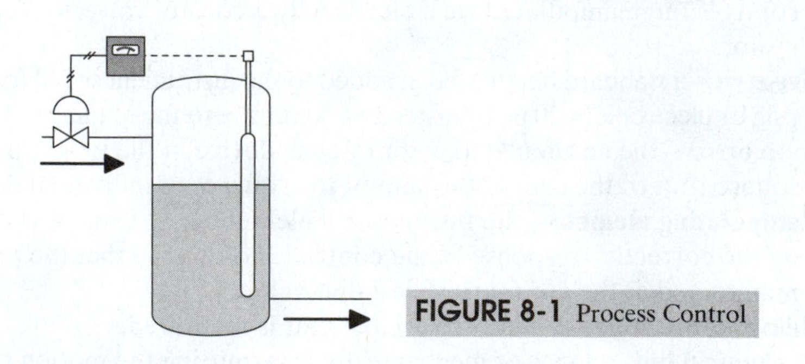

FIGURE 8-1 Process Control

The most common instruments (Figure 8-2) connected together to form a loop are the sensor or **transmitter**, the **controller**, and the **final control element**. All the **basic control functions** (e.g., sensing, measuring, converting, transmitting, etc.) are identified and associated with one of these instruments in a control loop. As shown in Figure 8.3, the arrows indicate the direction of information flow.

The PV itself is any measured property of a process other than an instrument signal. In a control loop, these measured properties are referred to as:

- Controlled: where a process variable is sensed to initiate the feedback signal
- Measured: where a process variable is measured to determine the actual condition of the controlled variable
- Manipulated: where the quantity or condition is adjusted as a function of the actuating error signal

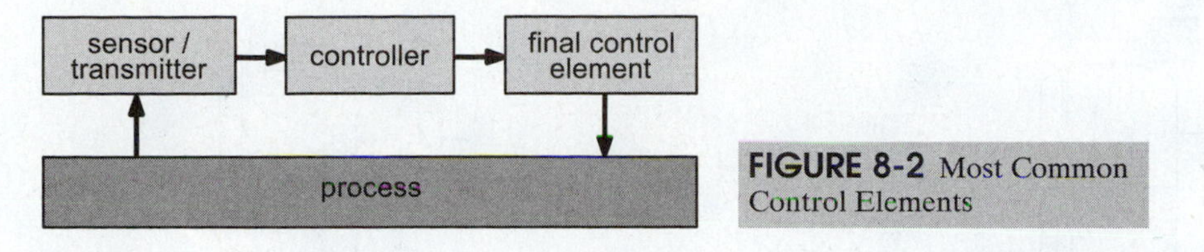

FIGURE 8-2 Most Common Control Elements

In many instances the **measured** and **controlled variables** are the same process variable. In some situations all three of these variables can be the same PV. For example, in a simple flow control loop, the flow is measured (PV), controlled, and manipulated.

Control Loops

A simple control loop is a group of instruments working together to control a single process variable such as pressure, temperature, level, flow, or some analytical or physical property. A typical instrument loop consists of the individual instruments necessary to control a process variable along with any additional instruments used to either convert signals or mathematically compute process variable measurements so that a signal can be manipulated.

Control Loop Types

There are different types of control loops in a processing plant. All process technicians should understand the fundamental concept of open-loop control versus closed-loop control. **Closed-loop control** is a combination of control units in which a process variable is measured and compared to setpoint. If there is a difference, a corrective signal is sent to the final control element to bring the process variable back to the desired value (setpoint). **Open-loop control** is a system of control that provides information necessary for control without comparing a measured value of a process variable to the desired value of a process variable. The difference between the two is based upon feedback. A control loop is closed if it has feedback (Figure 8-3) and open if it does NOT have feedback. Most control loops in process industries are closed-loop.

FIGURE 8-3 Function Block Diagram of a Feedback Control Loop

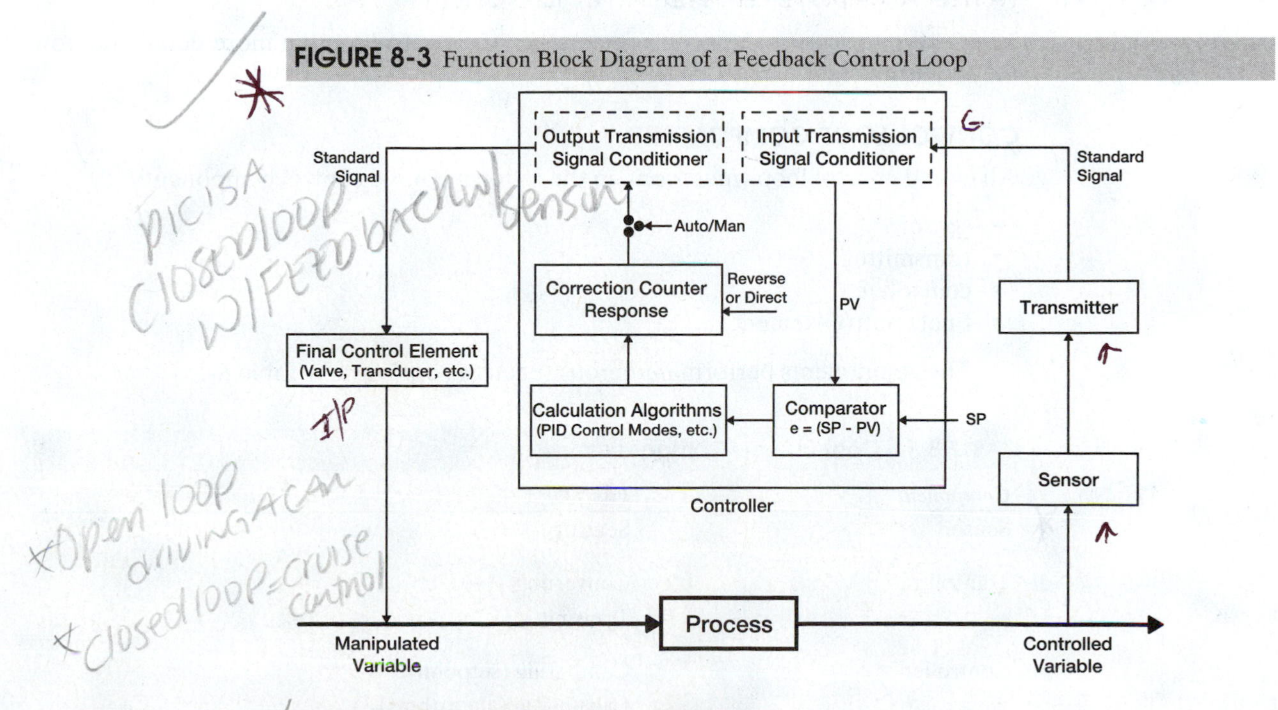

Closed-Loop Control (Feedback)

The most common control loop is the **feedback loop** (Figure 8-4). In feedback control, the change caused by the output of the controller is fed back into the input (process) providing a self-regulating action. As the error diminishes, so does the controller output.

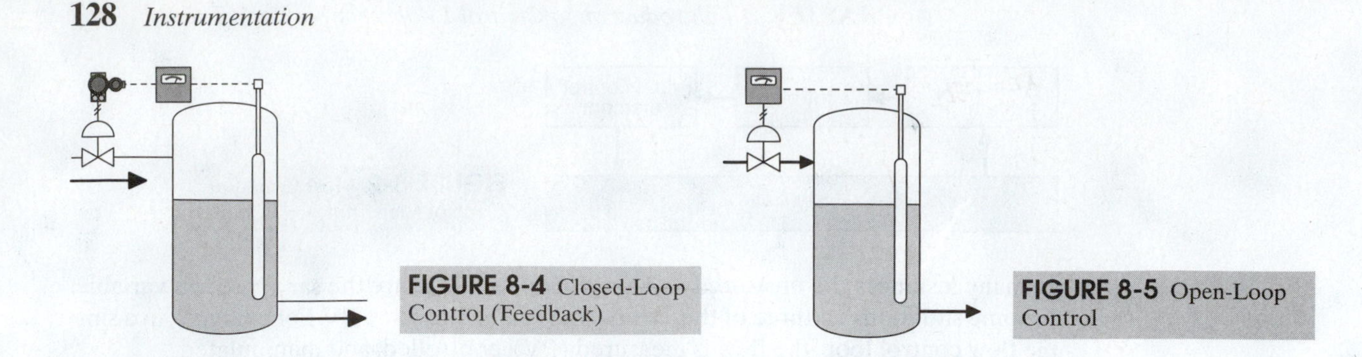
FIGURE 8-4 Closed-Loop Control (Feedback)

FIGURE 8-5 Open-Loop Control

A small amount of **process error** is always present in a feedback control loop since an error must exist before the error can be eliminated.

An air conditioner thermostat in a house acts as an ON/OFF controller. The air conditioner cools the air and the temperature measurement (PV) is the feedback signal compared to the setpoint. The air conditioner does not come on until the temperature has risen to a point above the setpoint. In a simple controller this corresponds to the automatic mode. Feedback control loops always work to eliminate error, not prevent it.

An example of closed-loop control is controlling the speed of a vehicle. If the vehicle has cruise control, a setpoint is entered and the cruise control mechanism automatically adjusts the fuel to the engine to adjust the desired speed. In a simple controller, this corresponds to the automatic (auto) mode.

Open-Loop Control

As stated earlier, open-loop control (Figure 8-5) is defined as a signal path without feedback. Open-loop control exists when a vehicle operator turns off the cruise control and physically manipulates the accelerator while monitoring the speedometer to achieve the desired speed. In a simple controller, this would correspond to the *manual* mode.

Control Loop Components and Signals

This section discusses the main components of a control loop in more detail and how they transmit signals between themselves.

CONTROL LOOP COMPONENTS

All closed control loops must contain the following basic control components:

- sensor
- transmitter
- controller
- final control element

The components perform numerous functions as shown in Table 8-1:

TABLE 8-1 Control Loop Components

Component	*Function*
Sensor	Sensing
Transmitter	Converting Transmitting
Controller	Comparing (setpoint) Calculating (algorithms) Correcting (counter response)
Transducer	Converting
Final Control Element	Manipulating

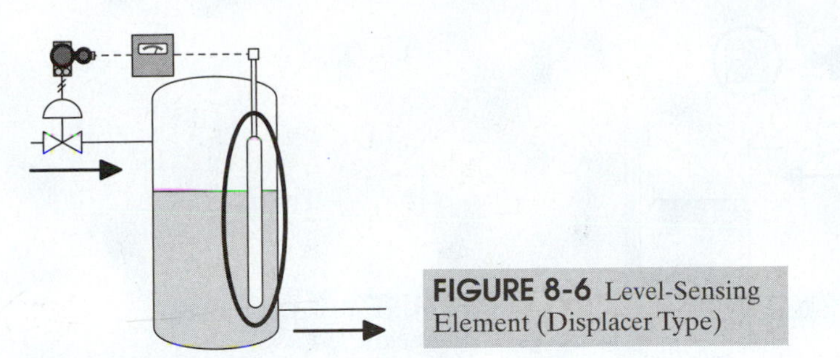

FIGURE 8-6 Level-Sensing Element (Displacer Type)

Sensor: Sensing

Sensing is the act of detecting and a sensor or **sensing element** (Figure 8-6) is the first device in a control loop. Obviously, the sensing element must be capable, in one way or another, of detecting the process variable. For example, a metal diaphragm can sense pressure when it moves as pressure is applied. A resistance temperature device (RTD) can sense temperature through a change in electrical resistance. Sensing the process variable is possibly the most important action taken by a control system. If the sensor is incapable of responding accurately, precisely (repeatable), and quickly enough to the process variable, then controlling the process variable may be difficult or even impossible.

Transmitter: Converting and Transmitting

Once the process variable has been sensed, it must be converted to a measurement. The measurement operation in an instrument is generally an assessment of the response of the sensor to the process variable rather than a direct measurement of the process variable itself. The device that transmits the measured signal is a transmitter (Figure 8-7). For example, a metal diaphragm responds to a change in position. The measurement transducer then responds electrically to the motion by producing an output. Effectively, the process variable is transduced (changed) into a standard instrument signal such as 3-15 psi, 4-20 mA, or digital.

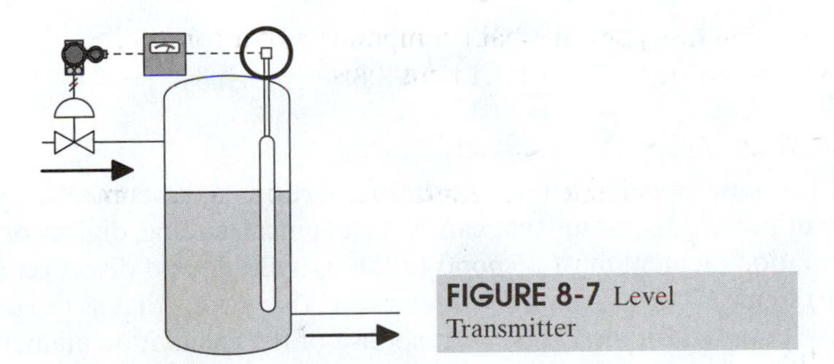

FIGURE 8-7 Level Transmitter

Prior to the use of transmitters, controllers were directly connected to the process, which meant that they had to be locally mounted in the field close to the process. The introduction of the transmitter into industry ushered in the small (6" x 6" and eventually the more modern 3" x 6") board-mounted controllers that allowed for the centralization of controllers (remote) into the more modern control rooms of today. The transmitter produces a standard instrument signal that can carry the measurement information to a remote location such as a control room. This was a very significant improvement in field instrumentation and a true mile marker on the evolutionary path to modern automatic process control, which today includes computers.

A measuring transducer must respond in an accurate and precise manner, just as the sensing element must. **Accuracy** is determined by how close the measurement corresponds to its true value (the process variable). **Precision** describes how repeatable the measurement is each time the same value is applied. Reproducing the same measurement results each time a specific value is applied to the sensor is as important, if not more important, than the actual accuracy of the measurement itself.

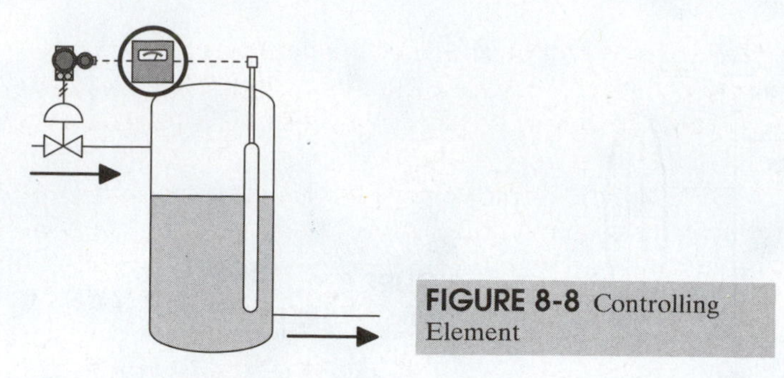

FIGURE 8-8 Controlling Element

For example, a person shooting at the bull's-eye on a target may hit the target every time. Precision is how close each of the holes is to each other or the tightness of the pattern. Accuracy would depend on how close (bias error) the holes are to the centermost point on the target. The possibility exists for there to be great precision and very poor accuracy.

Controller: Comparing, Calculating, and Correcting

Once the process variable (PV) has been measured and the appropriate signal has been transmitted to the controller (Figure 8-8), the PV must then be compared to a desired value called the setpoint (SP). The **comparator** part of the controller determines the difference between the SP and the incoming PV signal. If the measurement and SP are the same, the output from the comparator is zero and the controller output stays the same. If there is a difference between the SP and the PV, then the output of the comparator causes a change to the calculating function of the controller.

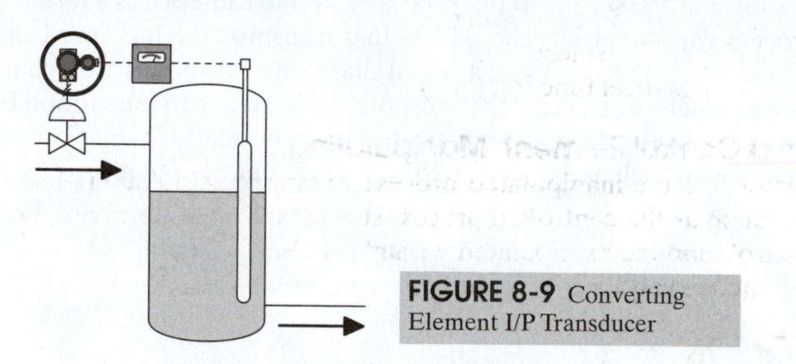

FIGURE 8-9 Converting Element I/P Transducer

The next function is the calculating portion of the controller performed by some type of preset **algorithms** that can be mechanical, analog, digital, or mathematical. The calculation element must respond to the magnitude and direction of the error (difference) signal received from the comparator. This means that as the difference generated by the comparator increases, the response of the calculation element increases as well. Also, the direction that the corrective element drives the final control element is directly related to whether the measurement is above or below the setpoint. In either case, an appropriate response is produced to reduce or eliminate the error; this is the corrective output.

The controller then acts upon this deviation by calculating a corrective output signal. These calculations are performed using one or more algorithms. The three main controller algorithms are usually referred to as proportional (P), integral (I), and derivative (D) modes. Other common accepted names for these functions are gain (G), reset (R), and rate (Rt). Digital controllers may be programmed with any combination of mathematical functions to produce a desired output. A controller usually contains the comparing, calculating, and corrective output elements. The controller action can be chosen direct (+) or reverse (−) acting to counteract the process direction.

Transducer: Converting

While not always present in a control loop, **converting devices** (Figure 8-9) may be used to receive information in one form of an instrument signal and change it into another.

These devices may also be called converters, transducers, and/or relays. Converting instruments are placed between the main control loop elements to convert one element signal type to another type of signal that must be read or understood by the next element in the loop.

The most prominent converting device is the current-to-pneumatic converter (I/P). An I/P converter is used in most electronic control loops because the final control element is usually a pneumatically actuated control valve. Many final control elements have several components; an I/P transducer, positioner, actuator, and valve integrally mounted as a single unit. The following are examples of other signal converters include:

- pneumatic-to-current converter (P/I)
- pneumatic-to-voltage converter (P/E)
- voltage-to-pneumatic converter (E/P)
- analog-to-digital converters (A/D)
- digital-to-analog converters (D/A)

Computing relays are those devices that are capable of computing values based on instrument signals. A square root extractor, for example, is a computing device that takes the linear output signal from a d/p transmitter that measures the pressure drop across an orifice plate and changes it to a square root output that is proportional to the flow rate. There are numerous types of *computing relays* and mathematical functions used in applications within control loops to include the following:

- adding and subtracting
- multiplying and dividing
- inverting
- scaling and proportioning
- high and low select
- mathematical function ($\sqrt{\ }$)

Final Control Element: Manipulating

Sometimes the manipulated process stream **(manipulating element)** (Figure 8-10) is the same as the controlled process stream and sometimes not. For example, in a flow control loop the manipulated variable is also the controlled variable. In this situation, the control valve is placed into the same process stream as the flow transmitter.

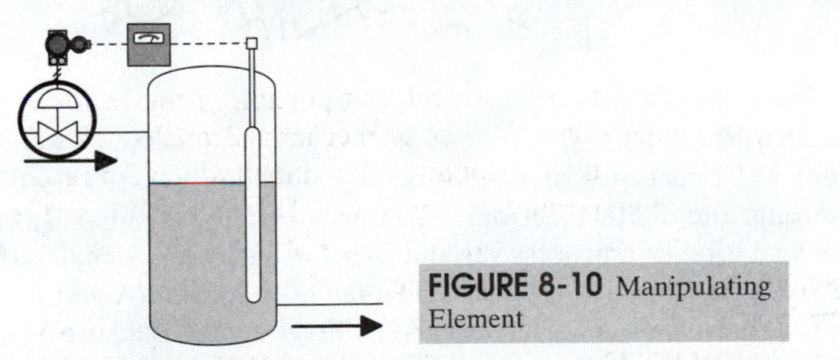

FIGURE 8-10 Manipulating Element

For other situations, one PV may have a controlling effect on another PV. For example, the flow of steam into the tube side of a heat exchanger is used to raise (control) the temperature of the product flowing through the shell side of the heat exchanger–two totally separate streams–one containing steam and the other containing the process fluid. In this case, the product temperature is the important process variable and this temperature is controlled by a steam valve opening and closing in the steam.

SIGNAL TRANSMISSION

All instrument signals can be thought of as a language that instruments use to communicate information between one another. There are four basic types of signal transmission used for instrumentation communication within processing facilities:

- pneumatic (analog)
- electronic (analog)

- digital
- mechanical

Pneumatic (Analog)

In the United States, the standard **pneumatic signal** has a range of 3–15 psig. Instrument signals always represent 0–100 percent of the information moving between loop devices. For example, a 3–15 psig signal leaving a pneumatic transmitter represents 0–100 percent of the measured variable; whereas the 3–15-psig signal leaving the controller represents 0–100 percent of the response of a controller to error (the output).

Any instrument capable of producing a pneumatic signal (Figure 8-11) must have an air supply connected. The air supply is usually about 20 psig for standard 3–15-psig signal devices. The air supply of a pneumatic instrument can be equated to a power supply for a piece of electronic equipment. After leaving an instrument, the signal travels through stainless steel, copper or plastic tubing. The tubing connects the individual instruments together in a loop.

Unlike electronic signals (discussed below), pneumatic signals are slower. Imagine the changing pressure propagating down a one-quarter to one-half inch tube that may be several hundred feet long. Pneumatic signal lines are kept to relatively short distances, usually no more than a couple of hundred feet (the longest are up to 500–600 feet) because of this potential transmission lag problem. Even though a pneumatic signal travels at about the speed of sound, the resistance to flow in the tubing and the size of the signal-producing element may cause the signal to take several seconds to reach its full pressure even in a short run of tubing. For long distances, and for large volume runs, a device called a booster relay may be added.

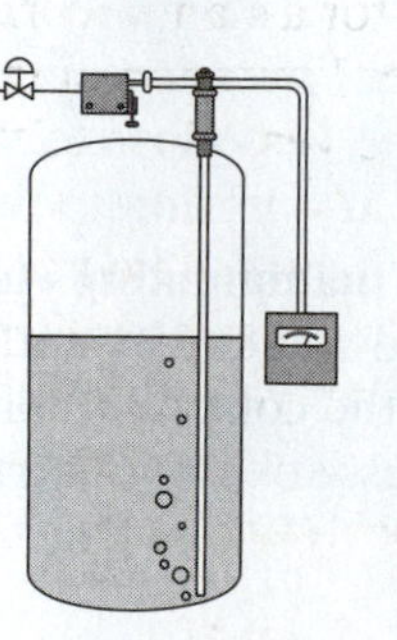

FIGURE 8-11 Pneumatic Signal

Electronic (Analog)

There are two types of **electronic signals** (Figure 8-12) used in industry: analog electronic and digital electronic. An analog electronic signal is a continuously variable representation of a process variable. Analog electronic signals usually have a range of 4–20 mA. Earlier analog signals had a range of 10–50 mA and 1–5 VDC.

Current signals are preferred over voltage signals because of their higher immunity to electrical and electromagnetic interference. Voltage signals are decreased by the resistance in the wires and also lose voltage to every device in the loop. Thus, the voltage signal sent is not the same as the one that is received. A current loop is much better because it is similar to water flowing in a pipe (water in = water out; current sent = current received). Most field transmitters have two wires connecting them to the controller. Power is supplied to the transmitter through the same two wires that carry the signal. In a sense, the *current-generating circuit* of the transmitter acts more like a *current-limiting circuit* in that it controls the current flow through the two wires to a range of between 4 mA and 20 mA. The actual output circuit is usually designed with operational amplifiers and other discrete components.

Electronic (Digital)

A **digital signal** (Figure 8-13) is characterized by data that is represented as coded information in the form of binary numbers. Digital signals are used to transmit data to and

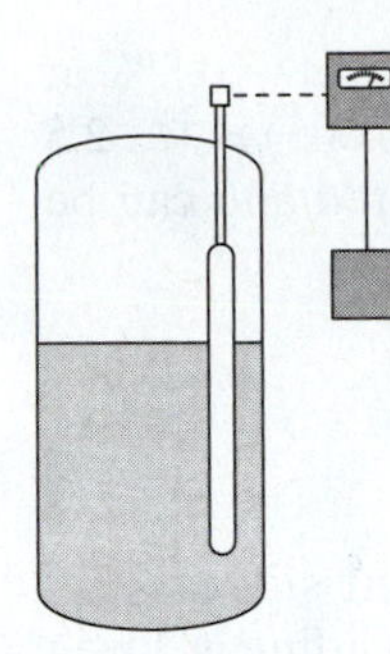

FIGURE 8-12 Electronic Signal

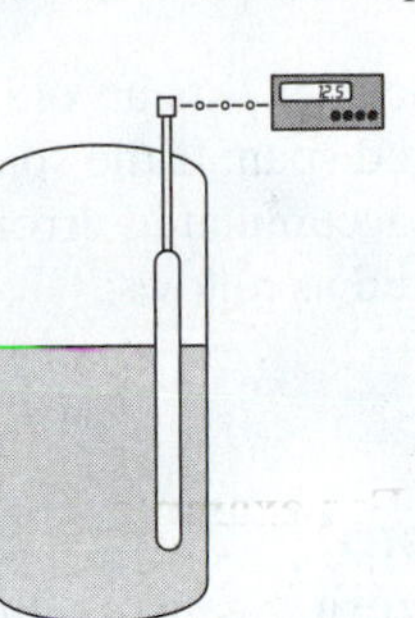

FIGURE 8-13 Digital Signal

from field transmitters on a twisted pair of wires, between computers and computer components on a data highway, and in a new generation of instruments systems called field buses. Digital information is transmitted as either serial or parallel data. **Parallel data communication** requires one wire per bit or 64+ wires for a 64-bit binary word, whereas a serial communication device requires only two wires (data). Parallel communication becomes impractical to use beyond a few feet. That is why serial data transmission is the most common means of communication (serial bus) used between plant equipment.

Unlike the standard analog pneumatic and electronic signals, digital signals vary immensely between manufacturers. There have been attempts over the years to standardize, but a lot of variation still exists. The Instrumentation, Systems, and Automation Society (ISA), a worldwide instrumentation organization, has come forth with standards for field buses.

Mechanical (Link or Linkage)

A **mechanical link** transmits the motion of a primary sensor to a controlling mechanism. By providing a mechanical link, the need for a standard instrument signal such as a 4–20 mA or 3–15 psig is eliminated. The major limitation of this type of transmission device is that the controlling device needs to be very close to the process sensor. This type of link may convey linear motion or rotary motion by using a pivoting crank. A fulcrum is sometimes used to produce a mechanical advantage.

An example of a mechanical link is found in a displacer level device. A displacer is a device that can infer level by measuring the buoyant force of the process fluid. The displacer is attached to the controlling device by a mechanical link (Figure 8-14) called a torque tube. The torque tube transmits the measured buoyancy force into a controlling device.

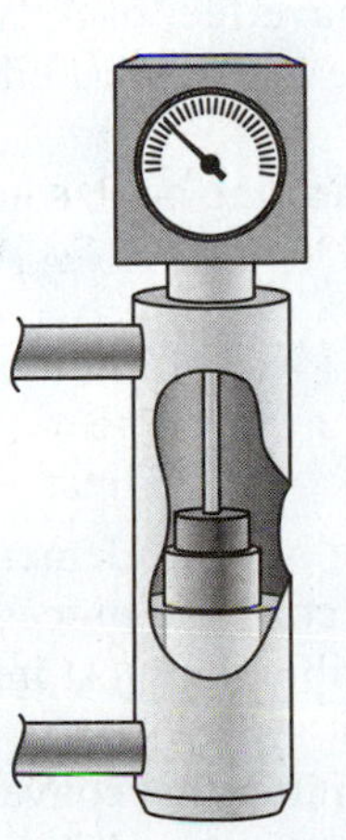

FIGURE 8-14 Mechanical Link

Another very common device is a control valve positioner that has a mechanical link to the valve that senses the position of the valve and ensures that the position corresponds to the position signaled by the controller.

LOOP ERROR

Each device in a control loop has an **error loop** (accuracy) associated with it. The device manufacturer usually lists the error or accuracy for their own device. For example, a manufacturer lists an old style temperature transmitter accuracy as E = +/− 0.5% of calibrated span. If the span is 500 degrees F, then E = (500 °F) X (+/− 0.005) = +/− 2.5 °F. The accumulated error for a control loop can be higher than expected and can be calculated as follows:

$$E_{loop} = +/- \sqrt{(E_{transmitter})^2 + (E_{valve})^2 + \ldots}$$

Live Zero

A **live zero** means a standard bias has been added to the instrument signal (e.g., pneumatic 3–15 psig; electronic 4–20 mA) instead of reading zero the reading is 3 psig or 4 mA. This is done for two reasons:

1. Calibration is much easier to perform on an instrument around a positive output like 3 psig or 4 mA rather than on zero (0) since a negative pressure would imply the need for a vacuum source or a negative current would require a polarity change. Either of these would be harder to do.
2. Process technicians in essence are given more information with a live zero condition. If the signal is truly indicating zero, then the tube or wire is broken or the transmitter or controller instrument has failed, whereas at 3 psig or 4 mA, functionality still exists but the instrument is at 0% of scale.

Summary

To maintain process control, one or more process variables must be sensed so that the desired product quality can be maintained and produced. Each process has a desired setpoint for comparison. The difference from the sensed process variable to the desired setpoint causes an output to move the measured value to approach the setpoint. For simple control situations, the process variable itself may be the sensed (controlled), measured to determine the condition of the variable, and also manipulated as a function of the actuating error signal.

Simple control loops consist of individual instruments necessary to control process variables in addition to any instruments required to convert signals or mathematically compute process variable measurements so that a signal can be manipulated. Control loops are basically considered either open- or closed-loop types with the primary difference between the two based on whether or not the loop has feedback. Open-loops do NOT have feedback and closed-loops DO have feedback. Most control loops in process industries are closed-loop and this type is often called feedback control or feedback control loop.

All closed control loops must contain the following basic control component:

- sensor
- transmitter
- controller
- final control element

Sensing is the act of detecting, and a sensor or sensing element is the first device in a control loop. A sensor must be capable of detecting the process variable accurately, precisely, and quickly. This function is quite possibly the most important action taken by a control system.

The next control loop component is the transmitter that converts the sensed process variable electrical response and produces a standard instrument signal output (3–15 psi, 4–20 mA, or digital) from the transducer. Measuring transducers must also be accurate and precise. Next, the controller receives the signal from the transmitter and compares the signal to a desired value (setpoint) in the comparator part of the controller. If there is no difference, then there is no output. If there is a difference, then the output of the comparator causes a change in the calculating function of the controller. Preset algorithms (mechanical, analog, digital, or mathematical) respond to the magnitude and direction of the error (difference) signal received from the comparator. An appropriate response, called the corrective output, is produced to reduce or eliminate the error.

Converting devices may be used at any place in a control loop to receive information in one form of an instrument signal and change it into another. These devices may also be called converters, transducers, and/or relays. The most prominent converting device is the current-to-pneumatic converter (I/P) since the final control element is usually a pneumatically actuated control valve.

Computing relays may also be used in control loops to compute values based on instrument signals. These are used anywhere mathematical functionality is required within the control loop.

For most simple control loops, the manipulated process stream is the same as the controlled process stream. In other situations, one process variable may have a controlling effect on another. The final control element in a control system is the instrument that is manipulated to maintain the desired value of the process variable. In most cases, the final control element is a control valve.

Signals are the language of instruments. The following basic types of signals are used for transmitting information within processing facilities:

- *pneumatic*: range of 3–15 psig; requires an air supply of 20–25 psig to operate; are characteristically slow
- *electronic*: two types are analog electronic and digital electronic
- *analog electronic signals*: continuously representing a process variable and have a range of 4–20 mA or 1–5 VDC
- *digital electronic signals*: have coded information (binary numbers) that characterize or simulate the analog PV and are transmitted to and from field transmitters using twisted pair wires, computer components on a data highway, or a field bus; transmission may be either serial (more common) or parallel (short distances only) data; digital signals vary greatly between manufacturers; field buses have a standard set by the ISA
- *mechanical*: motion by a primary sensor is transmitted to a controlling mechanism through a mechanical link (linkage); no standard signal is needed; controlling devices must be very close to the process sensor for this signal to operate properly; conveys linear or rotary motion when using a pivoting crank; may use a fulcrum to produce a mechanical advantage

Control loop devices have accuracy error associations. The respective manufacturer identifies these. Instruments transmitting signals are calibrated using a live zero that adds a small amount of positive output (e.g., 3 psig or 4 mA) to the signal scale zero point.

Checking Your Knowledge

1. Process control is the act of _____ one or more process variables.
   a. Sensing
   b. Measuring
   c. Tweaking
   d. Regulating
   e. Manipulating
2. Which of the following is NOT a component of a simple control loop?
   a. Transmitter
   b. Controller
   c. Final control element
   d. Vessel
3. *True or False* A closed loop is a signal path WITH feedback
4. Match the type of signal with the description of how the signal is transmitted.

Signal	*Description*
1. Pneumatic B	a. Transmits motion
2. Electronic analog D	b. Transmits pounds per square inch
3. Electronic digital C	c. Transmits binary code
4. Mechanical link A	d. Transmits milliamps

5. Write OPEN or CLOSED in the Loop column for the following devices and list the appropriate process variable (PV) for each.

		Loop	*PV*
a.	Light switch		
b.	Toilet flush		
c.	Water faucet		
d.	Automobile cruise control		

6. A control loop is composed of a transmitter, controller, I/P transducer, and a valve. Find the loop error if the individual accuracies (+ or −) are: transmitter accuracy = 0.5%; controller accuracy = 0.25%; I/P accuracy = 0.5%; and valve accuracy = 1.5%.
   a. 2.75
   b. 2.00
   c. 1.75
   d. 1.25
7. If accuracy is defined as [(measured value − true value)/true value)] X 100%, find the accuracy of a pressure gauge if the true value is 100 psig and it reads 98 psig.
   a. .98
   b. 2%
   c. 20%
   d. 1.02
8. Identify each instrument loop pictured below as either OPEN-LOOP or CLOSED-LOOP control.

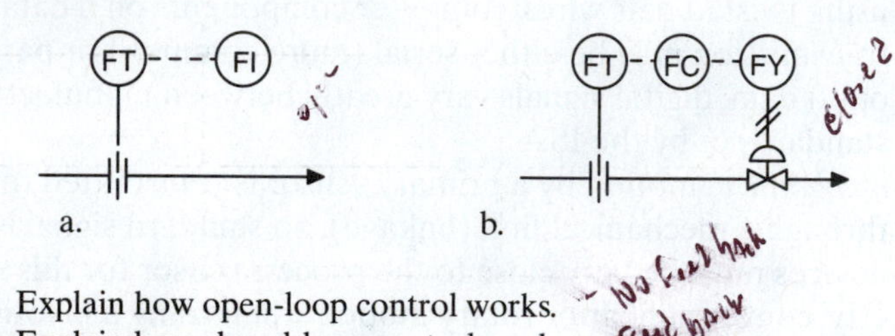

9. Explain how open-loop control works.
10. Explain how closed-loop control works.
11. For the instrument control loop illustrated below, fill in the appropriate information in the table below.

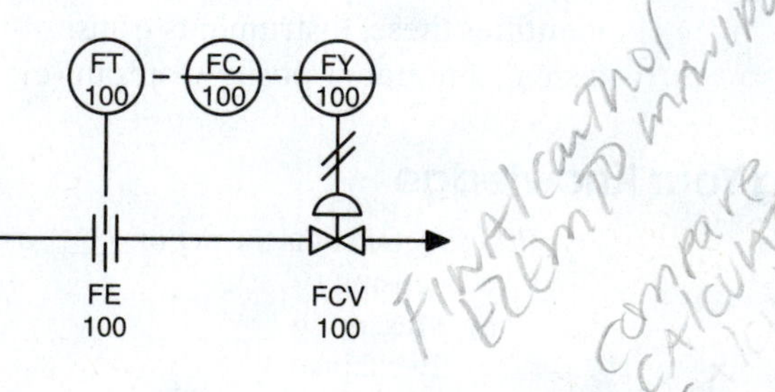

INSTRUMENT CONTROL LOOP			
Component	*Element Type*	*Process Variable Being Controlled*	*Component Function*

12. For the instrument control loop illustrated below, fill in the appropriate information in the table below.

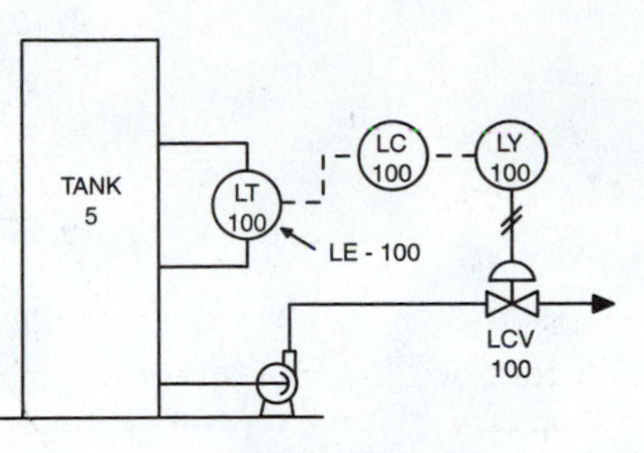

INSTRUMENT CONTROL LOOP			
Component	***Element Type***	***Process Variable Being Controlled***	***Component Function***

13. For the instrument control loop illustrated below, fill in the appropriate information in the table below.

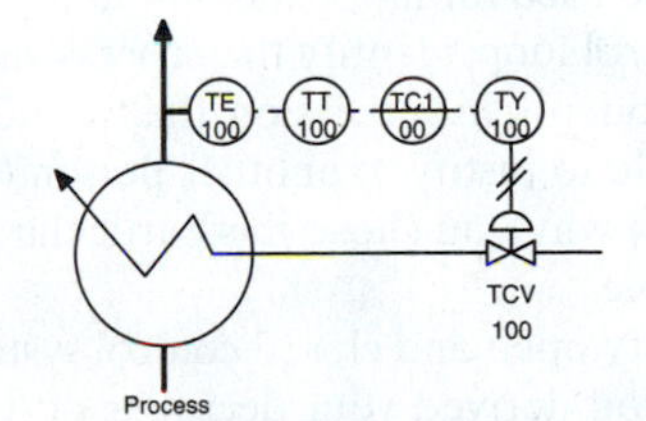

INSTRUMENT CONTROL LOOP			
Component	***Element Type***	***Process Variable Being Controlled***	***Component Function***

14. For the control loop illustrated below, fill in the appropriate information in the table below.

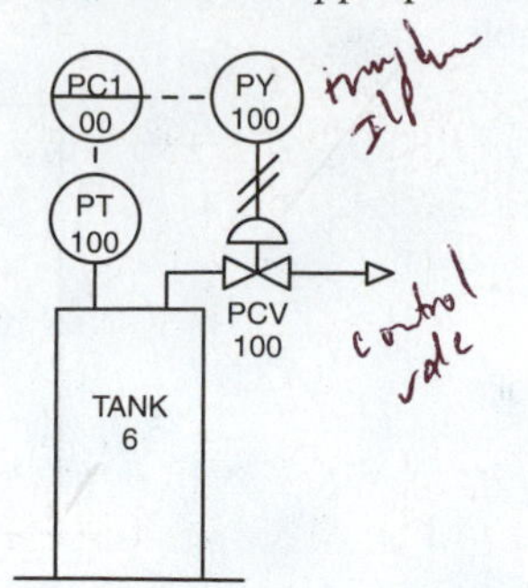

INSTRUMENT CONTROL LOOP			
Component	***Element Type***	***Process Variable Being Controlled***	***Component Function***

Student Activities

1. Photocopy individual control loop components from this textbook or other reputable sources (e.g., instructor classroom materials). Write the name of the component on the bottom front of each copy. On the back of each copy, write the types of signal combinations that may be used for input and output to the component. Using the copies as actual placement in a control loop, identify the process variable and then select the combination of control loop components, in succession, that would form an appropriate control loop for the application. Be able to justify to another person (e.g., your instructor or another member of your small group) why you chose this particular arrangement of components for the specified process variable.
2. Identify open and closed control systems around your home and school. Be sure to explain how you derived your decisions on why the system is open or closed. Share your findings with other class members to see how your lists compare.
3. Identify control loops and/or control loop components in the lab.
4. Use pilot plant equipment and instrumentation and locate the control loop components.

CHAPTER 9

Control Loops: Primary Sensors, Transmitters, and Transducers

Objectives

After completing this chapter, you will be able to:

- Describe the relationship between the measuring instruments (pressure, temperature, level, and flow) and their role in the overall control loop process.
- Describe the purpose and operation of the transmitter in a control loop.
- Discuss differential pressure in relation to the process input to the transmitter.
- Compare and contrast the transmitter input and output signals.
- Describe the function of a current to pneumatic transducer (signal converter).
- Describe the relationship between a 3-psig to 15-psig air signal and a 4mA to 20-mA electric signal.
- Given a process control scheme, explain how a control loop functions.

Key Terms

Discrete sensing element—a sensing element that can stand alone or is individually distinct; connected to the transmitter by sensor wires.

Impulse tubing—a tube that is usually made of stainless steel and allows the process variable to be sensed by the sensor located in the transmitter.

Instrument scale—a range of ordered marks that indicate the numerical values of the process variable.

Integrally mounted sensing element—where the sensing element is a physical part of the transmitter.

Linear scaling—a linear relationship between two scales (input versus output).

Lower range value (LRV)—the number at the bottom of the scale.

Operating range—one number that is the difference between the upper and lower range values (URV and LRV) on a scale.

Scaling—the act of equating the numerical value of one scale to its mathematically proportional value on another scale.

Sensor—detects the process variable; can be an integral part of a transmitter.

Signal converter transducer—Part of a transmitter that effectively converts the process variable into a standard instrument signal; a device that converts one energy form into another.

Span—the algebraic difference between the URV minus the LRV of a scale; expressed as one number.

Standard signal—the language that instruments use to communicate between one another (4–20 mA, 3–15 psig, or digital).

Upper range value (URV)—the number at the top of the scale.

Introduction

In this chapter, the individual control loop component's primary sensors, transmitters, and transducers are discussed in more detail as to their purpose and operation. Most transmitters house both the sensing and measuring functions that ultimately produce a signal that is transmitted to the next control loop element that may be connected to a controller, recorder, indicator, programmable logic controller (PLC) or digital control system (DCS), as well as a combination of these items. In this chapter, each function of sensing, measuring, and transmitting is discussed in more detail to include the purpose and operation of each portion of the process, and also how each component relates to others in the loop based on the process variable (PV).

Figure 9-1 shows a simple control loop in a liquid-level-measuring system. This drawing is used as the basis in this chapter for identifying the location of a transmitter and its functionality within a control loop.

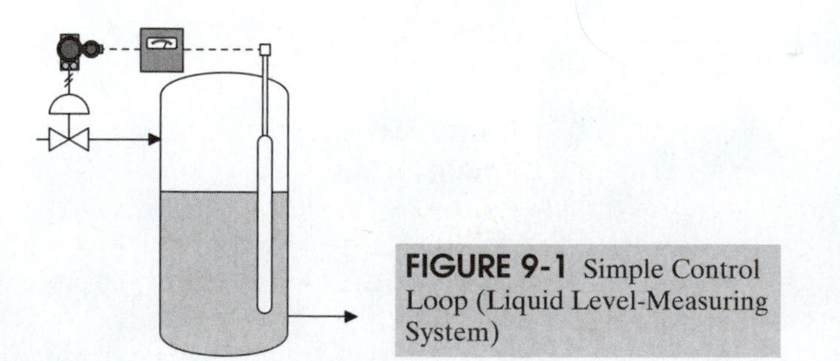

FIGURE 9-1 Simple Control Loop (Liquid Level-Measuring System)

Component Purpose and Operation

When control loops were introduced in the previous chapter, the control loop functions identified as sensing and measuring were shown as separate functions. Some transmitter instruments (Figure 9-2) are capable of both sensing and measuring the

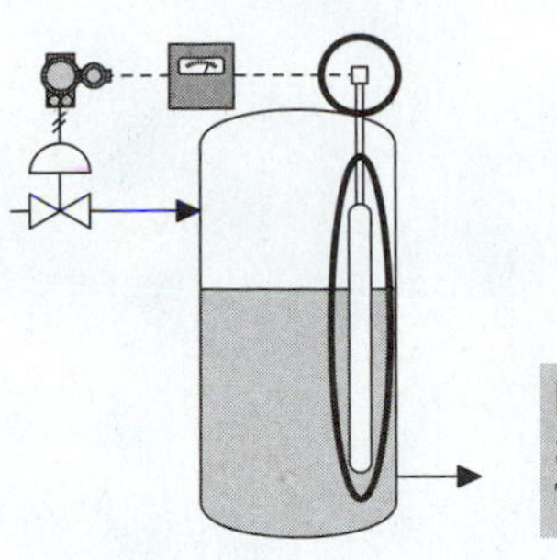

FIGURE 9-2 Primary Sensors, Transmitters, and Transducers

process variable. The transmitter produces an output signal that carries the measurement information to the next instrument in the loop.

There are actually several elements that perform the sensing and measuring functions, or act in tandem with the transmitter to perform these functions. These elements include the **sensor** and the *transducer*. The sensor detects the process variable and the transducer converts one energy form into another.

If a control loop is to function properly, then the sensors and transmitters must provide the process variable measurement in a usable form to the loop. In a sensor-driven loop, the *sensor* serves as a transducer and provides an input directly to the controller. In a transmitter-driven loop, the *transducer* is part of the transmitter and effectively converts the process variable into a standard instrument signal.

SENSORS

Sensors (Figure 9-3) can be mechanical or electronic. A thermocouple is an example of an electronic sensor that can be directly connected to a controller. The controller in a sensor-driven loop must be specifically designed to accept the *nonstandard input* rather than a standard instrument signal. For the thermocouple, the nonstandard input signal is in millivolts. These dedicated input controllers are only used when they clearly provide a significant advantage functionally or financially.

The role of a primary sensor is to detect a process variable. A sensor must also be capable of responding accurately, precisely, and quickly if good control is to be accomplished by the loop. **Discrete sensing elements** (stand alone or individually distinct) are devices connected to the transmitter by sensor wires. **Integrally mounted sensing** elements are a physical part of the transmitter.

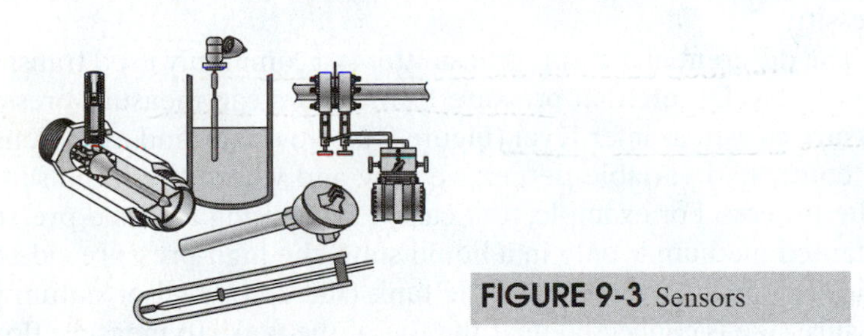

FIGURE 9-3 Sensors

Thermocouples and RTDs are examples of discrete temperature sensors that are usually installed in a thermowell extending into the process. The thermocouple wires, or the lead wires extending out of the sheathing of the RTD, are used to connect these sensors to the transmitter. The transmitter may be physically connected to the open end of the thermowell or located a short distance away. In either case, the sensors are separate entities from the transmitter.

A differential pressure transmitter, however, has an integrally mounted sensor. Since the d/p-sensing cell (Figure 9-4) is part of the transmitter, the process variable has to be brought to it. This is accomplished by installing **impulse tubing** (impulse means a wave transmitting through something) between the process and the transmitter. The tubing is usually made of stainless steel and allows the process variable to be sensed by the d/p cell located in the transmitter. The impulse tubing as a minimum must meet the material specification of the process where it is connected.

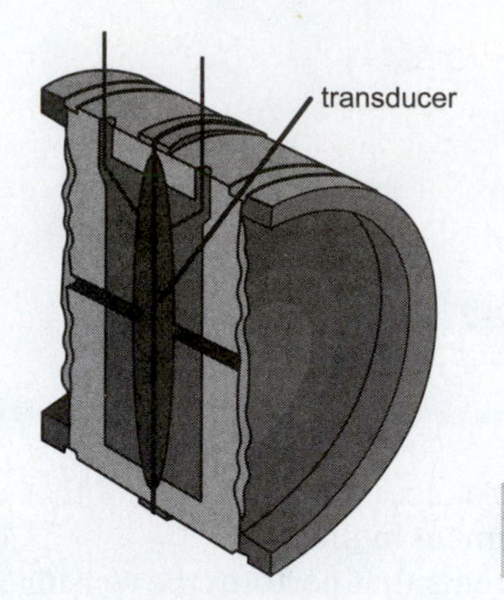

FIGURE 9-4 Differential Pressure Cell Cross Section

TRANSMITTERS

The most common method for providing a measured input to a control loop is through a transmitter. As stated previously, transmitters either have a primary sensor mounted within them or connected to them. After being sensed and measured, the process variable measurement is transduced (converted) by the transmitter circuit into a standard instrument signal. The standard instrument signals are 4–20 mA (electronic), 3–15 psig (pneumatic), or digital.

Unlike a directly connected sensor where the signal is nonstandard, transmitters provide a standardized signal that is accepted by all other instruments in the loop. For example, a differential pressure (d/p) transmitter contains a special type of pressure-sensing and measuring component called a d/p cell. The d/p cell is designed to measure the difference between two pressures and then produce an output representing that difference. Inside the transmitter is a transducer that transforms the output of the d/p cell into a standard instrument signal such as 4–20 mA or 3–15 psig.

To elaborate further, one type of electronic transmitter uses a capacitance d/p cell. As pressure increases, the distance between the capacitance plates decreases thus increasing the capacitance. The increased capacitance is transduced to an increased 4–20-mA output signal. Thus, the 4–20-mA output signal is directly proportional to the differential pressure. Since the output of the transmitter is also the input to the next instrument in the loop, a meaningful and recognizable **standard signal** is an absolute necessity.

The differential pressure transmitter is a commonly used transmitter in the processing industry. Differential pressure transmitters can measure pressure and differential pressure as well as infer level (Figure 9-5), flow rate, and even density. The application and controlled variable determine how and where the impulse tubing is connected to the process. For example, to measure level using a closed pressure vessel when the contained medium is only in a liquid state, the high-pressure side of the transmitter is connected near the bottom of the tank (the zero level or datum point) and the low-pressure side is connected near the top of the tank. To measure flow, the high-pressure side is connected upstream of the orifice plate and the low-pressure side is connected

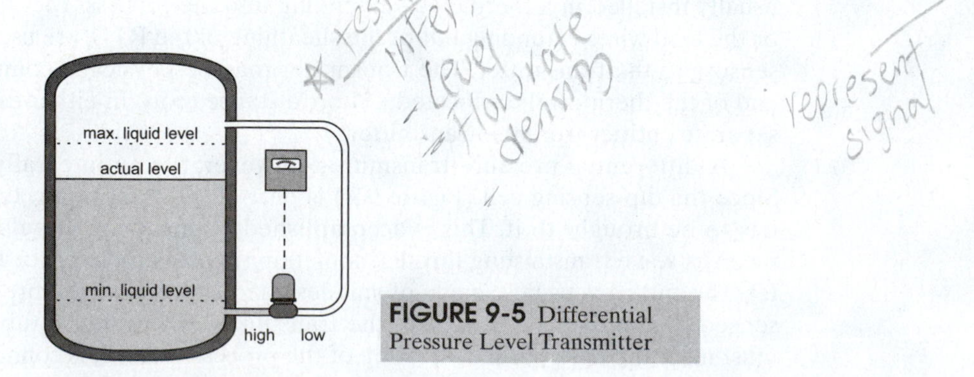

FIGURE 9-5 Differential Pressure Level Transmitter

downstream. Connecting the high-pressure side to the process and allowing the low-pressure side to be vented to atmosphere provides a gauge pressure measurement. D/P transmitters are very flexible, but they are not simply used as a gauge or pressure transmitter since simple Bourdon-tube-type pressure gauges will suffice in such applications. D/P transmitters are used where control applications are required to maintain process conditions at optimum levels.

Transmitter Signals and Scaling

Regardless of the type or magnitude of the process variable, the output of the transmitter is typically converted into one of the common standard instrument signals (Figure 9-6). In an analog electronic control loop, a thermocouple with a millivolt input to the transmitter has a 4–20-mA output. An RTD with its resistance input to the transmitter also has a 4–20-mA output signal. Pressure inputs, as well as analytical measurements applied to transmitters, also have a corresponding 4–20-mA output signal. All these process variables are converted into a standard instrument signal representing 0-100 percent of their respective measurements. That is why a standard instrument signal is so important. The process variables are different, but the output signals are the same.

DIFF FLOW TRANSMITTER
FLOW TRANSMITTER
FT 101 measures Flow

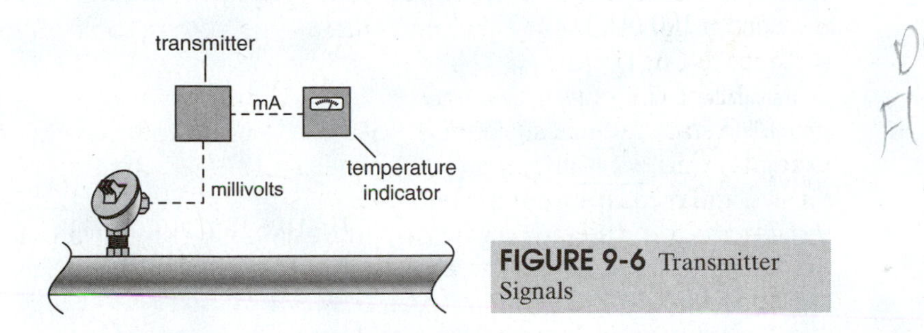

FIGURE 9-6 Transmitter Signals

Standard signals are the language that instruments speak when communicating with one another. For every input value applied to an analog electronic transmitter, there is a unique and corresponding milliampere output value. For example, since the output of a transmitter represents 0–100 percent of the measured process variable, then 4.0 milliamperes of a 4–20-mA standard output signal (in a direct acting transmitter) would represent 0 percent of the measured process variable and 20.0 milliamperes would represent 100 percent. That means that 12.0 milliamperes represents 50 percent.

For the example, given what is the percent span for the 25-psig value?

$$\%\text{Span} = \frac{\text{Value} - \text{LRV}}{\text{Span}} \times 100\% = \frac{(25 - 0)\text{ psig}}{50\text{ psig}} \times 100\% = 50\%$$

To better understand the relationship between the input of a transmitter and its output, calculating the input of a transmitter by observing its output is useful. The concept of instrument scaling must be addressed before calculations related to the input/output of a transmitter. (Table 9-1)

An **instrument scale** is a range of ordered marks at fixed intervals inscribed on an indicator plate. Scales are used to indicate actual measurement values. When calibrating a transmitter, an instrument technician observes two scales: the input scale and the output scale. The act of calibrating a transmitter involves applying a known set of input values to the measuring component and then adjusting the transmitter output to correlate the measurement range to a standard signal range. Scaling converts process signals so that they are compatible with instrument and control systems. The conversion process (scaling) is discussed in greater detail in Chapter 15 of this textbook.

Scaling is the act of equating the numerical value of one scale to its mathematically proportional value on another scale. For example, the measurement applied to a standard analog electronic pressure transmitter is represented on an appropriate pressure scale while the output signal is represented on a milliampere scale.

TABLE 9-1 Point Calibration on a Transmitter

Percent of Scale	*Input*	*Output*
URV 0%	500 °F	4 mA
25%	625 °F	8 mA
50%	750 °F	12 mA
75%	875 °F	16 mA
LRV 100%	1000 °F	20 mA

Process technicians may use formulas (Figure 9-7) to convert the numerical value of one scale to another. To know how to plug in the appropriate numbers in the formula, several scaling terms must be understood, to include:

- **Upper range value (URV):** The number at the top of the scale
- **Lower range value (LRV):** The number at the bottom of the scale
- **Operating range:** The set of values that exist between the LRV and the URV of a scale; expressed as two numbers (e.g., a calibration range may be expressed as 50 psig to 150 psig)
- **Span:** The algebraic difference between the URV minus the LRV of a scale; expressed as one number (e.g., If URV = 150 psig and LRV = 50 psig, then span = 100 psig)

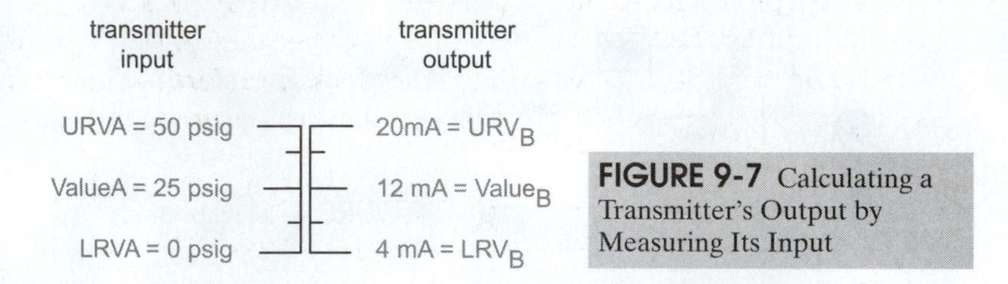

FIGURE 9-7 Calculating a Transmitter's Output by Measuring Its Input

The input-versus-output relationship of a transmitter can be either linear or nonlinear. This discussion is limited to **linear scaling**. When there is a linear relationship between two scales (transmitter input versus output), the following formula works:

$$\text{VALUE}_\text{B} = \frac{\text{VALUE}_\text{A} - \text{LRV}_\text{A}}{\text{SPAN}_\text{A}} \times \text{SPAN}_\text{B} + \text{LRV}_\text{B}$$

Where:
A = Original scale
B = New scale
Proportioning factor = Decimal representation of the original value on scale "A"

Example:
Refer to Figure 9-7. If a pressure transmitter is calibrated to measure 0–50 psig and is measuring 25 psig, what is the output of a standard 4–20 mA transmitter?

$$\text{VALUE}_\text{B} = \frac{\text{VALUE}_\text{A} - \text{LRV}_\text{A}}{\text{SPAN}_\text{A}} \times \text{SPAN}_\text{B} + \text{LRV}_\text{B}$$

$$\text{VALUE}_\text{B} = \frac{25\ \cancel{\text{psig}} - 0}{50\ \cancel{\text{psig}}} \times 16\text{ mA} + 4\text{ mA}$$

$$\text{VALUE}_\text{B} = 12\text{ mA}$$

Transducers and Signals

Generally speaking, a transducer (Figure 9-8) is a device that converts one energy form into another. Transducers, for example, can convert quantities such as temperature and pressure into an electronic form. More specifically, a primary sensor such as a diaphragm

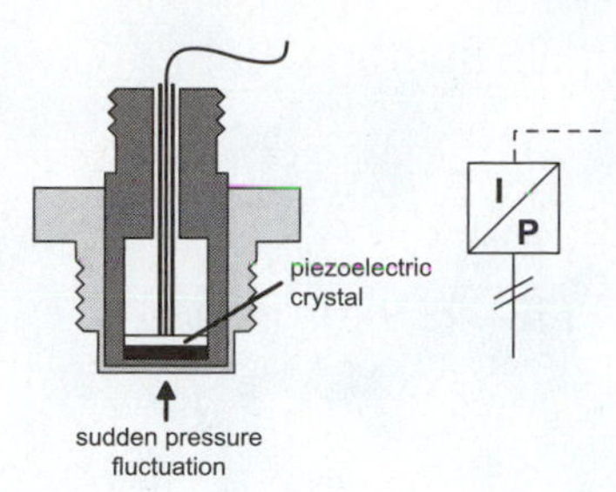

FIGURE 9-8 Transducers

produces motion in response to pressure and then a transducer converts the motion into a measurable electrical quantity. Sometimes the sensor and the transducer are one and the same. For example, a piezoelectric transducer is a crystalline wafer that responds to an applied pressure by producing an electrical response. Another example is a thermocouple that responds to a difference in temperature by generating a millivolt signal. These sensor-related transducers may produce any form or level of electrical impulse unlike their macro counterparts such as current-to-pneumatic transducers (converters).

In the previous chapter, the discussion covering instrument converters was in fact a discussion about transducers. In the strictest sense, any device that converts one signal form into another is a transducer.

A common instrument signal transducer converts an analog electronic signal (4–20 mA) into a pneumatic signal (3–15 psig). This device is called an I/P (current-to-pneumatic) transducer (Figure 9-9). This type of transducer is required in electronic control loops that use pneumatically actuated control valves. Therefore, a **signal converter (transducer)** is necessary to convert the electrical signal into a pneumatic signal.

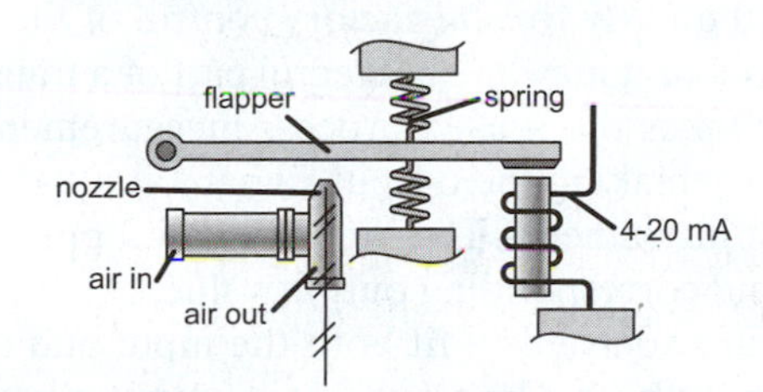

FIGURE 9-9 Current-to-Pneumatic Transducers

Pneumatic and Electrical Signals

Pneumatic and electronic signals are a very simple language that instruments use to communicate information between one another. Commonly used instrument signals in the process industry are the 4–20-mA electronic signal and the 3–15-psig pneumatic signal. Most modern smart transmitters, which are microprocessor based, usually produce a 4–20-mA output signal as well as a digital signal.

Both of these analog signals have what is referred to as a live zero, which means that the zero value of the signal is greater than true zero (e.g., 3 psig instead of 0 psig). This is an important point because verifying that the zero output of a transmitter is correct would be much more difficult without a live zero. This is why if the output of a pneumatic transmitter is reading 3.2 psig when its measurement is 0 percent, it is easier to correct the zero value by adjusting the pressure downward. Even if the value of the signal is 2.8 psig on the live zero scale, the problem is easy to correct by adjusting the value upward. But, if the output of the transmitter were below true zero on a scale without a live zero, the output signal would go unnoticed and consequently continue to provide an incorrect measurement.

Control Schemes and Control Loops

When observing a control loop in the field or in a schematic (Figure 9-10), the best place to start is at the measurement point. The information (measurement) is carried

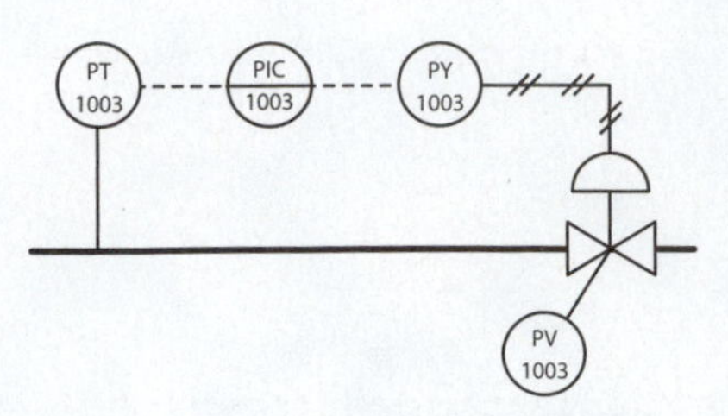

FIGURE 9-10 Pressure Control Loop Diagram

and dealt with throughout the control loop. Each instrument in the control loop functions according to how the incoming signal is affected. The sensor or transmitter processes the signal received from the measurement transducer and produces a standard instrument signal representing that measurement.

Summary

This chapter covered the first components of a control loop: the primary sensor, the transmitter, and the transducer. These components can be housed in the same device. Their definitions and operational purposes have specifics that make each of them unique and clearly distinct from each other.

Sensors and transmitters are capable of both sensing and measuring the process variable and then producing an output signal that carries the measurement information to the next instrument within the loop arrangement. The sensor performs the detection of the process variable such as flow rate, level, and/or pressure. Sensors come in various shapes and sizes depending on the particular application involved. Sensors may be directly connected to a controller such as in a thermocouple arrangement. This type of direct connection is not always either financially or functionally possible. The primary sensor must perform accurately, precisely, and quickly for satisfactory control of the process. The sensor itself may be a stand-alone unit, or it may be an integral part of a transmitter.

A sensor or transmitter takes the sensed process measurement and transduces it within the transmitter circuitry making the output signal a standard instrument signal. This signal is the language of instruments. The input value is applied to the transmitter and the output is a unique and corresponding output value.

Scales on a transmitter are adjusted to fit both the input and output of an instrument so there is a correlation to both. The transmitter output must correlate from the measurement range to a standard signal range. Simply stated, this means that one scale is equated numerically to its mathematically proportional value on another scale. While input to output relationships of a transmitter can be either linear or nonlinear, this chapter deals exclusively with a linear relationship.

The most common instrument signal transducer, an I/P (current-to-pneumatic) transducer, converts an analog electronic signal (4–20 mA) into a pneumatic signal (3–15 psig). This type of transducer is the most common since almost all the final control elements, such as a control valve within a control loop, are pneumatically actuated. Both the 4–20-mA and 3–15-psig signals have a live zero attribute. The zero is greater than actual zero.

When observing an instrument control loop, the best place to start is at the measurement point or the primary sensor of process variable. Each component of the control loop functions according to this measurement. In most control loops, the process variable measurement must be changed to a standard instrument signal before being processed by the next control loop component.

Checking Your Knowledge

1. The primary role of a sensor in a control loop is to detect and _____ the process variable.
   a. Transmit
   b. Transduce
   c. Measure
   d. None of the above

2. The d/p cell transmitter converts a process variable measurement into a standard instrument _____.
   a. Value
   b. Signal
   c. Pressure
   d. Difference
   e. None of the above
3. A differential pressure transmitter can be used to measure differential pressure as well as *infer* _____.
   a. Temperature
   b. Density
   c. Level
   d. Flow
   e. Pressure
   f. All of the above
4. The most common electronic instrument signal used in process industries is expressed in _____.
   a. Millivolt
   b. Milliamps
   c. Pounds per square inch
   d. Span
   e. Range
5. I/P transducers are commonly used in the process industry with _____ actuated valves.
   a. Electronically
   b. Digitally
   c. Electrically
   d. Pneumatically
   e. None of the above
6. What is the term used to describe the conditions where the LRV of a standard instrument signal is greater than true zero?
   a. Positive zero
   b. Live zero
   c. Integral zero
   d. Zero span
7. In a standard I/P transducer, an 8-mA input corresponds to _____ of output?
   a. 4 psig
   b. 6 psig
   c. 9 psig
   d. 7.2 psig
8. A temperature transmitter uses a thermocouple sensor and is calibrated 100 °F – 300 °F as a 4–20-mA output signal. If the fluid temperature is 200 °F, what is the output signal in mA?
   a. 8
   b. 10
   c. 12
   d. 18
9. In order to increase production rate, a reactor setpoint was increased from 160 psig to 190 psig. The reactor pressure transmitter was recalibrated from 0–200 psig to 0–300 psig. What is the percent span for the old and new setpoints?
   a. 50%
   b. 63%
   c. 80%
   d. 90%

   Remember: In an operating unit, when a transmitter is recalibrated, the scale must be changed on the controller as well as the controller retuned.

   New gain = (old gain) X (new span) / (old span)
10. In a standard I/P transducer, an 8-mA input corresponds to _____ of output?
    a. 4 psig
    b. 6 psig
    c. 9 psig
    d. 7.2 psig

11. Convert the following to find the missing factors in the items below:

Process Range	*Output Signal*	*Current Process Reading*	*Current Signal Reading*
200–800 F	4–20 mA	**625** F	14 mA
0–1000 gpm	4–20 mA	250 GPM	**8** mA
100–300 psi	3–15 psi	**150** psi	6 psi

Student Activities

1. Using a telephone handset, describe how it compares to a process primary sensor, transmitter, transducer configuration in a control loop.
2. Using a home computer network, describe how it compares to a process primary sensor, transmitter, transducer configuration in a control loop.
3. Identify or locate primary sensors, transmitters, and transducers in the lab.
4. Use pilot plant equipment and instrumentation to locate primary sensor, transmitter, and transducer components.
5. Using an I/P converter, explain the relationship between the psig air signal and the mA electric signal.
6. View a photograph of a process control loop from one of the local industries or make an actual visit to view a predetermined control loop at a particular site.
7. Perform a bench calibration on a transmitter and signal converter or transducer, or at minimum watch a calibration being performed or demonstrated.
8. IN CLASS ACTIVITY–Process Control Loop Operation

Instructions

a. Review the following process control loop.

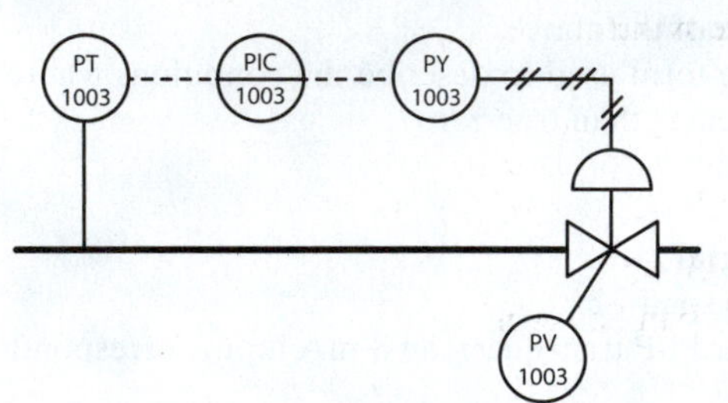

Electronic Pressure Control Loop

b. Write a description of how the control loop operates starting in the field at the measurement point. Explain how the information (measurement) is carried and dealt with throughout the control loop by addressing each instrument in the control loop according to the function of the instrument within the control loop and how it affects the incoming signal.
c. Give examples and any reasoning for justification of where this type of control loop could be found in process operations.
d. Follow your specific instructor guidelines for submitting responses to this activity.

CHAPTER 10

Control Loops: Controllers and Final Control Element Overview

Objectives

After completing this chapter, you will be able to:

- Define terms associated with controllers:
  - auto/manual switch
  - local/remote switch
  - setpoint
  - tuning
  - direct acting
  - reverse acting
  - proportional band or gain
  - integral and reset
  - derivative and rate
- Define bumpless transfer of auto to manual and manual to auto control.
- Describe the process for switching from auto control to manual control on a local controller.
- Describe the process for switching from manual control to automatic control on a local pneumatic controller without bumping the process.
- Demonstrate various control skills, such as:
  - make setpoint adjustments on a local controller
  - operate a local controller in manual mode
  - make setpoint adjustments on a remote pneumatic controller
  - switch from manual to automatic control on a remote pneumatic controller without bumping the process
- Given a drawing or actual local pneumatic controller, read the chart and state the high and low-range values.
- Given a simulator or actual device, identify if a control loop is in control or out of control and identify the information used to make the decision.
- Given a drawing or actual device, identify and describe the operation of the following:
  - local controller
  - remote controller
  - split-range controller
  - cascade or remote setpoint (RSP) controller
  - ratio controller

- Provide an application requiring the following devices:
  local controller
  remote controller
  split-range controller
  cascade or remote setpoint (RSP) controller
  ratio controller
- Describe the role of the final control element as it relates to the process and the control loop.
- Describe three types of final control elements and provide an application for each type:
  control valve: manipulates a process flow (liquid or gas) in response to a control signal
  damper or louver: manipulates an air flow to control draft setting or temperature setting
  motor: starts or stops in response to a control signal
- Given a drawing or actual instrument, identify and describe the operation of the following:
  louver or damper final control element
  variable speed motor used as a final control element
  instrument air regulator

Key Terms

Auto/manual switch—a switch that allows a process technician to select either automatic or manual control from the front of a controller.

Automatic to manual and manual to automatic switching—when controller action is moved from automatic to manual or vice versa by adjusting the setpoint of the controller to the actual controlled point and then switching the mode; or switching from automatic to manual by simply repositioning the mode selector.

Bump—a process upset occurring when a controller is switched from auto to manual mode.

Bumpless transfer—the act of changing the controller from manual to automatic (or vice versa) without a significant change in controller output.

Cascade control loop—where the control loop is characterized by the output of one controller becoming the setpoint of another.

Control mode—the control action or the control algorithm response such as PID or a programmed function.

Control valve—the most common final control element in the processing industry; has an actuating device mounted to it; drives the flow controlling mechanism (the plug or disc) in a valve.

Derivative action—a controller output response that is proportional to the rate at which the controlled variable deviates from the setpoint.

Fail-last (fail-in-place)—the condition of the valve seat upon loss of instrument air or power failure.

Gain—change in output divided by the change in input.

Integral action (reset)—a controller output response that is proportional to the length of time the controlled variable has been away from the setpoint.

Local controller—when the controller is physically mounted in the processing area near the other instruments in the loop.

Louvers (dampers)—devices used to control airflow.

Primary controller—perceives the secondary loop as a separate entity; responds to one process variable while the secondary controller responds to another.

Proportional band (proportional gain)—the amount of deviation of the controlled variable from the setpoint required to move the output of the controller through its entire range (expressed as a percent of span).

Rate action—the faster the rate of change of the process variable, the greater the output response (Derivative action).

Ratio controller—controllers that are designed to ratio (or proportion) flow rate between to separate flows entering a mixing point; designed so that its output represents the exact flow rate needed by the controlled flow loop to remain in alignment with the desired ratio to the uncontrolled flow; may also be capable of receiving two separate flow inputs and ratioing its output to the control valve located in the control line.

Remote controller—any controller that is not located in the processing area with the transmitter and control valve.

Remote setpoint (RSP)—a setpoint received from an external source.

Secondary controller—a special type of controller sometimes called a cascade controller or remote setpoint controller; a standard controller with the added capability of choosing to receive a remote setpoint from an external source or a local (internal) setpoint.

Setpoint knob—the mechanism by which a technician could manipulate the setpoint.

Split-range controller—where the output signal is divided between two final control elements.

Tuning—adjusting the control action settings so that they produce an appropriate dynamic response to the process resulting in good control.

Uncontrolled flow—flow in a process line where there is no control valve.

Introduction

The controller (Figure 10-1) is the device in a control loop that operates automatically to regulate a process variable. In a simple feedback control loop, the controller first compares the value of the measurement signal to a predetermined or desired setpoint value. The result of this comparison produces another value called the error. The error signal is then acted upon by one or more separate action components (control algorithms) in the controller to generate an appropriate output signal. The output signal of the controller then becomes the input to the final control element, or transducer that in turn is connected to the final control element. The final control element (usually a control valve) manipulates the process producing a change in the measurement sensed by the sensor or transmitter. The sensor or transmitter, once again, feeds the revised measurement information back to the controller and the cycle begins again.

The information in this chapter takes a closer look at controllers to include how they function, what type of responses that they may take to various signals, and what type of signal they output. The last and final element of the control loop is also introduced to include what types of final control elements are used and how they relate to the signals produced by the controller.

The theory of control is easier to understand by studying analog systems first. Digital control is covered in Chapter 19.

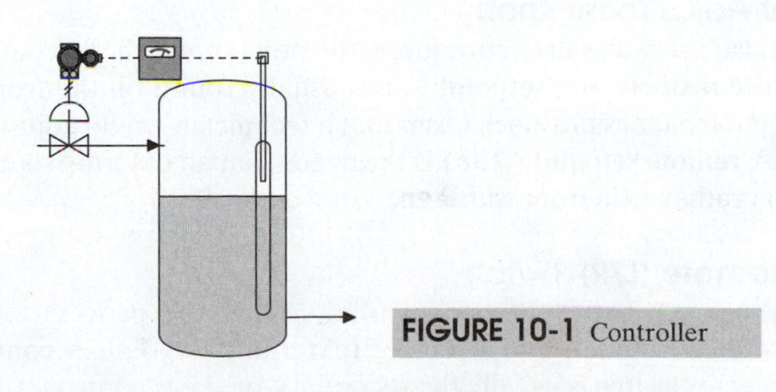

FIGURE 10-1 Controller

Controller Characteristics

The purpose of a controller is to ensure a process variable remains at a desired value or setpoint or is returned to that value/setpoint if there is a process change. The following discussion has been divided based on the characteristics of controllers as found on either the front or side panels of controllers.

FRONT PANEL

To perform as intended for the application, various controller configurations may be found on a typical front panel (Figure 10-2), to include the following:

- Auto/manual switch
- Local and/or remote setpoint indicator
- Local/remote switch
- Setpoint adjustment knob
- Manual output knob
- Manual output indicator
- Process variable indicator
- Setpoint indicator
- Setpoint variable
- Tag number

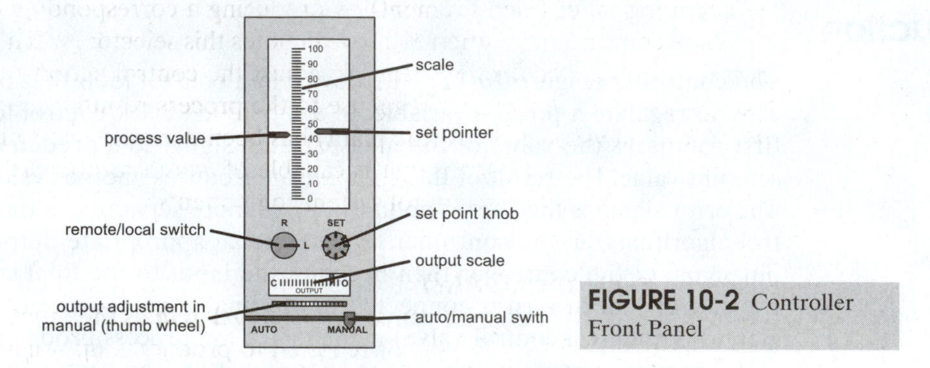

FIGURE 10-2 Controller Front Panel

Auto/Manual (A/M) Switch

The **auto/manual switch** enables the process technician to select either the automatic or manual mode from the front of a panel-mounted controller. When the process technician selects automatic mode, the controller automatically controls the process. When manual mode is selected, the process technician manually controls the process by adjusting the output of the controller accordingly. During startup, shutdown, and when certain process problems occur, it is necessary to use manual control. However, once the problem has been resolved or after the process has stabilized, the technician should switch the controller back to automatic. A manual output switch/knob is provided to send an output to the final element when the controller is in manual mode.

Setpoint Adjustment Knob

Setpoint, also known as desired value, is the point at which the controller is adjusted to regulate the process. The **setpoint knob**, usually found on the front plate of a panel-mounted controller, is the mechanism that a technician would adjust to change the local setpoint. A **remote setpoint (RSP)** is received from an external source (usually another controller) rather than from within the same controller.

Local/Remote (L/R) Switch

The local/remote switch on a controller allows the process technician to switch between a local setpoint and a remote (external) setpoint. A controller with a local/remote selector switch is usually the secondary, or slave, controller in a cascade control loop. The cascade control loop will be discussed later in this session.

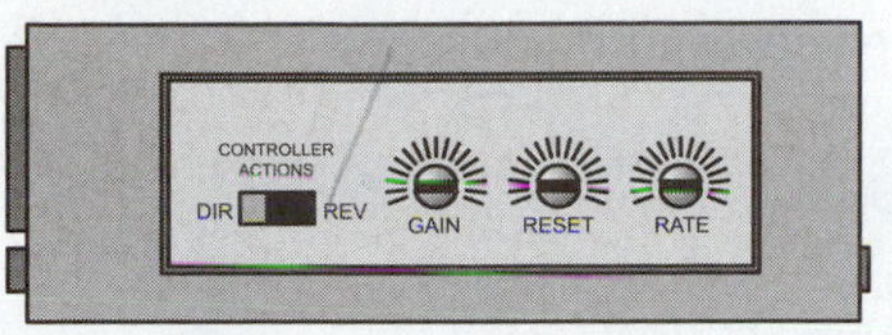

FIGURE 10-3 Controller Side Panel

SIDE PANEL

The components used for tuning a controller can be found on the side panel (Figure 10-3) of a typical pneumatic or electronic analog controller. These components for tuning include the various control actions (modes) as follows:

- direct/reverse acting (DIR and REV)
- proportional band (GAIN)
- integral action (RESET)
- derivative action (RATE)

Direct Acting and Reverse Acting (DIR and REV)

A controller selected for direct action responds to an increasing input (measurement) by producing a corresponding increasing output. One convention denotes this selector switch option as increase/decrease. A controller selected for reverse action responds to an increasing input (measurement) by producing a corresponding decreasing output. The same convention mentioned above denotes this selector switch option as increase/increase. **Tuning** a controller means to adjust the control action settings so they produce an appropriate dynamic response to the process resulting in *good* or *tuned control*. Once a physical process (the plant) has been designed and built, the controller is the only device in the system that is capable of counteracting the dynamics (process gain) of the process to the control system components.

Proportional Band (GAIN)

Proportional band (PB), also called **proportional gain**, is the input change to the controller (band or part of the range) required to produce a full-range change in output from a controller. A proportional band of 100 percent means that it takes a full range or 100 percent change in input to drive the output of the controller through its full range.

$$PB = \left(\frac{1}{\text{Gain}}\right)(100\%)$$

Gain is another term used to describe this input to output relationship. Gain is a unitless number and is calculated as:

$$\text{Gain} = \frac{(\text{change in output})/(\text{output transmitter span})}{(\text{change in input})/(\text{input transmitter span})}$$

$$\text{Note that (Process Gain)} \times \text{(Controller Gain)} = 1 = G_P \times G_C$$

Gain is a proportioning factor describing how the magnitude of the input relates to the magnitude of the output. For example, if a controller has a gain of 1, the controller will respond to an input change of 10 percent by producing an output change of 10 percent; if it had a gain of 2 instead, then the output change would be 20 percent. Regardless of how much change in input there is, the output of a controller can only produce a fixed output range (4–20 mA, which is 100 percent of the output range). Therefore, a gain of 1 is equal to a proportional band of 100 percent because it takes a full-range change in input (100% percent to produce a full-range output. Proportional-only controllers do exist; however, proportional action is usually teamed up with integral action (PI).

Example (relationship between PB and gain):

$$\% \text{PB} = \frac{1}{\text{gain}} \times 100$$

200% PB = 0.5 (gain)
100% PB = 1.0 (gain)
50% PB = 2.0 (gain)

Calculating Process Gain Cooling water is fed into a reactor jacket for controlling reactor temperature. When the input cooling water flow is changed from 60 gpm to 55 gpm, the reactor temperature changes from 250 degrees F to 275 degrees F. The cooling water flow transmitter is calibrated from 0 to 100 gpm and the temperature transmitter is calibrated from 0 to 400 degrees F.

In calculating the process gain for this process, one would find that it is reverse acting as shown below:

$$G_p = \frac{(275 - 250)°\text{F}/(400 - 0)°\text{F}}{(56 - 60)\ \text{gpm}/(100 - 0)\ \text{gpm}} = -1.25$$

Calculating Controller Gain For this formula, the controller must be selected as direct acting to counter the process.

$$G_c = \frac{1}{1.25} = 0.8$$

Integral Action (RESET)

The **integral action** (or **reset**) in a controller is designed to eliminate the offset inherent in feedback control loops. Reset describes the actual function of this **control mode**. Reset action repeats the proportional action in a given time period and is given as repeats per minute or repeats per second depending on the brand of the controller. Integral action is the inverse of reset and is given as minutes/repeat or seconds/repeat.

The controller senses a steady-state offset between the measurement and the setpoint and then produces a ramping response based on the proportional action. The output of the controller continues to change until the measurement is brought back to setpoint. Unlike proportional band, the integral action mode is never found alone and is combined with proportional action (PI) or in some cases with proportional and derivative action (PID).

Derivative Action (RATE)

The **derivative action** of a controller responds to the rate of change of the controlled variable (i.e., measurement or PV). This is why derivative action is also known as **rate action.** The faster the rate-of-change in the error signal, the greater the derivative response. Once the PV measurement stops moving away from setpoint, the derivative action drops to zero. Some controllers may be programmed with only derivative action, but derivative is usually found in combination with proportional and integral action (PID). Derivative action is added to a PI (proportional and integral) controller to provide a better response to slow loops, such as temperature loops.

Controller Switching

Controller switching makes the following types of mode transfers:

- bumpless
- auto to manual mode
- manual to automatic control

Bumpless Transfer

Bumpless transfer means to change the controller mode from manual to automatic, or vice versa, without the controller responding with a change in output. **Automatic to**

manual and manual to automatic switching procedures vary according to the manufacturer equipment specifications. The most modern controllers have a setpoint tracking circuit that makes switching between these modes as simple as selecting the desired mode position. However, if a controller does not have an automatic tracking circuit, then a more involved procedure must be performed. For these controllers, setpoint tracking is selected when the controller is in the manual mode, making SP = PV; hence the control error is zero and therefore no proportional or integral action is calculated. If SP = PV, then the integral or reset action would drive the internal output (called reset windup) to 0 percent or 100 percent output. Switching to auto mode at this time would create a process bump. Remember that manual mode would bypass these internal calculations.

Generally, switching from manual to automatic requires adjusting the setpoint of the controller to the actual controlled point (PV) and then switching the mode (e.g., aligning the setpoint indicator to the measurement indicator PV). On the other hand, switching from automatic to manual may require only a simple repositioning of the mode selector.

In the old pneumatic systems and some of the older electronic controllers, a **bump** or mild upset could occur when switching the controller back from manual to automatic if the process variable has drifted from the setpoint. The controller causes the process value to move quickly back to setpoint when the switch is made, causing a bump. By setting the setpoint so that the process variable matches the PV, the switch becomes bumpless. If the technician had changed the controller to manual for a specific reason, there should not be a reason for returning the controller to a previous (old) setpoint at this time.

Auto to Manual Mode

The process for switching from auto to manual mode includes the following steps:

1. With controller in auto, the PV matches the setpoint.
2. Note the output signal to the control valve.
3. Switch auto to manual.
4. Adjust manual output to achieve previous reading when it was in auto.
5. Check PV and tweak the output as needed if the PV is not at desired value. The output needs to be adjusted as often as necessary to maintain the desired process value.

Manual to Automatic Mode

The process for switching from manual mode to automatic mode without bumping the process includes the following steps:

1. Adjust the setpoint to match the PV.
2. Switch manual to auto.
3. Monitor the signal to see if the valve changes or if the PV drifts.
4. Once in auto, adjust the setpoint if it is not at the desired value.

Types of Controllers

Local Controllers

A controller is considered to be a **local controller** (Figure 10-4) if it is physically mounted in the processing area near the other instruments in the loop.

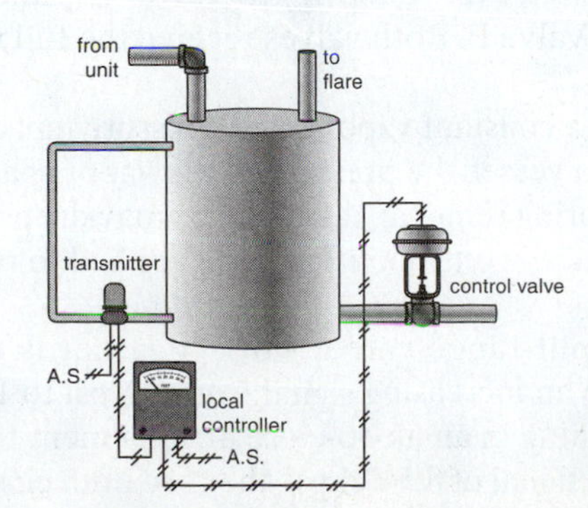

FIGURE 10-4 Local Controller

Local control systems are still found on remotely located equipment such as knockout pots (or drums) located at the base of flares. The knockout drum is the last point where liquids can be separated from burn-off gases before they rise up to the ignition point at the top of a flare stack. These vessels are generally remotely located and their levels are isolated and unrelated to other processes making them ideal candidates for local control.

Remote Controllers

Any controller that is not located in the processing area with the transmitter and control valve is considered to be remote (Figure 10-5). Most **remote controllers** are found in panels located in centralized control rooms. Remote controllers receive a signal from a transmitter located in the processing area, generate a corrective signal and then transmit an appropriate signal out to the final control element, which is also in the processing area.

With few exceptions, modern process control schemes place the controller away from the processing area in centralized control rooms. Controllers began moving out of the field and into centralized control rooms shortly after the development of the pneumatic transmitters around 1939. By the early 1950s, most controllers were located in control rooms where they have remained ever since. Figure 10-5 depicts a pressure loop with the transmitter and control valve located in the processing area and the controller in a centralized control room. JB-100 and JB-200 are collection and distribution points called junction boxes. A large multitube bundle connects each point (bulkhead fitting) in JB-100 to a corresponding point (bulkhead fitting) in JB-200.

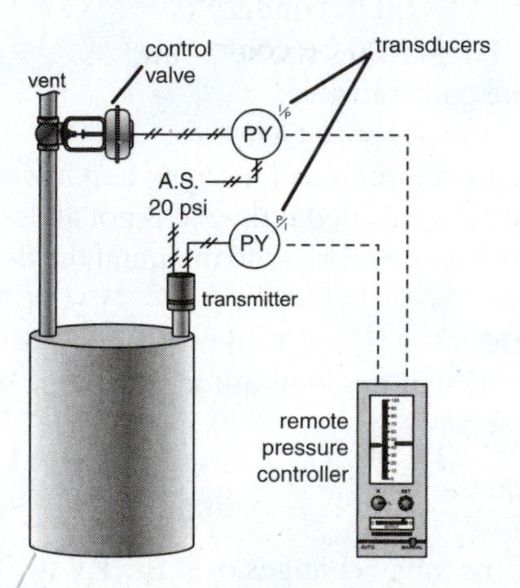

FIGURE 10-5 Remote Controller

Controllers with Split-Range Valves

The output of a controller may be divided between two final control elements. The most common scenario is where one final control element (Valve A) responds to one portion of the output signal where the other portion of the output signal (Valve B) responds to the upper half. Modern digital controllers can have more than one output; therefore, they can be **split-range controllers** (Figure 10-6). For example, the signal from output 1 could be 4 – 11.4 mA (3 – 8.5 psig) to Valve A, while output 2 could be 12.6 – 20 mA (9.5 – 15 psig) to Valve B. Both valves receive the full range output of the controller.

A processing tank requires a constant vapor space pressure (space above the liquid) of 10 psig. During the filling of a vessel, the pressure in the vapor space increases requiring the vent valve A to open. During removal, the pressure may drop requiring valve B to open. This type of situation requires control output to be applied to two separate control valves.

The most common way split-range valves work together is to have one valve, Valve B, respond by opening to an increasing signal from 9.5 psi to 15 psi and the other valve, Valve A, respond by opening in an air-to-close arrangement from 8.5 psi to 3 psi. When the valves are receiving a signal of 8.5–9.5 psi, they are both closed. As the pressure

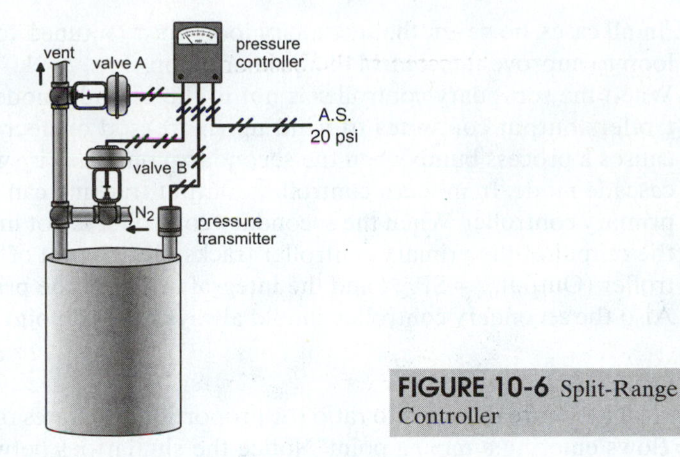

FIGURE 10-6 Split-Range Controller

drops in the tank, Valve B opens allowing pressure to enter the tank while the other remains closed. Conversely, when pressure rises in the tank, the Valve A opens allowing pressure to be relieved from the tank while Valve B remains closed.

Cascade/Remote Setpoint (RSP) Controllers

A **cascade control loop** is characterized by the output of one controller becoming the remote setpoint of another. As unusual as this type of control may seem, cascade control is very common. In Figure 10-7, the primary (master) controller is responding to the temperature of the product while the secondary (slave) controller is operating to control steam flow. The temperature could be controlled without the secondary loop, but it would not be as fast or as sure.

There are two important reasons for using a cascade loop:

- better control
- reduced lag times

The **primary controller** on the shell side is a temperature controller. The **secondary controller** is a controller with a remote setpoint option. The cascade controller is the same as a standard controller plus a cascade controller has the capability of choosing to receive a remote setpoint from an external source or a local (internal) setpoint.

In Figure 10-7, the primary (master) control loop is designed to control the temperature of the product leaving the shell side of the heat exchanger. This is done by controlling the amount of hot oil flow into the tube side of the heat exchanger. The secondary (slave), or intercontrol loop, is specifically controlling hot oil flow into the tube side of the heat exchanger. Since the secondary controller is a cascaded controller, it can be selected to control the hot oil flow locally (local position) or selected (remote position) to follow the needs of the primary controller and works to control the outlet temperature of the shell side of the heat exchanger.

Notice that the output of the primary controller is the remote setpoint of the secondary controller. This is just one of many different applications where a cascade control

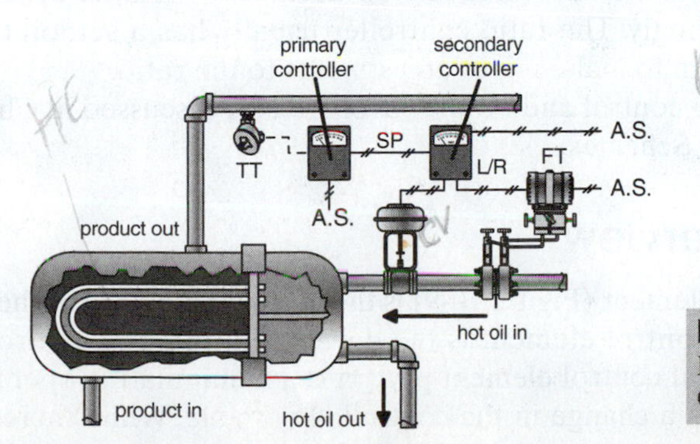

FIGURE 10-7 Cascade or Remote Setpoint (RSP) Controller

is found. In all cases, however, the secondary loop must be tuned to be faster than the primary loop to improve the control of a primary loop.

NOTE: When the secondary controller is not in the cascade mode, the primary controllers output continues to wind up (increased or decreased output) and causes a process bump when the secondary controller is switched back to the cascade mode. In modern controllers, output tracking can be selected for the primary controller. When the secondary controller is not in the cascade mode, the output of the primary controller tracks the setpoint of the secondary controller ($Output_{pri} = SP_{sec}$) and the integral action of the primary is turned off. Also the secondary controller should always have setpoint tracking selected.

Ratio Controllers

Ratio control loops are designed to ratio (or proportion) the rates of flow between two separate flows entering a mixing point. Notice the similarities between the ratio control loop and a standard cascade loop. The **ratio controller** (Figure 10-8) is designed so that its output represents the exact flow rate needed by the controlled flow loop to remain in alignment with the desired ratio to the uncontrolled flow. Another ratio controller is capable of receiving two separate flow inputs (the uncontrolled and the controlled) and ratioing the controller output to the control valve located in the controlled flow line. In either case, a ratio between the uncontrolled and controlled flows is maintained.

uncontrolled measure flow

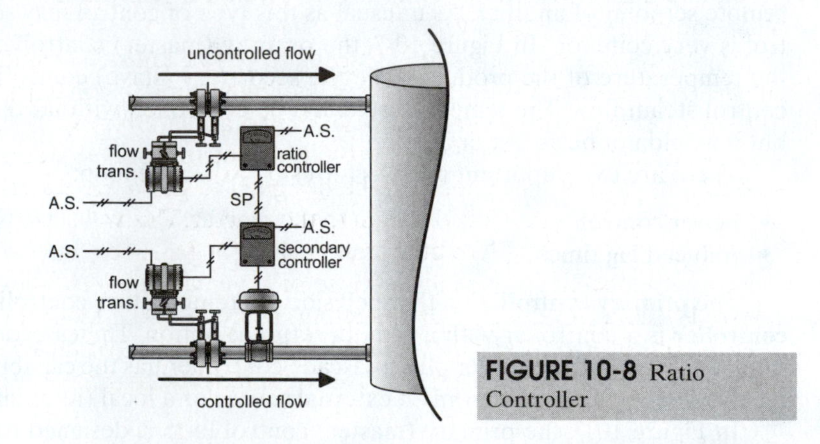

FIGURE 10-8 Ratio Controller

If this example represents a feedstock line (**uncontrolled flow**) and a catalyst feed line (controlled flow) entering a reactor, then the catalyst is proportioned according to how much feed is entering the reactor. Too much catalyst is dangerous and expensive and too little catalyst impacts the quality and quantity of the product. A proper ratio is imperative.

In the example, the uncontrolled flow transmitter produces an output that is received by the ratio controller where it produces an output representing the exact flow rate needed by the controlled flow loop. If the uncontrolled flow rate in the pipe increases, then the flow rate in the controlled pipe needs to increase proportionally as well. Conversely, if the uncontrolled flow decreases, then the controlled flow needs to decrease accordingly. The ratio controller usually has a setpoint range to allow the process technician to make minor adjustments to the ratio.

NOTE: Cascade control and ratio control are also discussed in Chapter 18: Advanced Control Schemes.

Final Control Element Overview

A final control element (Figure 10-9) is the last active device in the instrument control loop. The final control element is the device that directly controls the manipulated variable. This final control element performs a manipulating operation on the process that brings about a change in the controlled variable. While there are many different

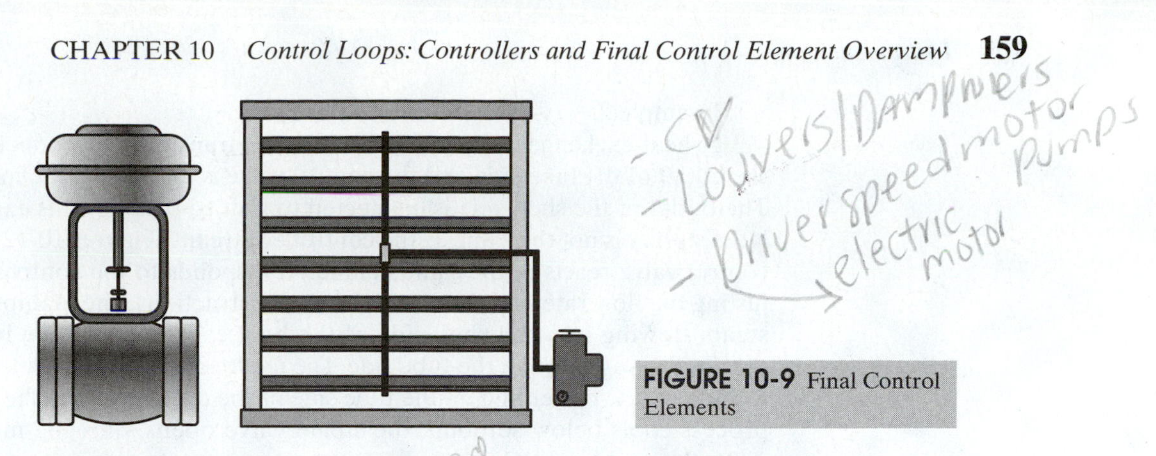

FIGURE 10-9 Final Control Elements

types of final control elements, the most common are control valves, louvers, and variable speed drives.

CONTROL VALVES

Control valves are the most common final control element in the processing industry. The function of a control valve is to manipulate the flow rate of some component in a process that will in some way affect the measured value of the controlled variable. Of course, this manipulated stream may also be the controlled variable stream. An actuating device is mounted to the control valve. The actuator changes an instrument signal into a linear or rotary motion. This motion drives the flow controlling mechanism (the plug or disc) in a valve.

Recall that control valves are designed to go to a fail-safe position upon air failure. For instance, a control valve controlling cooling water to a condenser may be designed to go to the open position upon air failure to allow full flow of cooling water. This would be referred to as a fail-open valve. If a control valve were controlling steam to a jacket of a reactor, the valve might be designed to go to the closed position upon air failure to shut down the flow of steam to the reactor. This valve would be referred to as a fail-closed valve. There are other terms used for fail-open and fail-closed. Fail-open might also be referred to as air-to-close or reverse-acting. Fail-closed might also be referred to as air-to-open or direct-acting. There is also a **fail-last** (also **fail-in-place**) feature for control valves that is not as common. Fail-last means when there is an instrument air failure, the valve stays in the same position that it was in at the time of the failure.

Example 1:

A simple feedback control loop controlling flow rate where the controlled variable is the same stream as the manipulated variable. In this example, the two streams are the same (Figures 10-10 and 10-11). The flow rate measured by the flow transmitter is the same flowing material that is being manipulated by the control valve. In this example, the controller adjusts the signal to the control valve, which manipulates the flow rate of the material moving through the process. The transmitter then measures the resulting flow rate and feeds the new measurement value back to the controller where the control cycle continues.

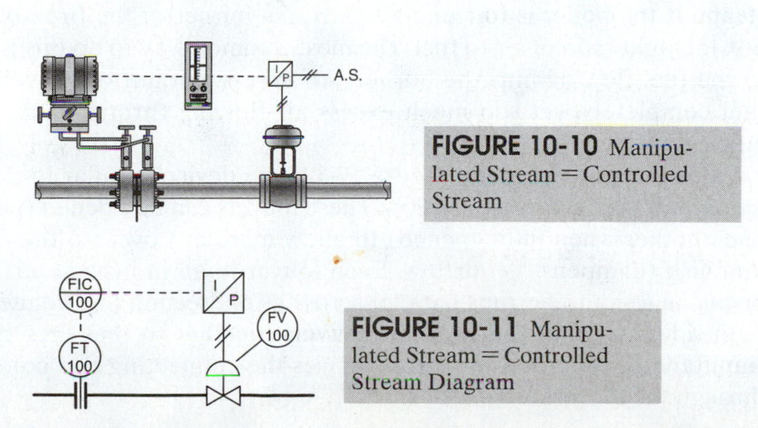

FIGURE 10-10 Manipulated Stream = Controlled Stream

FIGURE 10-11 Manipulated Stream = Controlled Stream Diagram

Example 2:

A heat exchanger where the temperature transmitter (product side) is located on the outlet of the tube side and the control valve is located on the inlet to the shell side. The outlet of the shell side is connected to a steam trap. In this example, the manipulated stream is not the same as the controlled stream (Figures 10-12 and 10-13). But the control valve reacts in the same manner: it responds to the controller output by readjusting the flow rate and then waits for new instructions. The manipulated stream is the steam flowing into the shell side of the heat exchanger where it heats the process stream flowing through the tube side. The feedback signal originates in the transmitter located on the outlet line on the tube side of the exchanger. As the temperature of the process cools below setpoint, the steam valve opens allowing more heat energy to enter the exchanger. Conversely, as the temperature of the process exceeds setpoint, the steam control valve reduces the flow of steam into the exchanger causing the temperature of the process material to move closer to setpoint.

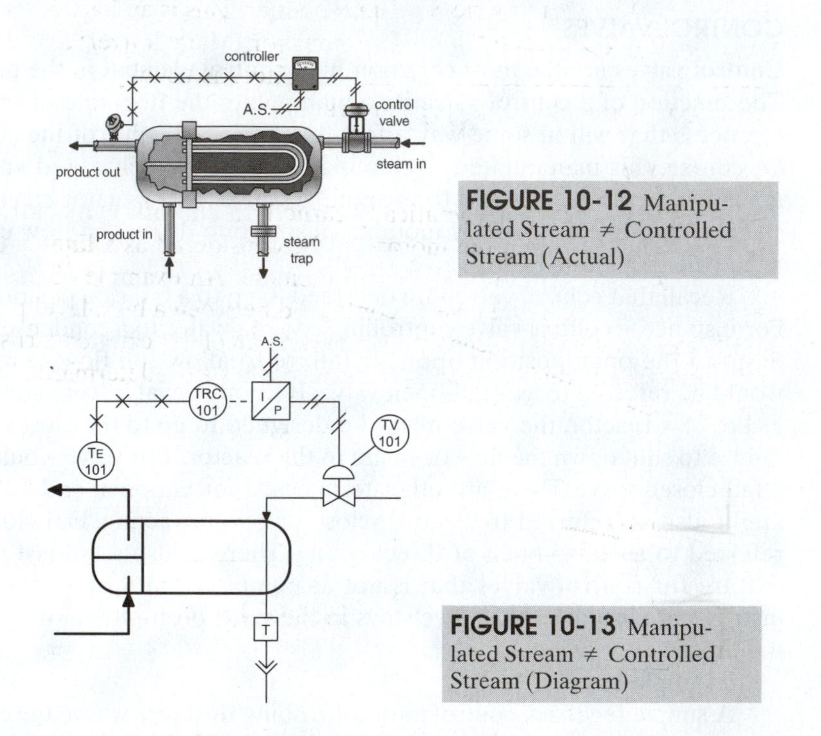

FIGURE 10-12 Manipulated Stream ≠ Controlled Stream (Actual)

FIGURE 10-13 Manipulated Stream ≠ Controlled Stream (Diagram)

Additionally discussed in Chapter 11 are control valve positioners that are attached to the control valve stem and used in conjunction with the controller to match the controller output signal to the desired valve position (open or closed).

LOUVER AND DAMPER

Many power plants, excluding nuclear plants, burn a hydrocarbon fuel to produce steam. If the boiler is to run efficiently, the burner in the firebox of the boiler must mix the right ratio of air to fuel. The most common way to do this is to ratio the airflow to the fuel flowing into the burner. An excess amount of air is needed to burn the fuel completely, yet too much excess air flowing through the system reduces the efficiency.

Louvers and **dampers** (Figure 10-14) are devices similar in design to shutters or miniblinds used to control airflow. These louvers can be opened (positioned parallel to one another when fully opened) to allow more airflow into the firebox or closed to diminish (dampen) the airflow. Each louver in an industrial airflow system usually pivots on a shaft that runs parallel across its midsection and connects to the frame on both sides. Another rod links the louvers together so that they can be manipulated simultaneously. An actuator then drives the connecting rod controlling the airflow through the louvers.

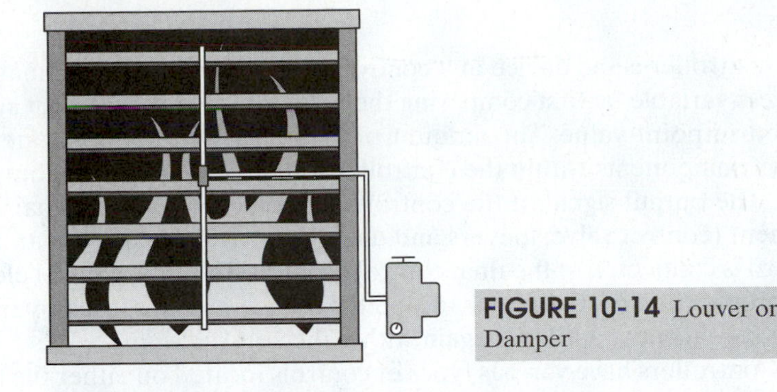

FIGURE 10-14 Louver or Damper

When the louvers are positioned parallel to the airflow, there is very little resistance. When positioned at an angle to the airflow, they can effectively regulate the draft through a firebox in a furnace or steam boiler. This is an important factor in adjusting the air-fuel ratio needed for good combustion. Many louvers and dampers have blowers to induce a force to provide airflow. There is usually an air-to-mechanical transducer (positioner) to adjust the louvers.

ELECTRIC MOTORS

If an electric motor is automatically turned on and off by a controller responding to a process variable, then the motor can be considered as a final control element or at least the actuator of the final control element. An example of this situation is a tank (Figure 10-15) where the level is controlled between a high-level point and a low-level point. When the level reaches the high mark, an electric motor turns on driving a pump that lowers the level in the tank. When it reaches a predetermined low point, the motor and pump are turned off.

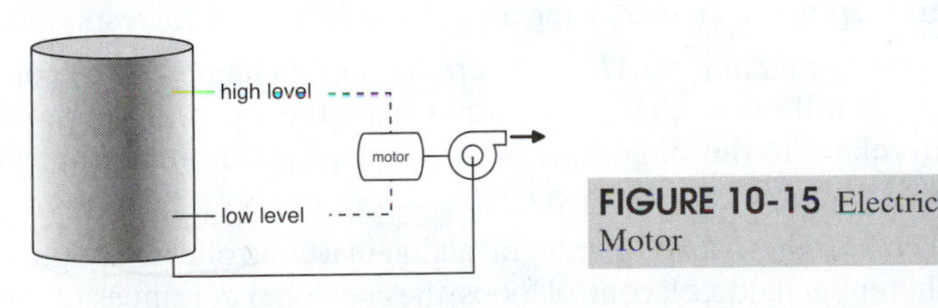

FIGURE 10-15 Electric Motor

Variable-speed electric motors can also be considered as the final control element. For example, if an electric motor drives a pump that is energized when the level reaches a certain height and is turned off at a designated low-level point, then, it is acting as an on-off final control element. Variable speed electric motors are capable of operating within a selectable revolution per minute (rpm) range. Usually the operating range is between zero revolutions per minute and a maximum upper range value. Variable-speed motor control may be achieved by controlling the level of direct current (DC) applied to the motor or by controlling the duty cycle of the applied alternating current (AC). Usually an mA signal from a speed controller is sent to a variable frequency drive (controller) that varies the frequency (0-70 Hz) supplied to the motor that in turn varies the rpm of the electric motor. An example (4–20 mA = 0–70 Hz = 0 to 2200 rpm) might be a combustion air blower on an incinerator.

Most people are familiar with a variable-speed electric drill, which is driven by a variable-speed motor. The concept is the same in larger industrial variable speed motors. Variable-speed motors are used to control fluid pumping rates, conveyor belt speeds, and other final controlling devices. If a variable-speed motor is driving a pump, then the flow rate can be modulated just like a pneumatically driven control valve.

A reversible electric motor with a gearbox is capable of actuating block valves or control valves just like a pneumatic actuator. Unlike the spring and diaphragm actuator with a loss of signal, the electric motor actuator typically fails in place.

Summary

The controller is the device in a control loop that operates automatically to regulate a process variable by first comparing the value of the measurement signal to a predetermined setpoint value. The amount of error is acted upon by one or more separate action components within the controller so that an appropriate output signal is generated. The output signal of the controller becomes the input signal to the final control element (control valve, louvers and dampers, or a motor), or to an I/P transducer that in turn is connected to the final control element. The final control element manipulates the process that produces a change in the measurement sensed by the transmitter. This cycle repeats over and over again within the control loop.

Controllers have various types of controls located on either the front or side panel. The front panel has various switches to include the following:

- *Auto/manual (A/M) switch* When in automatic control, the controller is in control of the process; when in manual control, the process technician is in control of the process by adjusting the output of the controller especially in situations such as startup or shutdown
- *Setpoint adjustment knob* The setpoint is the desired value that the controller is adjusted so that it regulates the process; if the setpoint indicator is local, it adjusts the readout from a nearby source; if the setpoint indicator is remote, then it receives information from an external source outside of the controller
- *Local/remote (L/R) switch* Allows technicians to switch between two different setpoint information sources (local and remote)

In addition, the side panel has tuning capability, to include the various mode settings as follows:

- *Direct acting/reverse acting* Direct acting controllers respond to an increasing input measurement by producing a corresponding increasing output; reverse action controllers respond to an increasing input by producing a corresponding decreasing output.
- *Proportional band (GAIN)* Where the input change to the controller is required to produce a full range change in output from the controller; how the magnitude of the output relates to the magnitude of the error; proportional action controllers are usually teamed up with integral action (PI).
- *Integral action (RESET)* Integral action in a controller is designed to eliminate the offset inherent in feedback control loops; the controller continues to reset the offset between the measurement and the setpoint until the measurement is brought back to setpoint.
- *Derivative action (RATE)* Also known as rate action; the controller responds to the rate of change of the controlled variable; the faster the rate of change, the greater the derivative response; derivative action stops when the rate of change stops; derivative action is most often found in combination with proportional and integral action (PID).

Controller modality changes or transfers include the following:

- *Bumpless*: When the controller is changed from automatic to manual control or vice versa and the controller does not have to respond with a change in output; usually performed by setting the output so that the process variable matches the setpoint before making the switch
- Auto to Manual control: see procedure
- Manual to Auto control: see procedure

Controllers are identified by their type, to include the following:

- *Local* The controller is physically mounted in the processing area near the other instruments in the loop; usually in a pneumatic control loop; most often used where the process is isolated and unrelated to other processes such as knockout drums.
- *Remote* Where the controller is not located in the processing area with the transmitter and control valve; pneumatic remote controllers are usually located no more than 300

feet away and electronic remote controllers may be significantly away from the other loop components.

- *Split-range* Where the output of the controller is divided between two final control elements as in the case where one valve responds to the lower half of an output signal and another valve responds to the upper half of the output signal; feed or product tankage are common places for this type of controller so that pressure is controlled on incoming and outgoing operations.
- *Cascade* Where the output of one controller becomes the setpoint of another as in where the primary (master) controller responds to temperature and the secondary (slave) controller operates steam flow; used where better control is needed or lag times need to be reduced.
- *Ratio* Where the rates of two separate flows enter a mixing point need to be proportionally introduced as in one part of stream A to four parts of stream B or a ration between one uncontrolled and one controlled flow is maintained.

A final control element is the last active device in the instrument control loop and directly controls the manipulated variable so that when one process is manipulated, then the other is changed. These are the three most common final control elements:

- *Control valves* The most common final control element; manipulates the flow rate of some component in a process so that the measured value of the controlled variable is affected; may have an actuating device that changes the instrument signal into a linear or rotary motion that drives the flow controlling mechanism such as a plug or disc in a valve.
- *Louvers or dampers* Commonly used to increase or decrease air flow to burners; when the slats are parallel (open), then airflow is increased and has little resistance; when the slats are closed, then no airflow is available and the burners may not function.
- *Electric motors* Where an electric motor drives a pump that is energized when the level reaches a certain height and is turned off at a designated low-level point; a reversible electric motor has a gearbox capable of operating block valves or control valves; variable speed motors are used to control how many rpms are allowed by controlling the level of DC applied to the motor or the duty cycle of an applied AC.

Checking Your Knowledge

1. *True of False* Controllers are usually placed in automatic mode during start up of the process or when process problems are encountered.
2. Match the following comments to the type of mode found in controllers:

Controller responds to the rate at which the variable changes	a. Direct
Controller responds to an increasing input with a decreasing output	b. Reverse
Controller responds to reset the measurement back to the setpoint	c. Proportional
Controller responds to an increasing input with an increasing output	d. Integral
Controller takes the full range of input to drive the output	e. Derivative

3. *True of False* A bumpless transfer is achieved if the controller mode is changed from manual to automatic, or vice versa, without the controller generating a change in output.
4. *True of False* When switching the Foxboro Model 43 controller from manual control to automatic control, the auto/manual switch should be flipped first and then the regulator knob turned.
5. *True of False* When switching the Foxboro Model 43 controller from automatic control to manual control, the regulator knob should be slowly moved so the indicator ball is within the central portion of the tube and then the auto/manual switch flipped.

6. Properly identify each type of controller by selecting the appropriate letter for the image of the controller:
   Local
   Remote
   Split-range
   Cascade
   Ratio

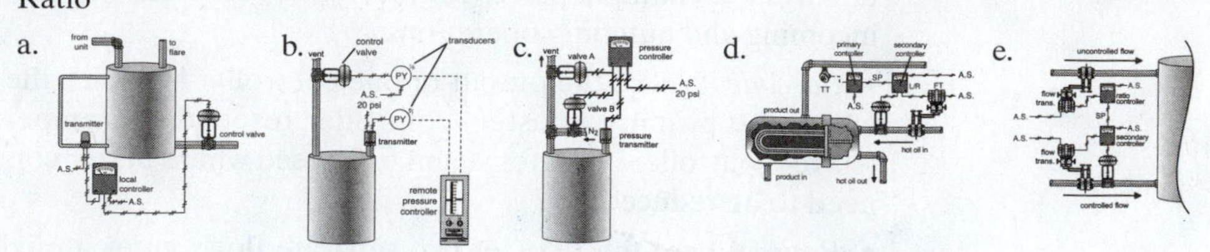

7. Match the following comments to the type of mode found in controllers:

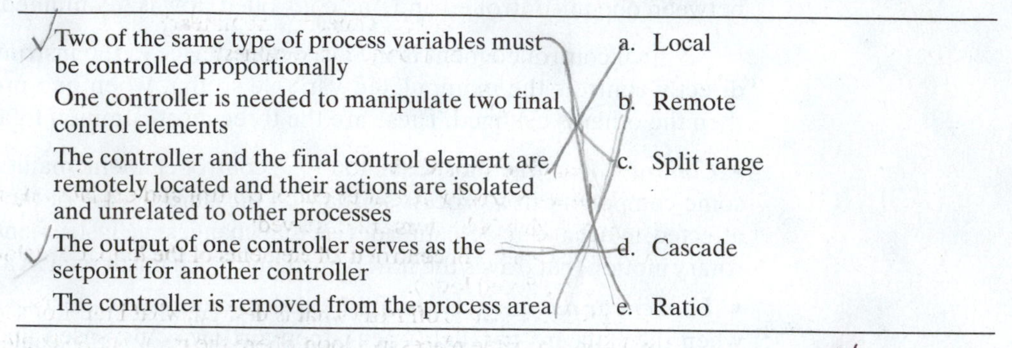

Two of the same type of process variables must be controlled proportionally	a. Local
One controller is needed to manipulate two final control elements	b. Remote
The controller and the final control element are remotely located and their actions are isolated and unrelated to other processes	c. Split range
The output of one controller serves as the setpoint for another controller	d. Cascade
The controller is removed from the process area	e. Ratio

8. Which of the following is considered the device that directly controls the manipulated variable?
   a. The measuring instrument
   b. The transmitting element
   c. The comparing and controlling element
   d. The final control element
   e. None of the above
9. Which of the following is the most common final control element in the processing industry?
   a. Control valves
   b. Electric motors
   c. Louver and dampers
   d. Compressors
   e. None of the above
10. Which of the following are variable-speed motors used to control?
    a. Fluid pumping rates
    b. Conveyor belt tension
    c. Internal timers
    d. None of the above

Student Activities

1. Photocopy the various control loop graphics found in this chapter. Use a white correction fluid or tape and *erase* ONLY the words. Photocopy the pages again making sure the words do not show. The last copy may be transferred to a transparency. Using a NON-permanent overhead transparency-type pen, write descriptors on each of the control loop elements. Check your responses against the original textbook copy. Erase your responses by the appropriate method as per your pen type. Use the transparency copy over and over again by yourself or with others to improve your visual recognition of each element and how it works within the control loop paying special attention to the controller functionality.
2. Identify or locate various types of controllers in the lab
3. Given a drawing or actual local pneumatic controller, read the chart and state the high and low-range values.
4. Perform the following control skills depending on available resources:
   - Make setpoint adjustments on a local controller.
   - Operate a local controller in manual mode.
   - Make setpoint adjustments on a remote pneumatic controller.
   - Switch from manual to automatic control remotely without bumping the process.

5. **CONTROLLER SWITCHING–A**

Using available instrumentation within your classroom facilities, perform the following in the presence of your instructor:

- Make setpoint adjustments on a local controller.
- Operate a local controller in manual mode.
- Make setpoint adjustments on a remote pneumatic controller.
- Switch from manual to automatic control on a remote pneumatic controller without bumping the process.

6. **CONTROLLER SWITCHING–B**

Obtain the *Pneumatic Controller High and Low-Range Values* handout from your instructor and explain how to read the chart.

7. **CONTROLLER SWITCHING–C**

Obtain the *Indicator Illustrations* handout from your instructor and determine the following:

- Does the illustration indicate linear or nonlinear?
- What are the upper and lower-range values?
- What is the range?
- What is the span?

8. **CONTROLLER SWITCHING–B**

Determine if a control loop is in or out of control and explain why it is in or out of control when the following indicators are observed:

- The control loop is in control if all elements of the loop are yielding the desired process result (auto, closed loop).
- If the process result is off from what is desired, what indicator shows an out of control condition? Possible places in a loop where the malfunction could exist include: the primary element, transmitter, controller, and/or final control element.
- The process technician must troubleshoot the cause of the problem.
- Placing the problem element on manual control can help regain control of the process.

9. **CONTROL LOOPS—CONTROLLERS**

Instructions

a. Obtain specific instructions from your instructor as to how to submit your activity responses.

b. Review the following indicator illustrations:

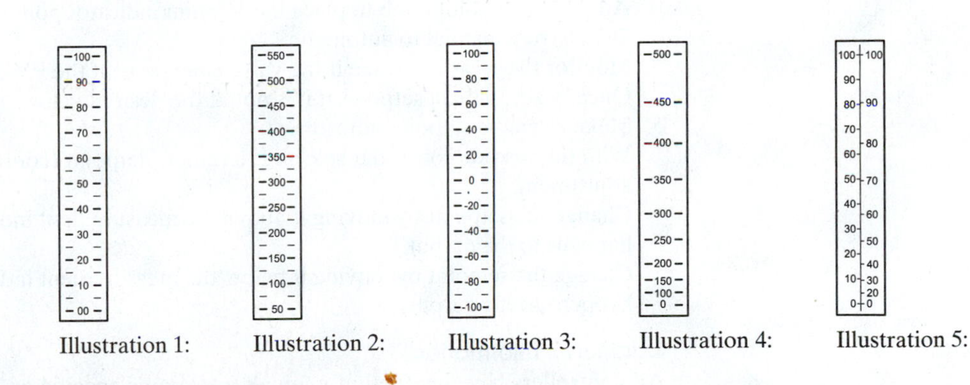

Illustration 1: Illustration 2: Illustration 3: Illustration 4: Illustration 5:

c. Answer the following questions:

- Is it linear or nonlinear?
- What are its upper and lower-range values?
- What is its range?
- What is its span?

d. Post your answers as per your instructions obtained in (a) above.

10. **CONTROLLER LABORATORY**

Background Information

The term *bumpless transfer* describes the manner in which the output of the controller responds to a change in setpoint value brought on by switching from manual control. In a cascade control system, the act of switching from a local setpoint to an external setpoint can cause a bump as well. Most modern controllers have a setpoint tracking circuit (or subroutine in computer systems) that allows switching between these modes without a bump in output. If a controller does not have an automatic tracking circuit, then a more involved procedure must be performed. Generally, switching from manual to automatic requires adjusting the setpoint of the controller to the actual controlled point and then switching the mode, i.e., aligning the setpoint indicator to the measurement indicator. On the other hand, switching from automatic to manual may only require a simple repositioning of the mode selector. It should be noted that the automatic-to-manual and manual-to-automatic switching procedure varies according to the manufacturer's equipment specifications.

When switching between manual and automatic control, a technician must first match the process variable to the setpoint and then switch to automatic or the controller will react aggressively to match the setpoint. If the PV and setpoint are significantly different, the controller will produce an output change to the final control element that will cause a spike in the process. In some cases, this is not a problem but in others it could be problematic.

Materials Needed

a control loop or process simulator
instruction manual for the automatic controller

Safety Requirements

Safety glasses should always be worn while in the lab.

Procedure

1. Start up the process plant in the lab and identify a specific control loop to manipulate.
2. Switch the controller from automatic to manual.
   Line out the process so that the process variable is in a steady-state condition.
   With the controller in Auto, the PV indicator pointer should match the setpoint indicator pointer.
   Note the output pressure to the control valve.
   Switch auto to manual.
3. Operate the controller in Manual.
   Under dynamic conditions, the controlled variable may have changed because the control valve is now frozen in position. Therefore, adjust the manual output knob to achieve the same desired value that you had while in automatic control.
   Periodically check the process PV and tweak the output knob as needed if the PV is not at desired value.
   Adjust the output as often as needed to maintain the desired process value.
4. Switch between manual and automatic.
   Adjust the setpoint knob to place the setpoint indicator pointer to match the process PV.
   Switch from manual to automatic.
   Monitor the pressure to see if the valve changes or if the PV drifts.
   Once in auto, adjust setpoint if it is not at the desired value.
5. Make a remote setpoint adjustment.
   With the cascade controller selected to remote setpoint (control), make a setpoint adjustment.
   Change the setpoint by moving it above the measurement indicator and record what happens to the output.
   Change the setpoint by moving it below the measurement indicator and record what happens to the output.

Additional Information

All controllers have instruction manuals containing specific operating instructions. If you have a cascade control system, transferring control from local (secondary controller's setpoint) to remote (primary controller's setpoint) requires the same logical approach as transferring from manual to automatic in a single automatic controller. With the primary controller in manual and the secondary controller in local automatic, line out the primary controller's PV. Make sure the output from the primary controller (its output) is equal to the secondary controller's setpoint. Do this by adjusting the primary controller's output manually. Then switch the secondary controller from local setpoint to remote setpoint. Now the primary controller should be directing the entire cascade loop in manual. Finally, switch the primary controller from manual to automatic in the same manner mentioned in the lab procedure.

Findings

Compile your observations and readings into a report that includes your recorded data, an explanation of the results of the experiment, and a conclusion. Include a paragraph explaining the differences between a local setpoint and a remote setpoint. Be sure to discuss the importance of bumpless transfer when switching an analog controller from automatic to manual control and from manual to automatic. Follow your instructor's guidelines on how to submit these findings.

CHAPTER 11

Control Loops: Control Valves and Regulators

Objectives

After completing this chapter, you will be able to:

- Given a drawing or actual device, identify the main components of a control valve:
  body
  bonnet
  disc
  actuator
  stem
  seat
  spring
  valve positioner
  handwheel
  I/P transducer
- Given a drawing or actual device, identify and describe the following:
  current-to-pneumatic transducer
  indications of a sticking control valve
- Define terms associated with valves and other final control elements:
  "air to close" (fail open)
  "air to open" (fail closed)
  fail last/in place/as is
- Describe operating scenarios in which fail-open, fail-closed, and fail-last positions are desirable.
- Discuss the purpose of diaphragm valve actuators and piston valve actuators.
- Compare and contrast a spring and diaphragm actuator to a cylinder actuator.
- Explain why the action of a valve actuator may not correspond with the action of the valve.
- Describe a valve positioner and explain three of its uses.
- Explain the function of each of the three gauges located on a pneumatic valve positioner.
- Given a pressure indication for each of the three gauges on a valve positioner, predict what the control valve movement will be.
- Describe two ways a controller's output signal can be reversed at the valve so the valve's action is opposite of the controller output signal.
- Describe a control scheme that utilizes reversing a controller's output signal at the valve.

- Explain how reversing a controller's output signal at the valve affects the valve's fail-safe position upon loss of air.
- Explain the purpose and operation of the following:
  globe valves
  three-way valves
  butterfly valves
- Given a drawing, picture, or actual device, identify and describe pressure regulators.
- Define the following terms associated with regulators:
  back-pressure regulator (self-actuated)
  pressure-reducing regulator
- Given a process flow diagram and/or piping and instrumentation diagram (P&ID), locate and identify pressure regulators used in process control.
- Given an instrument air pressure regulator, perform the following tasks:
  blow down the regulator to check for condensate or oil
  set specific pressure for operating the final control element

Key Terms

Actuator—a device that provides motion to a valve for controlling purposes.

Back-pressure regulator—a device used to regulate and/or control the pressure of a process fluid upstream of the device location.

Body—the housing component of a valve.

Bonnet—the top portion of the valve that connects the valve to the actuator; it can be removed to allow entry into the valve body cavity; usually contains the packing box and stem mechanism.

Disc or plug—the only moveable component in the valve that is actuated to open or close the flow path through the valve.

Handwheel—an actuator accessory used to manually override the actuator or to limit its motion.

I/P transducer (current-to-pneumatic transducer)—a device that converts a milliampere signal into a pneumatic pressure.

Pressure-reducing regulator—a device used to regulate and/or control the pressure of a process fluid downstream of the device location.

Regulator—a self-contained and self-actuating controlling device used to regulate variables such as pressure, flow, level, and temperature in a process.

Seat—the stationary part of the valve trim connected to the body that comes in contact with the valve plug; when the valve plug is fully seated, the flow through the valve ceases.

Spring—the device that provides the energy to move a valve in the opposite direction of the diaphragm loading motion so that the valve can be opened and closed proportionally with the instrument signal; also provides energy to return the valve back to its fail-safe condition.

Stem—the pushing rod that transfers the motion of the actuator to the valve plug.

Valve positioner—a device used to make the valve position match the controller output signal by positioning the moving parts of a valve in accordance to a predetermined relationship with the instrument signal received from the loop controller; may also be used to adjust the position of the valve according to the specific needs or change the amount of signal needed to fully stroke the valve as in a split range application.

Introduction

Recall in Chapter 10 that a final control element is the last active device in the instrument control loop. The final control element is the device that directly controls the manipulated variable by performing a manipulating operation on the process to bring about change in the controlled variable. Of the many different types of final control elements, the most common final control element is a control valve.

In this chapter, the various components of control valves as well as their purpose and operation are covered. Simple troubleshooting tips are given as well as what happens during valve failure.

Also covered are the various actuators and valve positioners. Additionally, input and output signal differences, fail-safe positions, and special valve types are also discussed to show how various arrangements and configurations can be made to enhance the safety and operability of the process.

Instrument air is commonly used to actuate the valve actuator. This air may need to be regulated, clean, and dry. Dust can plug small orifices found in all pneumatic instruments. Moisture or water in instrument air lines can interfere with control and can freeze in cold weather. A discussion about instrument air regulators and how they are used in the operation of control valves is also included. Most examples and discussions are about sliding stem valves. Other types such as butterfly valves have important differences.

The Control Valve

Control valves are the most common final control element in the processing industry. Control valves operate the process by producing a differential pressure drop across the valve. For good control, a throttling valve must drop at least 10 percent of the process pressure in the line. Control valves regulate the flow of materials in the process. For example: material may be added to or taken away from a vessel during the opening and closing operation of a control valve.

A control valve has an actuating device or actuator. The actuator changes an instrument signal into a linear or rotary motion. This motion drives the flow-controlling mechanism (e.g., plug or disc) in a valve.

Key components of a sliding stem control valve as shown in Figure 11-1 and described below.

- The **body** of a valve is the housing component. The body of the valve is usually joined to the process by a threaded or flanged connection but in a critical situation the

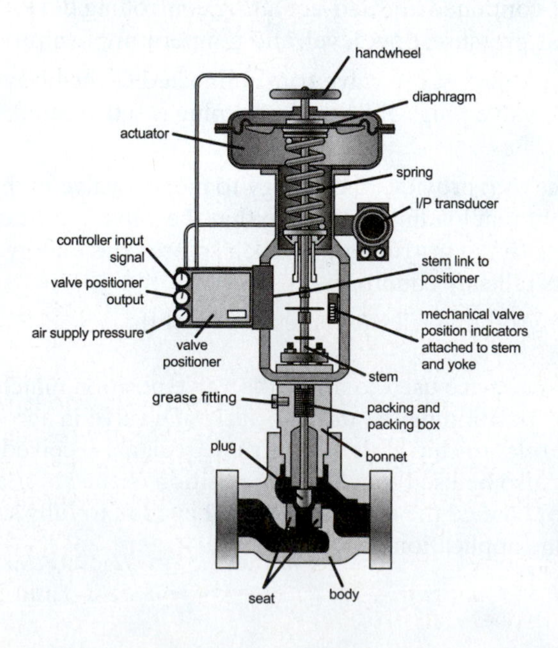

FIGURE 11-1 Control Valve Key Components

connection could be welded. Certain critical services may have welded connections that may complicate valve replacement.

- The valve **bonnet** is the top portion of the valve body and connects the valve body to the actuator and can be removed to allow entry into the valve body cavity.

- The bonnet usually contains the packing box that provides a seal around the sliding stem mechanism that connects the actuator to the valve plug.

- The valve **plug** assembly that includes the valve stem moves to open or close the flow path through the valve. As the plug moves away from the seat, the flow through the valve increases. As the plug nears the seat, flow decreases. Once the plug reaches the seat, a reasonably tight seal is formed stopping the flow through the valve. In other valves, such as the butterfly and ball valve, the plug component is referred to as a **disc** or ball.

- The **actuator** is the device that provides motion to the valve using a spring diaphragm, spring piston, or double-acting piston.

- The **stem** is the pushing and pulling rod that transfers the motion of the actuator to the valve plug.

- The **seat** in a valve is the stationary part of the valve trim connected to the body that comes in contact with the valve plug. When the plug is fully seated, the flow through the valve ceases.

- The **spring** provides the energy to move the valve in the opposite direction of the diaphragm loading motion. This provision is made so that the valve can be opened and closed proportionally with the instrument signal. The spring provides the energy to return the valve back to its fail-safe condition.

- The diaphragm is the flexible member that creates a force to move the stem.

- A **valve positioner** is actually a proportional-only controller. The position of the valve stem is sensed by a mechanical link that is directly connected to the positioner. The position of the valve stem is then compared to the value of the instrument signal and a response is produced to make the position of the valve and the signal equal.

- A **handwheel** is an actuator accessory that is used to manually override the actuator or to limit its motion. The handwheel may be located on the top of the actuator or on its side. Process technicians may manually limit or close a problem valve if it is equipped with a handwheel. Many control valves have associated handwheel inlet, outlet, and bypass valves to permit maintenance activities since these reduce installation costs.

- An **I/P** or **current-to-pneumatic transducer** is a device that converts a milliampere signal into a pneumatic pressure. The most common use for an I/P transducer is to provide the source of energy needed to drive a diaphragm or piston actuator. A current to pneumatic transducer typically receives a 4-20 mA current signal and converts it to a 3-15 psig pneumatic signal.

Control Valve Failure Conditions

Control valves are ultimately responsible for regulating the movement of fluids in a process. If there is a power or air failure, they should move to a safe position. If there is an I/P transducer before the valve, loop (current) failure should also be considered when determining valve fail position. Design engineers address these possibilities with each control valve placed into the process. Control valves are generally designed to fail in an open, closed, or last position. Actuators without return mechanisms (such as springs) typically fail in last position. External devices, such as solenoids and external air tanks, may be placed into the system to drive the control valve actuator to a predetermined intermediate position in the event of power failure or reduction in air pressure.

If a plant loses power or air supply, the design engineers must recognize the required fail-safe conditions during the design phase of a plant and choose each control valve response accordingly. Every process technician should be aware of how control valves fail under power or air supply loss. Since most control valves are pneumatically actuated, the following discussion assumes a pneumatic diaphragm actuator with a spring return device.

When using P&IDs, the control valve failure position should be indicated under the valve symbol. These symbols are further discussed in Chapter 12.

Fail Closed

When an air-to-open control valve (Figure 11-2) loses its instrument air signal or supply, the valve fails closed because a return spring provides more opposing force than the diminishing instrument air applied to the diaphragm and the force applied by the process pressure.

For example during a power failure, a valve in the controlling position that is allowing material to flow into a tank should close. If the valve stayed open, the tank could overfill if power were not restored. This requires the control valve to fail closed on loss of instrument air.

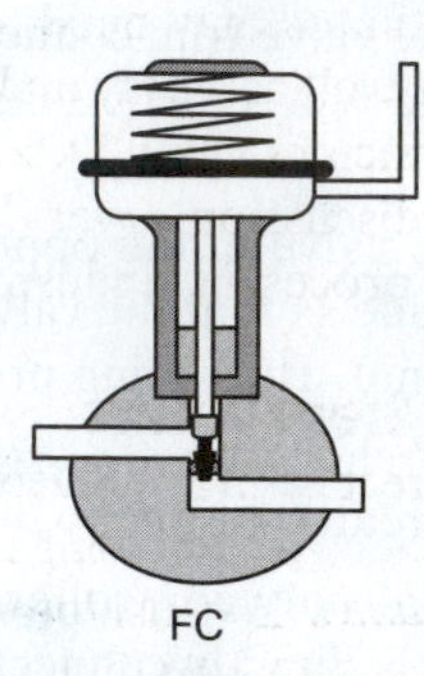

FIGURE 11-2 Fail Conditions–Air-to-Open or Fail Closed

Fail Open

When an air-to-close control valve (Figure 11-3) loses its instrument air signal or supply, the valve fails open because the return spring, once again, provides more opposing force than the diminishing instrument air applied to the diaphragm and the force applied by the process if any.

For example during a power failure, you would want a pressure-relieving valve on a reactor to fail in the open position preventing a pressure buildup that may rupture the vessel.

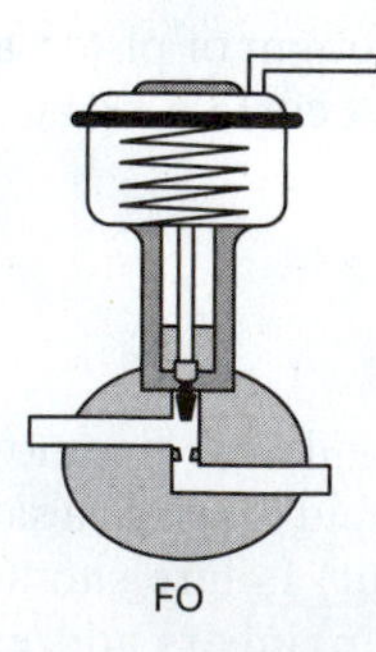

FIGURE 11-3 Fail Conditions–Air-to-Close or Fail Open

Fail-in-Place or Fail Last

Pneumatic actuators with opposing springs naturally fail in the direction of their spring tension, unless it has a lockup relay attached. A lockup relay seals in the existing signal applied to the actuator at the point of power loss. Actuators without a spring or other return mechanism usually fail in their last position (Figure 11-4) just prior to loss of power unless the process pressure is high enough to change the valve position. Electric motor actuators, for example, naturally fail in place.

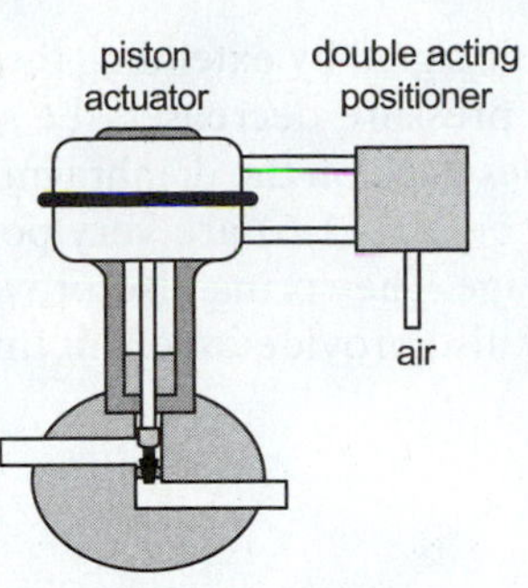

FIGURE 11-4 Fail Conditions–Fail-in-Place or Fail Last

Control Valve Actuators

An actuator is a device that responds to an applied instrument signal by creating a linear or rotational motion. An actuator is what makes a valve a control valve. By responding to an instrument signal, an actuator provides the motion necessary to throttle the valve.

In most control situations a valve and actuator may need to be selected so as to accommodate the needs of the control loop as well as safety and shutdown considerations. For this, and other reasons, process technicians should know what combinations could be expected in the field. The following discussion covers the possible combinations of actuator and valve actions used in the processing industry.

Actuator and Valve Actions for Sliding Stem Valves

To begin with, actuators and valves are either direct acting or reverse acting (Figure 11-5). An actuator is considered to be direct acting when an increasing application of pressure causes the stem to extend. Conversely, an actuator is considered to be reverse acting when an increasing application of pressure causes the stem to retract. A valve is considered to be direct acting when the valve stem is pushed down to close and reverse acting when the valve stem is pulled up to close.

There are four possible combinations of actuator and valve actions:

- direct/direct
- direct/reverse
- reverse/direct
- reverse/reverse

The easiest and probably the best way to determine what the actuator-valve combination is capable of doing is to look at the identification (ID) plate attached to the actuator. The ID plate provides information about the combined actuator and valve action (ATC/atc or ATO/ato) as well as other information.

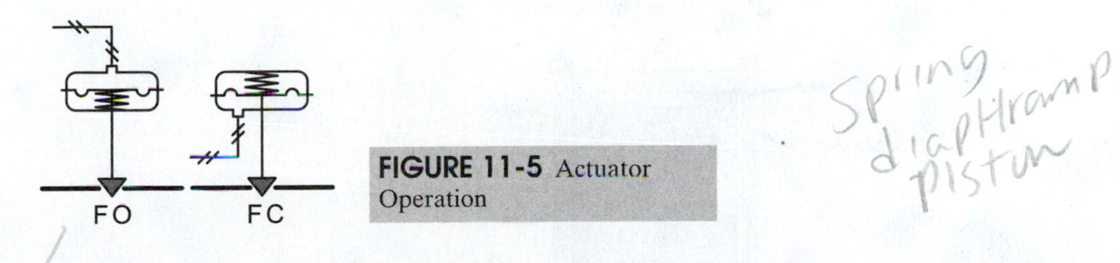

FIGURE 11-5 Actuator Operation

Pneumatically Driven Actuators

There are two major subcategories of pneumatically driven actuators: (1) the spring and diaphragm actuators, and (2) the piston type actuators.

The most common spring-and-diaphragm actuator has a single diaphragm supported by a diaphragm plate connected to a steel rod called a stem. A spring is placed on the opposite side of the plate to create an opposing return force. In a direct acting actuator, the diaphragm responds to the pneumatic signal by extending the stem component while compressing the return spring. As the pressure decreases, the spring tension pressure exceeds that of the diaphragm and pushes back on the diaphragm, retracting the stem.

Spring-and-diaphragm actuators (Figure 11-6) are very popular because of their low costs and high mechanical advantage. The spring also provides a mechanical fail-safe condition upon loss of signal. They also provide excellent throttling control with or without a positioner.

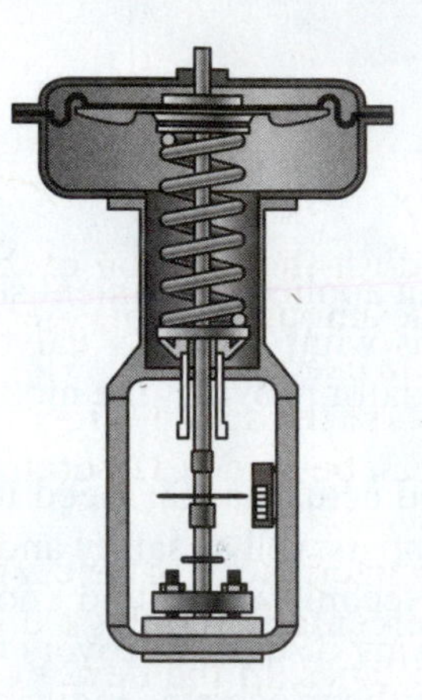

FIGURE 11-6 Spring-and-Diaphragm Actuator

Piston-type actuators fall into two major categories: single and double acting.

- In the single-acting category there are two subcategories: the spring-opposed actuators and the air-cushion actuators. In the single acting spring-opposed actuator, like the diaphragm type, the piston is opposed by a spring. In the air cushion type, the piston actuator has pressure trapped under the piston and the air compresses as the piston is pushed down. Then, as the instrument signal is reduced, the trapped compressed air pushes the piston back up.
- In the double-acting type, instrument air pressure is routed to both sides of the piston driving the stem to a required position by balancing the pressures on either side of the piston. A double-acting positioner is required to do this.

Piston actuators (Figure 11-7) tend to have longer strokes and can accept much higher input pressures. Higher pressure produces more force and provides more power, which can be important in some applications where there is a high pressure drop across the valve. Many piston-type actuators are double acting, which means that they need to be controlled with a positioner. Piston-type actuators may also be of the spring or air-cushion opposed types that provide a fail-safe condition.

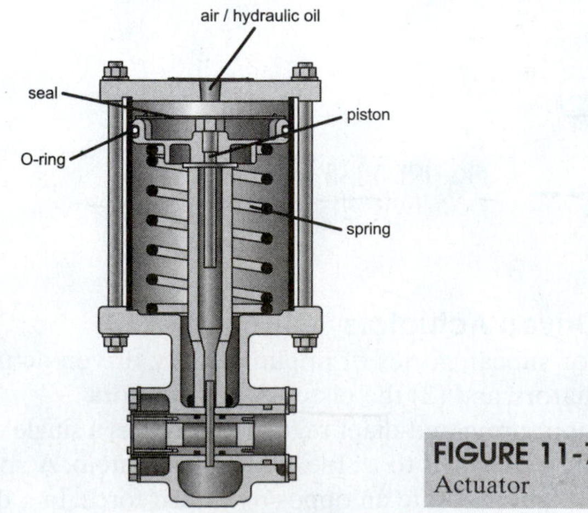

FIGURE 11-7 Piston-Type Actuator

Valve Positioners

The function of a positioner is to make the valve position match the controller output signal. The valve positioner positions the moving parts of a valve in accordance to a predetermined relationship with the instrument signal received from the loop controller. The positioner can be used to modify the relationship between the input and output instrument air signal.

The uses of a valve positioner are to:

- position the valve
- reverse the action
- mimic a valve trim type
- provide split-range control

A valve positioner may be used to adjust the position of the valve (Figure 11-8) according to specific needs or to change the amount of signal needed to fully stroke the valve. A split-range application would be an example.

A valve positioner may be used to reverse the action (Figure 11-9) of the valve. For example, an air-to-open/fail-closed valve can be made to operate as if it were an air-to-close/fail-open valve.

In old-style positioners, the trim characteristics can be changed by replacing the cam to quick opening, linear, or equal percentage. See Figure 11-10 to see the flow/percent open characteristic of these trim types. In the newer digital positioners, the trim characteristics can be programmed online and downloaded as needed.

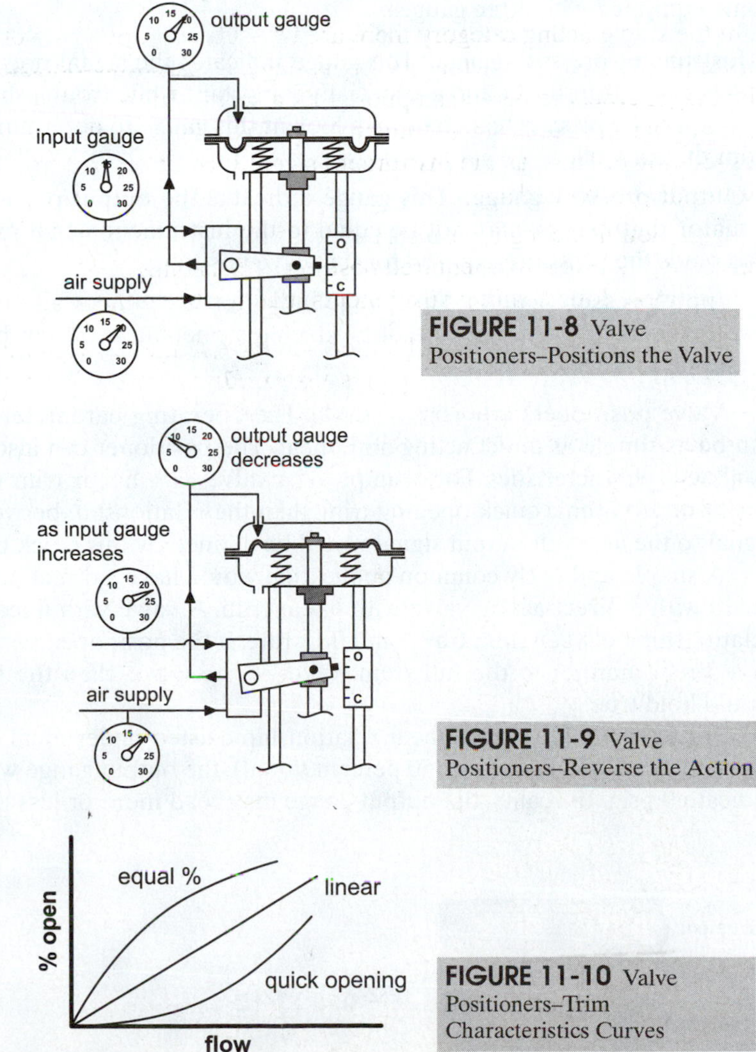

FIGURE 11-8 Valve Positioners–Positions the Valve

FIGURE 11-9 Valve Positioners–Reverse the Action

FIGURE 11-10 Valve Positioners–Trim Characteristics Curves

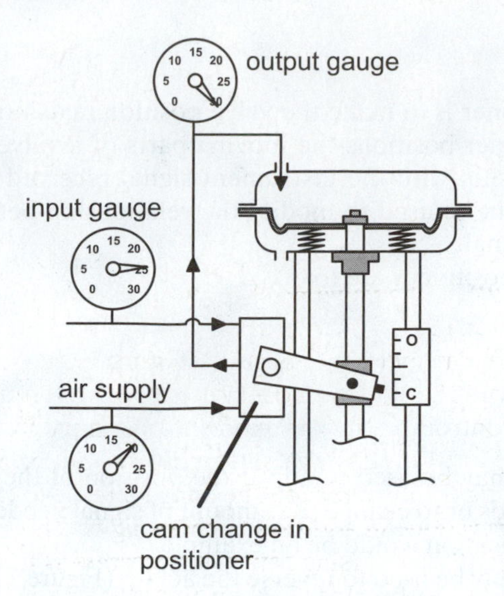

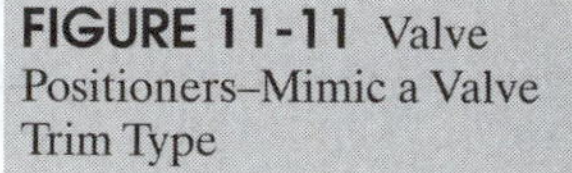
FIGURE 11-11 Valve Positioners–Mimic a Valve Trim Type

A valve positioner may also be selected to mimic (Figure 11-11) the characteristics of different valve trim types such as quick opening, linear, and equal percentage.

Valve Positioner Operation

Pneumatic valve positioners (Figure 11-12) on spring-and-diaphragm actuators may come equipped with three gauges:

- Instrument pressure gauge: This gauge indicates the signal pressure from the controller or I/P transducer and is especially important while troubleshooting the loop for problems. The process technician can look at this gauge to determine the output signal from the controller.
- Output pressure gauge: This gauge indicates the output pressure applied to the actuator that may or may not be equal to the instrument signal pressure. This gauge also reads the pressure required for the valve position.
- Supply pressure gauge: This gauge indicates instrument air supply pressure. All pneumatic instruments capable of producing a pneumatic output have an instrument air supply connected to them.

Valve positioners are very versatile. The operating parameters of the positioner can be as simple as direct acting and linear. The positioner can also be set up to have nonlinear characteristics. For example, if a valve with linear trim were set up by the positioner to mimic quick opening trim, then the relationship between the instrument signal to the actuator output signal would be defined by this quick opening property.

A simple and fairly common application would have a direct-acting actuator combined with a direct-acting valve with linear trim. A valve with linear trim has a linear relationship between stem travel and flow rate. If the positioner were set up to respond in a direct manner to the full stem travel of the valve, then the following scenario would hold true.

For example if a direct linear relationship existed under ideal conditions and the signal gauge indicated 9 psig (50 percent signal), the output gauge would probably also indicate 9 psig. In reality, the output gauge may read more or less than 9 psig because

instrument
output
supply

FIGURE 11-12 Valve Positioner Operation

the positioner drives the output to the actuator until the valve stem reaches the position equal to the instrument signal.

Output Signals

In some process control applications the output of the controller may need to be reversed at the control valve. This can be accomplished by configuring the current-to-pneumatic transducer (I/P) or the valve positioner to respond in reverse to the signal from the controller. For example, the I/P or the valve positioner would reverse an increasing signal from the controller to a decreasing signal to the control valve. In old-style controllers, this may make the controller output indication opposite to the valve closed and open position. Also note that in modern programmable controllers the output signal can be matched to the valve so that the output indicator shows 0 percent as fully closed and 100 percent as fully opened. Choosing increase to close YES or increase to close NO does this. This is done in addition to choosing reverse or direct acting.

One example of where this reversing action is necessary is in a split-range control system. If both valves need to be closed when the signal from the controller is at midrange, with one open with an increasing signal above mid-scale and the other open with a decreasing signal from mid-scale, then one valve has to respond in a reverse manner to the signal from the controller. Or, choose the first valve to fail closed (calibrated to 9.5–15 psig; close-to-open) and the second valve fail open (calibrated 8.5–3 psig); close-to-open).

Since a spring and diaphragm actuator is designed with a particular fail-safe position, this type of actuator is forced into the design fail-safe position if instrument air is lost. Therefore, whether or not the output of the controller is reversed does not matter since the actuator fails into the position demanded by the opposing spring.

Types of Control Valves

The following are the main types of valves used in control configurations:

- globe
- three-way
- butterfly
- ball or segmented ball

Globe Control Valves

The globe-style valve body is the most common type of valve used in the processing industry. The plug and seat, often called valve trim, are located within the inner cavity, or body, of the valve and provide an inlet and outlet connection. The globe valve gets its name from its globular-shaped appearance. Although a true globe shape is more like a sphere, any reasonably or somewhat rounded valve body style usually falls into this category. Flow through a globe valve body changes direction (Figure 11-13).

The globe valve, like all valves, controls the flow of material through its inner cavity with its plug and seat components. Control is accomplished with one or more ports.

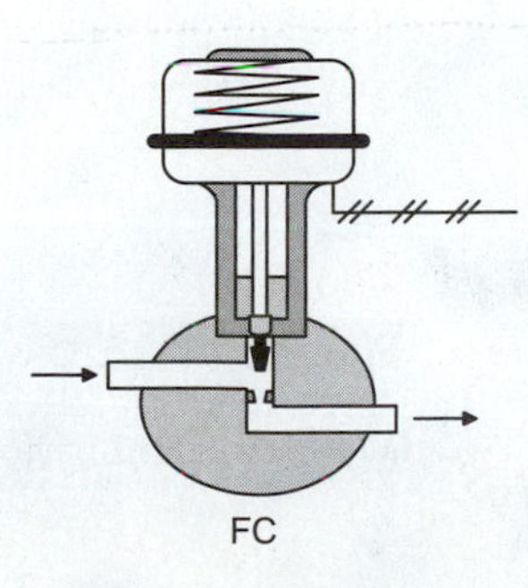

FIGURE 11-13 Globe Control Valve–Single Port

A valve port is the restrictive orifice in a valve defined by the diameter of the seat. The single-port valve is the most common of the globe body valves and a variation of this valve, a cage-style valve, is the single most popular control valve in the processing industry. Double-ported valves were developed to balance the high differential pressure acting on the single-port valves. The double-seated valve generally has a higher flow capacity than a single-ported valve of the same size and requires less actuator force to drive the stem. The problem with double-ported valves is that they tend to leak because there are two sets of plugs and seats that must shut at the same time. Unfortunately, over time with the inevitable uneven wearing across the two ports, this becomes almost impossible.

Three-Way Control Valves

A three-way valve is a special type of globe body valve that has three connecting ports instead of two. They are designed to either mix (Figure 11-14) two flowing streams together or divert one flowing stream between two output ports. In the diverting three-way valve, there is one inlet and two outlets and in the mixing three-way valve there are two inlets and one outlet.

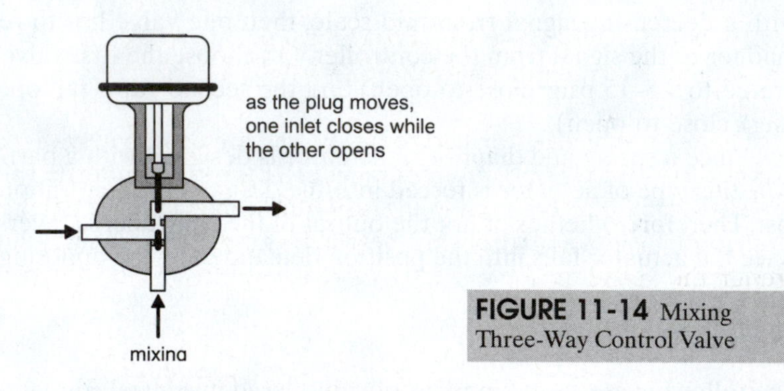

FIGURE 11-14 Mixing Three-Way Control Valve

The diverting valve (Figure 11-15) could be used as a switching valve diverting a flowing stream from one vessel to another, or as a temperature control valve diverting part of or the entire process stream into a heat exchanger. The mixing three-way valve could be used for blending two separate streams into one, producing a proportioned mixture of the two.

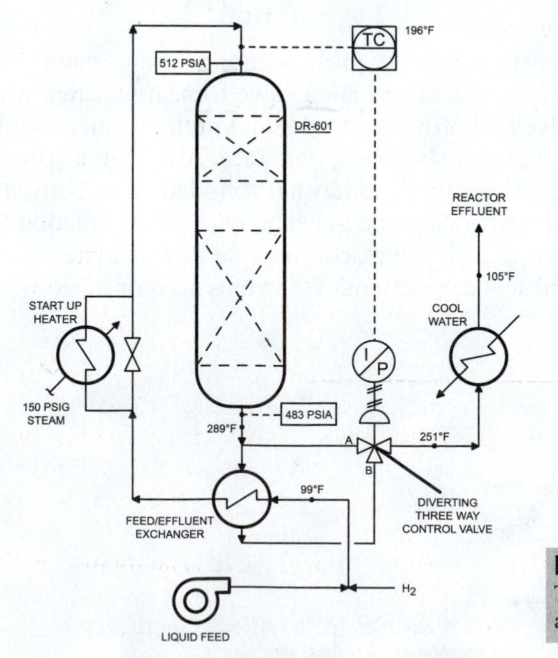

FIGURE 11-15 Diverting Three-Way Control Valve for a Hydrotreater Reactor

As shown in Figure 11-15, if the reactor inlet temperature drops, the diverting three-way control valve closes port A and opens port B to put hotter reactor effluent to the exchanger. This increases reactor inlet temperature.

Butterfly Control Valves

Butterfly valves (Figure 11-16) are on the rise as compared to the predominant globe-style valves. The reason for their greater acceptance in the processing industry is their lower manufacturing costs and higher flow capacities. Also, the rotary stem has less wear on the packing than the sliding stem. Actuators used with butterfly valves use a rotary motor.

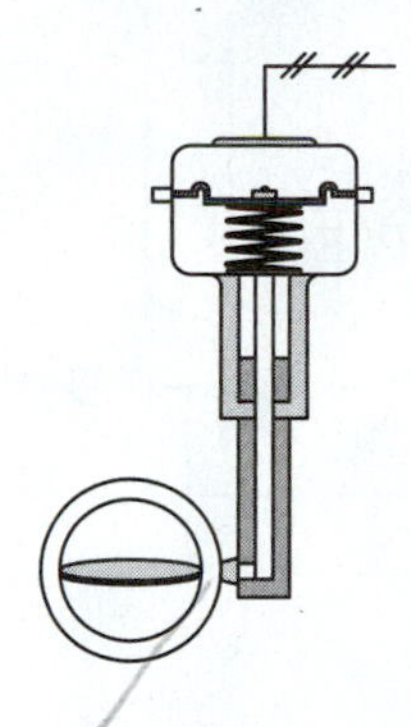

FIGURE 11-16 Butterfly Valve–Spring Diaphragm

Butterfly valves, also called rotary valves, are equipped with piston-type actuators (Figure 11-17) or spring-and-diaphragm actuators. The butterfly valve has flow characteristics somewhere between linear and quick opening which makes this type of valve exhibit a nonlinear relationship between the percent opening and the rate of flow through the opening. The butterfly valve is used to control all types of fluids including both liquids and gases.

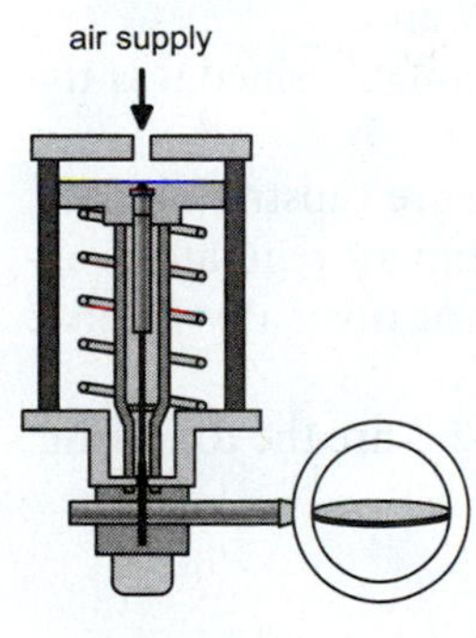

FIGURE 11-17 Butterfly Valve–Spring Piston

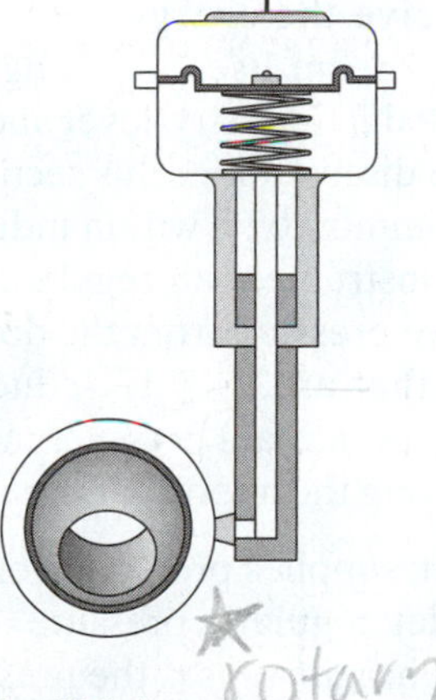

FIGURE 11-18 Ball Control Valve–Spring Diaphragm

Ball or Segmented Ball Control Valves

The full ball control valve (Figure 11-18) is a rotary valve that contains a spherical plug. The control valve actuator rotates the plug to control the flow of fluid through the valve body. By comparison, a segmented ball valve has one edge either contoured or having a V-shaped edge to yield a desired flow characteristic.

Ball control valves may be used as a tight shut-off or as a modulating valve offering high flow capacity because there are no internal obstructions when the valve is fully open. Ball valves are also often used in applications where the process fluid is slurry to minimize the settling and straining of these materials.

The spherical plug of the ball control valve is adaptable to function in three-way service (Figure 11-19). The control valve in the effluent from the Timtene reactor has a single, hot inlet with two outlets. One inlet admits hot effluent to heat up colder reactor

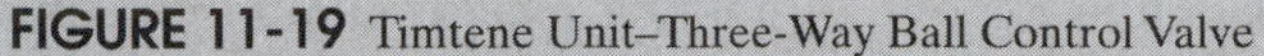

FIGURE 11-19 Timtene Unit–Three-Way Ball Control Valve

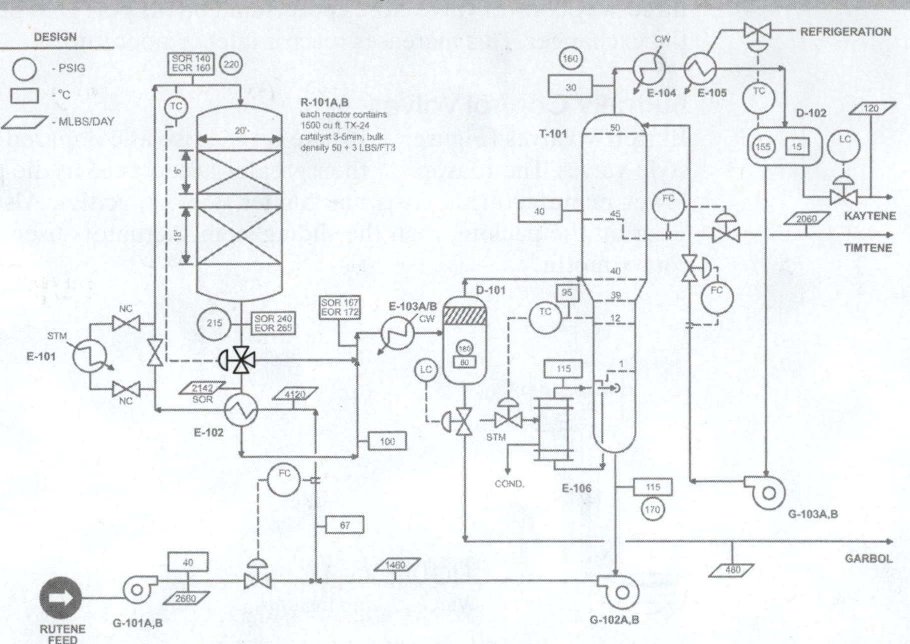

feed while the other bypasses unneeded hot material directly to cooling. Rotation of the ball allows more or less hot effluent to heat the feed (depending on the inlet temperature controller) while bypassing the remainder to cooling.

Instrument Air Regulators

A **regulator** is a self-contained and self-actuating controlling device used to regulate variables such as pressure, flow, level, and temperature in a process. Regulators generally receive the energy necessary to move the internal valve mechanism from the process stream itself. Although pressure and flow regulators are the most common types used in industry, level and temperature regulators are also found.

The discussion in this section is limited to the air pressure regulator since it is the more common type within industry.

An instrument air regulator can either reduce the supply pressure (upstream side), or relieve pressure from the downstream side. Unlike the instrument air regulator, regulators that are used to reduce pressure in a process line cannot normally release excess downstream pressure into the atmosphere.

Among the important parts of a pressure regulator (Figure 11-20) are the following:

- inlet: supplies pressure from a source
- outlet: regulates pressure
- diaphragm: senses the pressure on the outlet side of the regulator
- pilot valve assembly: working part of the regulator; entire mechanism works to both supply pressure or relieve pressure from outlet side

Instrument air regulators, such as a Fisher Model 67, drop the pressure received from the instrument air distribution system (usually 80–150 psig) down to a supply pressure range that most pneumatic instrumentation is capable of tolerating. The supply pressure applied to pneumatic instruments such as transmitters, controllers, and control valves cannot exceed about 25–30 psig and many of them recommend a supply pressure of around 20 psi. In an electronic control loop, the I/P transducer also needs a regulated supply pressure in this range.

An instrument air regulator action follows this sequence:

1. The handwheel is turned.
2. The spring compresses.

FIGURE 11-20 Regulator Cutaway

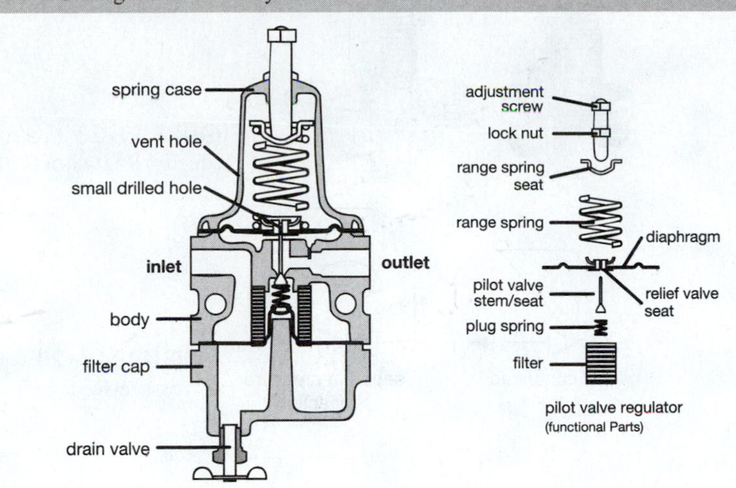

3. The diaphragm is pushed down.
4. The pen and plug move from their seat.
5. Air rushes in increasing downstream pressure (downstream of the regulator).
6. The output gauge responds accordingly.
7. Equilibrium is established.

A pressure control system would include a sensing device, a controlling device, and a final controlling element. Since the regulator is a self-contained controlling device, it has these elements integrally located inside.

The two major types of pressure regulators are the back-pressure regulator and the pressure-reducing regulator.

Back-Pressure Regulators

A **back-pressure regulator** (Figures 11-21 and 11-22) is a device used to regulate and/or control the pressure of a process fluid upstream of the location of the regulator. For example, this type of regulator is used to maintain the pressure in the vapor space of a vessel. Most sealed tanks in industry have a pressure control system associated with them.

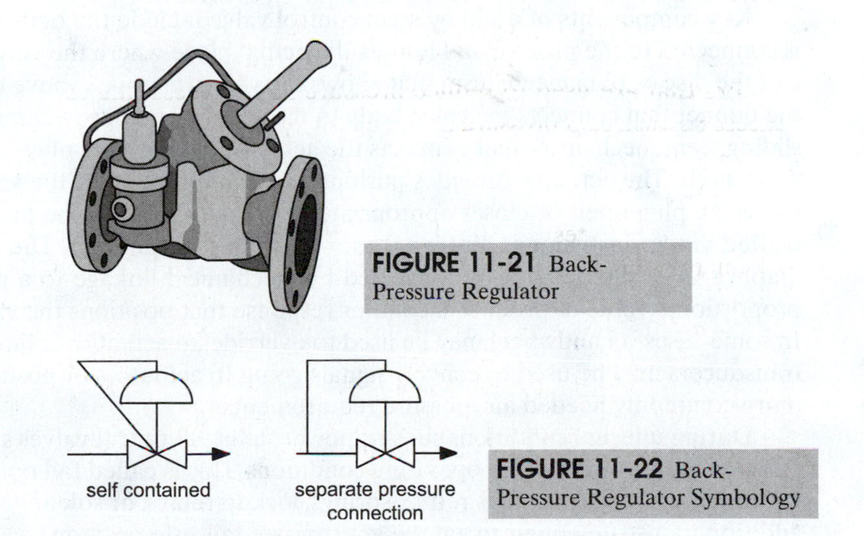

FIGURE 11-21 Back-Pressure Regulator

FIGURE 11-22 Back-Pressure Regulator Symbology

Pressure-Reducing Regulators

The other major type of pressure regulator is the **pressure-reducing regulator** (Figures 11-23 and 11-24). This device is used to control the pressure of a process fluid *downstream* of the location of the regulator.

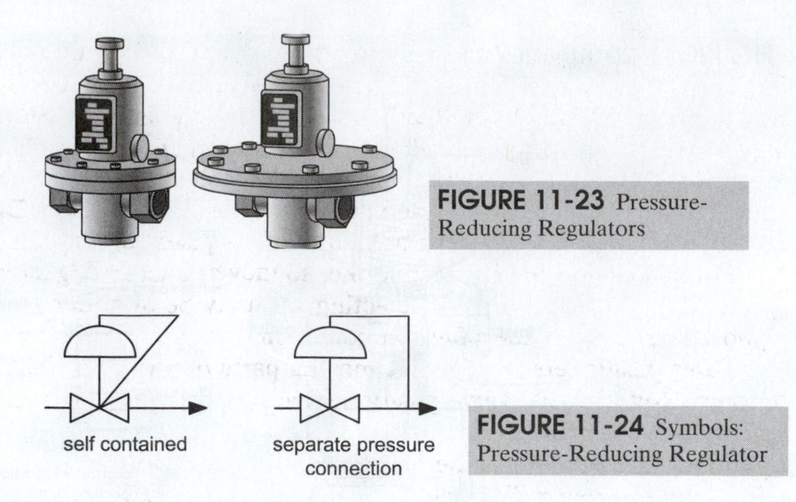

FIGURE 11-23 Pressure-Reducing Regulators

FIGURE 11-24 Symbols: Pressure-Reducing Regulator

An application where a pressure-reducing regulator might be used is where steam leaves a boiler and could potentially have hundreds of pounds of pressure that need to be reduced to specific unit requirements that may only be 150 psi. This type of regulator is called a steam pressure-reducing regulator. An instrument air regulator is also a pressure-reducing regulator.

Notice that in both the back-pressure and pressure-reducing regulator symbols there are two distinct sensor points. One is internally sensed and the other is externally sensed. In both cases, the pressure reduction occurs at the regulator.

Earlier discussions in this chapter mentioned that a regulator is self-contained. The external sensing point does not affect the basis for determining the regulator self-contained characteristic since the regulator still gets its manipulating power from within the regulating system.

Summary

The valve itself is not controlled unless it has some type of actuator and positioning arrangement. Actuators begin the movement of the valve either to an open or closed position while the positioner determines the percent open. The movement of the valve may be predicted by observing the three associated pressure gauges.

Key components of a sliding stem control valve include the body of the valve that is connected to the process and houses the actual place where the valve seat is located and the disc or plug mechanism that allows the valve to move. Above the valve body is the bonnet that connects the valve body to the actuator and provides a seal around the sliding stem mechanism that connects the actuator to the valve plug located inside the valve body. The actuator provides pushing or pulling motion to the valve stem so that the valve plug open or closes appropriately. The actuator may be pneumatically controlled via an instrument air signal to a moveable diaphragm. The pressure on the diaphragm in the actuator is converted by mechanical linkage to a positioner into a proportional value to produce a counter response that positions the valve accordingly. In some cases, a handwheel may be used to override an actuator or limit its motion. I/P transducers may be used to convert signals going to actuators or positioners into their more commonly needed air pressure requirements.

During unusual conditions such as power failures, control valves should move to a position that promotes safe operating conditions. This is called fail open, fail closed, or fail in place. Devices such as return springs, lockup relays, or solenoids may be used in addition to instrument air to aid the appropriate fail-safe position.

Actuators respond to instrument air signals by creating either a linear or rotational motion on the stem that makes a valve either open or close. How the stem moves, either direct or reverse, when increasing pressure is applied, determines the type of actuator. Since there are four potential types of actuator-valve combinations, the best way to determine their capability is to look at the valve ID plate.

The two major subcategories of pneumatically driven actuators are the (1) spring-and-diaphragm actuators and the (2) piston-type actuators. Spring-and-diaphragm actuators are popular due to their lower costs and higher mechanical advantage allowing them to provide a mechanical fail-safe condition upon loss of signal with or without a positioner. Piston-type actuators may be either single or double-acting. Single-acting may be either of the spring-opposed or air-cushion subcategories. Double-acting types allow instrument air pressure to both sides of the piston driving the stem.

Piston actuators have longer strokes so they can accept higher input process pressures. Many of these are double acting and may be of either spring or air-cushion opposed types to provide a fail-safe condition.

Valve positioners position the moving parts of a valve in relation to the control loop controller signal. Valve positioners are used to position the valve, reverse the action, or mimic a valve trim type.

Pneumatic valve positioners may come equipped with three gauges: (1) instrument pressure gauge that indicates the signal pressure from the controller, (2) output pressure gauge that may or may not be equal to the instrument pressure, and (3) the supply pressure gauge that indicates the instrument air supply pressure.

In some conditions, the output of a controller may need to be reversed at the control valve. The valve positioner can be configured to respond in reverse to the signal from the controller as in a split-range control system. The fail-safe condition due to failure of the instrument air is controlled by the opposing spring in the actuator.

The main types of valves used in controlling operations are globe, three-way, butterfly, and ball or segmented ball. Globe valves may have one or two ports depending on the particular design and are usually shaped like a globe. Three-way valves are also globe valves, but they have three ports. This type is used for mixing (two inlets–one outlet) or diverting (one inlet–two outlets). Butterfly valves have lower manufacturing costs and allow higher flow capacities for both liquids and gases. Ball and segmented ball may be used where a tight shut-off or modulation is needed.

While there are many types of regulators used within processing operations, this chapter limited the discussion to instrument air regulators, which have the ability to regulate the amount of supply pressure on the upstream side of the regulator as well as discharge pressure on the downstream side. The two main types are the back-pressure regulator and the pressure-reducing regulator and both types are self-contained mechanisms.

Checking Your Knowledge

1. On the following diagram, identify the following:
   a. Handwheel
   b. Actuator
   c. Diaphragm
   d. Spring
   e. I/P Transducer
   f. Stem
   g. Bonnet
   h. Plug or disc
   i. Body
   j. Seat
   k. Valve Positioner
   l. Packing and Packing Box

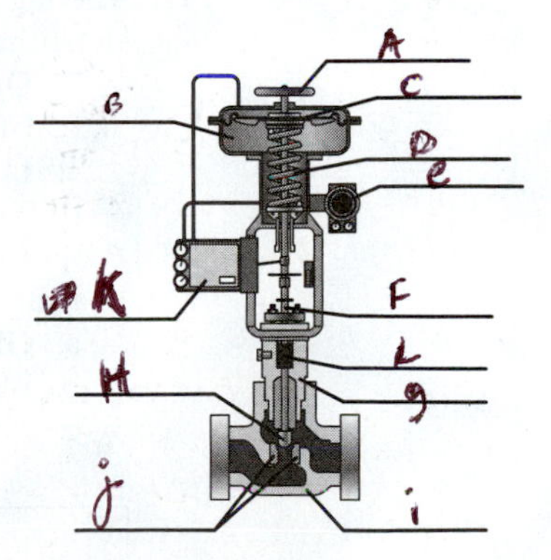

2. The stem on a control valve that is sticking _____ rather than moves smoothly.
   a. jumps
   b. twists
   c. pivots
   d. bends
3. A control valve with a spring-and-diaphragm actuator is said to be air-to-close; if there is an air or power loss the valve will:
   a. Be forced completely closed by the spring
   b. Be forced completely open by the spring
   c. Remain in its last position
   d. Hunt for the best option

4. During a power failure, you want a valve that is allowing the addition of material to a tank to stop the flow. Therefore, this valve should be a:
   a. Fail-closed valve
   b. Fail-open valve
   c. Fail-in-place valve
   d. It doesn't matter; the power is off.
5. A(n) _____ is a device that reacts to an instrument signal by creating linear or rotational motion.
   a. modulator
   b. agitator
   c. controller
   d. actuator
6. *True or False* A double-acting cylinder-type actuator always requires a positioner.
7. If a control valve and its spring-and-diaphragm actuator are both direct acting, then the control valve will _____ if air is lost to it.
   a. Fail open
   b. Fail closed
   c. Fail intermediate
   d. None of the above
8. Which of the following can a valve positioner do?
   a. Position the valve stem in reference to the instrument signal
   b. Reverse the direction of flow through the valve
   c. Reverse the action of the signal received from the controller
   d. None of the above
9. What are the three pressure gauges on a pneumatic valve positioner?
   a. Signal, output, and supply
   b. Signal, input, and supply
   c. Input, output, and valve
   d. Input, output, and position
10. *True or False* A direct-acting positioner on a direct actuator and valve with linear trim will have a linear relationship between the instrument signal and the valve stem position. That is, under ideal conditions, an instrument signal of 9 psi applied to the positioner produces an output pressure of 9 psi to the actuator.
11. The output signal from a controller can be reversed by which of the following?
    a. A regulator
    b. Reversing the lead wires
    c. The valve positioner
    d. None of the above
12. The tag on a pneumatically actuated control valve identifies it as air-to-open. If the positioner has been configured to reverse the signal, then the valve will fail _____.
    a. Open
    b. Closed
    c. In its last position prior to the loss of air
    d. None of the above
13. Properly identify each valve by putting the correct letter of appropriate name with its image.
    a. Globe body valve
    b. Butterfly valve
    c. Three-way valve

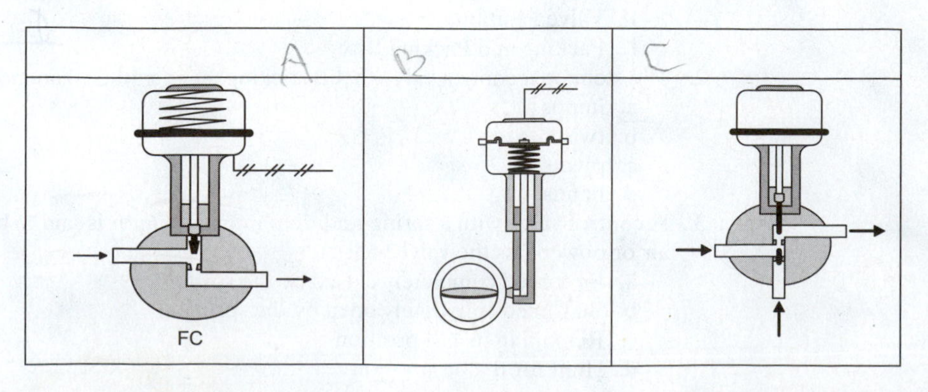

14. In the processing industry, the most common control valve body style is the _____.
    a. Butterfly
    b. Three-way
    c. Globe
    d. Double-port
15. A butterfly valve is opened with a _____.
    a. Linear motion
    b. Rotary motion
    c. Sliding-stem
    d. Flipper modulator
16. A three-way valve can be used to _____. (Select all that apply)
    a. Divert a flowing stream into two separate pipes
    b. Mix two separate flowing streams together into one pipe
    c. Control the amount of air flowing into a furnace
    d. Replace two valves

Student Activities

1. Using control valve final control element photograph copies from Chapter 10 Activities, identify the various types of control valves, actuators, positioners, etc. on each picture. If no activities of this sort were accomplished during the Activities section of Chapter 10, then using your previous textbooks or library books, photocopy several control valves and perform the same. Other sources include internet downloads, control valve vendor brochures, and/or engineering firm location photograph with appropriate instrumentation.
2. In small groups, make answer keys for the photographs used in the item above. Test each other's recognition of each of the instruments. Compare and contrast various manufacturers and configurations of control valves, actuators, positioners, and then instrument air regulators.
3. Identify/locate various types of control valves and regulators in the lab.
4. Stroke an assembled control valve with instrument air to check P versus % open.
5. Given several pressure indications for a valve positioner, work in small groups to determine how the valve will react.
6. Describe operating scenarios in which fail open, fail closed, and fail last positions are desirable.
7. Using a simple control loop system within a pilot plant, change a variable (e.g., tank level). Record the mA meter signal, then the controller output, then the air signal from the transducer, and record how the valve moves. Record the readings at several different milliamp and air points between 0 and 100%, 4 to 20 milliamps, and 3 to 20 psig on the air signal. If pilot equipment is not available, use graphics for each of the changes. Calculate the milliamp and air pressure values based on the relationship between the two and determine the expected valve movements and positions.

CHAPTER 12

Symbology: Process Diagrams and Instrument Sketching

Objectives

After completing this chapter, you will be able to:

Process Diagrams

- Describe the types of petrochemical and refining industry drawings that contain instrumentation.
- Describe the lettering and numbering standards based on ISA instrumentation symbols.
- Describe how to determine the instrument type from the symbol information.
- Describe the standards for line symbols.
- Using a legend, correctly identify instrumentation on a drawing.
- Compare and contrast P&IDs and PFDs.
- Given a PFD, a P&ID, and a symbols chart, locate and identify the following instruments:

  field-mounted instruments
  board-mounted instruments
  DCS-mounted instruments
  Diaphragm-actuated valve
  piston actuated valve
  motor-operated valve
  control valve
  three-way valve
  transducer
  equipment
  indicator
  recorder
  blind controller
  indicating controller
  recording controller
  instrument air supply and signal
  electronic signal
  capillary tubing
  process piping
  line marking
  revision block

continuation arrows
title block
clouds (revision/abandon-in-place)
notes
reference block
variables (measured and manipulated)
control loops

- Given a P&ID, trace a control loop.

Instrumentation Sketching

- Given a P&ID, explain the relationship of one piece of instrumentation to another.
- Given a drawing that has major system equipment, add control loops:

flow
level
temperature
pressure

- Sketch instrumentation control loops on available trainer resources.

Key Terms

Balloon—the basic instrumentation symbol.

Basic equipment symbol—common equipment such as pumps, towers, furnaces, etc. are basic pieces of equipment for most processing facilities and have commonly recognizable, or basic, equipment symbols.

Block flow diagram (BFD)—a flow scheme in a simple sequential block form.

International Society of Automation (ISA)—a global, nonprofit technical society that develops standards for automation, instrumentation, control, and measurement.

Legend—an explanation of what the symbols and codes on a drawing represent; usually located on an individual drawing in a framed area or on a page within a set of drawings.

Line symbols—connectors between the basic pieces of equipment without which process streams could not be moved.

Piping and instrumentation diagram (P&ID)—contains more detail than a PFD to include piping and instrumentation details and the entire control system.

Process flow diagram (PFD)—a pictorial description of an actual process including the major process equipment while providing process information including the heat and material balances; usually developed when initiating the design of a new plant.

Symbology—various graphical representations used to identify equipment, lines, instrumentation, or process configurations

Introduction to Process Diagrams

The three most likely drawings that a process technician uses are the following:

- block flow diagram (BFD)
- process flow diagram (PFD)
- piping and instrumentation diagram (P&ID)

A **block flow diagram (BFD)** (Figure 12-1) shows the flow scheme in a simple sequential block form. Not all, but most block flow diagrams show flow from left to right and tend not to cross over lines any more than necessary.

A **process flow diagram (PFD)** (Figure 12-2) pictorially describes the actual process, including the major process equipment, and may provide process variables as well as heat and material balance information. This is one of the first documents developed

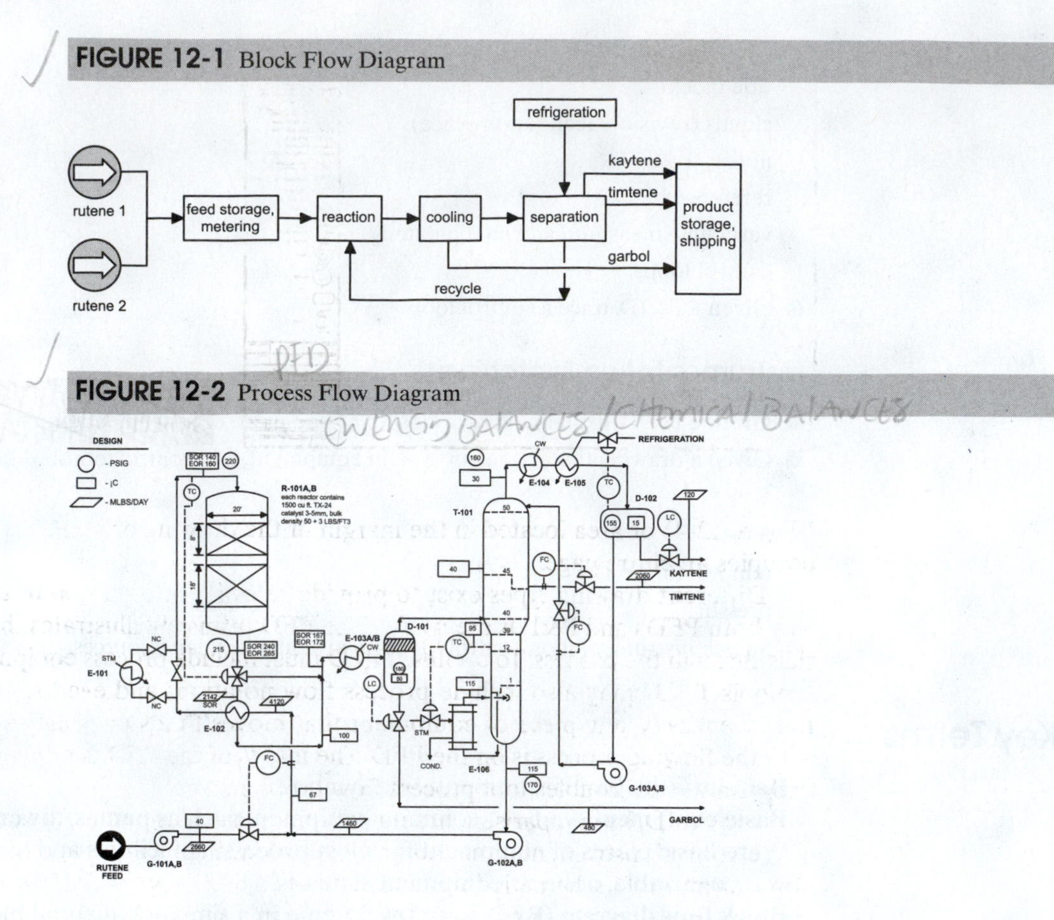

FIGURE 12-1 Block Flow Diagram

FIGURE 12-2 Process Flow Diagram

when initiating the design of a new plant. The material balance is used in all further flow calculations including main process pumps and compressors, vessel sizing, etc. The PFD contains major controls but only those block valves necessary for understanding (e.g., valves at startup heater E-101 in Figure 12-2).

A **piping and instrumentation diagram (P&ID)** (Figure 12-3) is similar to a PFD but contains no process information but much more detail including instrumentation and the entire control system. These drawings provide the basic mechanical design details and operating philosophy for the plant.

P&IDs are usually filed within a drawing package when a plant or new process is designed. Other drawings that come in the complete drawing package that may be useful to the process technician are the site plan and utility drawings as well as the more complete instrument drawings called the loop diagrams.

All PFDs and P&IDs should have an associated legend. A legend is an explanation of what the symbols and codes represent. The legend may be located in a small box

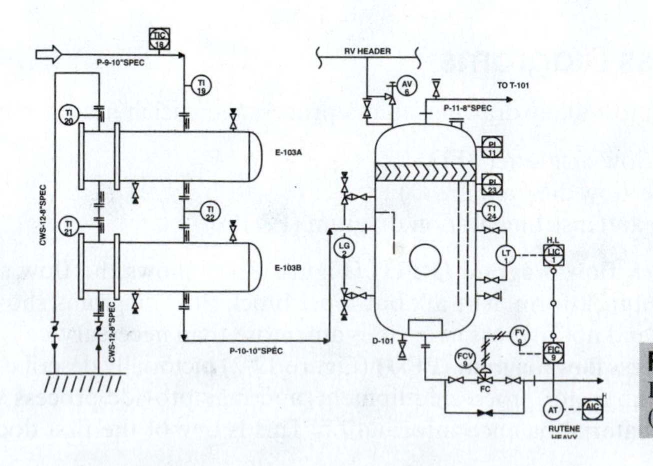

FIGURE 12-3 Piping and Instrumentation Diagram (P&ID)

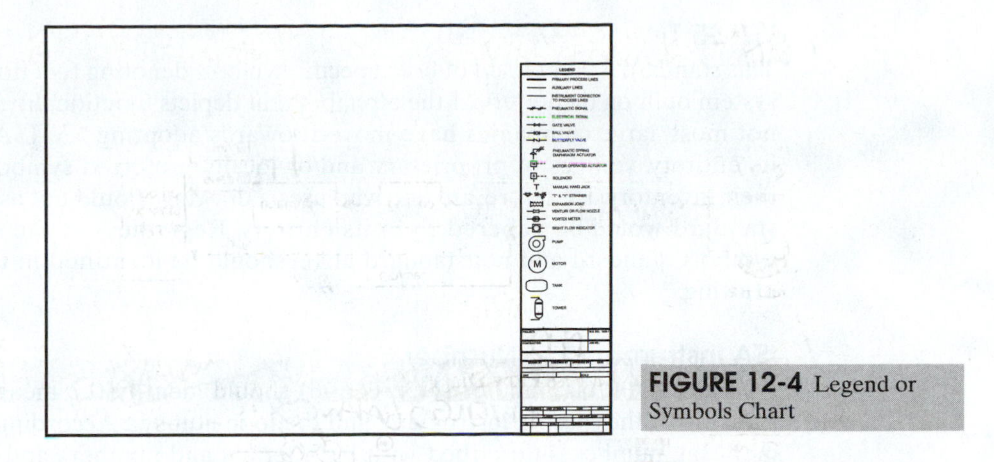

FIGURE 12-4 Legend or Symbols Chart

(Figure 12-4) or area located in the margin of the drawing or it may be so large that it occupies an entire page.

Different drawing types exist to provide relevant information to the user. That is why both PFDs and P&IDs are useful. The PFD primarily illustrates the flow of materials through the process. To do this, a PFD must include process equipment and piping symbols. PFDs may also include process flow notations and even some instrumentation. Generally, any piece of equipment that moves fluids or comes in direct contact with the flowing process is on the PFD. The intent of the PFD is to help the technician understand and troubleshoot process flow problems.

The P&ID, by comparison, has most of the same items as a PFD with the addition of the control instrumentation and considerable mechanical details. The instrumentation is drawn in a schematic form to illustrate functionality. A P&ID shows the entire control loop in proximity to the field instrumentation. Again, this is a schematic representation of the loop, not a drawing. A P&ID does not represent the actual physical placement of the components as they are situated within a plant or unit. P&IDs help technicians understand how the control instruments are interconnected and function together with respect to the process. By understanding how the control loop(s) function, a technician becomes a more capable operator and troubleshooter.

Introduction to Symbology

Throughout the years of modern processing design, a system of symbols has been utilized to streamline the various drawings that are used to describe how both equipment and its associated instrumentation are interconnected. While various engineering and/or chemical companies have striven to make these symbols standardized within their own facilities or companies, there have been others that have formed organizations or societies to address issues across various process industries. For instrumentation, the **International Society of Automation** (ISA) (Figure 12-5) is the dominant source and specifically its **symbology** Standard 5.1.

FIGURE 12-5 ISA–International Society of Automation

ISA S5.1

This standard is comprised of both specific symbols denoting functionality and a coded system built on the letters of the alphabet that depicts functionality. Although many, if not most, large companies have moved towards adopting the ISA S5.1 standard in its entirety, vestiges of proprietary and/or locally preferred symbols may be kept in their inventory. Therefore, anyone who uses a drawing should not assume that the ISA standard would be adhered to in its entirety. Regardless of the standard used, all symbols, standard and nonstandard alike, should be identified in the legend of each drawing.

ISA Instrument Tag Number

An instrument tag number (Figure 12-6) should identify the measured variable, the function of the specific instrument, and the loop number. Accordingly, the ISA instrument tag number is described with both letters and numbers and should be unique since most plants now use a global database to identify devices.

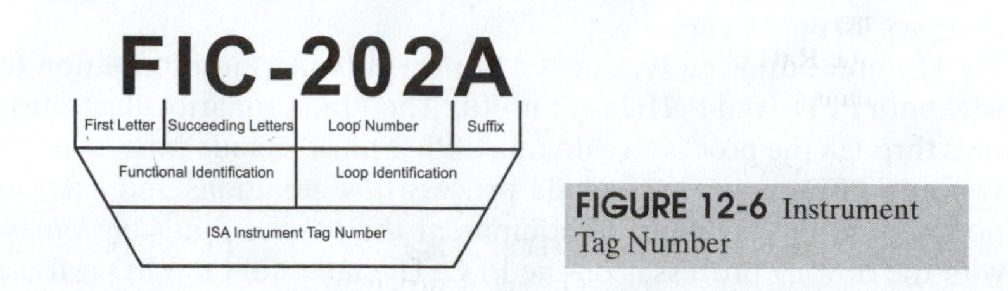

FIGURE 12-6 Instrument Tag Number

The first letter identifies the measured or initiating variable and the following or succeeding letters describe the function of the instrument. For example, in Figure 12-6 above, "F" stands for flow, "I" for indicate, and "C" for control. That makes this instrument a "flow-indicating controller," or a controller that is controlling flow and has an indicator on its faceplate.

Loop numbers are unique numbers assigned locally by the plant-engineering group or by the engineering firm that has been hired to produce the drawings. In either case, every instrument in the loop exclusively shares the loop number. If a loop has more than one instrument with the same functional identification, then a suffix is added to the end of the loop number. For example, a split-range control loop has two control valves. One should be designated with the suffix "A" (TV-202A) and the other with the suffix "B" (TV-202B). A multipoint temperature recorder (a temperature-recording device that records numerous temperature trends according to a printed number and/or color) would instead have numerical suffix identifications, such as TE-43-1, TE-43-2, etc. Suffix identifiers are the exception rather than the rule. When encountered, you need to know what they are expressing in terms of the physical loop instrumentation.

The ISA Functional Identification Table (Table 1 in the ISA S5.1 standard) lists the first letter with its possible modifiers and the succeeding letters including the passive or readout function column, the output function column, and possible modifiers associated with the succeeding letters (Table 12-1). The following information is taken from that table and should always be referenced back to the original ISA table for verification. The original table in the ISA publication also lists reference numbers where additional explanatory information may be found within Section 5.1.

Instrument tag examples can be interpreted by using the functional identification table shown in Table 12-2.

Instrumentation Symbols Chart

A review of **basic equipment symbols** is important since instruments are used to determine pressure, temperature, level, flow, speed, and position as well as various analytical activities on process streams going into or out of these basic pieces of equipment. (Table 12-3) Instrumentation is the control systems used to monitor, maintain, and

TABLE 12-1 ISA Table 1–Identification Letters

	FIRST-LETTER		*SUCCEEDING-LETTERS*		
	Measured or Initiating Variable	***Modifier***	***Readout or Passive Function***	***Output Function***	***Modifier***
A	Analysis		Alarm		
B	Burner, Combustion		User's Choice	User's Choice	User's Choice
C	User's Choice			Control	
D	User's Choice	Differential			
E	Voltage		Sensor (Primary Element)		
F	Flow Rate	Ratio (Fraction)			
G	User's Choice		Glass, Viewing Device		
H	Hand				High
I	Current (Electric)		Indicate		
J	Power	Scan			
K	Time, Time Schedule	Time Rate of Change		Control Station	
L	Level		Light		Low
M	User's Choice	Momentary			Middle, Intermediate
N	User's Choice		User's Choice	User's Choice	User's Choice
O	User's Choice		Orifice, Restriction		
P	Pressure, Vacuum		Point (Test) Connection		
Q	Quantity	Integrate, Totalize			
R	Radiation		Record		
S	Speed, Frequency	Safety		Switch	
T	Temperature			Transmit	
U	Multivariable		Multifunction	Multifunction	Multifunction
V	Vibration, Mechanical Analysis			Valve, Damper, Louver	
W	Weight, Force		Well		
X	Unclassified	X Axis	Unclassified	Unclassified	Unclassified
Y	Event, State or Presence	Y Axis		Relay, Compute, Convert	
Z	Position, Dimension	Z Axis		Driver, Actuator, Unclassified Final Control Element	

TABLE 12-2 Instrument Tag Examples

Letters	*Functional Interpretation*
P	Pressure
T	Temperature
F	Flow
L	Level
E	Element
I	Indicator
C	Controller
CV	Control Valve
Y	Transmitter/Transducer
R	Recorder
PT	Pressure Transmitter
TT	Temperature Transmitter
FRC	Flow Recording Controller

TABLE 12-2 Continued

PIC	Pressure Indicating Controller
LV	Level Valve (preferred way of identifying a control valve in a loop; may also be expressed as PV, FV, TV)
PY	Pressure Relay or Compute (convert) (e.g., could be an I/P transducer in a pressure loop)
TE	Temperature Element (e.g., could be a thermocouple, RTD, or filled thermal system)
LI	Level Indicator
PC	Pressure Controller (since this controller does not have an indicator or recorder function, it would probably be behind the panel out of the sight of the operator)
FFIC	A Flow (Ratio) Indicating Controller

TABLE 12-3 Basic Equipment Symbols

Equipment Symbol	*Equipment Name*
	Tank
	Heat Exchanger
Motor	Motor
	Pump and Motor
	Tower or Column
	Compressor

manipulate the process variables. The symbols are given as a review of the basic building blocks of a process flow diagram.

Connecting each of the basic pieces of processing equipment are the various piping arrangements and/or signal paths that communicate between the instruments controlling the processes. **Line symbols** are more important than most technicians realize.

Not only do they show how loop instrumentation is related and connected, together they also identify the type of energy the instrument uses.

An analog pneumatic transmitter produces a pneumatic signal and an analog electronic transmitter produces an electrical signal. Most analog control systems are fairly consistent with their signal forms. Pneumatic loops tend to be pneumatic from one end to the other whereas analog electronic usually switches to pneumatic at the control valve.

If a digital control system (DCS) is used, then another line symbol, the software link, may be introduced into an electronic analog loop. One possible scenario: an electronic signal can be accepted into a DCS by first converting into a binary number. This is done with an analog-to-digital (A/D) converter. Once the signal value has been converted into a binary number, it moves through the computer programming as a software link. The output of the computer leaves through a digital-to-analog (D/A) converter and reenters the loop as an analog signal before manipulating the process. Without a system of symbols depicting these instrument signal lines, understanding would be very difficult as to how a controlled variable signal moves through the computer system. A digital signal is like a bar code.

Other line symbols include instrument connections to the process. The electromagnetic or sonic symbol includes radar, radiation, and even a video camera focused on a flare. The capillary tubing usually indicates a liquid-filled system and the hydraulic line symbol may be any line that carries a fluid signal, such as a piston actuator.

Standards for line symbols, as indicated on most symbol charts, include those shown in Table 12-4.

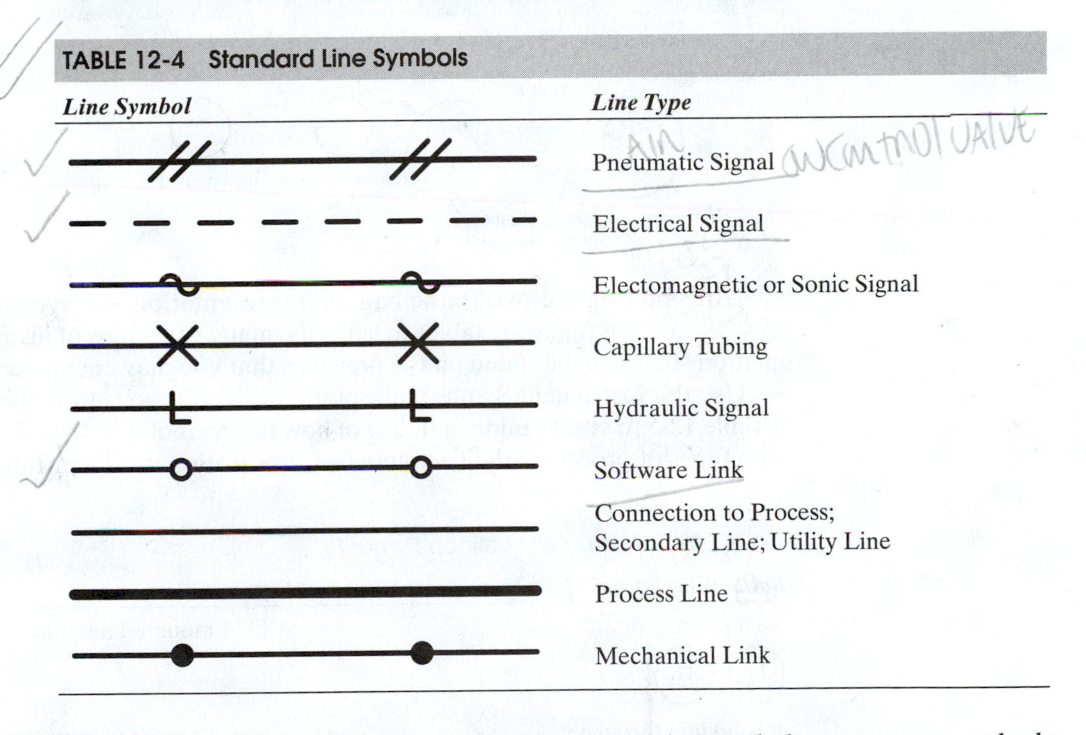

TABLE 12-4 Standard Line Symbols

Line Symbol	*Line Type*
	Pneumatic Signal
	Electrical Signal
	Electomagnetic or Sonic Signal
	Capillary Tubing
	Hydraulic Signal
	Software Link
	Connection to Process; Secondary Line; Utility Line
	Process Line
	Mechanical Link

Unlike process equipment symbols, instrumentation symbols may or may not look like the physical device they represent. For example, a 7/16-inch diameter circle, called a balloon, is commonly used to represent any number of functionally different instruments. The only distinguishing difference from one balloon to another is its unique alphanumeric tag number. Considering the complexity of many control systems this schematic approach works very well. The tag number, covered earlier in this chapter, is the primary key to defining the functionality of the instrument whereas slight modifications of the balloon depict where the instrument is physically located.

In Figure 12-7, a typical legend may show how the various instrument balloons are represented for a particular drawing. One balloon (Figure 12-8) is further explained as to what each of the letters or numbers that may be found on the tag may represent. Also indicated in Figure 12-8 are two variations of the balloon.

FIGURE 12-7 Legend Example for Instrument Balloon Interpretation

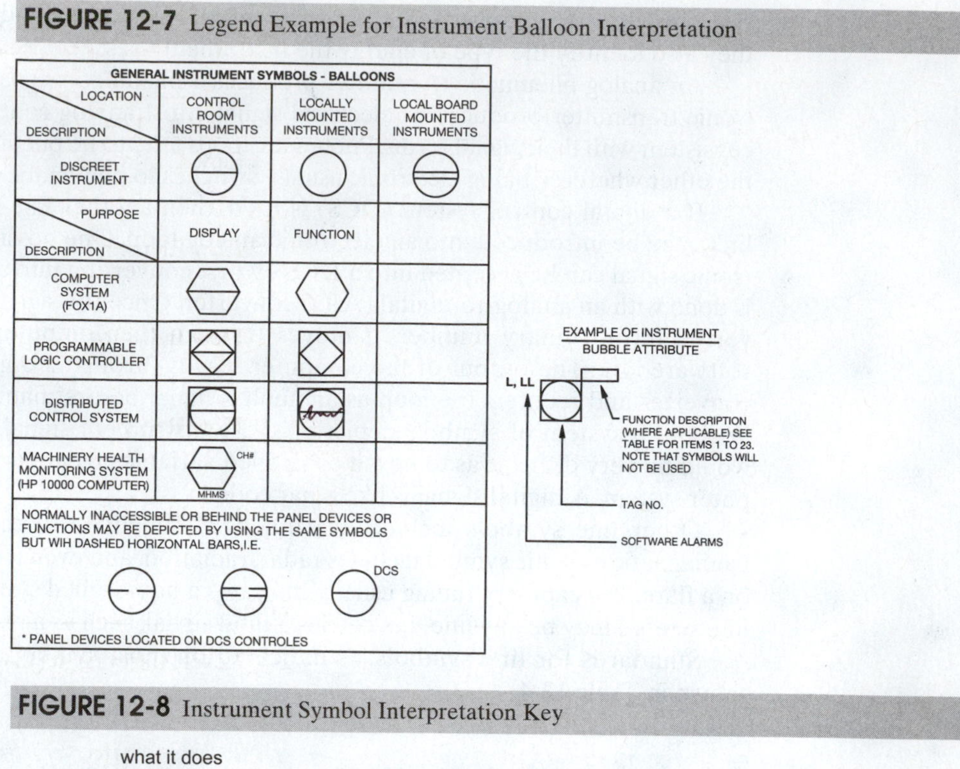

FIGURE 12-8 Instrument Symbol Interpretation Key

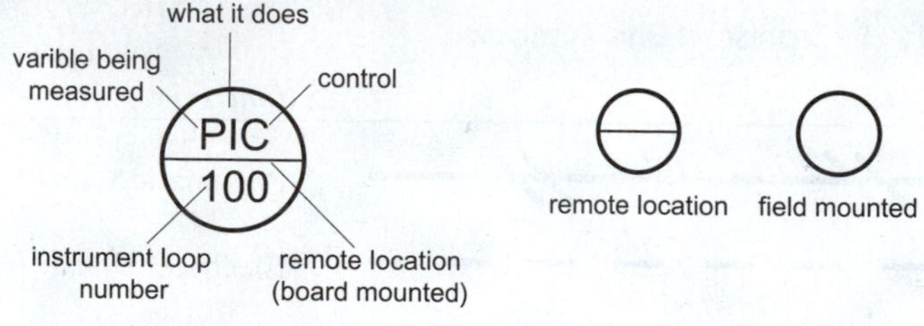

After putting the two (basic balloon representation and symbol interpretation) together, use the following tables to learn the many variations of instrumentation combinations that may be found on the drawings that you may encounter.

Use the Instrument Symbol Interpretation Key as you study the balloon symbols in Table 12-5 to ensure understanding of how the symbol was developed for each of the main types of process variables where instrumentation is used in Table 12-6.

TABLE 12-5 Instrument Balloon Symbols

Balloon Variation	*Interpretation*
No lines in balloon (FT 1280, LT 1024)	A field-mounted instrument.
A solid line through it (FRC 288, PT 768, PIC 121)	A board-mounted instrument
Two parallel lines through it (TT 72)	Located in an auxiliary location, usually on a control panel located in the processing area
A broken line through it (FY 360)	Located behind the panel board, or at least is not readily accessible
A box around it (TC 144)	Digital Control Systems (DCS) or computer interface

TABLE 12-6 Instrument Balloon Symbol Examples

FLOW

Balloon Variation	*Interpretation*
FC	Flow Controller
FE	Flow Element
FI	Flow Indicator
FR	Flow Recorder
FT	Flow Transmitter

LEVEL

Balloon Variation	*Interpretation*
LAH / 15	Level Alarm High (LAH) ("H" for High)
LIC / 30	Level Indicator Controller
LG	Level Gauge, field mounted
LI	Level Indicator, field mounted
LR / 20	Level Recorder, board mounted
LT / 25	Level Transmitter

PRESSURE

Balloon Variation	*Interpretation*
PAH	Pressure Alarm High (PAH or PHA), panel mounted
PIC / 60	Pressure Controller, board mounted
PI	Pressure Indicator, field mounted
PIC / 100	Pressure Indicating Controller, board mounted
PR / 50	Pressure Recorder, board mounted
PCR / 90	Pressure Recording Controller, board mounted
PT / 45	Pressure Transmitter, board mounted

TABLE 12-6 *Continued*

TEMPERATURE	
TC	Temperature Controller
TE	Temperature Element
TI	Temperature Indicator
TR	Temperature Recorder
TT	Temperature Transmitter
MISCELLANEOUS	
AT pH	pH Analyzer Transmitter
I/P 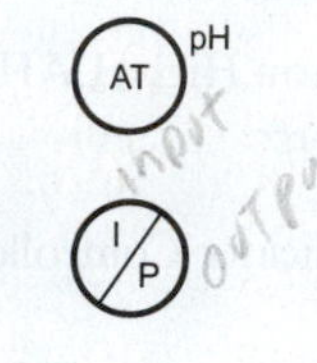	I/P and P/I Converters convert a current signal to a pneumatic signal so a Distributed Control System (DCS), Programmable Logic Control (PLC), or Personal Computer (PC) can control a valve or actuator; may also convert a pneumatic signal to current so remote pneumatic devices can interface with electronic instruments and computer based monitoring systems.
E/P 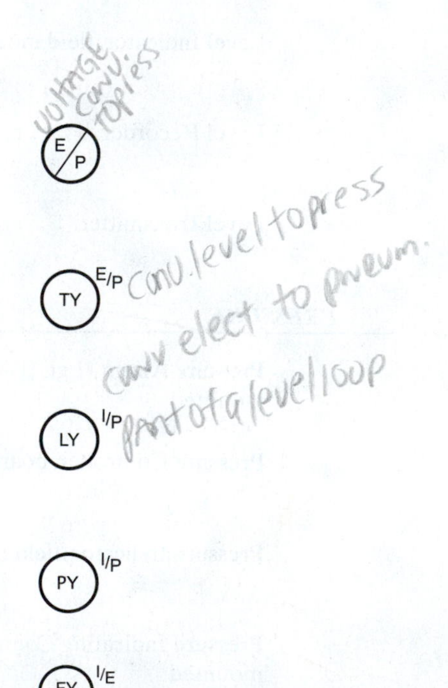	E/P transducers convert electrical signals to equivalent pneumatic signals and they are commonly used in the field to supply instrument air to field control elements.
TY E/P	Temperature transducer—transducer electronic to pneumatic symbol used to activate a pneumatic positioner on a valve for a temperature loop.
LY I/P	Level transducer—transducer electronic to pneumatic symbol used to activate a pneumatic positioner on a valve for a pressure loop.
PY I/P	Pressure transducer—transducer to pneumatic symbol used to activate a pneumatic positioner on a valve for a pressure loop.
FY I/E	Flow transducer—transducer, isolator and converter; current to voltage.

Using information from the previous chapters on pressure, temperature, level and flow, the following tables (Tables 12-7 through 12-14) are given to show how the instrumentation balloon symbols, line types, and equipment symbols are used together to represent process components. In addition, a table of final control elements is shown with corresponding instrument symbol configurations.

TABLE 12-7 Pressure

Symbol Variation	*Interpretation*
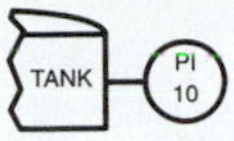 TANK, PI 10	A Pressure Indicator directly connected to a tank/vessel
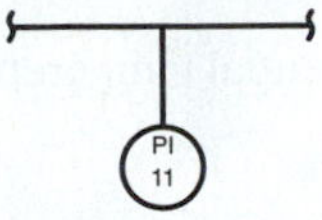 PI 11	A Pressure Indicator connected to process piping
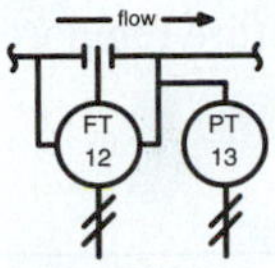 flow, FT 12, PT 13	A Pressure Transmitter connected via piping to a low-pressure lead coming off a Flow Transmitter
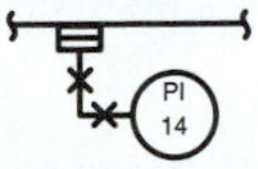 PI 14	A Pressure Indicator connected to a special type of chemical seal to protect the instrument from the process fluid (capillary)
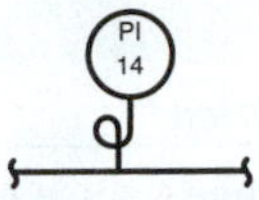 PI 14	A Pressure Indicator connected to process piping where a siphon is installed
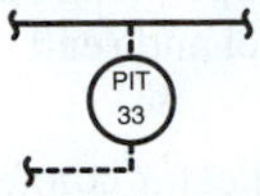 PIT 33	Pressure element, strain-gauge type, connected to pressure indicating transmitter (TAG Strain Gage PE-33)
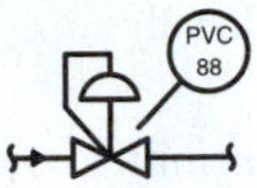 PVC 88	Back-pressure regulator, self-contained with handwheel adjustable setpoint
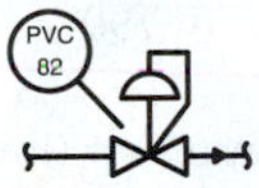 PVC 82	Pressure-reducing regulator, self-contained
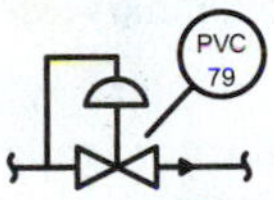 PVC 79	Backpressure regulator with external pressure tap

TABLE 12-8 Temperature

Symbol Variation	*Interpretation*
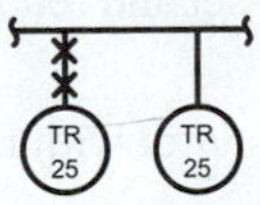 TR 25, TR 25	A local Temperature Recorder, filled thermal system and thermowell connected to process piping
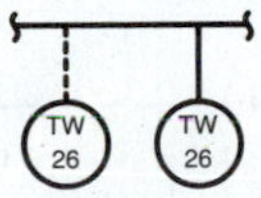 TW 26, TW 26	A local Temperature Recorder, thermocouple, or RTD and thermowell connected to process piping
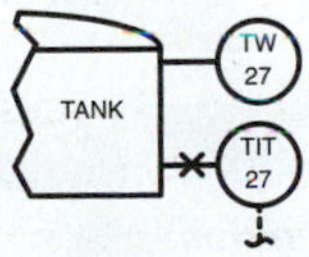 TANK, TW 27, TIT 27	A Temperature Indicating Transmitter and filled thermal system connected to a tank/vessel and thermowell
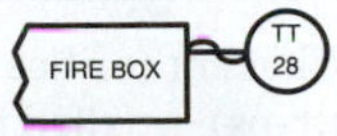 FIRE BOX, TT 28	A Temperature Transmitter of the thermal radiation type using an optical pyrometer connected to a furnace fire box
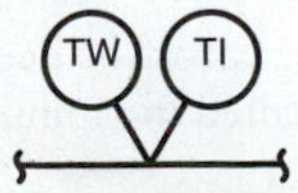 TW, TI	A bimetallic thermometer (Temperature Indicator) inserted in a thermowell in process piping

TABLE 12-8 Continued

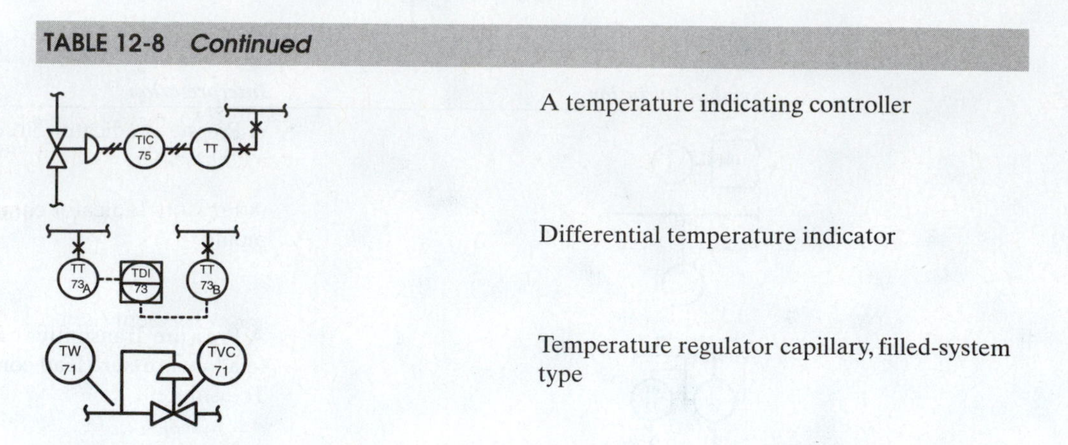

Symbol Variation	Interpretation
TIC 75, TT	A temperature indicating controller
TT 73A, TDI 73, TT 73B	Differential temperature indicator
TW 71, TVC 71	Temperature regulator capillary, filled-system type

TABLE 12-9 Level

Symbol Variation	*Interpretation*
TANK, LG 10	Level Gauge (gauge glass) connected to a tank/vessel and read visually
LI 11	Level Indicator connected to a tank/vessel and read locally
LT 12	Level Transmitter connected to a tank/vessel and read remotely
TANK, LT 13	Level Transmitter (low side vented) connected to a tank/vessel and read remotely
TANK, LI 15	Level Indicator (gauge board—float actuated) connected to a tank/vessel and read locally
LI 17	Level Recorder/Level Electronic (bubble tube direct connect to final device) connected to a tank/vessel and read remotely
TANK, LC 19	Local Controller (piped direct) connected to a tank/vessel and read remotely
LCV 70, TANK	Level regulator with mechanical linkage

TABLE 12-10 Flow

Symbol Variation	*Interpretation*
FE 10	Flow Element (orifice plate with flange/corner taps) installed in piping
FI 11	Flow Indicator (orifice plate with flow indicator) installed in piping

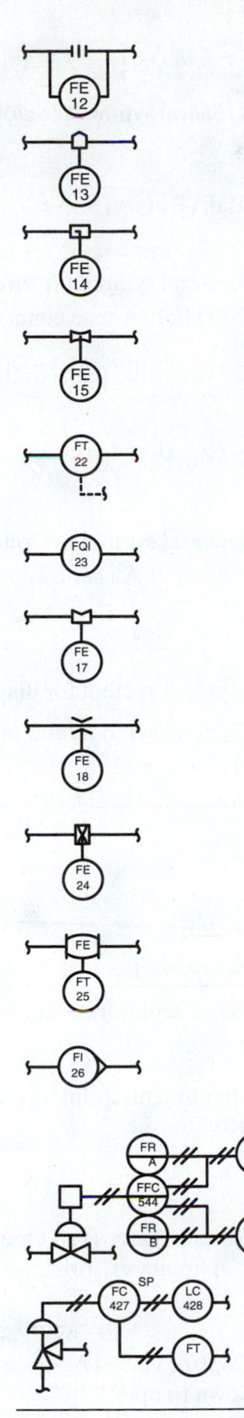

Flow Element (orifice plate with vena contracta radius or pipe taps) installed in piping

Flow Element (orifice plate in quick change fitting) installed in piping

Flow Element (pitot tube) installed in piping

Flow Element (venturi or flow nozzle) installed in piping

Flow Transmitter installed in piping

Flow Quantity Indicator

Flow Element (weir) installed in piping

Flow Element (flume) installed in piping

Flow Element (turbine or propeller type primary element) installed in piping

Flow Target (meter) installed in piping

Rotameter (variable area flow indicator) installed in piping

Flow-ratio controller with two pens to record flow

Cascade control

TABLE 12-11 Final Control Elements (Valve Bodies)

Symbol Variation	*Interpretation*
	General symbol for valve (ON/OFF only)
	General symbol for angle valve NOTE: ISA uses connected triangles.
	General symbol for butterfly valve

TABLE 12-11 Continued

Symbol	Interpretation
	General symbol for globe valve
	Ball (Rotary) Valve
	General symbol for three-way valve NOTE: ISA uses connected triangles.
	3-Way Valve (Fails to Bottom)
	3-Way Valve (Fails Straight)
	General symbol for four-way valve NOTE: ISA uses connected triangles.
	General symbol for diaphragm valve
M	General symbol for a motor-operated valve

TABLE 12-12 Actuator Symbols

Symbol Variation	*Interpretation*
or	Hand actuator or handwheel
	Diaphragm, spring-opposed, or unspecified actuator
FO	Control Valve (Fail Open) (straight through, diaphragm vs. spring actuator)
	Control Valve (Alternate) (Fails Open) (push down to open)
FC	Control Valve (Fail Closed) (straight through, diaphragm vs spring actuator)
	Control Valve (Alternate) (Fails Closed) (push down to open)
FO	Butterfly Control Valve (Fails Open)

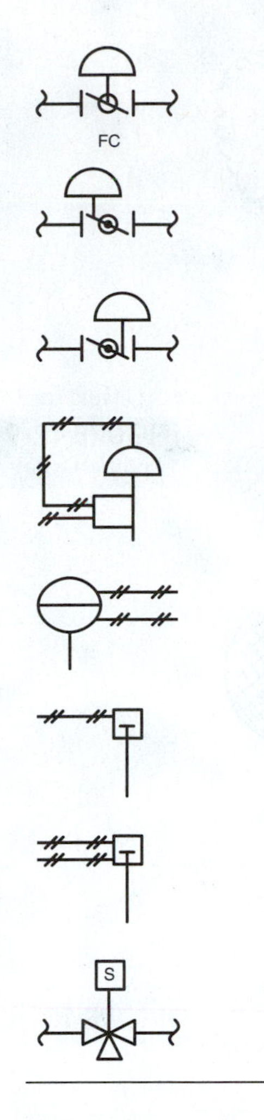

Butterfly Control Valve (Fails Closed)

Butterfly Control Valve (Alternate) (Fails Closed) (push down to open)

Butterfly Control Valve (Alternate) (Fails Open) (push down to close)

Control Valve Actuator with Positioner

Actuator (Diaphragm vs diaphragm)

Actuator (Piston)
NOTE: ISA draws the piston from side to side on the box.

Actuator (Double-acting Piston)
NOTE: ISA draws the piston from side to side on the box.

Three-way Solenoid Valve

TABLE 12-13 Final Control Elements

Symbol Variation	*Interpretation*
	Damper/Louver (position open)
	Damper/Louver (position closed)

There are many other types of symbols used on drawings frequently or infrequently. For example, between updates, hand-drawn clouds (Figure 12-9) may be used to signify a revision of some type such as an addition.

Equipment that has been abandoned or dismantled is bordered and filled with crosshatched lines (Figure 12-10) indicating abandonment in place or dismantlement.

Still other miscellaneous symbols are common to drawings that have continuations (e.g., connector lines) similar to key maps used for driving across large distances where you have to connect one map drawing to the next.

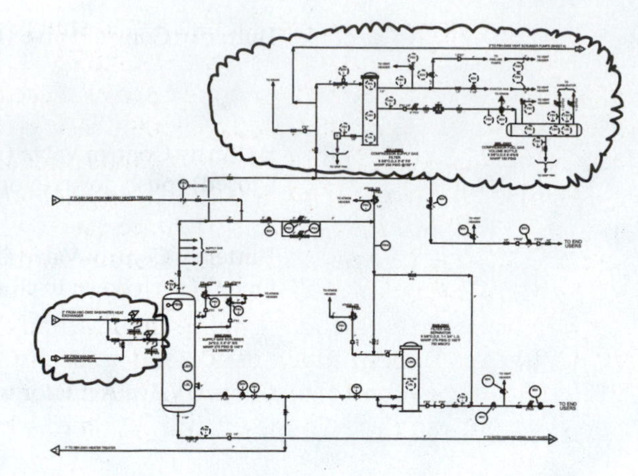

FIGURE 12-9 Revision or Addition–Clouds

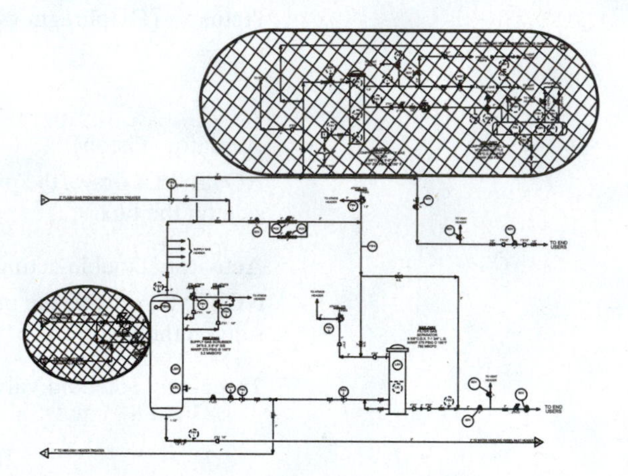

FIGURE 12-10 Cross-Hatching

TABLE 12-14 Miscellaneous Symbols

Symbol Variation	*Interpretation*
or	Connector lines (one-way or two-way) used when going from drawing to drawing NOTE: From . . . and To information is usually noted inside the arrow designations.

Summary

There are several types of drawings that process technicians come in contact with frequently. Of those, the three most common are the block flow diagram (BFD), the process flow diagram (PFD), and the piping and instrumentation diagram (P&ID). Block flow diagrams show simple sequential flow from left to right while process flow diagrams include major process equipment and provide heat and material balances of the chemical processes represented within the drawing. Piping and instrumentation diagrams are similar to process flow diagrams but contain much greater amounts of detail about mechanical design, instrumentation, and primary control systems. More detailed drawings such as PFDs and P&IDs usually have legends that explain how various symbols and codes are used. When process technicians understand the interrelationships of the various pieces of equipment and instruments that monitor and maintain the processes, they can relate to their job responsibilities such as troubleshooting and quality control with more qualified judgment capability.

People have always used symbols to communicate, but standardization is the key to understanding between facilities and location-to-location. The International Society of Automation (ISA) has attempted to bring together across these boundaries a set of drawing symbols to standardize instrumentation and other basic symbols so that errors between engineering companies and facility design and implementation personnel are able to understand and replicate drawing parameters. Even so, there are still variations being used. To alleviate some of this problem, legends are used on drawings for explanations of symbols used. These legends should be carefully studied when observing and/or studying PFDs and P&IDs.

Instrument tag numbers are also standardized by the ISA and contain an alphanumeric code to include the instrument functionality and a specific numeric identifier specific to a particular drawing or control loop. The ISA has a Functional Identification Table to help identify what the specific letters indicate and what their modifiers may mean.

Symbols can be quickly divided into several types to include the following:

- basic equipment
- lines
- instrument balloons
- combinations by process variables for pressure, temperature, flow, level, etc.

Checking Your Knowledge

1. Which of the following is NOT one of the most common types of drawings used by process technicians in their daily activities?
   a. BFD
   b. PFD
   c. PRD
   d. P&ID
2. Which of the following types of diagrams provides heat and material balances of the chemical process?
   a. BFD
   b. PFD
   c. PRD
   d. P&ID
3. Which of the following types of drawings contains the most detail to include instrumentation and the entire primary control system?
   a. BFD
   b. PFD
   c. PRD
   d. P&ID
4. What is the part of a drawing called that contains a compilation of the symbols used as well as any codes or other notes?
   a. Margin
   b. Legend
   c. Perimeter
   d. Cloud
5. What does ISA stand for?
   a. Instrumentation Society of America
   b. International Society of Automation
   c. International Standards and Automation Society
   d. Instrumentation Standards Association
6. The ISA standard for specific drawing symbols denoting functionality and a coded system built on the letters of the alphabet are in which of the following standards?
   a. ISA-S5.1
   b. ISA-S5.2
   c. ISA-S5.3
   d. ISA-S5.4
7. *True of False* An ISA instrument tag number is described with both letters and numbers and should identify the measured variable, the function of the specific instrument and the loop number.

8. Instrumentation symbols commonly use a _____ to represent the physical device.
   a. Square
   b. Triangle
   c. Circle or balloon
   d. None of the above
9. Clouds on a P&ID represent which of the following?
   a. Revision or addition
   b. Abandon in place
   c. Continuation of process
   d. Equipment for removal
10. The purpose of the arrows attached to process lines on the P&ID is to:
    a. Show the direction of the flow
    b. Show the continuation of the process on another drawing
    c. Show the process in an order step
    d. Show the difference between a utility line and a process line

Student Activities

1. Photocopy symbol pages or draw symbols on repositionable sticky paper found at most craft supply stores. Cut out symbols and temporarily mount them on plastic such as transparency film or plastic protective page covers. Print Drawing 1 in this section scaling to 11 × 17 if possible. Place symbol stickers in appropriate places on the drawing where instrumentation would be found. When complete, bring to class and present your work to your instructor and classmates.
2. Repeat the above procedure for Drawing 2.
3. Given a simplified flow diagram of a section of your pilot plant or table top model, use the symbols legend to insert instrumentation on the diagram.
4. Create an original PFD of the pilot plant section or tabletop model located at your facility. Add control loops to the diagram and write a short paragraph for each control loop added describing why the loop was added and how the process would be affected by having the control system in place.
5. Using the symbols in this chapter, create a simple diagram with process equipment and its associated instrumentation. If trainer resources are available, then these drawings may be created there. Drawings and printouts should be passed around to other groups for discussion and improvement.
6. Complete the following
   - Given a P&ID, explain the relationship of one piece of instrumentation to another.
   - Given a drawing that has major system equipment, add control loops:
     flow
     level
     temperature
     pressure
   - Sketch instrumentation control loops on available trainer resources.

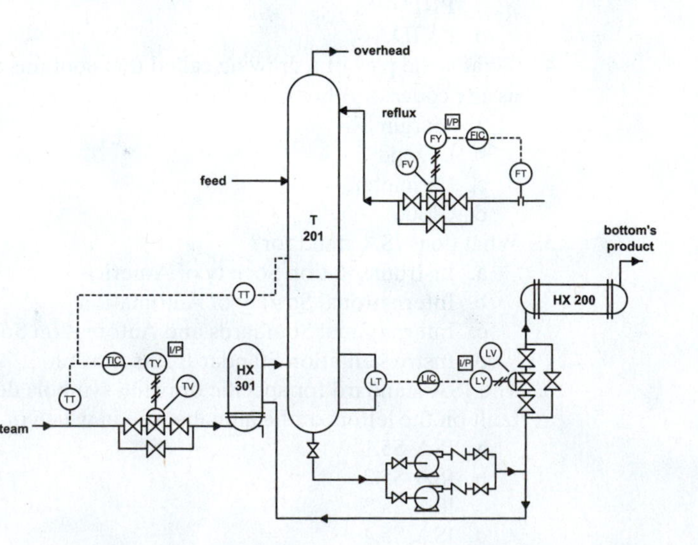

7. **Diagrams, Symbols and Sketching**
   Instructions
   a. Review the following P&ID.
   b. Visually trace the control loop from the steam input to the heat exchanger on the left side of the P&ID.
   c. Write a description of all the control elements for the loop and the purpose of the loop.
   d. Follow your specific instructor guidelines as to submission of your work.
8. **USING LEGENDS**
   Instructions
   a. Review the following P&ID and Symbols Legend.
   b. Identify all the listed control elements.
   c. Write the answers to what each control elements is.
   d. Follow your specific instructor guidelines as to submission of your work.

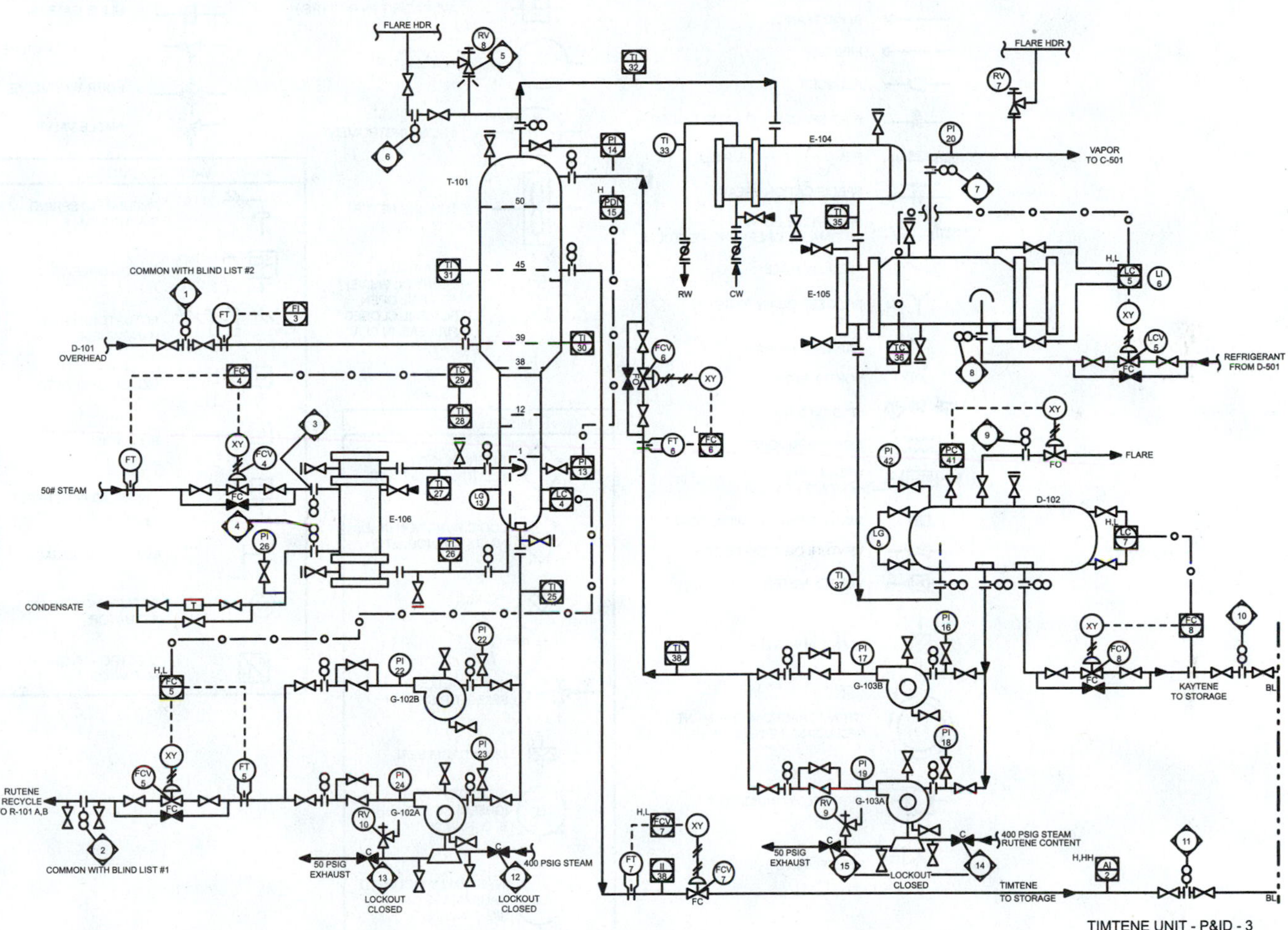

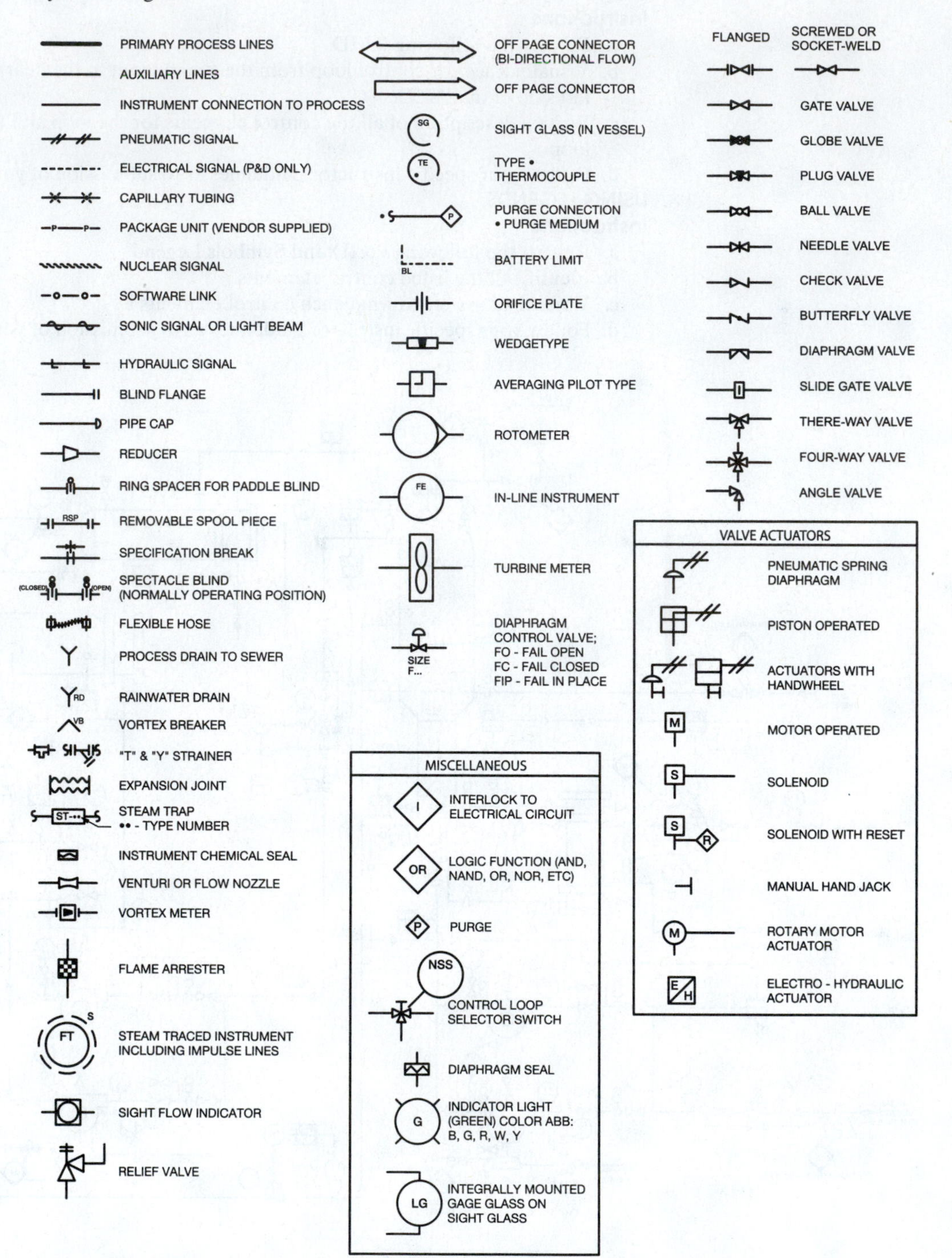

9. **LABORATORY PROCEDURE: Symbols and Diagrams**
Background Information
Students need to know how to read a PFD, P&ID, and a symbols legend. Students also need to understand how instrumentation fits into the diagram. Keep in mind that this is a diagram, not an actual picture of the process. The finished diagram should represent the field and panel mounted instruments properly located and connected in a control loop. Look at several P&IDs before attempting this lab procedure.

Materials Needed

- Sketching equipment (pencil, straightedge, eraser, paper, template, ruler, etc)

Safety Requirements

None required for this lab.

Procedure

a. Review the simplified process flow diagram below.

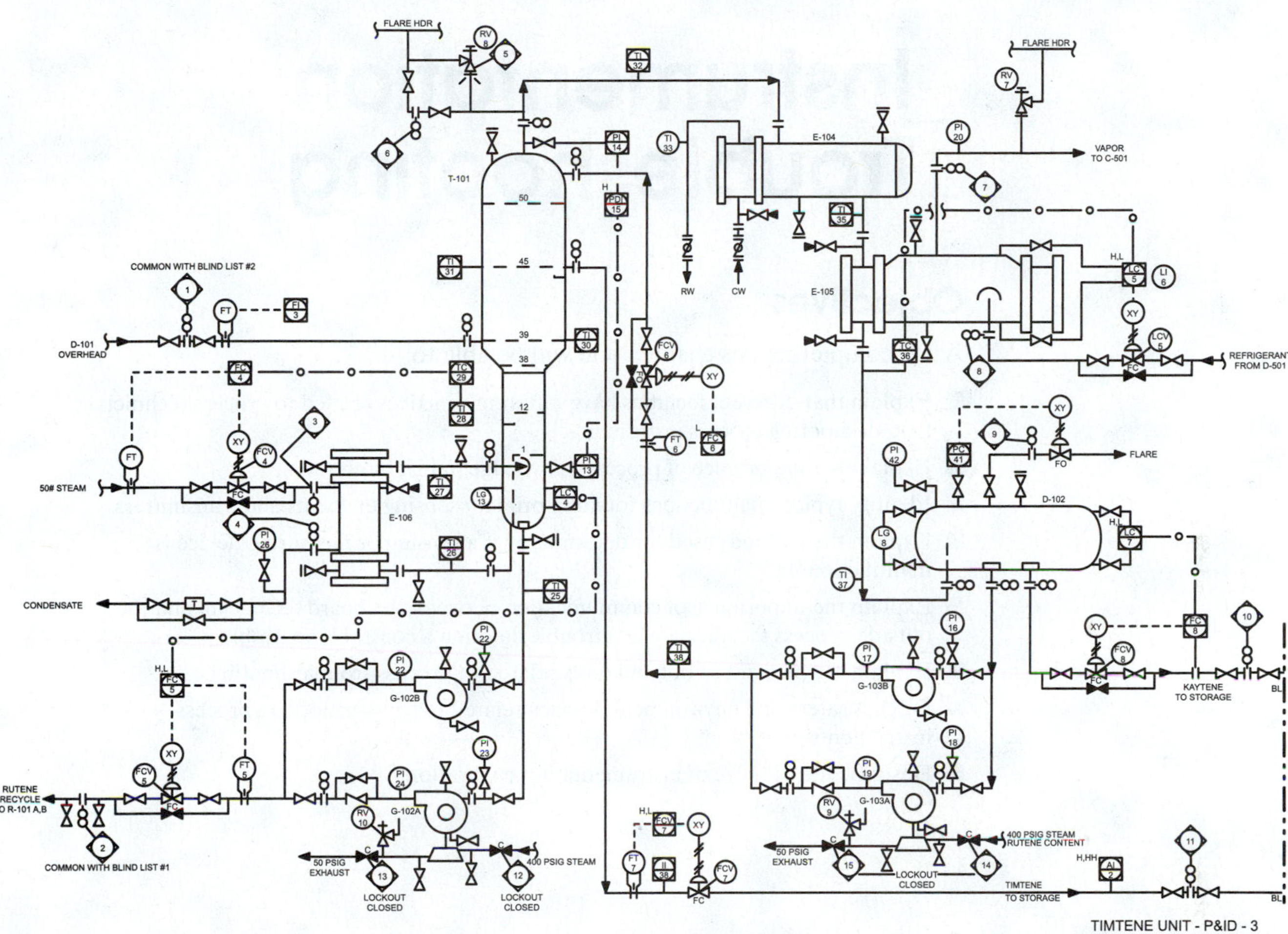

b. Study the Symbols Legend provided previously.

c. Complete the drawing by including:
   - a level control loop
   - a temperature control loop
   - a tower feed flow control loop

d. Determine where the sensing point and the final control element for each loop will be placed on the Simplified PFD and draw them first. Complete the loop showing the controller and transmission lines. Complete one loop at a time.

e. Determine whether the loop is a simple feedback loop or a more complex control scheme such as cascade or ratio.

f. Complete the final report for this lab.

Additional Information

None

Findings

Write a short paper on each control loop included in the drawing to include how each control loop affects the process. All non-textbook drawings should be turned in with the write-up as per your specific instructor guidelines.

CHAPTER 13

Instrumentation Troubleshooting

Objectives

After completing this chapter, you will be able to:

- Explain that different facilities have different practices related to process technicians troubleshooting process instruments.
- Explain the importance of process knowledge in troubleshooting.
- Identify typical malfunctions found in primary sensing elements and transmitters.
- Explain the methods used for determining if a sensing or measuring device is malfunctioning.
- Explain the importance of communication between the board technician and the outside process technician when troubleshooting a control loop problem.
- Explain the proper use of hand tools related to process troubleshooting.
- Discuss safety and environmental issues related to troubleshooting process instruments.
- Describe the purpose of instrumentation calibration.

Key Terms

Calibration—the act of applying a known input span to an instrument and adjusting the device so that it provides an indication or output corresponding to the known values.

Smart instruments—instruments that have one or more microprocessors or "smart chips" included in their electronic circuitry so they may be programmed and have diagnostic capability.

Troubleshooting—the process of systematically examining, localizing, and diagnosing equipment malfunctions or anomalies.

Introduction

Process technicians are the first line of defense against all process unit–related problems. They should be aware of general equipment problems, such as the occasional failure of the control instrumentation as well as process problems (Figure 13-1).

Generally, process technicians are expected to report all maintenance problems, and in some cases fix the problems. To do either requires a certain level of knowledge associated with normal plant operations.

A technician must first know what is right in order to recognize what is wrong. If technicians are only expected to report instrumentation problems, then they must be able to accurately describe the problem from a symptomatic standpoint. However, if a technician is expected to go one step further and identify the malfunctioning instrument, then additional training on instrumentation system troubleshooting must be made available.

Most, if not all, facilities expect process technicians to troubleshoot process problems, while some facilities expect them to do more. The level of troubleshooting involvement expected of the process technician varies from one company to another. For that matter, the level of involvement may even vary between plants in the same company and possibly from one unit to another in the same plant. With so much variation in the industry, the process technician's role as an instrument troubleshooter is less defined than other broadly accepted tasks. Therefore, in this chapter, troubleshooting is discussed in general terms so that it can be applied to any processing plant.

The importance of understanding the troubleshooting process cannot be overstated. A process technician who knows the process recognizes a problem as it occurs,

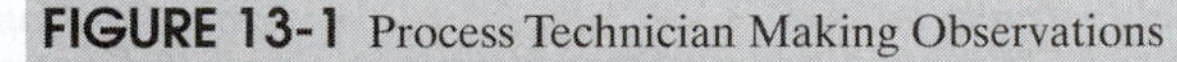

FIGURE 13-1 Process Technician Making Observations

Credit: Emerson

or soon after, depending upon the ramifications. Some problems are obvious, while others are not so easy to recognize and are only caught when some other part of the process or another controlled variable becomes abnormal. The more a process technician knows about the process and the equipment associated with it, the better equipped that technician will be when troubleshooting a malfunction.

Typical Malfunctions

All equipment eventually malfunctions or fails. Normal equipment failures are most likely to occur soon after installation or at the end of the equipment's expected life span. Failures caused by environmental problems such as exceeding temperature limits, excessive internal or external corrosion, over-ranging, or humidity can happen at any time. Other situational problems, such as a slow and regular accumulation of solids in the impulse tubing connecting the pressure transmitter to the process, can be avoided through a regular preventive maintenance schedule.

Preventive maintenance is specifically intended to prevent instrumentation failures from occurring by addressing some aspect of maintenance that prevents malfunctions. This is important because scheduled maintenance reduces unscheduled downtime and saves the company money.

The following are common environmental factors that cause instrument malfunctions:

- excessive temperatures
- corrosion
- sudden change in temperature
- over-ranging
- inclement weather
- high humidity
- high vibration
- mechanical damage

The following is a list of various sensor types and the malfunctions normally associated with each:

Sensor-Related Instrument	*Common Malfunctions*
Impulse tubing	• Plugged or partially plugged tubing • Leakage • Improperly sized tubing • Excessive vibration • Heat tracing too hot • Heat tracing too cold • Mechanical damage
Thermocouples	• Burnout • Corrosion • Bad connection • Delayed response time if used with thermowell without thermo conductive fluid
Signal wire	• Bad connection • Broken wire • Too much loop resistance • Shielding problems

	Transmitters	• Low voltage from power supply (or air supply with pneumatic instruments) • Loss of power • Environmental problems (e.g., corrosion or excessive humidity) • Calibration shifts
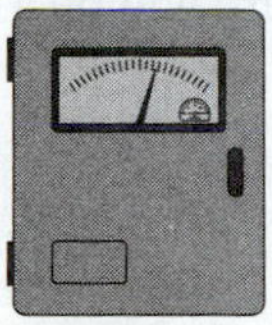 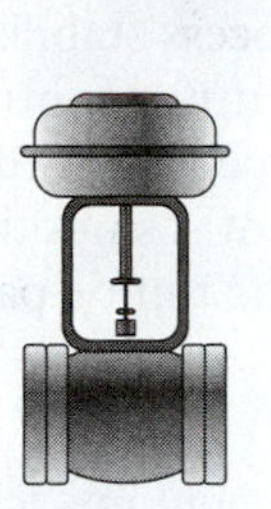	Controller	• Plugging • Water in the air • Power loss (in electronic) • Water in the device itself • Air or electrical power failure • Controller needs re-tuning due to change in transmitter span
	Final control element	• Process plugging • Corrosion or erosion in the valve • Packing leaks • Sticking valve • Mechanical failure of tubing or valve components • Water in the air line • Water in the device itself • Air or electrical power failure • I/P or positioner out of calibration

Troubleshooting Methods

Troubleshooting is an art based on sound reasoning and technical knowledge. There are a number of troubleshooting steps and guidelines in publication today, each with their own unique strengths and similarities. The following troubleshooting steps provide a general framework of thought and action that a technician can use to find and repair an instrument or process problem.

1. Interpret and analyze—Determine the source of problem.
2. Evaluate and infer—Use the information obtained in the analysis phase to identify a solution for the problem.
3. Recognize—Be able to determine if the problem is an instrument problem.
4. Repair—Fix the problem.
5. Reevaluate—Make sure the problem is corrected.
6. Document—Document the repair or change for historical purposes or regulatory reasons.

CHECKING THE CONTROLLER

A controller is the most common remote instrument in a loop (Figure 13-2). In an analog loop, the controller is a stand-alone device that can have any number of problems

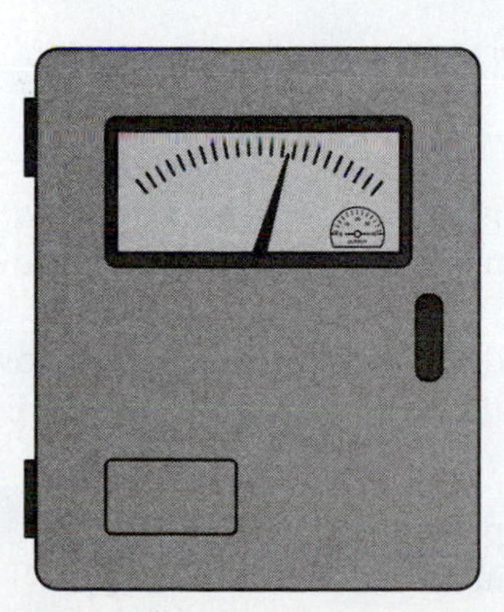

FIGURE 13-2 Controller

normally associated with hardware. If the controller is a part of a DCS system, however, then the problem is probably associated with the analog or digital interface (PLC) components as opposed to the hardware.

Software, unlike hardware problems, has a failure curve that is rather steep initially and then diminishes over the remainder of its useful life as errors are found and corrected. Therefore, outside of a problematic programming change, the problem is not likely to be in the process computer. If it is, then there will be a common problem associated with more than one loop.

When troubleshooting a controller:

1. Make sure that the PV indication on the controller corresponds to the true output of the transmitter by having an instrument tech simulate signals on the input to the controller or from the transmitter.
2. Place the controller in manual and see if the process stabilizes.
3. With the controller still in manual, determine if the controller moves the final control element properly (e.g., 0 percent, 50 percent, and 100 percent).
4. Confirm that the final control element movement is smooth.
5. Ensure that the loop is properly tuned (check the tuning parameters).

DIVIDING THE LOOP IN HALF

One of the best ways to find a malfunctioning instrument in a loop is to divide the loop into halves until the problem has been isolated to one component. Typically only one device fails at any given time; however, there are times when multiple problems occur simultaneously. Since most loops have a sensor or transmitter, a controller, and a final control element, one should start at the controller (Figure 13-3).

When troubleshooting a controller, place the controller in manual and observe the response. Based on your observations you should be able to determine that the problem is either:

1. In the instrument(s) before the controller,
2. In the controller itself, or,
3. In the instrument(s) beyond the controller.

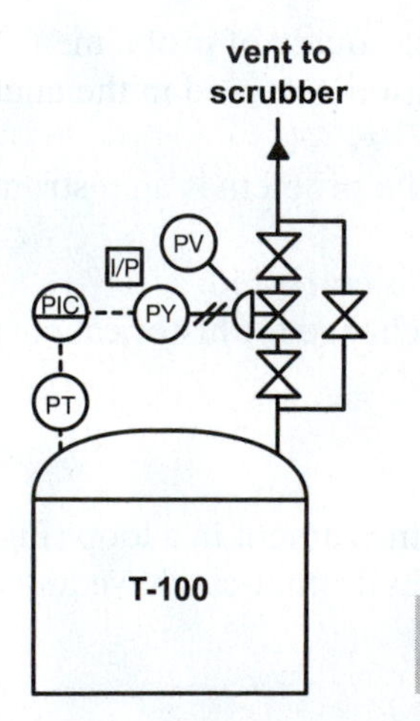

FIGURE 13-3 Typical Pressure Control Loop

Process to Controller

If a determination is made that an instrument or sensor located between the process and the controller is the problem, then checking the output of the sensor or transmitter is the next step.

A transmitter is an input/output device, so if the input is known then the expected output should be known as well. If the input and output are different from what is expected, a technician would need to do the following:

1. Check the transmitter zero and span.
2. Check process connection or piping.

3. Check the sensor and calibrate or replace the transmitter as needed.
4. Reconnect the transmitter to the controller and check for proper reading.

Controller to Process

If a determination is made that an instrument located between the controller and the process is the problem, then checking the final control element, usually a control valve, is the next step. The most common instruments on the output side of the controller are the I/P converter and the control valve or final control element. The I/P converter and control valve can be checked by having a technician go outside and look for problems with the valve as the inside technician maneuvers the valve open and closed. The outside technician should look for a smooth stem motion and a complete stroke as the valve is driven open and closed. If only one technician is available then that person can simulate controller output with a handheld instrument.

Communication During Troubleshooting

During the troubleshooting process, communication between the board technician and the outside technician is of utmost importance. These two technicians must work as a team, usually communicating using radios, as one technician instructs the other technician to either initiate a change or watch for a response.

Troubleshooting Tools

Proper tool selection is important when troubleshooting instrumentation. Technicians commonly use the following tools during a typical troubleshooting procedure (Figure 13-4):

- slotted and Phillips head screw drivers
- adjustable wrench
- multimeter to test the electrical side of the loop
- pressure and/or temperature testing gauges
- two-way radio

Failure to use the proper tools can make solving problems more difficult or create new problems. For example, channel-locks are a convenient tool; however, they are usually not the proper choice for the job since channel-locks tend to round off the edges of hex nuts or bolt heads making them difficult to retighten. Furthermore, using the wrong sized screwdriver is also a problem. If a screwdriver is too large or too small it may strip or damage the slot or mating surface and compromise the mechanical integrity of the screw.

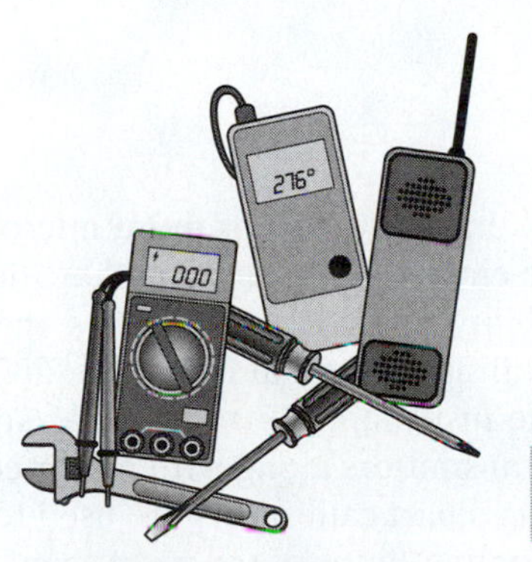

FIGURE 13-4 Basic Troubleshooting Tools

Troubleshooting Safety and Environmental Issues

Troubleshooting usually involves the manipulation one or more instruments. If a transmitter or control valve is to be taken out of service then safety and environmental precautions must be addressed. Prior to opening any process line proper personal protective equipment should be donned and proper safety procedures should be followed.

Calibration and Troubleshooting

Over time, even under normal conditions, instruments vary from their original calibrated spans. Because of this, instrumentation must be checked and adjusted periodically against a standard. This action is called **calibration** (Figure 13-5).

Calibration is the act of applying known input values to an instrument and adjusting the device so that it provides an indication or output corresponding to the known values.

While the proper use of calibration equipment requires a substantial amount of training, using a meter to check a signal value is fairly simple. Some transmitters have a test terminal under the cover. However, the technician needs to know what type of signal to expect (e.g., a 4–20-mA transmitter may have a millivolt test signal). The best way to determine which type of signal to expect is to read the instrument instruction manual.

If instruments are calibrated according to a regular preventive maintenance schedule, the overall number of malfunctions decreases, and the need for troubleshooting declines as well.

FIGURE 13-5 Instrument Calibration

Credit: Emerson

Troubleshooting "Smart" Instruments

Smart instruments are instruments that have one or more microprocessors (or smart chips) included in their electronic circuitry. This means that the device can be programmed and has diagnostic capability. In smart instruments, the microprocessor can compute and store analog and digital signals and at least 64 additional pieces of digital data pertaining to the status of the instrument. Furthermore, some of these devices include digital control valves and transmitters along with motor control circuitry.

In addition to smart instruments, smart calibrators are used to calibrate the instrument and check on the "health" or online status of the instrument. An instrument technician can call up this data from anywhere in the loop (junction box, at the device, or from any set of loop screw terminals). An added feature of these instruments is that all the data can be sent to a DCS operator station. This allows the process or instrument technician to analyze the data when the loop is in the diagnostic mode. (Note: Loop control functions have priority over diagnostics.)

Some of this digital data includes manufacturing specifications, materials of construction, sensor calibration graphs, and online status. For example, a smart flow transmitter can report a plugged impulse line, reverse flow, an empty pipe, or a bad sensor as well as many other statuses. A smart temperature transmitter can detect the difference between a bad sensor, open loop circuitry, RTD drift, or calibration error.

While smart field devices have made instrument troubleshooting easier, it is still important for process or instrument technician to be able to analyze this data correctly.

Summary

Process technicians are the first line of defense against all unit-related problems, especially the occasional failure of control instrumentation. Depending on the particular facility requirements, process technicians may fix or repair problems or simply report them so maintenance can perform the repairs. In either event, technicians must know both what is right and what is wrong with the instruments in their area of responsibility.

Since all equipment eventually malfunctions or fails, knowing the factors that increase malfunction and failure rates helps prevent or reduce the frequency of those problems. Environmental issues such as excessive temperatures, corrosive atmospheres, over-ranging conditions, or excessive humidity may happen spontaneously, while other problems, such as the slow accumulation of solids in pressure transmitter tubing, may happen over time and may not be as noticeable in the short term. To prevent excessive wear and tear on process instrumentation it is important to conduct daily inspections and perform routine preventative maintenance.

Troubleshooting is an art based on sound reasoning and technical knowledge. The following basic steps are required for troubleshooting:

1. Making sure that the PV indication on the controller corresponds to the true output of the transmitter by having an instrument tech simulate signals on the input to the controller or from the transmitter.
2. Placing the controller in manual and see if the process stabilizes.
3. With the controller still in manual, determining if the controller moves the final control element properly (e.g., 0 percent, 50 percent, and 100 percent).
4. Confirming that the final control element movement is smooth.
5. Ensuring that the loop is properly tuned (check the tuning parameters).

For control loops, the controller is the first and most common device to check. From the controller, the loop can be divided into halves (process to controller and controller to process) and the various instruments checked until the problem is discovered. Many times this process requires communication between inside and outside personnel and the coordination of various activities.

Various electronic and hand tools may be used during the troubleshooting process. These include screwdrivers, multimeters, pressure and temperature-testing gauges, and communication devices such as two-way radios. During these activities it is important to use the proper tools for the job since instruments are easier to damage than other types of process equipment.

When troubleshooting instrumentation on certain hazardous process streams, special types of safety or environmental conditions must be met or given special consideration. All necessary precautions should be taken during these activities.

When instruments are installed, they are calibrated to certain ranges or setpoints. Over time these instruments may lose their calibration due to vibration or other processing conditions. Therefore, each instrument should be tested and calibrated periodically so the reading matches the standard known value. If the reading is within the range, then the instrument does not need to be calibrated.

Checking Your Knowledge

1. *True or False* Process technicians have a clear set of duties in the area of troubleshooting instrumentation that are universal from plant to plant and company to company.
2. *True or False* In order for a process technician to be fully equipped to troubleshoot a process, they must first recognize that it is really an instrument problem.

3. Match the instruments below with the malfunctions normally associated with each.

Instrument	*Common Malfunctions*
Signal wire	a. • Burnout • Corrosion • Bad connection • Delayed response time if used with thermowell without thermoconductive fluid
Impulse tubing	b. • Bad connection • Broken wire • Too much loop resistance • Shielding problems
Transmitters	c. • Plugging • Water in the air • Lose power (in electronic) • Water in the device itself • Air or electrical power failure • Controller needs re-tuning due to change in transmitter span
Thermocouples	d. • Low voltage from power supply (or air supply with pneumatic instruments) • Loss of power • Environmental problems (e.g., corrosion or excessive humidity) • Calibration shifts
Controller	e. • Plugged or partially plugged tubing • Leakage • Improperly sized tubing • Excessive vibration • Heat tracing too hot • Heat tracing too cold • Mechanical damage
Final control element	f. • Process plugging • Corrosion or erosion in the valve • Packing leaks • Sticking valve • Mechanical failure of tubing or valve components • Water in the air line • Water in the device itself • Air or electrical power failure • I/P or positioner out of calibration

4. One of the best ways to begin troubleshooting an instrument loop is to first _____.
   a. Identify if the problem is between the process and the controller
   b. Identify if the problem is between the controller and the process
   c. Divide the loop in half
   d. Check the controller since it is usually the most common remote instrument in a loop
5. During troubleshooting, radio communication usually takes place between the field technician and the _____.
   a. Board technician
   b. Foreman

   c. I&E technician
   d. Maintenance supervisor
6. Which tool(s) are used regularly when troubleshooting instrumentation? (Select all that apply.)
   a. Channel locks
   b. Multimeter
   c. Slotted screwdriver
   d. Temperature-testing gauges
   e. Sledgehammer
7. *True or False* Technicians wear only basic PPE (hard hat, safety glasses, gloves, safety boots) when troubleshooting instrumentation.
8. When calibrating an instrument, the technician applies a known _____ span to the instrument in order to adjust indication correspondingly.
   a. Temperature
   b. Pressure
   c. Variable
   d. Level
   e. Input

Student Activities

1. In your institution or local city library, research various types of troubleshooting methodology. Apply the techniques to instrumentation as used in this chapter and discuss how they would or would not be applicable.
2. Given a control loop malfunction scenario, correctly apply a deductive troubleshooting methodology. Successfully deduce the source(s) of the malfunction and explain what actions are necessary to return the variable to within acceptable limits.
3. Role-play an inside and outside technician communicating during the troubleshooting of a control valve action.
4. Given a simulation trainer that utilizes a flow and level loop, and instructor-induced malfunctions, determine whether the problem is an instrumentation problem or process-related problem.
5. Perform minor instrumentation calibrations.

CHAPTER 14

Switches, Relays, and Annunciators

Objectives

After completing this chapter, you will be able to:

- Define the function of a switch:
  - energizes alarms, interlocks, safety systems, equipment, or other devices when a process condition meets a preset value
  - may be operated by hand, actuated by a mechanical signal, or actuated by a process or electrical signal
- Describe three common uses for switches in process control:
  - alarms
  - shutdown
  - autostart
- Given a drawing, picture, or actual device, identify and describe basic switch devices.
- Define the following terms associated with switches used in process control:
  - autostart switch
  - bypass
  - on/off/auto switch
  - limit switch
  - proximity switch
  - vibration switch
  - process variable switch
- Given a PFD or P&ID and a legend, locate and identify switches used in process control.
- Explain how relays are used in the process industry:
  - maintains signal
  - used in redundant systems
- Given a ladder diagram and a legend, locate a relay.
- Explain the purpose of annunciator systems.
- Define terms associated with annunciator systems:
  - first out (first shutdown conditions)
  - acknowledge, reset, and test modes

Key Terms

Alarm switch—used to notify an operator when a process variable enters an abnormal range (e.g., high or low) by triggering an alarm (a light or annunciator).

Bypass switch—used to override the normal operation of a system or device.

Limit switch—used to verify the state or presence of a condition that exists in the process.

Process variable switch—a type of switch that actuates when a predetermined value of a process variable (e.g., pressure, temperature, level, flow, and analytical) is present.

Proximity switch—a type of switch that requires the presence of an object or device to facilitate its operation such as a magnetic coupling or decoupling.

Relay—a device that "boosts," maintains, or controls the flow of a signal so it can be properly received.

Shutdown switch—used to actuate a circuit that shuts down a process.

Switch—a mechanical or electrical device that is used to operate or energize mechanical or electrical circuits for alarm, shutdown, or control purposes using a predetermined operating point or setpoint.

Vibration switch—used to determine the velocity, acceleration, displacement or any combination of these characteristics, for the purpose of predicting wear or impending failure.

Introduction

Switches, relays, and annunciators are important elements in the world of process technology. All three of these items are related and are intended to work together.

Switches allow electrical devices to operate. For example, switches can be used to initiate a startup or a shutdown or some other change, and can even trigger alarms. The passing of information to a switch is often facilitated by a relay. Relays help pass or "boost" signals so they are received properly. An annunciator system announces, via an audible or visual alarm, once a signal is received that is outside of normal range. Annunciators let the process technician know there is a problem.

Switches

A **switch** is a mechanical or electrical device that is used to operate, energize, or de-energize mechanical or electrical circuits for alarm, shutdown, or control purposes using a predetermined operating point or setpoint.

Switches may be operated by hand, actuated by a mechanical signal, or actuated by a process or electrical signal. Switches may be used to control pressure, temperature, level, or flow.

Process field switches can be purchased as normally open (N.O.) contacts, normally closed (N.C.) contacts, or both. The N.O. switch closes its contacts when powered, while an N.C. switch opens when powered. When the process condition is met, the contacts open and create an open circuit. This type of switch is commonly used in safety systems since it also alarms or shuts down the process when electrical power is lost or when the circuit is cut. This type of switch is easy to troubleshoot.

TYPES OF SWITCHES

There are many different types of switches. The following are some of the more common types:

toggle/HOA
alarm
shutdown
autostart
bypass
limit
proximity
vibration
process variable

These types of switches are really of two different abstractions. One abstraction is how the switch is physically constructed and what it senses, while the other is how the switch is integrated into the system (what the system response to that switch will be). If these switches were to be divided logically, they would be categorized as follows:

Physical Switch Types	*System Responses Due to Switch Action*
Toggle/HOA	Alarm
Limit	Shutdown
Proximity	Autostart
Vibration	Bypass
Process variable	

The nine switch types listed here are not mutually exclusive. Any given physical switch type may be used to initiate a given system response. For example, a limit switch sensing mechanical motion may be use to sound an alarm, initiate a shutdown, autostart a process, and/or bypass something. Just to pick one of the system responses, a shutdown switch may have any number of different physical forms: shutdown on vibration or shutdown on process variable (e.g., temp, pressure, flow).

While there are many different kinds of switches in process control (e.g., autostart, bypass, limit, proximity, vibration, and process variable), the most common are alarm, shutdown, and autostart.

Toggle or HOA switches

On/off/auto switches may also be referred to as a toggle, rotary, or HOA switches (Figure 14-1). These types of switches may be set to one of three positions:

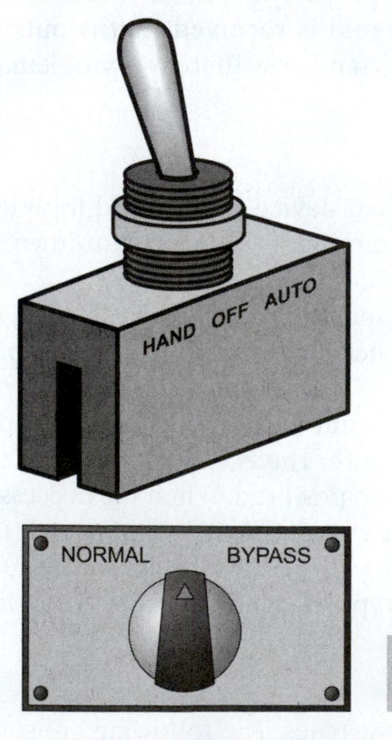

FIGURE 14-1 Toggle and HOA Switch Examples

- hand (manual)—An on switch that allows a process to proceed unless interrupted by safety interlocks
- off—An off switch that prevents a process from proceeding
- auto—A switch that allows for the automatic start of a process at a preset condition

Figure 14-2 shows an example of an autostart.

FIGURE 14-2 AutoStart Example

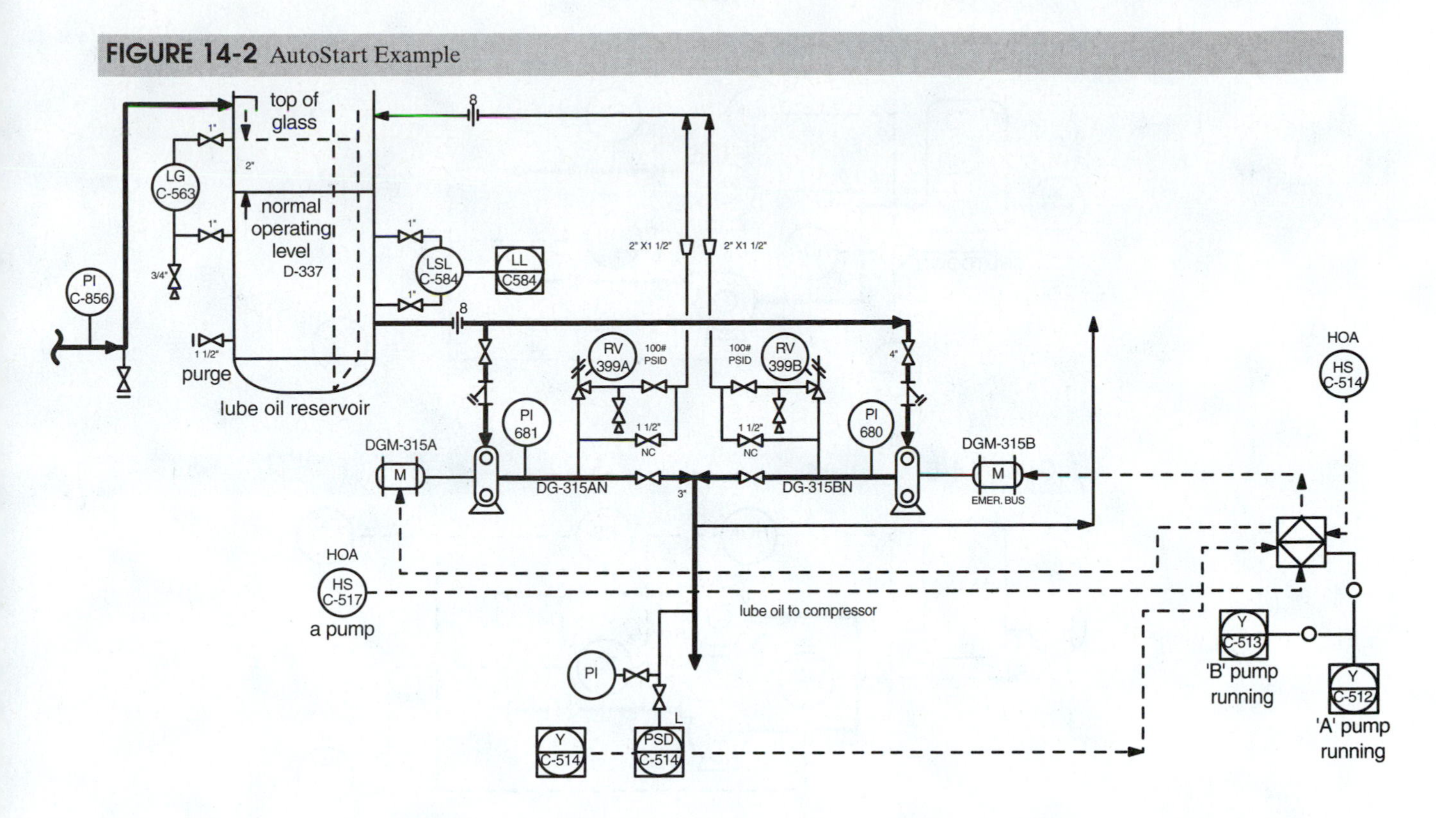

Alarm Switches

Alarm switches, like the one shown in Figure 14-3, are used to notify when a process variable enters an abnormal range (high or low). The switch triggers an alarm (lights and/or a buzzer) that informs the process technician of the condition. For example, a high-pressure switch (PSH) may be calibrated to alarm when pressure increases to 150 psig and clear at 145 psig. The 5-psig difference is called the deadband of the pressure switch.

FIGURE 14-3 Alarm Switch

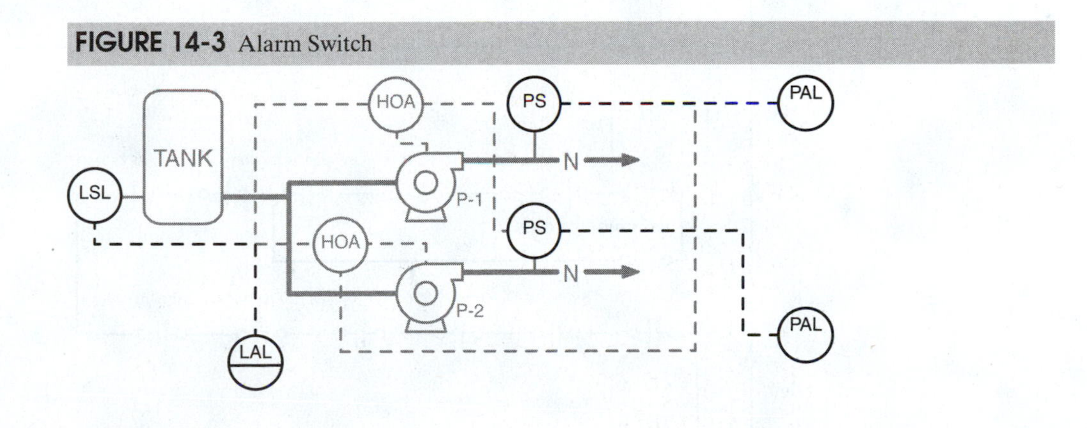

Shutdown Switches

Shutdown switches can be used to actuate a circuit that shuts down a process. In fail-safe situations they can be used to deactivate a circuit or detect an open circuit to shut-down part of the process. Figure 14-4 shows an example of a shutdown switch.

Autostart Switches

Autostart switches are switches that trigger an autostart sequence when predetermined process conditions are met. Figure 14-5 shows an example of an autostart switch.

FIGURE 14-4 Shutdown Switch Example

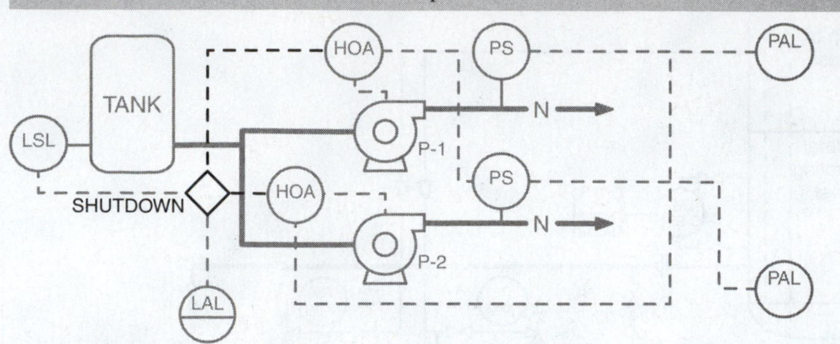

FIGURE 14-5 Autostart Switch Example

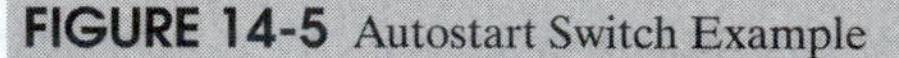

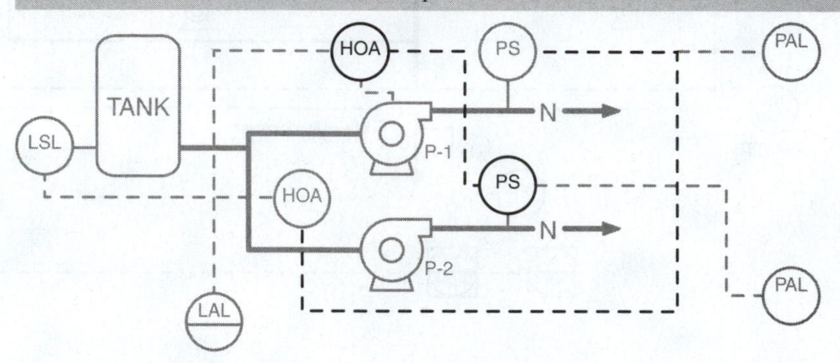

Bypass Switches

Bypass switches are used to override (bypass) the normal operation of a system or device. Figure 14-6 shows an example of a bypass switch.

FIGURE 14-6 Bypass Switch Example

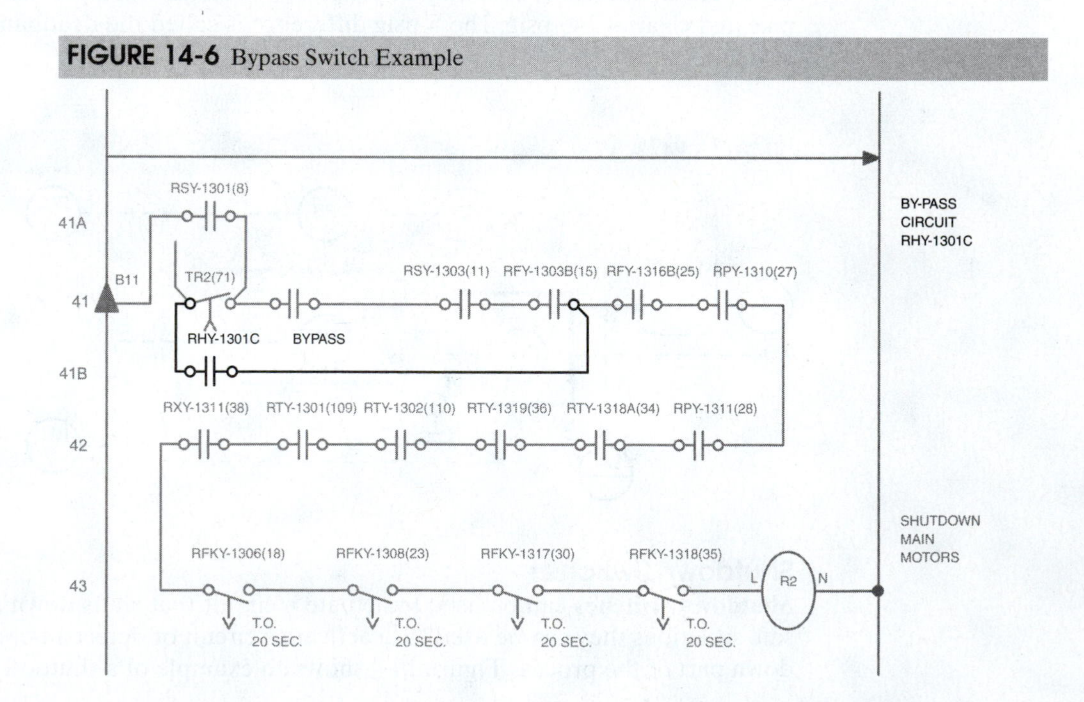

Limit Switches

A **limit switch** is normally used to verify the state or presence of a condition that exists in the process (e.g., if a valve is fully open or fully closed). Figure 14-7 shows an example of a limit switch.

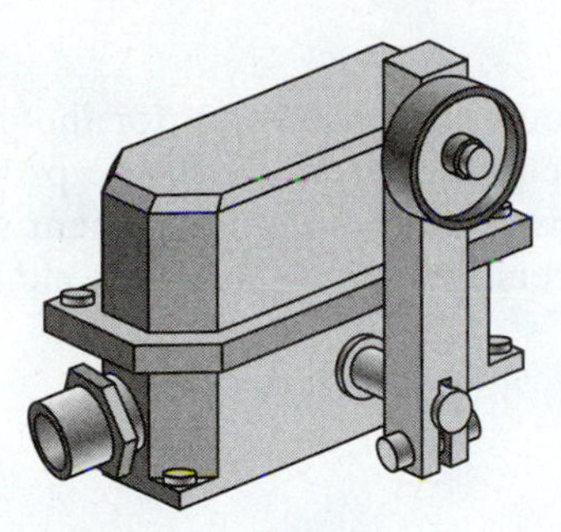

FIGURE 14-7 Mechanical Limit Switch

Proximity Switches

Proximity switches sense the closeness of an object without having to make physical contact as a limit switch does. Some proximity switches use magnetic fields to sense the position of an object, while others use beams of light. These switches are typically sealed against environmental and process conditions (liquid, dust, gases), as are their connecting cables. Figures 14-8, 14-9, and 14-10 show examples of proximity switches.

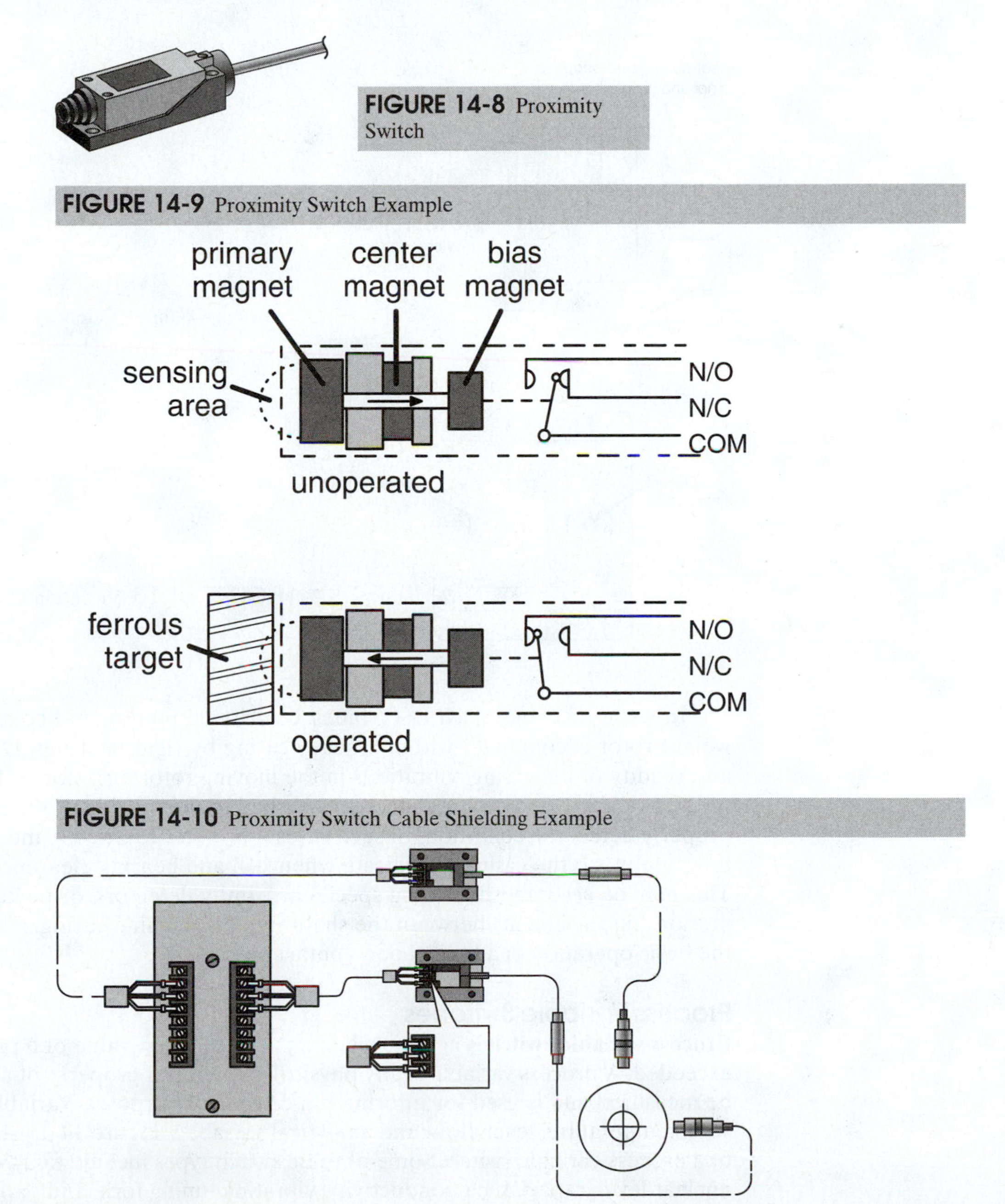

FIGURE 14-8 Proximity Switch

FIGURE 14-9 Proximity Switch Example

FIGURE 14-10 Proximity Switch Cable Shielding Example

Vibration Switches

Vibration switches may be used to determine vibration for the purpose of predicting wear or impending failure on the equipment in which the process is taking place. Vibration switches are commonly used on rotating equipment such as compressors, centrifuges, steam turbines, and blowers. Figures 14-11, 14-12, and 14-13 show examples of vibration switches.

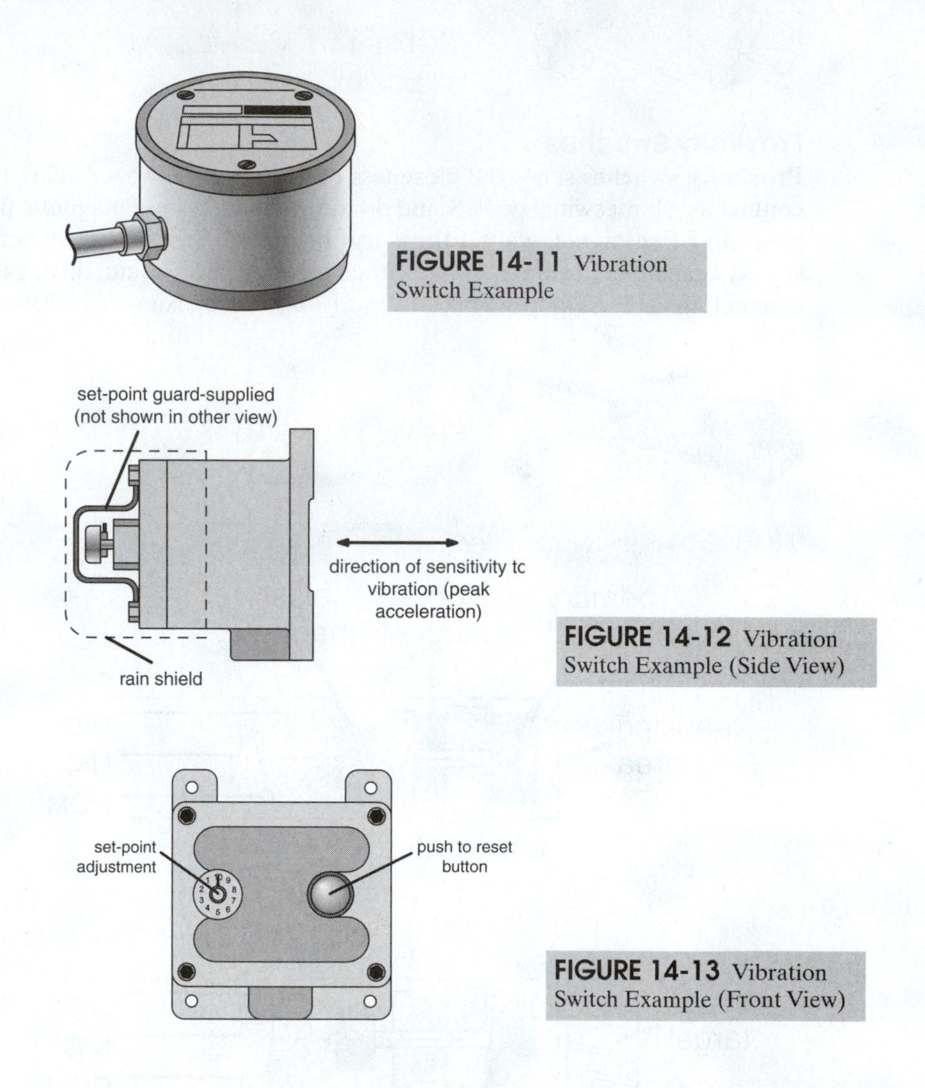

FIGURE 14-11 Vibration Switch Example

FIGURE 14-12 Vibration Switch Example (Side View)

FIGURE 14-13 Vibration Switch Example (Front View)

In some machines such as turbines, centrifugal pumps, and compressors, a lightweight rotor is contained within a heavy casing by rigid bearings. Due to the weight and rigidity of the casing, vibrations in the moving rotor may not be felt on the frame of the machine, so vibration detectors attached to the outside of the machine may not properly detect the condition. In such cases it is better to directly measure shaft vibration relative to the casing to indicate when seal and bearing clearances are in danger. This may be accomplished using special proximity detectors, or pickups, that indicate motion (displacement) between the shaft and the machine casing. Figure 14-14 shows the basic operation of a typical non-contact pickup.

Process Variable Switches

Process variable switches actuate when a predetermined value of a process variable is exceeded. A process variable is any physical or chemical property of a process that can be measured and is used for information or control purposes. Variables include: pressure, temperature, level, flow, and analytical variables. Figure 14-15 shows an example of a process variable switch. Some of these switch types include float, ultrasonic, radar, nuclear level, capacitance, conductivity, vibrating tuning fork, and paddle wheel.

FIGURE 14-14 Noncontact Pickup Example

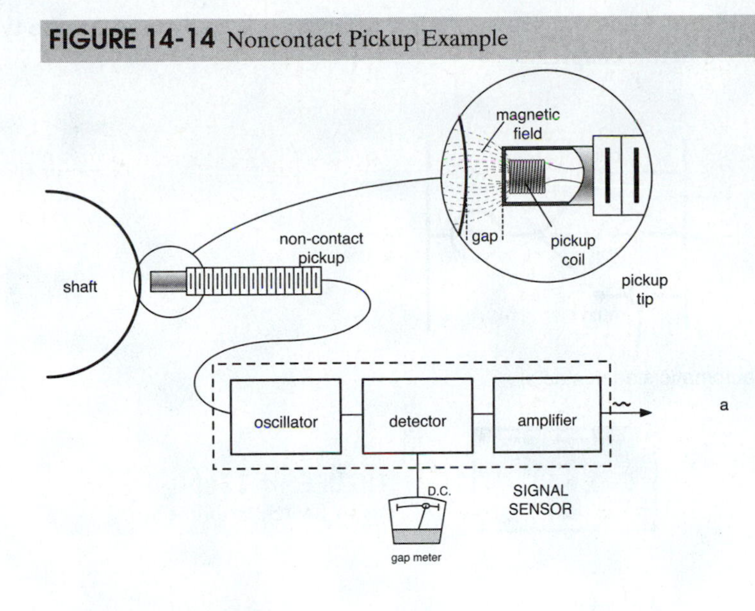

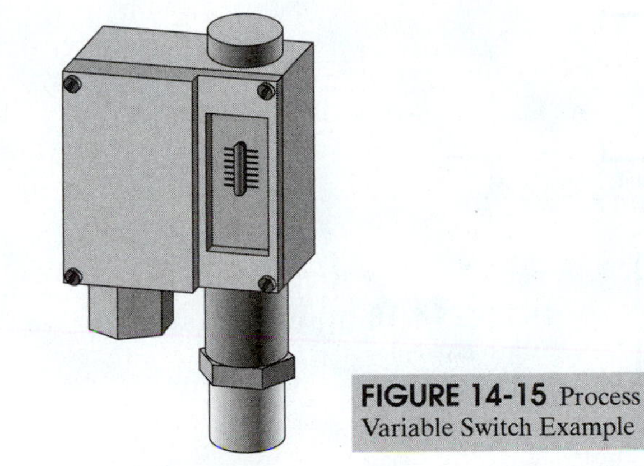

FIGURE 14-15 Process Variable Switch Example

SWITCH SYMBOLS

The piping and instrumentation diagram of a flow control loop shown in Figure 14-16 contains several examples of switches and other piping and instrument details (for more information on symbology, review symbols for switches presented in Chapter 12: Symbology: Process Diagrams and Instrument Sketching).

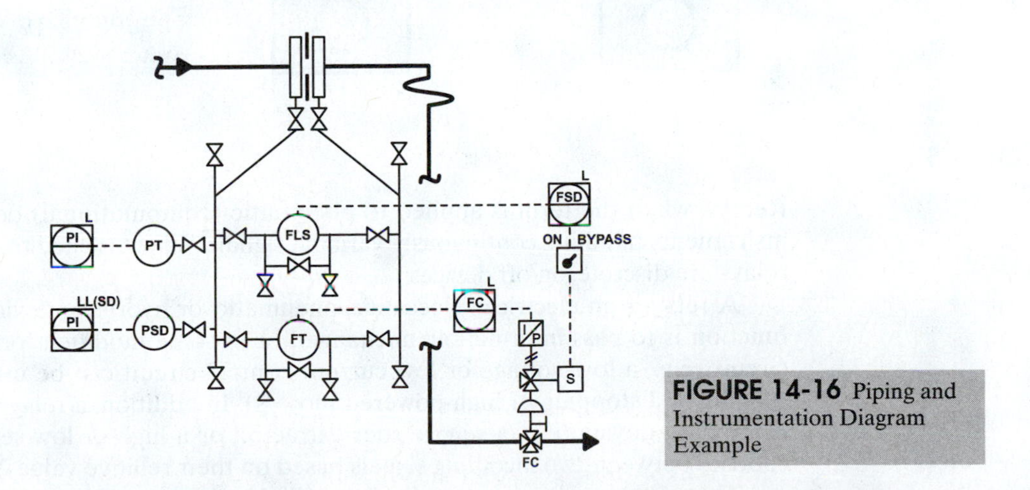

FIGURE 14-16 Piping and Instrumentation Diagram Example

Switches are represented on PFDs and P&IDs using standardized symbols and should be noted in the Legend of the drawing if used. The symbols shown in Figures 14-17,

14-18, and 14-19 are representative of ones commonly used for the types of switches discussed in this chapter.

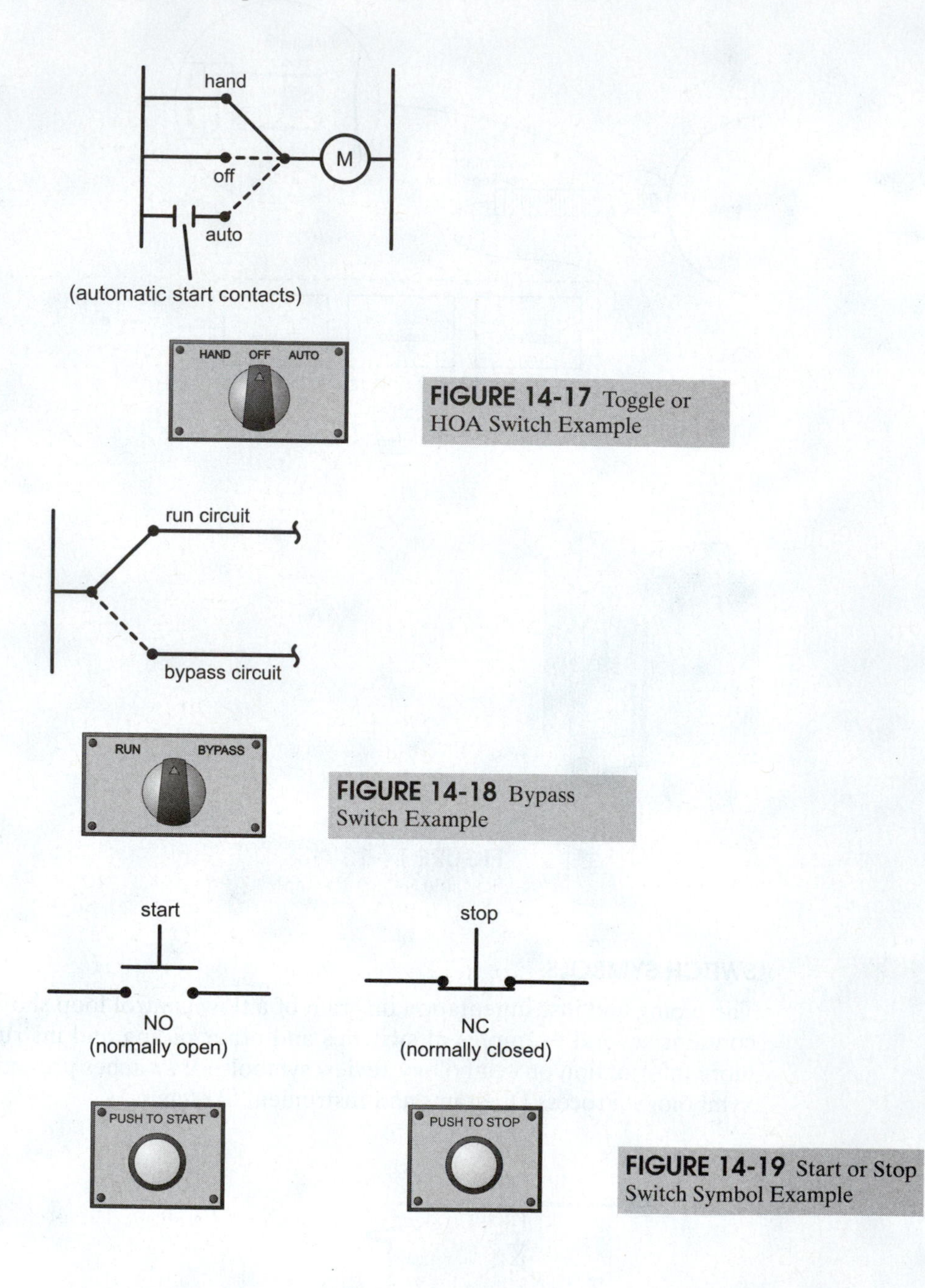

FIGURE 14-17 Toggle or HOA Switch Example

FIGURE 14-18 Bypass Switch Example

FIGURE 14-19 Start or Stop Switch Symbol Example

Relays

Relays, when the term is applied to pneumatic (computational) devices, are analog instruments handling continuously variable signals. When applied to electrical devices, relays are discrete, on/off devices.

A relay is an electrical, electronic, pneumatic, or hydraulic device whose primary function is to pass information, unchanged or in some modified form, to an external circuit (e.g., a low-voltage or low current control circuit can be used to control the starting and stopping of high-powered motors). In addition, a relay may be used as a computing device (e.g., a square root extractor, or a high or low selecting relay that chooses between two incoming signals based on their relative values).

The term relay is more specifically applied in this chapter to an electric, pneumatic, or hydraulic switch that is actuated by a signal from a control circuit. Generally, however, a relay describes an electromechanical device that is operated by a relatively low power signal and is used to control a higher power source through one or more sets of electrical contacts. This allows a control system to turn on high-power devices remotely.

TYPES OF RELAYS

There are several types of relays. These include pneumatic, hydraulic, electrical, electronic (on/off control), timing, pneumatic booster and selection. The following table lists each of these relay types and a brief description of each.

Relay Type	*Description*
Pneumatic and hydraulic	Used to perform mathematical operations, signal conditioning, and selection or modification operations.
Electronic (on/off control)	Used to control one or more circuits that are not normally part of the control circuit.
Timing	May be electrical, mechanical, or electromechanical; used to determine the elapsed time between the start and stop of related operations or events (e.g., a timing relay may be used to start an agitator 15 seconds after the pump starts).
Pneumatic booster	Contains a high-pressure or volume source that increases the signal value or volume.
Selection	Can take two or more inputs and chose a predetermined value (e.g., high, medium, or low).

RELAY APPLICATIONS

Relays can be used for several applications. These applications include maintenance, modifications, selection, and computation.

Maintenance

Many electromechanical relays are used to control the flow of a signal through a logical sequence to ensure safety and/or product quality.

In a pneumatic control system, where there is an unusually long distance between the controller and the control valve, an amplifying or "booster" relay may be necessary. A booster relay maintains a pressure signal by producing a greater volume than that which a controller is capable of supplying to the control valve. In essence, it matches the pressure but increases the quantity of air, allowing an actuator to move faster, producing better control.

Modifications

Signals sometimes need to be modified (e.g., some control valve actuators require a 6–30 psi signal. A typical pneumatic signal is 3–15 psi, so the signal must be converted). Although a current to pneumatic transducer is called a converter (4–20 mA to 3–15 psi), it is also considered to be a relay (ISA identifies it as a relay or compute function). A modifying relay can easily accomplish this task.

Selection

Selection relays are used to select a predetermined signal by its value relative to the other signal inputs. For example, consider a scenario where three redundant transmitters are being used to measure a very important process pressure. If a single transmitter's signal were to be lost or inaccurate, the product quality could be in serious trouble. In a situation such as this, a middle select relay would be the most appropriate solution since it would average the two signals whose values were the closest together and consider the third signal to be an inaccurate or failed signal.

Computation

Computations are an important part of control systems. Controllers have components called comparators that provide different mathematical functions. Since they are part of the controller, they are not considered to be relays. However, standalone devices such as multiplying, adding, or subtracting devices are considered to be relays (e.g., square root extractors used to linearize the flow rate signal produced by the d/p drop across an orifice).

Annunciator Systems

An annunciator is a type of instrument that provides audible and/or visual alarms in reference to some measured variable. An annunciator can be a stand-alone panel board device or configured as the alarm package of a distributed control system (DCS) (e.g., computer). Newer annunciator systems are programmable so they can reside on the Ethernet or process variable input and use LCD displays. Figure 14-20 shows an example of an annunciator system.

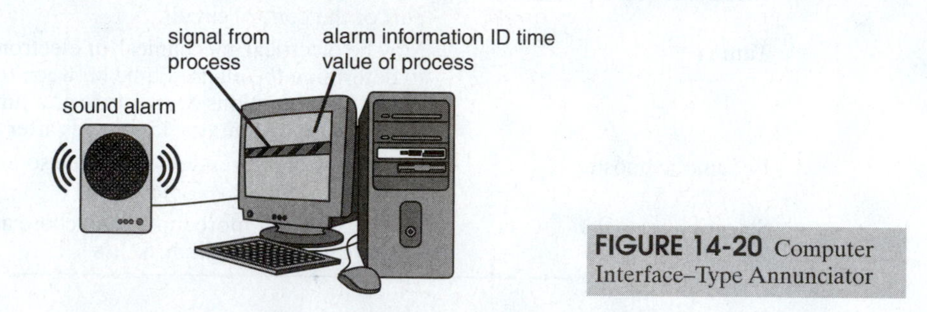

FIGURE 14-20 Computer Interface–Type Annunciator

In a typical annunciator system an alarm panel, is designed to operate as follows:

Operating Condition	*Description*
Normal operating condition	Alarm circuits are not activated (no lights) and the horn circuit is not activated (no audible signal).
Alarm Condition	Alarm circuit is de-energized; flashing lights (normally white) and horn sounds. NOTE: A number of alarm points have a "first-out" provision to indicate which of several conditions occurred first causing a machine to trip off. When a first alarm occurs, these points are audible and display an intermittent fast flash. Subsequent first out alarm points are audible and display a steady fast flash.
First out (first shutdown)	Used to show the order in which the alarms came on. This is very useful for troubleshooting.
Acknowledge	Audible is silenced. The first out alarm lights flash slowly and all subsequent alarms remain steady on. NOTE: Common trouble alarms have two windows. The signal comes from a local panel. Both lights come on and the audible sounds in the alarm condition. "Acknowledge" silences audible in the control room, but both lights continue to show until local acknowledgement is made. Once acknowledgement is made in the control room, the first window in the control room will go off, while the second window remains steady on.
Reset	Used to put the panel indicator back in a ready position after a problem has been corrected.
Test	On panel board systems this functional test, which is usually conducted at the beginning of a shift, triggers an audible signal and intermittent fast flash. To conduct a test, depress **test**, **acknowledge**, and **reset** in that order. Any alarm triggered during a functional test remains locked in until acknowledge and reset are depressed, then the alarm sounds. NOTE: On a DCS or PC system the *test* functionality will differ.

FIGURE 14-21 Alarm Panel Example

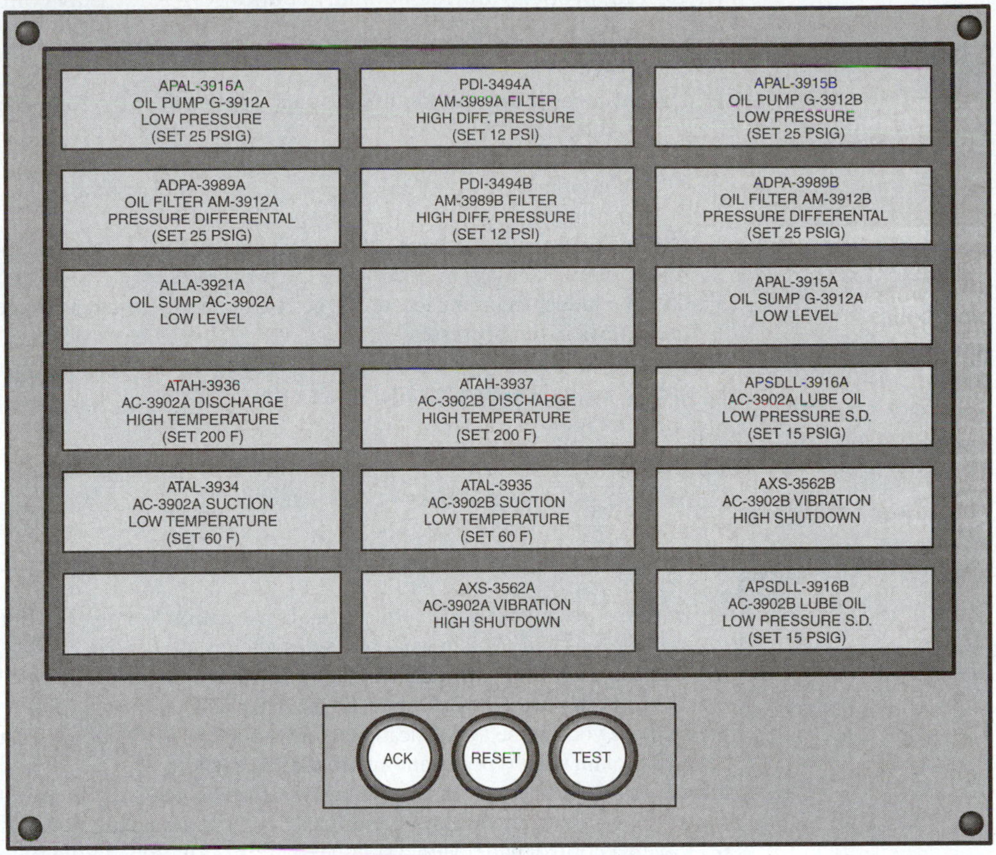

APAL-3915A OIL PUMP G-3912A LOW PRESSURE (SET 25 PSIG)	PDI-3494A AM-3989A FILTER HIGH DIFF. PRESSURE (SET 12 PSI)	APAL-3915B OIL PUMP G-3912B LOW PRESSURE (SET 25 PSIG)
ADPA-3989A OIL FILTER AM-3912A PRESSURE DIFFERENTAL (SET 25 PSIG)	PDI-3494B AM-3989B FILTER HIGH DIFF. PRESSURE (SET 12 PSI)	ADPA-3989B OIL FILTER AM-3912B PRESSURE DIFFERENTAL (SET 25 PSIG)
ALLA-3921A OIL SUMP AC-3902A LOW LEVEL		APAL-3915A OIL SUMP G-3912A LOW LEVEL
ATAH-3936 AC-3902A DISCHARGE HIGH TEMPERATURE (SET 200 F)	ATAH-3937 AC-3902B DISCHARGE HIGH TEMPERATURE (SET 200 F)	APSDLL-3916A AC-3902A LUBE OIL LOW PRESSURE S.D. (SET 15 PSIG)
ATAL-3934 AC-3902A SUCTION LOW TEMPERATURE (SET 60 F)	ATAL-3935 AC-3902B SUCTION LOW TEMPERATURE (SET 60 F)	AXS-3562B AC-3902B VIBRATION HIGH SHUTDOWN
	AXS-3562A AC-3902A VIBRATION HIGH SHUTDOWN	APSDLL-3916B AC-3902B LUBE OIL LOW PRESSURE S.D. (SET 15 PSIG)

ACK RESET TEST

When working with alarm panels like the one in Figure 14-21, it is important to remember that the backlit panels on the display will light up, flash and sound an audible alarm when a condition reaches alarm status. Pushing the Acknowledge (ACK) button shuts off the audible alarm and triggers the flashing lights to go steady on.

Summary

Switches, relays, and annunciators are important in process technology.

Switches, which may be operated by hand, actuated by a mechanical signal, or actuated by a process or electrical signal, may be used for many things including energizing alarms, interlocks, safety systems, equipment, or other devices when a process condition meets a preset value.

While there are many different kinds of switches in process control (e.g., autostart, bypass, HOA, limit, proximity, vibration, and process variable), the most common are alarm, shutdown, and autostart (HOA).

In process technology there are many different kinds of relays including pneumatic, hydraulic, electrical, electronic, timing, pneumatic booster, and selection. Relays assist in signal transfer by increasing (boosting), passing, or maintaining a signal.

If a signal is sent that is out of an acceptable range or that meets preset variable limits, an annunciator may be triggered. Annunciators are instruments that provide audible and/or visual alarms in reference to some measured variable. Annunciators can be standalone panel board devices or configured as part of a distributed control system (DCS).

A typical annunciator system is designed to notify the process technician that the system is operating under normal operating conditions or that an alarm condition is present. If an alarm condition is present, the *First Out* feature on the annunciator lets process technicians know which light came on first. From there they can acknowledge the alarm and/or reset the system. If process technicians wish to verify the system is functioning properly during nonalarm conditions, they may conduct a system test.

Checking Your Knowledge

1. Draw and label the symbols used to represent each of the following switch types:
   a. Autostart
   b. Bypass
   c. On/off/auto
   d. Limit
   e. Proximity
   f. Vibration
   g. Process variable
2. Given a PFD or P&ID, locate all the switches and identify each type of switch. Describe the role each switch plays in controlling the processes.
3. Given a ladder diagram, locate all the relays and identify each type. Describe the role each relay plays in the processes.
4. Explain the purpose of an annunciator system.
5. Define each the following annunciator system terms:
   a. First out
   b. Acknowledge
   c. Reset
   d. Test

Student Activities

1. Explore your home and educational institution for applications of switches. List the applications and explain what types of switches are used in each.
2. Explore your home and educational institution for applications of relays. List the applications and explain what types of relays are used in each.
3. Explore your home and educational institution for applications of alarms. List the applications and explain what types of alarms are used in each.
4. Given a set of actual switches, identify each type and explain how it is used.
5. Given a set of actual relays, identify each type and explain how it is used.
6. Given a control panel with a simulated alarm situation, do the following:
   a. Identify the alarm condition.
   b. Tell which system item was the first out.
   c. Acknowledge the alarm.
   d. Reset the system.
   e. Conduct a test of the system.

CHAPTER 15

Signal Transmission and Conversion

Objectives

After completing this chapter, you will be able to:

- Recall the purpose and operation of transmitters.
- Describe methods for protecting integrity and reliability of signal transmission:
  - heat tracing
  - shielding
  - insulation
  - materials of construction
  - other safety concerns
- Recall types of common transmissions:
  - mechanical
  - pneumatic
  - electrical
  - digital
- Recall purpose and operation of signal converter equipment.
- Describe how wet-leg instrument lines are "zeroed out."
- Describe the effect on a primary sensing element that uses a wet-leg/sealing liquid when the density of the sealing liquid changes.
- Discuss methods used to ensure a dry-leg remains dry.
- Perform scaling calculations:
  - linear signal
  - I/P conversion
  - E/P conversion
  - I/E conversion
  - nonlinear signal
  - square root to linear signal

Key Terms

Digital transmission—a signal transmission method that uses digital devices to create a signal.
E/I conversion—the conversion of a signal from voltage to current.
E/P conversion—the conversion of a signal from voltage to pneumatic.
Electrical transmission—a signal transmission method that uses electrical devices to create a signal.
Heat tracing—the technique of adding heat to process piping, instrument piping, or an instrument in order to keep the process fluids at a constant temperature so signal transmission is not impacted.
I/E conversion—the conversion of a signal from current to voltage.
I/P conversion—the conversion of a signal from current to pneumatic.
Insulation—a protective covering placed around the steam and electric tracing tubes to prevent heat loss and to protect personnel from contact with hot surfaces.
Mechanical transmission—a signal transmission method that uses mechanical methods to create a signal.
P/I conversion—the conversion of a signal from pneumatic to current.
Pneumatic transmission—a signal transmission method that uses compressed air/gas to create a signal.
Shielding—a technique used to control external electromagnetic interference (EMI) by preventing transmission of noise or static signals from the source to the receiver; consists of foil, mesh, or woven wire.
Zeroing out—adjusting a measuring instrument to the proper output value for a zero measurement signal.

Introduction

Transmitters and conversion equipment are integral parts of signal transmission. They send process information or data in different signal formats to displays, controllers, or other transmitters to utilize this information to control a process or part of a process, or sometimes just to convey information for human decisions or verifications.

There are many different types of transmitters and most are categorized by the type of information they convey. When transmitting information, it is important to protect the integrity and reliability of signal transmission since the accuracy of this signal and its transmission determine how well the final control elements perform their intended function.

Signal integrity can be improved or protected using several methods including heat tracing, shielding, insulation, error correction, signal conditioning, and proper construction material selection.

Transmitted information can be in either a mechanical, pneumatic, electrical (analog or digital), or optical format. When sending information between two or more devices, it is sometimes necessary to convert the signal(s) from one format to another. This type of conversion is performed through a piece of equipment called a signal converter.

Transducers are devices that convert energy from one form to another—for example, sound to electrical. So whether the variable is measured and converted or just simply measured the information or data will need to be transmitted in order to be useful for process control.

Signal Transmission

Signal transmission is an important part of data transfer. In order to transmit signals properly, it is important to ensure the integrity and reliability of the signal through such methods as heat tracing, grounding and shielding, insulation, error correction, signal conditioning, and construction material selection.

TRANSMITTER PURPOSE AND OPERATION

Transmitters are devices that encode information or data in a format that is suitable for a transmission medium. In process control applications this information corresponds to process variable data (e.g., flow, pressure, and level). The type of transmitter and transmission medium used are determined by the kind of process variable and its sensor type, physical conditions or environment, and the degree of precision required.

The most common of all transmission modes is the electrical signal. In **electrical transmission** information or data is sampled or collected by a sensor or transducer, then mixed with or encoded onto an electrical carrier signal. (NOTE: Sometimes the carrier signal may simply be a power supply.) In Figure 15-1 the microphone is the sensor-transducer. Sound waves are sampled or collected by the microphone diaphragm that in turn presses on small crystals that cause a voltage change in proportion to the magnitude and frequency of the sound waves hitting the microphone diaphragm. This small voltage signal is the information data or process variable. This signal is applied to a voltage source (the battery) and together they act as a transmitter to convey the data to the speaker (the receiver or final control element).

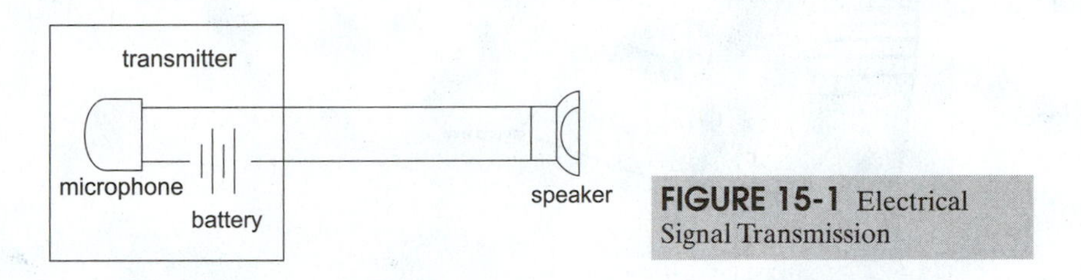

FIGURE 15-1 Electrical Signal Transmission

Although this is a simple illustration, the principle for utilizing any electrical field sensor is similar. If the field sensor is polarity sensitive, then sinking or sourcing supply connections are applicable. Also in the above example, the transmission medium is wire. If this medium was inappropriate, the signal could easily be converted (mixed) to RF (radio frequency) and transmitted in AM or FM. It could also be digitized (analog to digital–A/D) and transmitted over radio or optics.

PROTECTING INTEGRITY AND RELIABILITY OF SIGNAL TRANSMISSION

During a process it is important to maintain the integrity and reliability of signal transmission. To accomplish this, several methods may be employed. These include the incorporation of heat tracing, shielding, insulation, error correction, signal conditioning, and the selection of proper construction materials.

HEAT TRACING

Heat tracing is the technique of adding heat to process piping, instrument piping or an instrument by placing a steam line or electric heating element adjacent to, or coiled around, the device or piping that is to be heated. While the heat tracing itself does not produce a transmittable signal it is an important part of signal transmission because it keeps process fluids at a constant temperature. If the fluid were to change states (e.g., freezing or vaporizing) the impulse of the signal would be incorrect because signal transmission is directly related to fluid flow through the pipe. Figure 15-2 shows an example of a pipe with heat tracing attached.

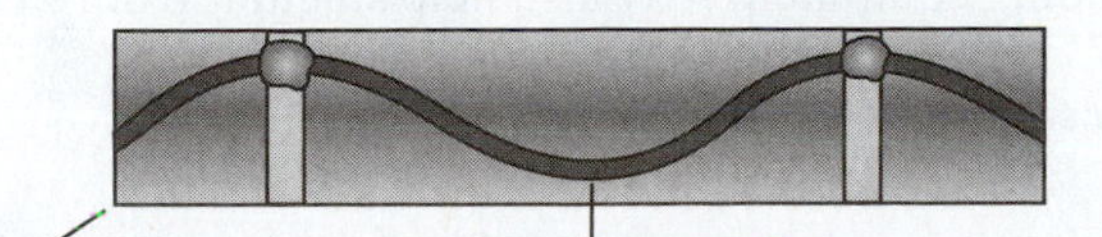

FIGURE 15-2 Heat Tracing

GROUNDING AND SHIELDING

Two of the problems associated with signal transmission are radio frequency interference (RFI) and electromagnetic interference (EMI). RFI and EMI are induced, radiated, or conducted electrical disturbances that can interfere with other signals and cause undesirable responses or malfunctions in electrical or electronic equipment. Failure to reduce or eliminate these RFI and EMI interferences can lead to erroneous signal transmission or signal noise. One way to prevent signal noise is through proper grounding and shielding.

Grounding is the establishment of a conductive connection so harmful current (from an electrical circuit or equipment) is diverted to the earth or some other conducting body that serves in place of the earth. Proper grounding redirects rogue signals away from the intended transmission, thereby reducing interference. Grounding is typically accomplished through a single grounding wire connected directly to the housing of the unit being grounded. Figure 15-3 shows an example of grounding.

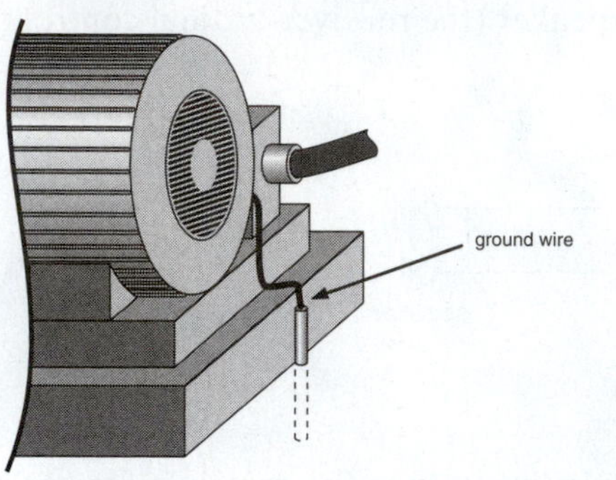

FIGURE 15-3 Grounding

Shielding is another technique for controlling external EMI in signal transmission. In shielding a wire mesh or foil is used to encase the transmission wire. This mesh or foil is then connected to a grounding wire which diverts erroneous signals and reduces interference. Figures 15-4 and 15-5 show examples of shielding and how it is incorporated into a transmission wire.

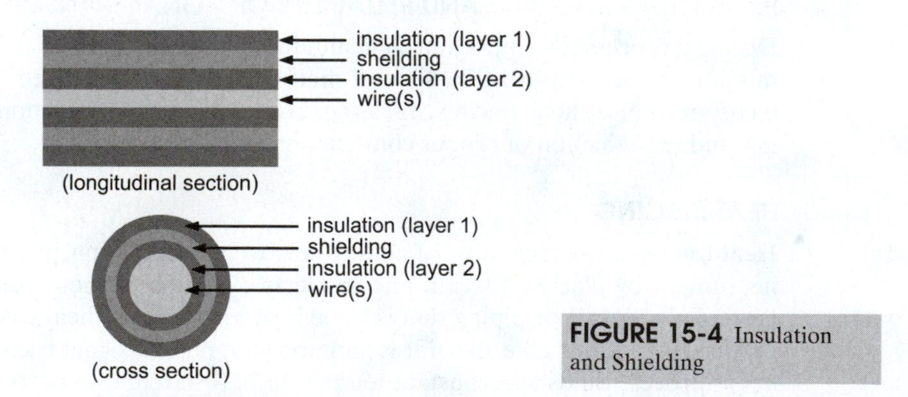

FIGURE 15-4 Insulation and Shielding

Although we incorporate grounding techniques in shielding, actual grounding of equipment is primarily for safety purposes for both people and equipment. It allows for conductivity to earth, thereby protecting equipment personnel from high-voltage shorts. Grounding equipment may also help shield it from signal noise.

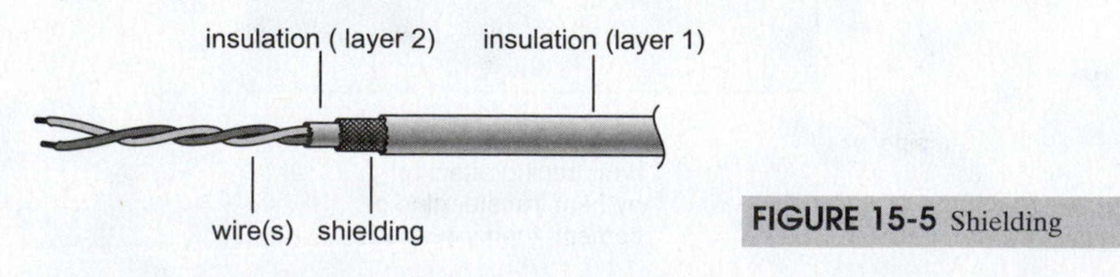

FIGURE 15-5 Shielding

Instrument DC shielded cable should have the shield or ground wire grounded in the control room or the junction box—not in the field. The ground or shield should not be connected in the field to prevent ground loops. Shielding ensures that the control signal is transmitted without interference (e.g., static).

INSULATION

Insulation is a protective covering placed around steam and electric tracing tubes to prevent heat loss and to protect personnel from contact with hot surfaces (Figure 15-6). Insulation may be used on process piping, instrument piping, and instruments.

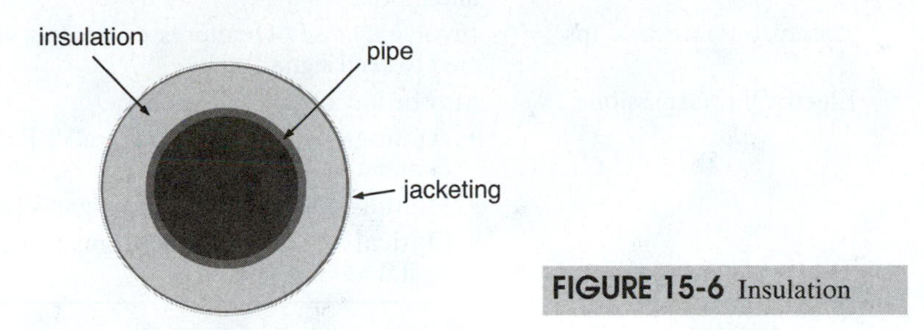

FIGURE 15-6 Insulation

ERROR CORRECTION

In digital transmission information is sampled or converted to binary data (the signal exists in two states, either on (a one) or off (a zero). This on-off signal is usually very reliable and does not require any correction to the data. If, however, there are large noise sources, error correction can be employed using two methods: (1) binary signal transmission, or (2) the addition of error checking bits to the transmission itself.

SIGNAL CONDITIONING

In gathering data with various sensors and transducers, the converted energy or signal may be suitable for the receiver (the controller or final control element) but may need amplification, filtering, or signal conversion (to be discussed later).

Amplification involves increasing the magnitude of a signal voltage or current level increase to a level that allows the transmission to arrive safely at its destination. Signals that have considerable amounts of noise may require filtering to eliminate the noise and enhance data reliability. And finally, to ensure reliable data transmission, the signal may need to be converted to another type more suitable to its environment.

CONSTRUCTION MATERIAL SELECTION

One of the greatest enemies of signal transmission is friction or heat that may be due to the resistive conductive properties of wire joints, splices or connection, or the two physical surfaces rubbing or sliding against each other as in liquid or gas flow. In any case, material selection is critical to maintaining proper signal conductivity or propagation by reducing heat loss or friction. The quality of physical joints, splices, and connections is also critical for ensuring proper conductivity.

OTHER SAFETY CONCERNS

It is very important that proper safety precautions and equipment are implemented and in good working order (e.g., heat tracing is installed and working properly, insulation is not torn or removed, and proper construction materials have been selected). Failure to use or maintain proper safety precautions could result in human and/or equipment loss or injury.

The following are just a few safety issues that could arise as a result of improper safety precautions:

equipment fires
thermal burns (personal)
sensor damage (loss of measurement and control signals)
process piping damage (erosion and corrosion)

Common Types of Transmissions

Transmissions can be in a variety of forms. The most common, however, are **mechanical transmissions**, **pneumatic transmissions**, electrical, and optical. The table below lists each of these transmission types and a brief description of each.

Transmission Type	*Description*
Mechanical transmissions	Include levers, arms, torque tubes, cables, and pulleys and more.
Pneumatic transmissions	Involve the use of compressed gas (air or some other inert gas) to send signals.
Electrical transmissions	May be any of several types: • Analog—Varying signals from VLF (very low) to Gamma ray frequency. • Digital—Discrete or binary signals (having two states). • Optical—Light frequency signals of either analog or digital modes.

ANALOG TRANSMISSION

Analog signals have an infinite number of values and can vary both in magnitude or frequency depending on the mode. Amplitude-varying signals (the usual process control type) contain the intelligence in the magnitude of the signal. For transmission through the air amplitude modulation (AM) is used, which is the mixing of a low-frequency signal (the data) with a higher frequency (the carrier); the LF modulating signal "rides" on the HF carrier signal. Frequency-varying signals (less susceptible to lower-frequency noise) contain the intelligence in the frequency of the signal. Frequency modulation (FM), used for through the air transmission, mixes the lower-frequency data information signal with a higher carrier frequency but differs by having the carrier frequency increase or decrease in proportion to the lower data signal magnitude.

Simple analog transmission in which the voltage or current amount (magnitude) is transmitted or conducted to a controller or receiver is the usual method for process control signals. The method is reliable and inexpensive and satisfactory for low-noise, nonhostile environments. It uses simple electrical circuits. The main disadvantage is the length of transmission–typically 2,500 feet—and its susceptibility to noisy environments. Standard 4–20mA, 0–10V, and 0–50mV loops utilize this method of transmission but usually also utilize some kind of noise filtering and/or signal amplification or buffering.

DIGITAL TRANSMISSION

Digital transmission consists of either retransmitting digital pulses that are already produced by the field sensor (rotary encoders, etc.) or by converting the analog signal containing the process control variable data or information into a digital representation. The latter is normally achieved by an analog-to-digital or A/D converter.

The advantage of converting signals to binary for transmission is the signal has very good noise immunity and simple buffering filters noise. The disadvantage is that it is more expensive and generates its own noise, which is harmful to other signals. There are standards (IEEE-488, RS-232, Ethernet, Field-bus, etc.) for the various digital transmission methods, but compatibility between the transmitting field device and the controller, final control element, or receiver must be considered.

OPTICAL TRANSMISSION

Fiber-optic transmission is becoming more and more common in process control because of cost reductions, noise immunity, and intrinsically safe operating characteristics. Light transmitters (LEDs or laser diodes) and receivers (photo diodes, photo transistors, or photo-resistive cells) used with fiber-optic cabling make up a complete circuit. The transmitter is modulated with the process variable information that is conveyed over the optical cable to the receiver, which senses or detects the modulated signal contained in the light signal. The process variable data may either be analog or digital but is typically digital.

Signal Conversion

In order for a signal to be transmitted and interpreted properly, it is sometimes necessary to convert the signal from its original form to something the receiving device can understand. Transducers are the devices used to perform this type of conversion.

PURPOSE AND OPERATION OF SIGNAL CONVERTER EQUIPMENT

The purpose of signal conversion is to change the form of one signal value into another signal value to ensure device or signal compatibility. Transducers perform this type of conversion.

The following are common signal conversions:

- **I/P** = current to pneumatic
- **P/I** = pneumatic to current
- **I/E** = current to voltage
- **E/I** = voltage to current
- **E/P** = voltage to pneumatic
- I/F = current to frequency
- F/I = frequency to current
- E/F = voltage to frequency
- F/E = frequency to voltage

"ZEROING OUT" WET LEG INSTRUMENTS

Wet leg is a reference leg containing a liquid. It is a reverse transmission instrument.

Zeroing Out

Zeroing out means adjusting a measuring instrument to the proper output value for a zero measurement signal.

Zero Suppression

Zero suppression is the compensation applied to the lower range value when the actual zero reference point is below the calibrated lower range value.

A differential pressure level transmitter is located below the tank with impulse tubing connecting them together.

Zero Elevation

Zero elevation is the compensation applied to the lower range value when the actual zero reference point is above the calibrated lower range value.

A differential pressure transmitter with a wet leg is used to measure level.

SCALING CALCULATIONS

Scaling calculations convert process signals so they are compatible with instrument and control systems. Through scaling it is possible to equate the numerical value of one scale to its mathematically proportional value on another scale.

For example, measurements applied to a standard analog electronic temperature transmitter are represented on an appropriate temperature scale, while the output signal is represented on a milliampere scale.

Figure 15-7 shows an example of various scale types and how they relate to one another.

Instrument Scale Definitions

In order to properly interpret and discuss various scales, it is important to know a few key terms—specifically, upper range value (URV), lower range value (LRV), range, and span.

FIGURE 15-7 Scaling Chart Comparison

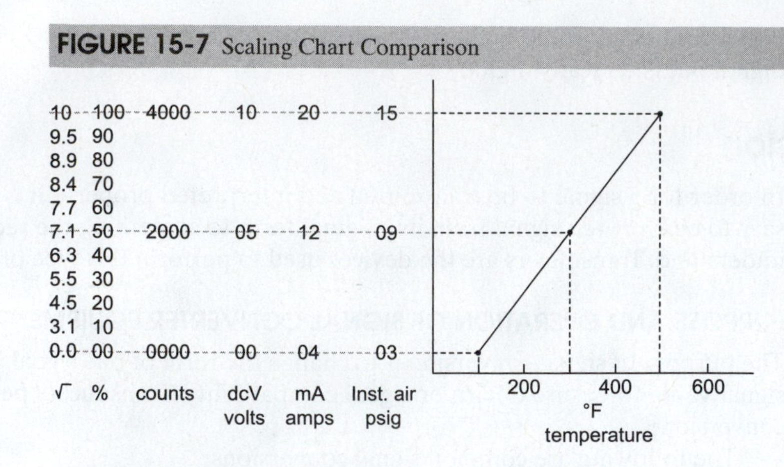

Upper range value (URV) is the number at the top of the scale expressed as one number.

Lower range value (LRV) is the number at the bottom of the scale expressed as one number.

Range is the set of values that exist between the upper and lower range values (URV and LRV) of a scale. Range is expressed as two numbers (e.g., a calibration range of 50–150 psig).

Span is the algebraic difference between the upper range value of a scale minus the lower range value of a scale expressed as one number (e.g., span = 100 psig)

Figure 15-8 provides a visual representation of upper range value, lower range value, range, and span.

FIGURE 15-8 Example of URV, LRV, Range, and Span

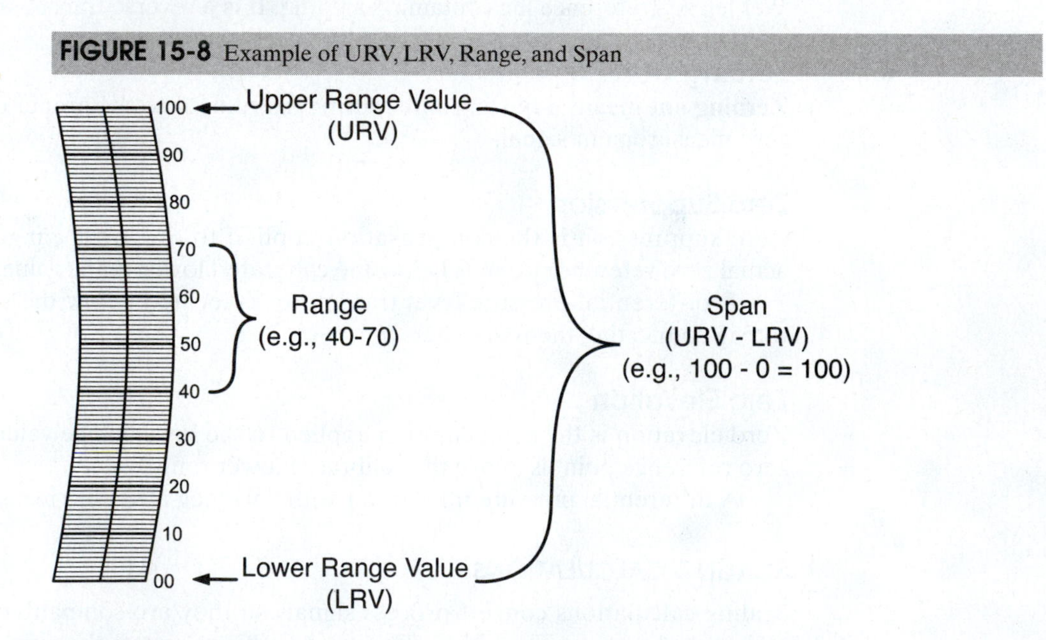

Linear Signal Conversion

Linear scaling is used when there is a linear relationship between two scales (e.g., transmitter input versus output). The following is a formula for conducting a linear conversion:

$$\text{VALUE}_B = \frac{\text{VALUE}_A - \text{LRV}_A}{\text{SPAN}_A} \times \text{SPAN}_B + \text{LRV}_B$$

I/P Conversion

I/P conversion is the conversion from current to pneumatic.

E/P Conversion

E/P conversion is the conversion from voltage to pneumatic.

I/E Conversion

I/E conversion is the conversion from current to voltage.

Nonlinear Signal Conversion

Flow rate is the flow rate through an orifice meter run is proportional to the square root of the differential pressure created by the orifice. Consequently, the need to calculate flow rate as a result of a square root number is required.

Percent d/p is a measurement of the actual differential pressure drop compared to the d/p at 100 percent flow. If a d/p transmitter is calibrated to measure 0–100 inches water column (in. w.c.), then a measurement of 25 in. w.c. would also be 25 percent d/p. In order to convert the differential pressure into percent flow rate, use the following equation:

$$\text{Percent flow rate} = \sqrt{\text{d/p\%}}$$

Step	*Equation*
Insert the % d/p into the equation.	Percent flow rate $= \sqrt{25\%}$
Change percent into a decimal form.	Percent flow rate $= \sqrt{0.25}$
Take the square root.	Percent flow rate = 0.50
Make the decimal a percent again.	Percent flow rate $= 0.50 \times 100$
Record your answer.	Percent flow rate = 50%

Square Root to Linear Signal Conversion

Refer to Figure 15-7 for the conversion.

Summary

Transmitters and conversion equipment are integral parts of signal transmission. Transmission is the conveying of information from one place (the sample point) to another (the endpoint) via some kind of transmission medium (transmitter).

Information from the real world (physical data) is imperative in process control.

Transmitters send process information or data in different type signal formats to displays, controllers, or other transmitters to utilize this information to control a process or part of a process, or sometimes just to convey information for human decisions or verifications.

There are many different types of transmitters and most are categorized by the type of information they convey (e.g., temperature, pressure, level, flow, and computer machine or controller status). The operation of these various process control transmitters is relatively simple in that they all convey the sampled data, either in the same format it is gathered in, or possibly the format signal is converted to a more appropriate one because of the environment the transmission is occurring in.

When transmitting information, it is important to protect the integrity and reliability of signal transmission. The accuracy of this signal and its transmission determine how well the final control elements perform their intended function. The integrity of these signals or data must be maintained by transmitters and converters, or additional error compensation is required. This may be accomplished using several methods including heat tracing, shielding, insulation, error correction, signal conditioning, and proper construction material selection. Transmitted information can be in either a mechanical, pneumatic, electrical (analog or digital), or optical format. When sending information between two or more devices, it is sometimes necessary to convert the signal(s) from one format to another (e.g., from pneumatic to electrical). This type of conversion is performed through a piece of equipment called a signal converter. Some of the most common conversions are P/I, I/P, and I/E (or E/I).

The actual sampling or measurement of a process variable is performed by a sensor. This sensor is specifically designed for the type of variable to be measured and has operating characteristics unique to that design and is usually of two categories, contacting and noncontacting. These variables include temperature, color, light, density, chemical properties, force, weight, motion, position, flow, level, pressure, sound, particle radiation, electromagnetic radiation, and electrical.

Transducers are devices that convert energy from one form to another. For example, sound to electrical (e.g., a microphone even though this device also serves as a sensor). So whether the variable is measured and converted or just simply measured the information or data will need to be transmitted in order to be useful for process control.

Checking Your Knowledge

1. Explain the purpose of transmitters and how they operate.
2. The following are methods used to protect the integrity and reliability of signal transmission. Briefly describe each method and explain what each one is for. Then, identify possible safety concerns if these methods are not employed properly.
   a. Heat tracing
   b. Shielding
   c. Insulation
3. Explain how improper selection of construction materials can impact the integrity and reliability of signal transmission.
4. List four examples of safety issues that could arise if heat tracing, shielding, and insulation are not installed or implemented properly.
5. The following are four common transmission types. Briefly explain each type and what it is used for.
   a. Mechanical
   b. Pneumatic
   c. Electrical
   d. Digital
6. Explain the purpose of signal conversion equipment.
7. Explain how to "zero out" a wet leg instruments.
8. Explain the effect of when one impulse line is required for measurement versus when two impulse lines are required?
9. Discuss methods used to ensure a dry-leg remains dry.
10. Match the abbreviations below to the type of signal conversion they represent.

Abbreviation	*Conversion Type*
a P/I ______	1. Current to pressure
b. I/P ______	2. Voltage to current
c. E/I ______	3. Pressure to current
d. I/E ______	4. Current to voltage

11. On the scale below; match the letters to the appropriate value type, then record the numerical values for each:

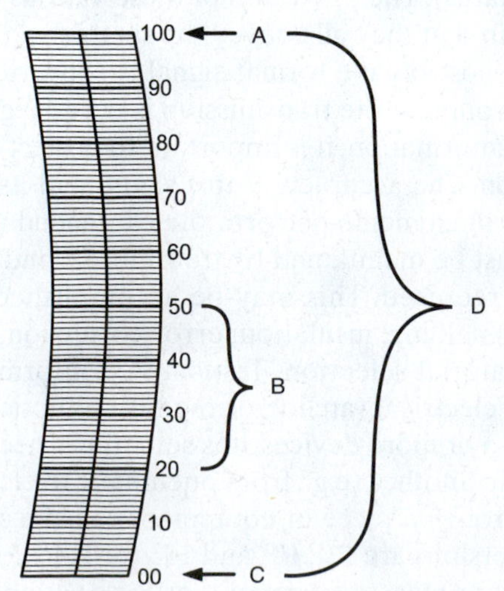

12. Perform a linear signal conversion.
13. Work the following I/P signal conversion equations alone or as a team:
    a. A 4–20-mA to 3–15-psig converter has an output of 8 psig. What is the input value in milliamperes?
    b. A 4–20-mA to 3–15-psig converter has an output of 11.8 psig. What is the input value in milliamperes?
14. Work the following E/P signal conversion equations alone or as a team.
    a. Due to the temperature range in a tank, a specific thermocouple produces between 2.547 mV and 4.868 mV. The output of this E/P converter is 3–15 psig. If the input is 3.75 mV, what is the converter's output?
    b. Following from the previous question, if the output of this E/P converter is 13.2 psig, what is its input in mV?
15. Work the following I/E signal conversion equations alone or as a team.
    a. A 4–20-mA to 1–5-VDC converter has an output of 2.5 VDC. What is its input value in milliamperes?
    b. A 4–20-mA to 1–5-VDC converter has an output of 4 VDC. What is its input value in milliamperes?
16. Work the following non-linear signal conversion equations alone or as a team.
    a. The differential pressure measured at an orifice meter run equates to 38 percent d/p. What is the percent flow rate through the meter?
    b. A 4–20-mA to 1–5-VDC converter has an output of 2.5 VDC. What is its input value in milliamperes?
    c. A 4–20-mA to 1–5-VDC converter has an output of 4 VDC. What is its input value in milliamperes?
       HINT: The output of a standard transmitter would have a linear relation to its input (measurement); therefore, take its output and change it into a percent.
    d. If the flow indicator in a control room is reading 120 gpm on a scale marked 0–250 gpm, what is the mA output of the standard d/p transmitter in the field?
       HINT: Change 120 into a percent number first.
    e. The output of a standard differential pressure transmitter measuring the d/p across an orifice is reading 7.5 mA. What is the percent flow rate through the meter run?
       HINT: The output of a standard transmitter has a linear relation to its input (measurement).
17. Perform a square root to linear signal conversion.

Student Activities

1. Perform a zero-based calibration of a differential pressure, gauge pressure, or absolute pressure transmitter. It is recommended that a calibration graph, calibration report, and a lab report be included with this exercise. If a lab exercise has not been developed, use the calibration and documentation procedures in the manufacturer's instruction manual.
2. Perform a zero suppression calibration of a differential pressure, gauge pressure, or absolute pressure transmitter. It is recommended that a calibration graph, calibration report, and a lab report be included with this exercise. If a lab exercise has not been developed, use the calibration and documentation procedures in the manufacturer's instruction manual.
3. Perform a zero elevation calibration of a differential pressure, gauge pressure, or absolute pressure transmitter. It is recommended that a calibration graph, calibration report, and a lab report be included with this exercise. If a lab exercise has not been developed, use the calibration and documentation procedures in the manufacturer's instruction manual.
4. Using a table with an mA-input column and a psig-output column (corresponding to an I/P transducer). Determine the input values based on 10%, 30%, 50%, 70%, and 90% of the analog input signal (4–20 mA). Then calculate the output values based on a 3–15-psig signal. Finally, set up an I/P converter (transducer) in the lab and verify the answers by applying the mA-input values to the instrument and reading its output.

5. Set up an I/P converter (transducer) in the lab and check its calibration. To do this, you will have to understand its input/output relationship. Then calculate several additional output values based on random input values verifying each result with the I/P transducer.
6. Using examples of signal transmission lines, locate and identify the various transmission signal types. It is recommended that this activity be done independently.

CHAPTER 16

Controllers

Objectives

- Define the following terms associated with process dynamics:
  process equilibrium (material balance)
  load change
  dead time or lag time
  over range
  spanning
- Define the following terms associated with tuning in process control controllers:
  indirect controller action
  on/off (narrow band proportional) control
  feedback control scheme
  feedforward control scheme
- Define the following terms associated with process control and controllers:
  proportional (gain) and proportional band
  integral (reset)
  derivative (rate) or preact
  windup
  offset
  tuning
- Define the following terms associated with process control and controllers:
  trend
  hunting
  flatline
- Given a drawing, picture, or actual trend, distinguish between effects of proportional, integral, and derivative control.
- Discuss the necessity for tuning a controller.
- Explain how to differentiate between an upset caused by a controller malfunction, as opposed to a process change, some external cause, etc.
- Demonstrate a correct method for tuning a loop.

Key Terms

Chart recorder—a device the physically or electronically records data that process and instrument technicians can use to troubleshoot process trends and instrumentation problems.

Cycling—the moving or shifting of a process variable above and/or below the setpoint.

Dead time—the elapsed time between the initiation of an input (measured, controlled, or manipulated variable), change, or other stimulus and the point at which the resulting response can be measured or observed.

Deviation—the difference between the setpoint and the controlled variable under dynamic (changing) conditions.

Direct acting controller—a controller whose output signal value increases or decreases as the controller input signal value increases (increased input = increased output) or decreases.

Feedback control—a closed loop control strategy where the difference in the setpoint and measurement drives the output of the controller to the final control element.

Feedforward control—an open loop control strategy where the controller output is based on knowledge (usually mathematical modeling) of the relationship between the output of the controller and its input received from the point in the process where the disturbance occurs.

Hunting—a condition that exists when a control loop is improperly designed, installed or calibrated, or when the controller itself is not properly aligned; this condition will be observed by the process technician as a random behavior or cycling above and/or below the setpoint (desired control point).

Lag time—a relative measure of the elapsed time between two events, states, or processes.

Load change—a change in any variable in the process that affects the value or state of the controlled variable.

Offset—the difference between the setpoint and the controlled variable under steady-state (not changing) conditions.

On/off control—control used when controlled variable cycling above and below the setpoint is considered acceptable, or when the process quickly responds to an energy change, when the energy change itself is not too quick.

Over range—the condition that exists when the signal value of a device or system exceeds the maximum allowable value.

Process equilibrium—the condition that exists when there is a balance of material and energy within a given system or process; also referred to as "steady state."

Proportional action—a controller output response that is proportional to the amount of deviation of the controlled variable from the setpoint.

Reverse acting controller—a controller whose output signal value decreases as the controller input signal value increases (i.e., increased input = decreased output).

Spanning—the process of setting or calibrating the lower and upper range values of a device; also the process of testing a device or system by traversing up and down the scale through the entire calibrated or operational range.

Steady state—a trend condition in which the process is in equilibrium and running smoothly producing a relatively straight line on a graph or chart.

Trend—the plotting of a process variable over time.

Windup—the saturation of the controller output signal to its minimum or maximum value if the process variable does not return to the setpoint.

Introduction

When studying controllers and process dynamics, it is important to understand that processes have certain behaviors that are common and predictable. For example, process can be in either a state of equilibrium (steady state) or in a state of change. During states of change, a process may experience load change, dead time, or spanning, or be over range.

When tuning process controllers, you may experience direct or indirect controller action and the control schemes may include on/off control, feedback control, or feed-forward control.

Other key terms that may be encountered with regard to process control and controllers include, gain, reset, rate, windup, offset, and tuning.

As processes occur, chart recorders record and document operations and maintenance data, which can be a very important troubleshooting tool. Through the use of historical graphs and charts, important observations can be made with regard to trends. From these trends, a process technician can determine if there are problems in the control loop (hunting) or if the process is running smoothly in a state of equilibrium (flatline).

Process Dynamics and Control

Processes behave in certain recognizable ways. Process instruments are used to maintain a desired condition or to alter a process in order attain a desired condition. There are several terms associated with process dynamics. These include process equilibrium, load change, dead time or lag time, over range, and spanning.

PROCESS EQUILIBRIUM

Process equilibrium (Figure 16-1) is the condition that exists when there is a balance of material and energy within a given system or process. This condition is also referred to as **steady state**.

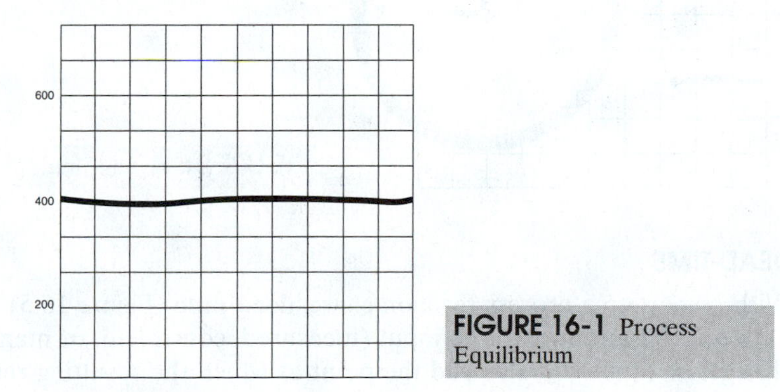

FIGURE 16-1 Process Equilibrium

LOAD CHANGE

Load change (Figure 16-2) is a change in any variable in the process that affects the value or state of the controlled variable (e.g., more cooling, more heat, higher flow rate). More specifically, when a load change occurs, the process will move to a new steady state value as show in Figure 16-3.

LAG TIME

Lag time (Figure 16-4) is a measure of the elapsed time between two events, states or processes. For example, a process lag time is the elapsed time from when the process change can be measured or observed to the time it reaches a new equilibrium value. Process examples include measurement lag time, transmission lag time, and controller lag time.

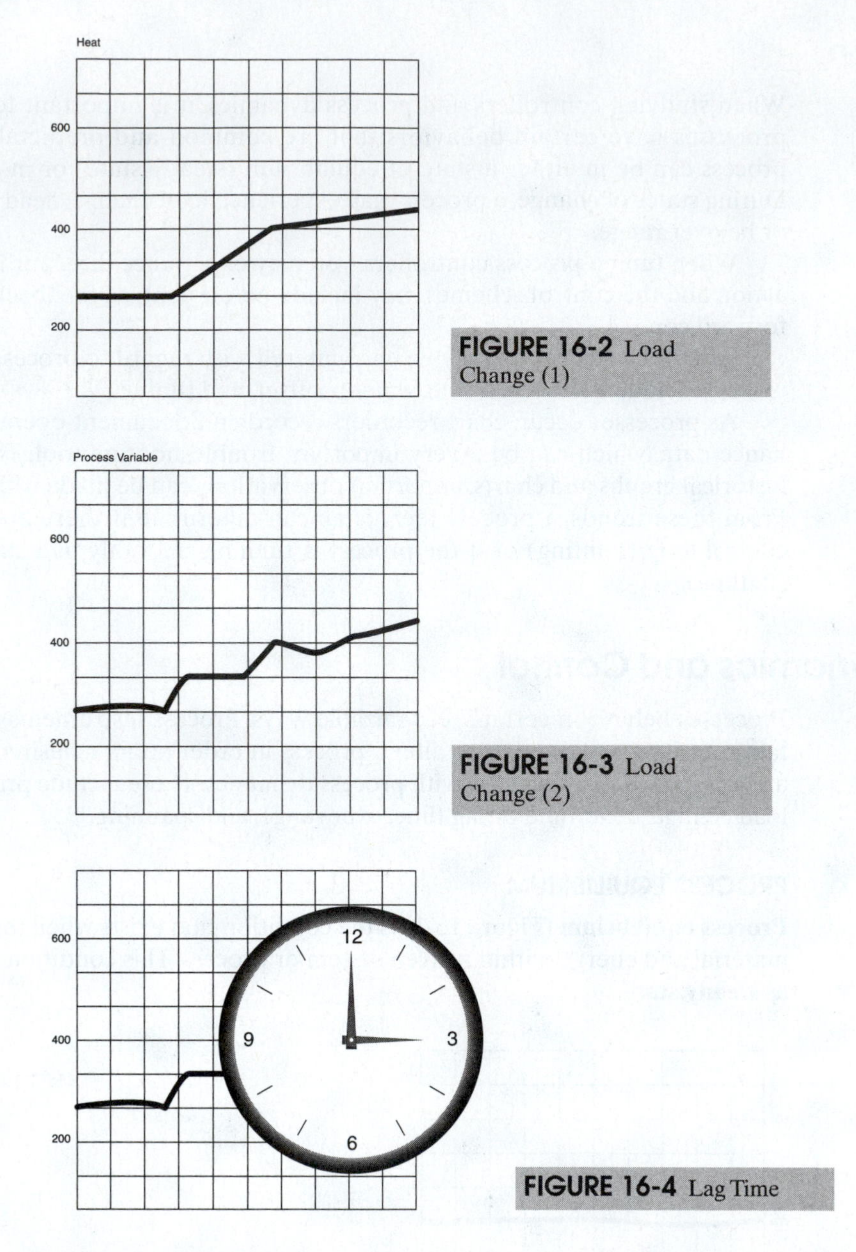

FIGURE 16-2 Load Change (1)

FIGURE 16-3 Load Change (2)

FIGURE 16-4 Lag Time

DEAD TIME

With respect to a process reaction curve, **dead time** (Figure 16-5) is the elapsed time between the initiation of an input (measured, controlled, or manipulated variable), change, or other stimulus and the point at which the resulting response can first be measured or observed.

OVER RANGE

Over range (Figure 16-6) refers to the maximum or minimum allowable values. For example, most transmitters can output less than 4 mA and more than 20 mA even though they have been calibrated to specific process variable to 4 to 20 mA. The same over or under range values can occur with controller input and output signals. These values may or may not be displayed on the controller.

- device range limits (e.g., transmitter ABC may have a range of 0 to 62.5 feet)
- device span limits (e.g., transmitter ABC could be calibrated (spanned) 0 to 30 feet for a 33-foot-high storage tank)
- operational limits (e.g., the tank mentioned above may be operated only from 5 to 25 feet)

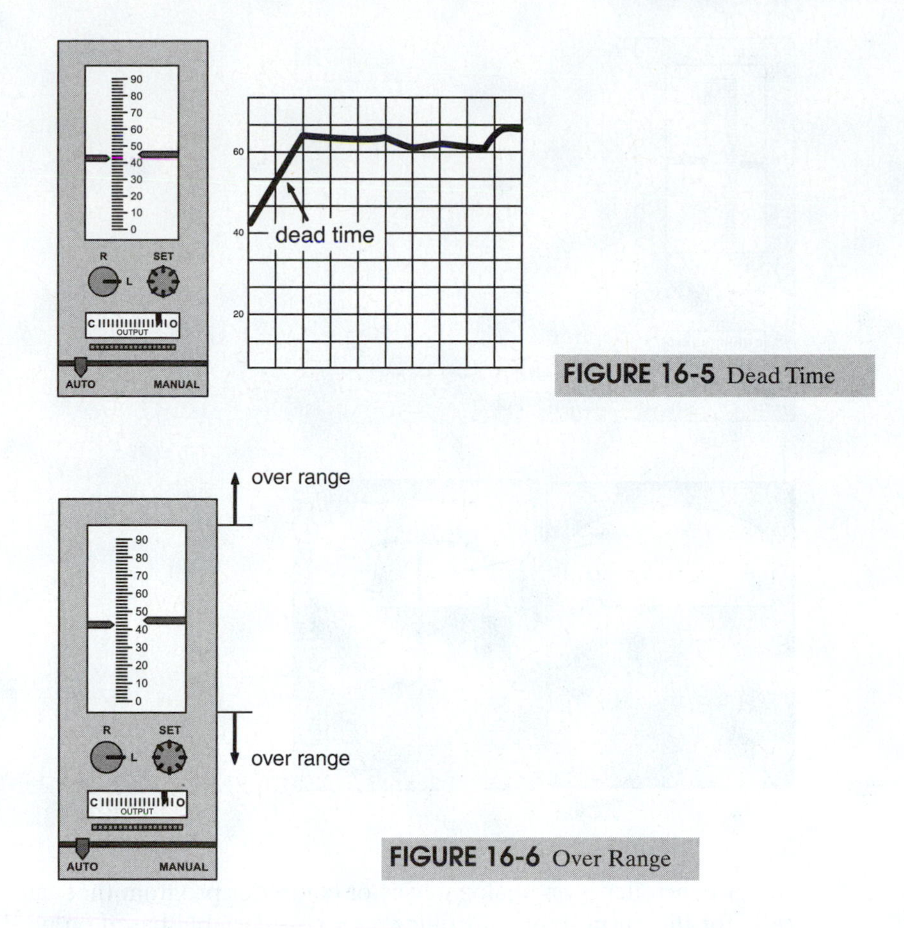

FIGURE 16-5 Dead Time

FIGURE 16-6 Over Range

SPANNING

Spanning (Figure 16-7) is the process of setting or calibrating the lower and upper range values of a device. Testing a device or system (transmitter, controller, etc.) can be done by traversing up and down the scale through the entire calibrated range.

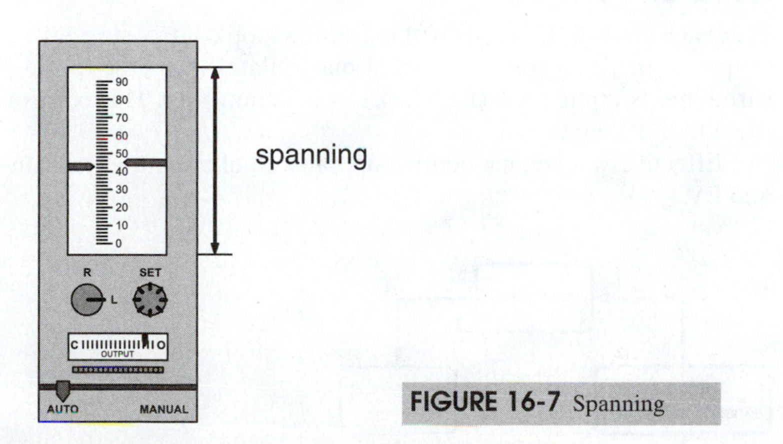

FIGURE 16-7 Spanning

Controllers and Control Actions or Modes

Controllers receive measurements of process variables from instrumentation and produce output signals to final control elements (Figures 16-8 and 16-9). Once a physical process (the plant) has been designed and built, the controller is the only device in the system that is capable of matching the dynamics (loop gain) of the process to the control system components. Different brands of controllers may be in use (Foxboro, Taylor, etc.). However, the principles of operation will be the same.

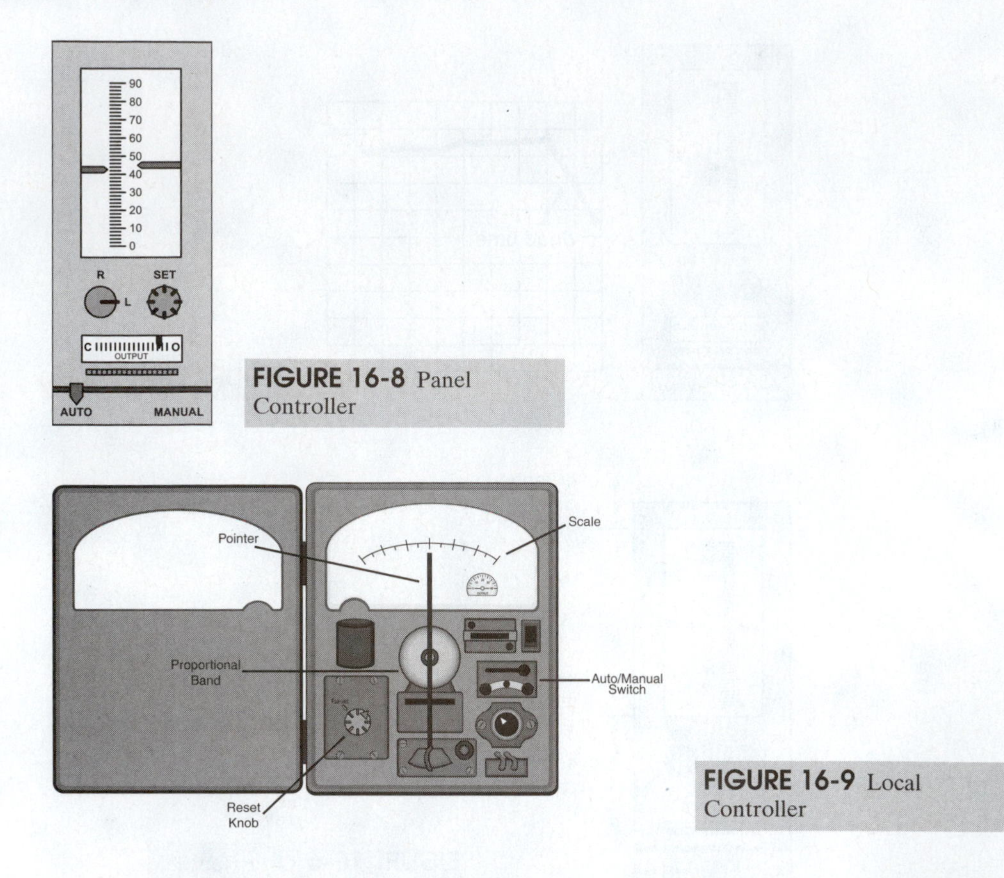

FIGURE 16-8 Panel Controller

FIGURE 16-9 Local Controller

A controller is an analog device or computer program that can operate automatically for the purpose of controlling a process variable based on predetermined conditions. All controllers compare a process variable to a setpoint, perform some type of calculation, and then produce an output. This output drives a control valve (or other final control element) that manipulates the process in a way that moves the measurement to the setpoint.

FEEDBACK CONTROL

Feedback control (Figure 16-10) is a closed loop control strategy where the controller's output to the final control element manipulates a change to the process that is measured and is input back (feedback) to the controller. This control action should continue until PV equals SP.

Effectively, feedback control operates to eliminate an existing error between SP and PV.

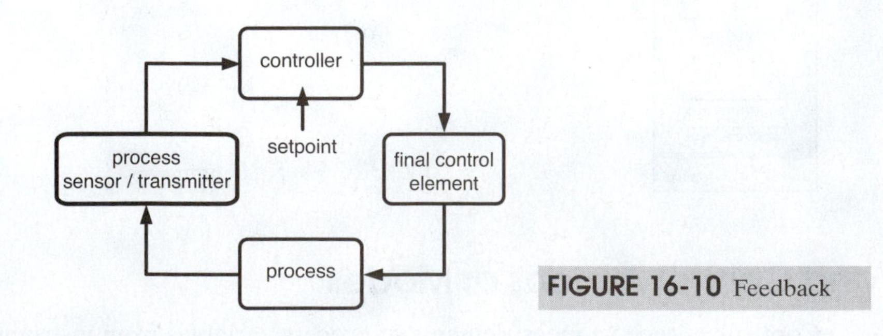

FIGURE 16-10 Feedback

FEEDFORWARD CONTROL

Feedforward control is an open loop control strategy where the controller's output is based on knowledge (usually mathematical modeling) of the relationship between the output of the controller and its input received from the point in the process where the disturbance occurs. In other words, feedforward control is used to prevent an error from occurring in the first place.

DIRECT OR REVERSE ACTING CONTROLLERS

A controller may be selected as "direct acting" or "reverse acting" to counteract the process dynamics. This can be accomplished through programming on newer controllers or by a selector switch on other types of controllers. All controllers must be set to act in one of these two modes (Figure 16-11).

FIGURE 16-11 Direct and Indirect Acting

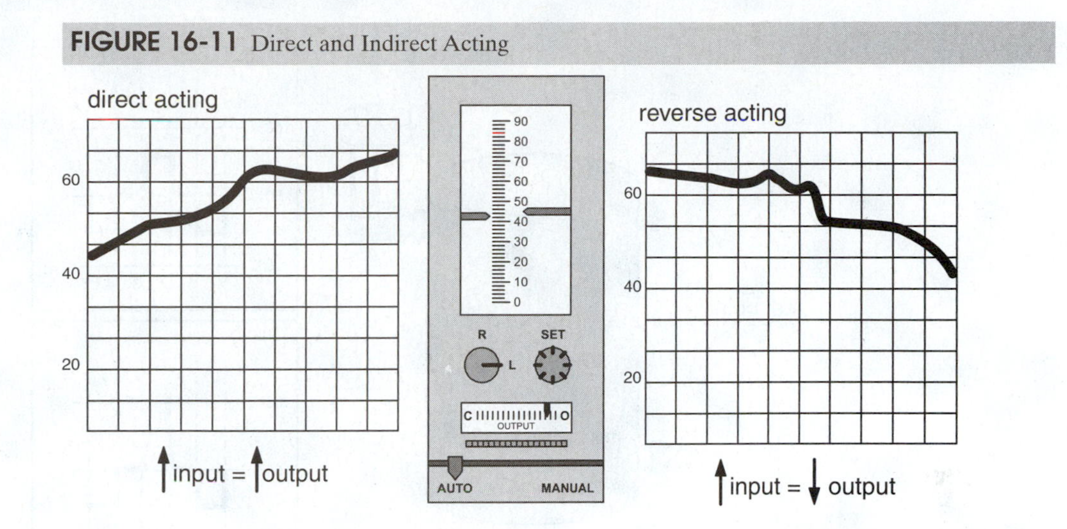

Direct Acting

A **direct acting controller** is when as the controller's input signal value increases, the controller's output signal value increases (increased input = increased output). Direct acting also refers to decreased input/decreased output.

Direct-acting controllers are designed to increase the output signal when the input signal increases (Figure 16-12). For example, a controller receives an increasing signal as a tank fills above setpoint. The controller responds by sending an increasing output signal to the drain valve. The drain valve opens incrementally in response.

FIGURE 16-12 Direct Action

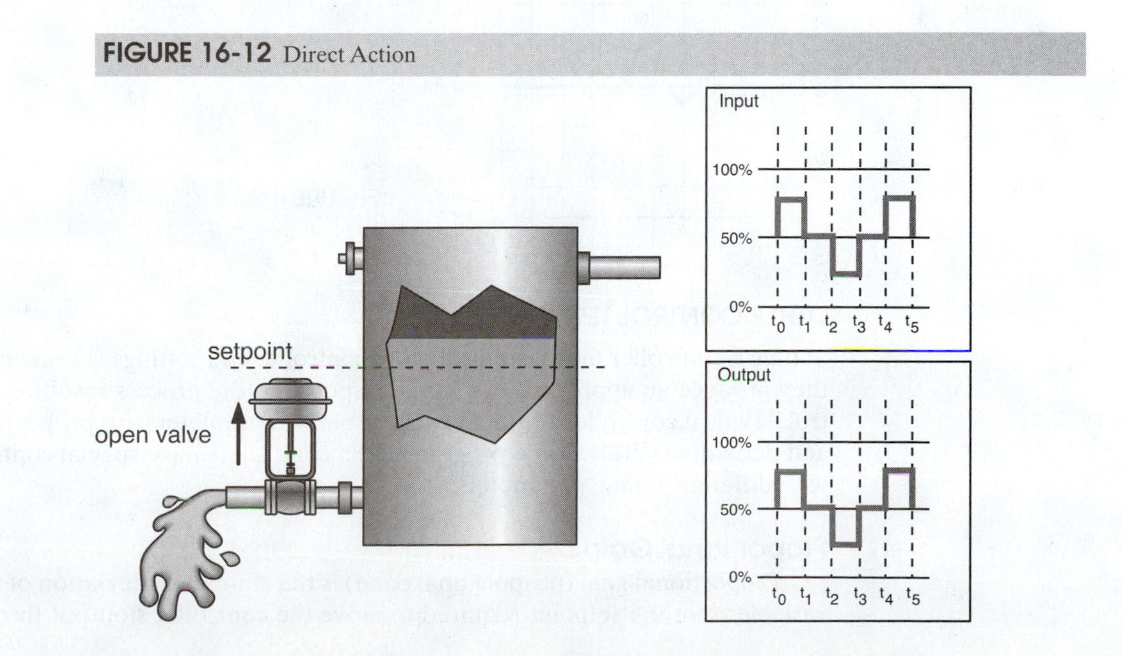

Reverse Acting

A **reverse acting controller** is when the output signal decreases when the controller's input signal value increases (increased input = decreased output).

Reverse-acting controllers are designed to decrease the output signal when the input signal increases (Figure 16-13). For example, a controller receives an increasing signal as the tank level rises above setpoint. The controller responds by sending a decreasing output signal to the fill valve. The fill valve closes incrementally in response.

FIGURE 16-13 Reverse Action

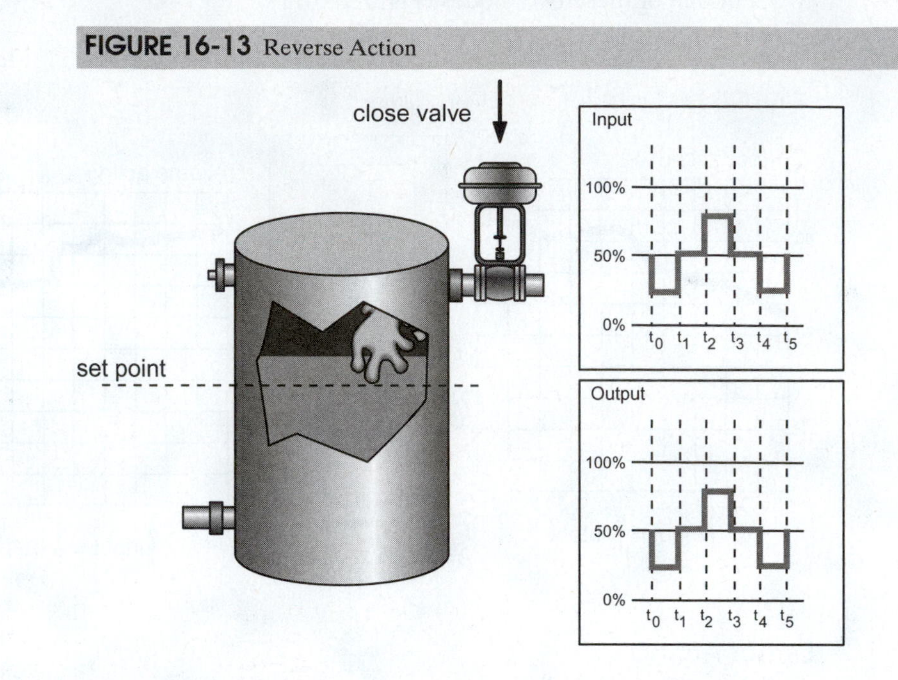

ON/OFF CONTROL

On/off control may be used when controlled variable cycling above and below the setpoint is considered acceptable, or when the process quickly responds to an energy change, when the energy change itself is not too quick (Figure 16-14).

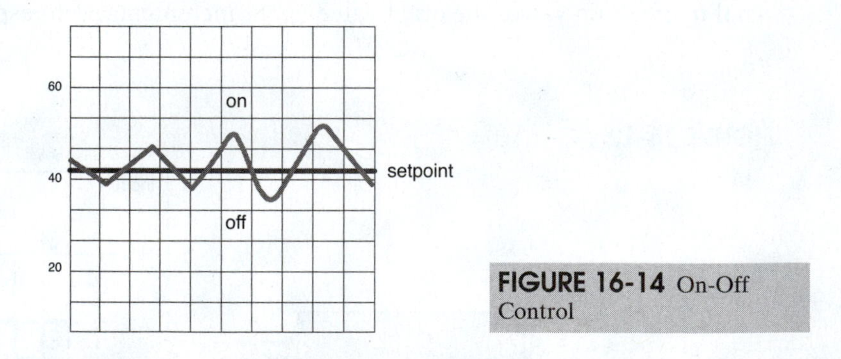

FIGURE 16-14 On-Off Control

OTHER CONTROL TERMS

Tuning a controller means adjusting the control action settings (tuning parameters) so they produce an appropriate dynamic response to the process resulting in "good control." Typical controller action settings or tuning parameters are proportional, integral, and derivative (PID). Some programmable controllers have special control algorithms with different tuning parameters.

Proportional Gain

Proportional gain (proportional band) is the amount of deviation of the controlled variable from the setpoint required to move the controller's output through its entire range (expressed as 10 percent of span). The controller output for a standard feedback control scheme will be the signal value that is being transmitted to the final control element.

Proportional Action

Proportional action is a controller output response that is proportional to the amount of deviation of the controlled variable from the setpoint. The proportional action contribution to the controller's output responds once to the deviation of PV from SP. It acts in the present time, usually only once.

In other words, proportional gain asks: "How big is the error?" There is a reciprocal relationship between proportional band and gain:

$$100\%/\text{PB} = \text{Gain} \quad 100\%/\text{Gain} = \text{PB}$$

where PB is given in percent (%) and gain is a decimal number without units.

In other words, the wider the proportional band, the lower the gain.

Integral Action (Reset)

Integral action (reset) is a controller output response that is proportional to the length of time the controlled variable has been away from the setpoint and it repeats the proportional action a given number of times; it acts on the past error. The integral action contribution to the controller's output is the controller's gain divided by the integral time multiplied by the sum of the (PV – SP) error. In other words, integral action asks: "How long has the error existed?"

Windup

Windup is the saturation of the controller output signal to its minimum or maximum value if the process variable does not return to the setpoint. The elapsed time that it takes for this to occur will be based on the process characteristics and the gain and integral or reset modes of the controller. In many controllers anti-reset windup parameters can be entered to prevent the output from going to a maximum or minimum.

Derivative Action (Rate, Preact)

Derivative action (rate, preact) is a controller output response that is proportional to the rate at which the controlled variable deviates from the setpoint. In other words, derivative action asks: "How fast is the error changing?" (See Figures 16-15, 16-16, and 16-17.)

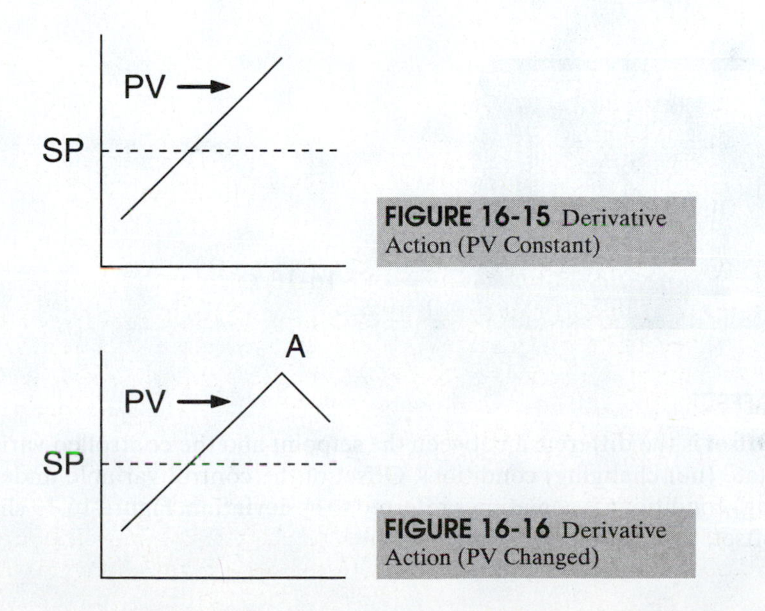

FIGURE 16-15 Derivative Action (PV Constant)

FIGURE 16-16 Derivative Action (PV Changed)

FIGURE 16-17 Derivative Action

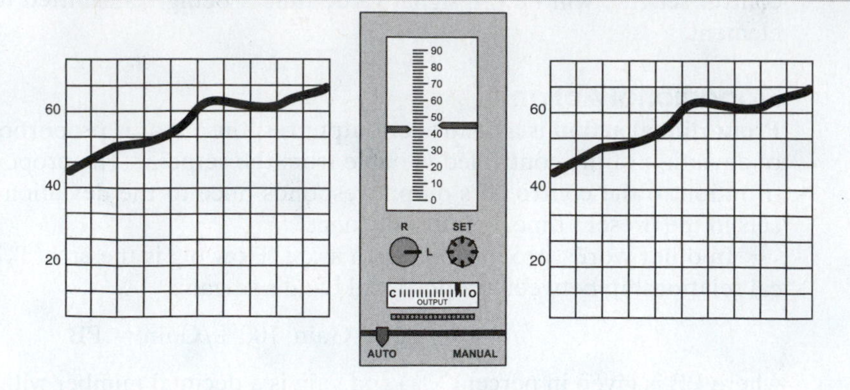

This control action acts strictly on the rate of change that the PV deviates from SP. Note that derivative does not act if the process technician makes a setpoint change: that would be a rate of change to the setpoint. This is shown on the graph in Figure 16-15.

Process Recording

Chart recorders or historical files in a computer system provide both operations and maintenance with an important troubleshooting tool (Figure 16-18). Process technicians use chart-recorded data to spot both good and bad process trends. Instrument technicians use these trends to help troubleshoot problems in the instrumentation control loop and to calculate controller tuning parameters.

The following are a few of the trends that can be observed through chart recordings.

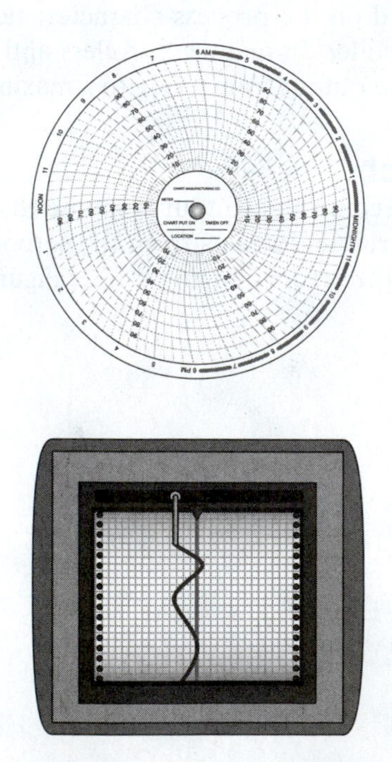

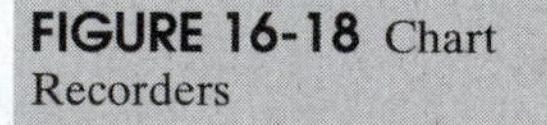

FIGURE 16-18 Chart Recorders

OFFSET

Offset is the difference between the setpoint and the controlled variable under steady-state (not changing) conditions. Offset of the control variable under dynamic (changing) conditions is sometimes referred to as **deviation**. Figure 16-19 shows an example of offset.

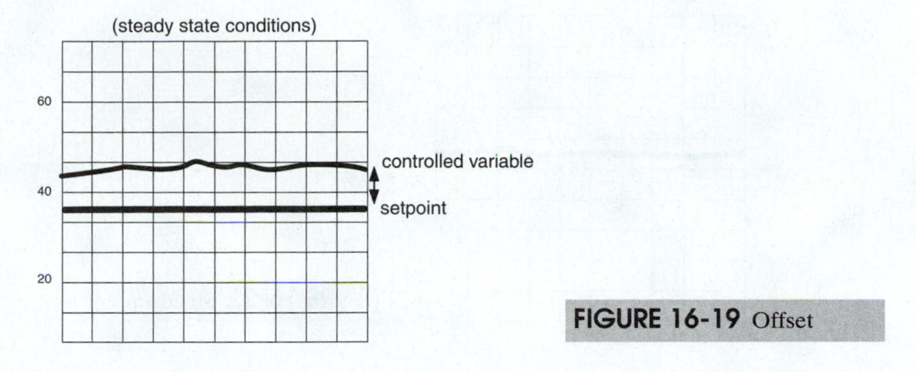

FIGURE 16-19 Offset

TREND

Trend is the plotting of a process variable over time. In a DCS any output or input, keystroke, alarm, or calculated value can be trended. Trends can be recorded on media such as a hard drive, tape drive or optical platter and can be displayed in different increments (e.g., hours, minutes, days). Figure 16-20 shows and example of a trend.

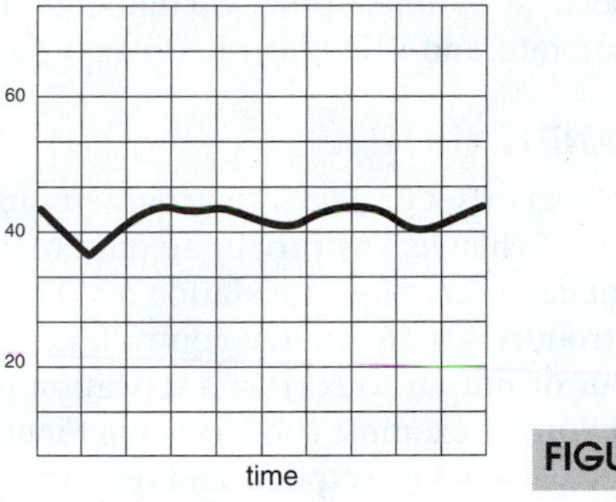

FIGURE 16-20 Trend

HUNTING

Hunting is a condition that exists when a control loop is improperly designed, installed or calibrated, or when the controller itself is not properly aligned. Process technicians will observe this condition as a random behavior or **cycling** above and/or below the setpoint (desired control point). Figure 16-21 shows an example of hunting.

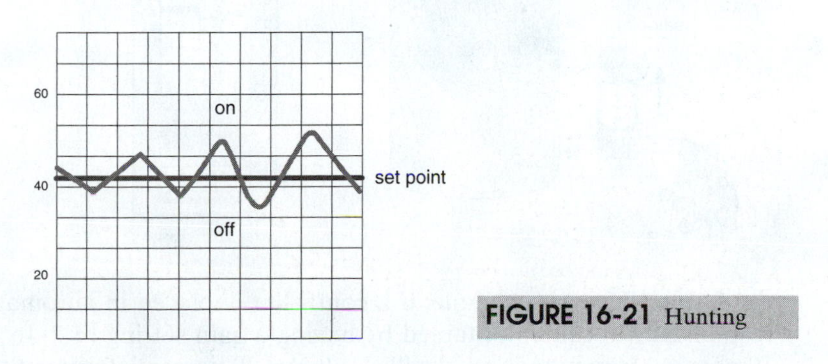

FIGURE 16-21 Hunting

STEADY STATE

Steady state is sometimes used to describe the trend condition in which the process is in equilibrium and running smoothly producing a relatively straight line. Figure 16-22 shows an example of steady state.

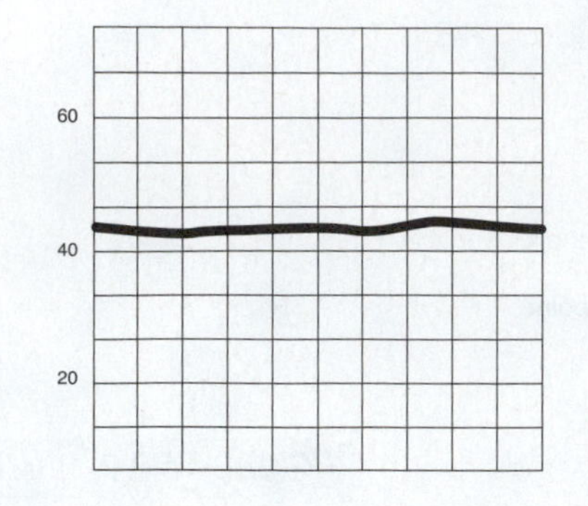

FIGURE 16-22 Steady State

Tuning Modes (Gain and Reset)

Control may be exercised using one or a combination of three modes: gain, reset, and rate. [Note: These three modes may also be referred to as proportional, integral, and derivative (PID). This will be explained later in the session.]

These modes must be selected and set correctly for effective process control. This is called tuning a control loop.

Tuning seeks to optimize process control in automatic mode much the same way a process technician would seek to optimize process control in manual mode. The following sections on gain, reset, rate, and PID illustrate this point.

GAIN (PROPORTIONAL BAND)

The required response to correct offset or deviation from setpoint may not be the same percentage as the percentage of change. The proper amount of correction may require some factor of the amount of deviation (e.g., a deviation may require a multiplier). This may be accomplished by introducing gain into the controller.

In order to change input or output to correct a deviation using the proportional band setting, consider the following example (illustrated in Figure 16-23): if the level in a particular tank drops 10 percent below setpoint and remains there, it may be necessary for the outlet valve to close 20 percent to compensate. Experience dictates that a 20 percent output change will correct a 10 percent offset.

FIGURE 16-23 Manual Correction

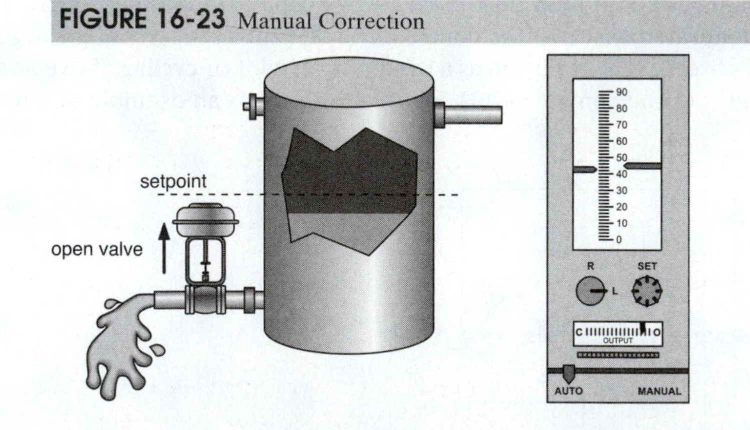

Using the same example: if a controller is placed in automatic mode this same response ratio can be obtained by having a gain setting of 2. In this case as well, a 10 percent drop in tank level will result in a 20 percent closing of the drain valve (see Figure 16-24). Thus, gain consists of a ratio established between process deviation and controller response.

If the percentage response required is exactly equal to the percentage change in the measured variable, a gain of 1 would be established (see Figure 16-25).

What if the drain valve needs only to close 5 percent in response to a 10 percent drop in tank level? A gain set at .5 would accomplish this (see Figure 16-26).

FIGURE 16-24 Gain Of 2

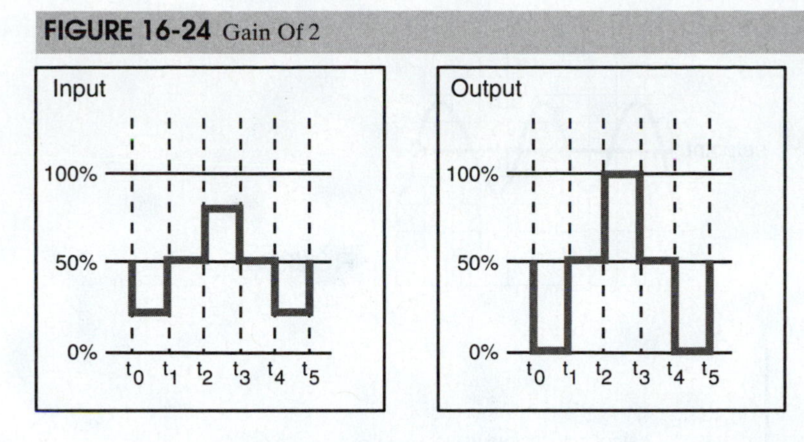

FIGURE 16-25 Gain Of 1

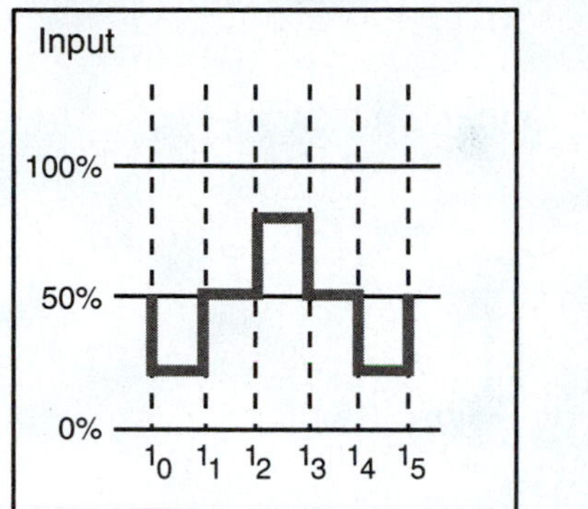

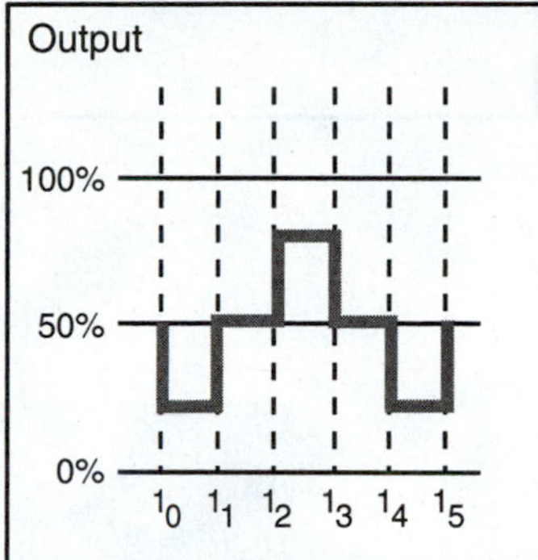

FIGURE 16-26 Gain Of .5

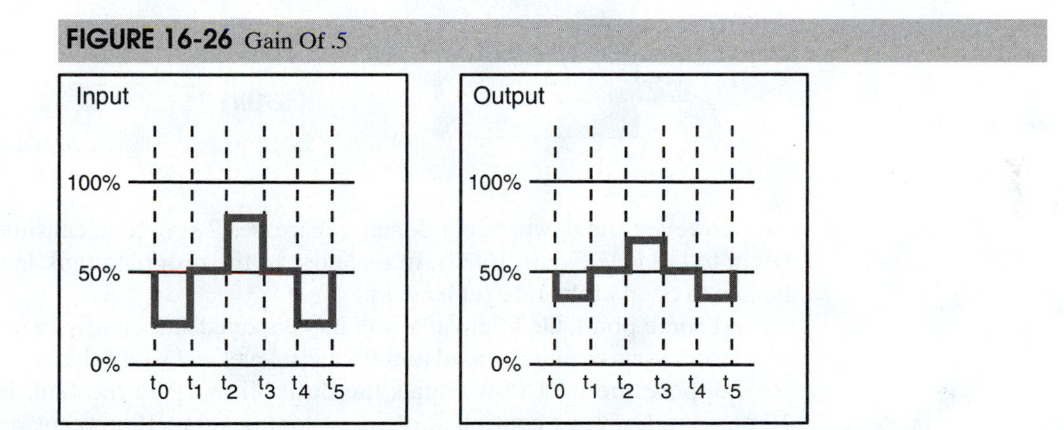

What is the optimum gain setting? This may depend upon several varying factors.

If the gain is set too high, the controller will over-compensate for each deviation detected and the process will begin to cycle (see Figure 16-27).

If the gain is set too low the controller will not provide enough correction to reach setpoint and a constant offset will result (see Figure 16-28). However, even a very careful gain setting is unlikely to maintain a process stable at setpoint.

Suppose that gain has been set so accurately for a given process under a particular set of conditions that setpoint is maintained. What would happen when the process conditions change? Consider the following example:

The level in a tank is maintained at setpoint by a controller with a gain of 1 (see Figure 16-29).

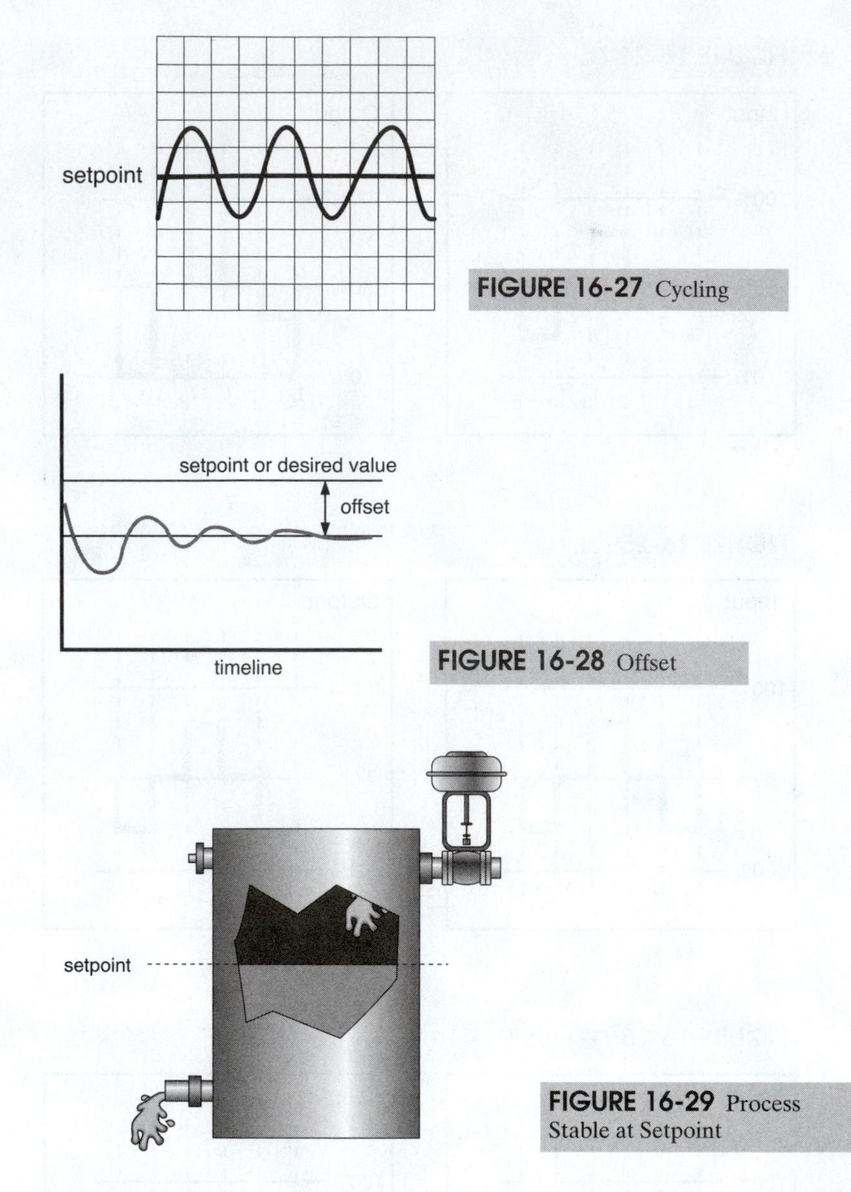

FIGURE 16-27 Cycling

FIGURE 16-28 Offset

FIGURE 16-29 Process Stable at Setpoint

However, the downstream demand increases 20 percent, causing the tank level to begin to fall (see Figure 16-30). In response to the dropping tank level, the inlet valve begins to open at the rate set by gain.

At some point, the inlet valve will have opened sufficiently to compensate for the increased downstream demand and the tank level will stop falling.

Suppose the inlet flow equals the outlet flow when the tank level has dropped 10 percent. Only the continued offset of tank level at 10 percent below setpoint will keep the inlet valve open the additional 10 percent necessary to compensate for the increased downstream demand (see Figure 16-31).

As this example illustrates, any change in product flow will render gain inadequate. In the real world, product flow is always fluctuating. This is a major reason why gain is rarely used alone. Gain only provides a ratio response to a deviation. Nothing in the gain mode will induce the controller to seek setpoint. Thus, reset mode is applied along with gain to solve this problem.

RESET (INTEGRAL)

Before discussing how reset works, let us consider how a process technician would react to continual offset using manual mode.

Suppose a process technician is monitoring a tank and notices the product level is continually below an acceptable level. In manual mode the process technician gently

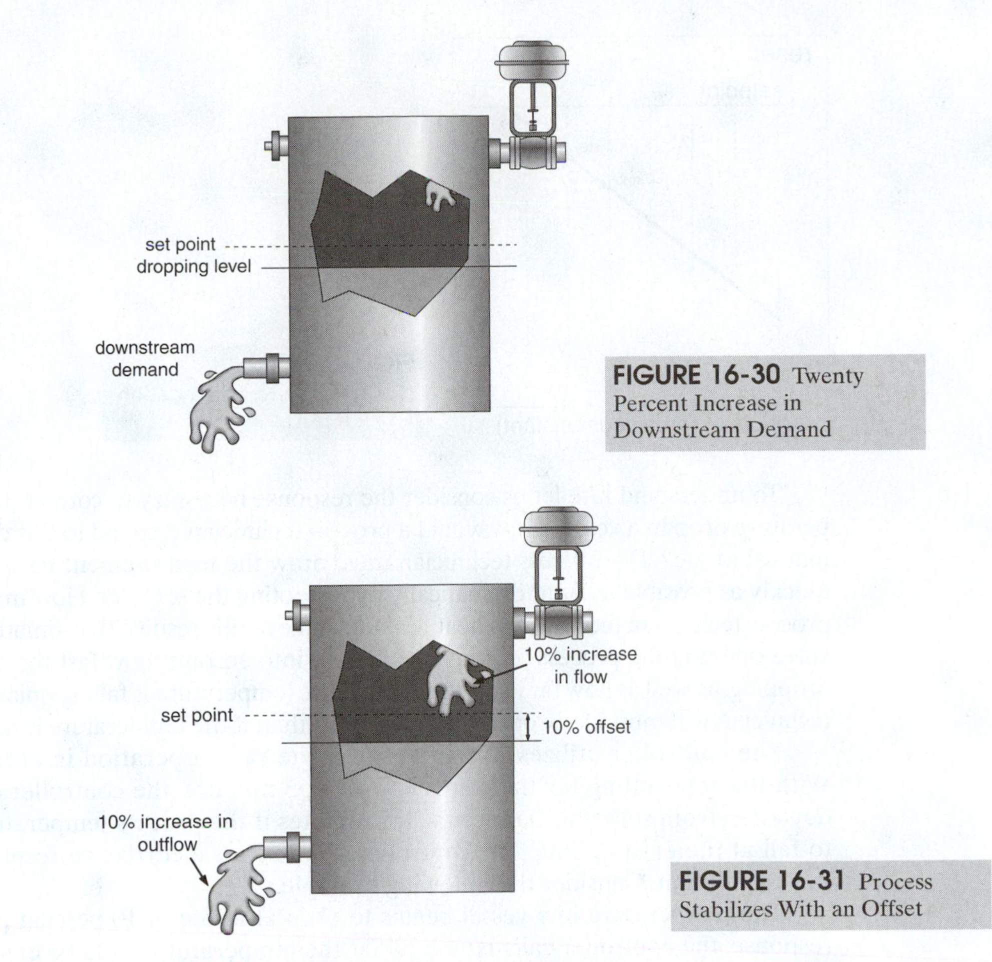

FIGURE 16-30 Twenty Percent Increase in Downstream Demand

FIGURE 16-31 Process Stabilizes With an Offset

bumps the inlet valve open until the product stabilizes near the desired level. How would the controller in automatic mode use the reset function to accomplish this same result? Reset opens the valve at a steady rate until setpoint is reached. But how rapidly should the valve be opened? Throwing the valve open quickly to 100 percent is likely induce cycling.

In order to establish the correction necessary to regain setpoint quickly without overshoot and cycling, the reset mode calculates a slope to draw the process variable up to setpoint. This slope is defined by gain across time. Typically reset is defined in repeats per minute and refers to the gain correction command.

Let us consider a tank with a direct action level controller set with a gain of 1 and a reset of 1 minute. The level in the tank rises 10 percent above setpoint resulting in a 10 percent increase in signal to the controller. The controller establishes a correction slope of 10 percent per minute. This means the controller signals the inlet valve to close at 10 percent per minute. The controller continues to decrease the valve opening at the rate of actual deviation multiplied by gain until the process has regained setpoint.

The chart in Figure 16-32 illustrates how reset corrects an offset if the deviation remains constant.

Gain is seldom used alone as a control mode since this mode is only capable of providing a ratio response to a deviation. It has no capacity to automatically adjust an overcorrection or an undercorrection. For this reason, most controllers utilize gain and reset together. Gain puts the correction closer to where it should be while reset fine-tunes the correction until the measured variable returns to setpoint.

RATE (DERIVATIVE OR PREACT)

The third tuning mode, rate, is used far less frequently than gain or reset and is most commonly used in temperature related applications.

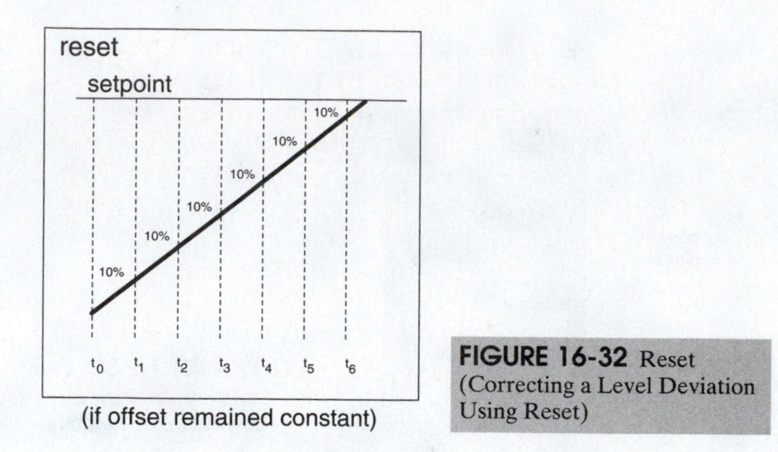

FIGURE 16-32 Reset (Correcting a Level Deviation Using Reset)

To understand rate let us consider the response necessary to correct a dramatic temperature drop in a vessel. How would a process technician respond to this deviation using manual mode? The process technician must draw the measurement back to setpoint as quickly as possible without dramatically overshooting the setpoint. How much should the process technician increase the heat input to achieve this result? To estimate the optimum valve opening the process technician will take into account how fast the temperature is dropping, as well as how far it has dropped. If the temperature is falling quickly, the process technician will most likely open the valve more than if the temperature is falling slowly.

The controller utilizes rate to perform the same operation in automatic mode. With the rate setting for the controller set at 5 minutes, the controller calculates the deviation from setpoint that occurs in 5 minutes if the process temperature is allowed to fall at the existing rate. The controller then makes a corrective response based on this calculation. Consider the following example.

The temperature in a vessel begins to drop at a rate of 10 percent per minute. In response, the controller calculates how far the temperature could be expected to drop in 5 minutes at the present rate. A 10 percent drop over 5 minutes would result in a 50 percent total drop (see Figure 16-33).

Based on this calculation, the controller immediately signals a 50 percent increase in heat input. This results in a quick, hard kick back toward setpoint (see Figure 16-34).

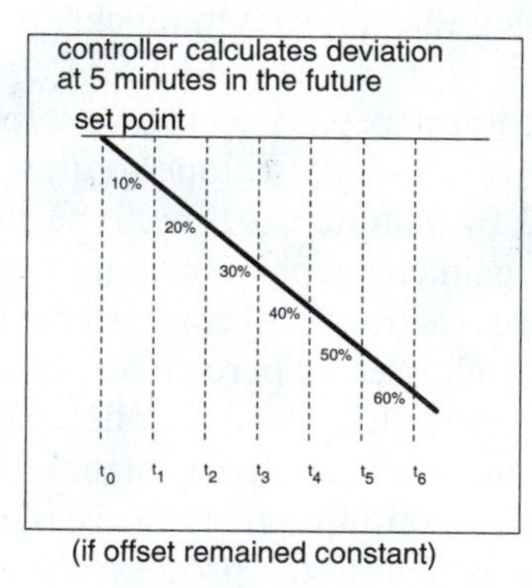

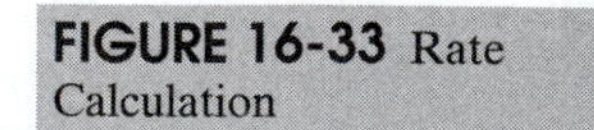
FIGURE 16-33 Rate Calculation

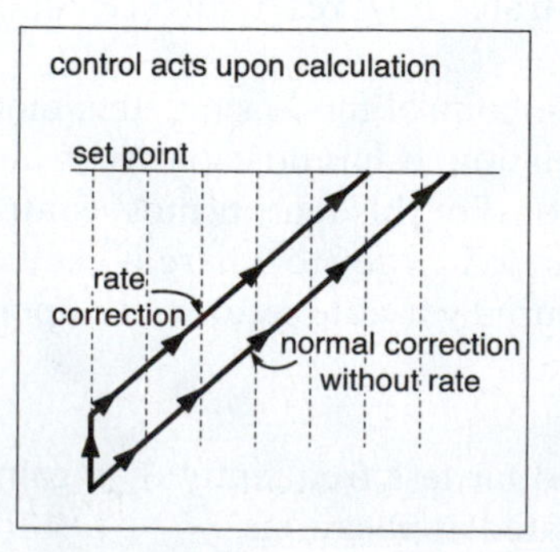

FIGURE 16-34 Rate Correction

Therefore, rate calculates correction based on an estimated future deviation (in this example, how far from setpoint the measurement will have fallen 5 minutes in the future). Because of this, rate is often referred to as *preact.*

At what value should rate be set? This depends on a number of factors. A rate that is set too low will not optimize the time necessary to return to setpoint. An excessive rate setting could kick the temperature up too hard, resulting in a large overshoot of setpoint and cycling.

PID

Earlier in the session it was mentioned that the three controller modes are sometimes referred to as proportional, integral, and derivative (PID). To clear up any possible confusion: integral is the same as reset and derivative is the same as rate. Proportional, on the other hand, is not the equivalent but rather the reciprocal of gain.

Proportional control refers to proportional band, an older term used to explain the ratio between the input and the output of a controller. Proportional band is a process technician's way of explaining how much offset is necessary between a process measurement and setpoint to move a controller across its entire range of measurement.

Pretend the two-headed arrow shown in Figure 16-35 is a solid bar supported in the middle by a fulcrum (like a child's teeter-totter). The left head of the arrow represents the input signal into a controller. The right head represents the output signal from a controller.

In this example, moving the right head of the arrow from 0 percent to 100 percent requires the movement of the left head of the arrow from −50 percent to +50 percent. This movement (from −50 percent to +50 percent) equates to travel of 100 percent. In this case the proportional band is said to be equal to 100 percent. Also, because an increase in the input signal results in an equal (1-1) increase in the output signal, the gain is said to be 1.

In the example in Figure 16-36, the fulcrum has been moved to the left. Thus, the left arrow only needs to move half the range of the previous example to accomplish

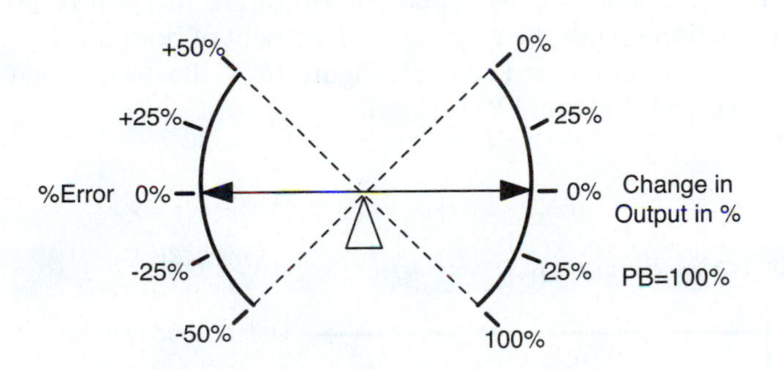

A 100% change in output is caused by a 100% change in error. The proportional band is therefore 100% or the gain is 1.

FIGURE 16-35 Gain Of 1

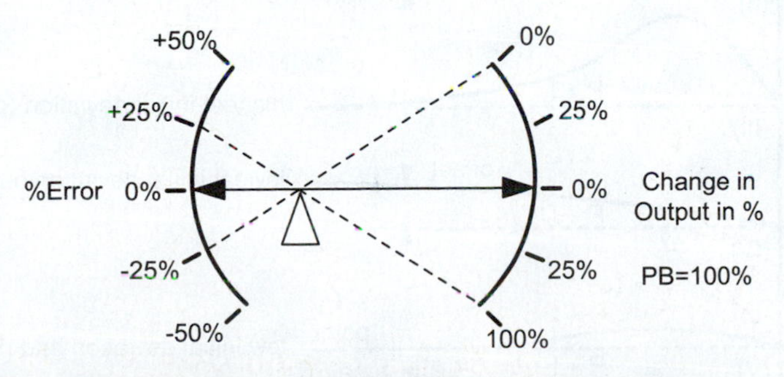

A 100% change in output is caused by a 50% change in error. The proportional band is therefore 50% or the gain is 2.

FIGURE 16-36 Gain Of 2

the same result. In essence, it only needs to travel from −25 percent to +25 percent to move the right arrow the full range of 0 percent to 100 percent. In this case the proportional band is said to be only 50 percent . Also, because movement of the left arrow results in a movement twice the distance by the right arrow (1-2), the gain is said to be 2.

The relationship between proportional band and gain can also be illustrated with the following chart.

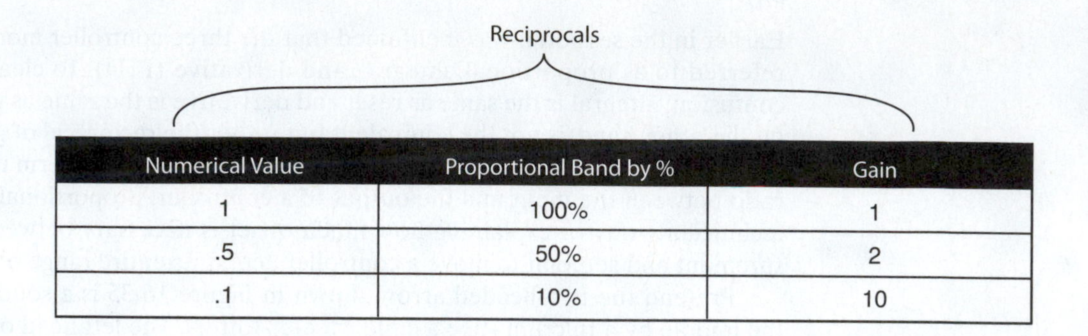

Numerical Value	Proportional Band by %	Gain
1	100%	1
.5	50%	2
.1	10%	10

Combined Control Action

It has already been shown that offset is inevitable with proportional only control. However, there are controls available that reduce offset. A proportional plus integral (P+I) control reacts to the size of the change in the process variable. Proportional plus derivative (P+D) control detects the speed with which the deviation occurs and speeds up or slows down the valve reaction accordingly.

Derivative action is usually combined with a proportional plus integral control to give P+I+D, the most sophisticated control of all. Integral removes the offset while derivative action speeds the response. Adjustment of both the integral and derivative is provided on most combined controls. Figure 16-37 illustrates each of the various types of control (P, P+I, P+D, and P+I+D).

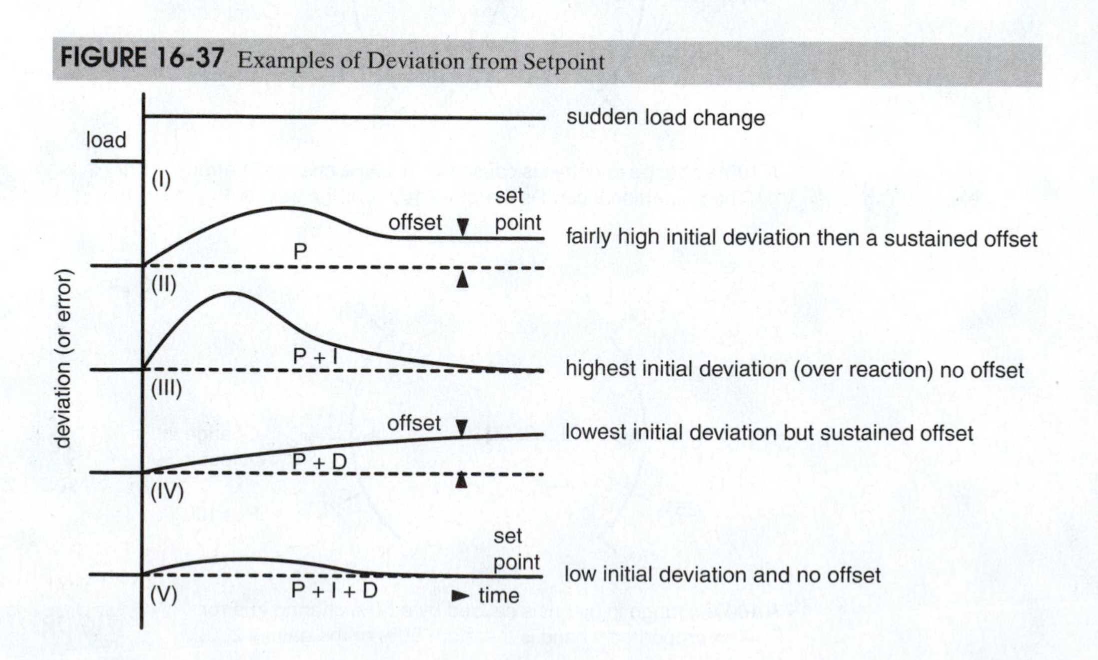

FIGURE 16-37 Examples of Deviation from Setpoint

Troubleshooting

This section discusses how tuning errors can create process problems. However, a process technician should never assume that cycling or some other problem is caused by a poorly tuned loop. A number of symptoms that appear to be tuning related can also be caused by other components of the control loop. Cycling and offset are two of those examples.

CYCLING

Earlier in this chapter it was stated that cycling can be caused by a tuning error. For example, consider the following:

- If the gain is set too high the controller will overreact to measurement deviation.
- If the time increment for reset is too small the controller will also overreact establishing a correction slope that is too steep. Overshoot and cycling will result.
- If the time increment for rate is too large the controller may also exaggerate the initial "kick" back toward setpoint, also resulting in overshoot and cycling.

However, cycling can also be caused by a process problem such as the following:

- A sticking valve (the valve may fail to react until the process has already deviated dramatically)
- Pump cavitations
- Faulty wiring (can result in an intermittent signal transmission, which leads to cycling)

OFFSET

Continual offset can be caused by a gain that is set too low. It can also be caused, however, by a malfunctioning final control element (e.g., a valve that is unable to open fully may prevent a process from ever reaching setpoint).

If a controller exhibits cycling or a continual offset, do not assume a tuning problem. Try correcting the problem in manual mode. If the problem persists in manual mode, tuning is not the culprit. In this case the process technician needs to try to isolate which element in the control loop could be at fault.

Only after determining that control can be reestablished in manual should it be concluded that a loop needs to be tuned.

Summary

Processes have certain behaviors that are common and predictable. During states of change a process may experience load change, dead time, or spanning, or may be over range.

During processes, controllers receive measurements of process variables from instrumentation and produce output signals to final control elements. In any given process there may be several types of control including feedback, feedforward, and on/off control.

In order to produce appropriate dynamic responses to a process and maintain "good control," controllers must be tuned. When tuning a controller, it is important to take into consideration proportional gain, proportional action, integral action, windup, and derivative action (rate, preact).

When running a process it is important to use process recording instrumentation (chart recorders) to document process trends. By recording process trends, process technicians and instrumentation technicians are better able to spot both good and bad process trends and troubleshoot problems.

When examining recorded process data, there are several things a technician might look for. These include offset, hunting, cycling, and steady states. In addition, technicians might also examine gain, and reset rate (derivative or preact). All this information helps technicians determine if the instruments are correctly tuned and everything is working properly.

Checking Your Knowledge

1. Match the following process dynamics terms with their correct definition.

Term	*Definition*
Dead time	a. The condition that exists when the signal value of a device or system exceeds the maximum allowable value.
Load change	b. The process of setting or calibrating the lower and upper range values of a device, or testing a device or system by traversing up and down the scale through the entire calibrated or operational range.
Over range	c. A change in any variable in the process that affects the value or state of the controlled variable.
Process equilibrium	d. The elapsed time between the initiation of an input and the point at which the resulting response can be measured or observed.
Spanning	e. The condition that exists when there is a balance of material and energy within a given system or process. This condition is also referred to as "steady state."

2. Match the following process control terms with their correct definition.

Term	*Definition*
Proportional gain	a. A condition that exists when a control loop is improperly designed, installed, or calibrated, or when the controller itself is not properly aligned. This condition will be observed by the process technician as a random behavior or cycling above and/or below the setpoint (desired control point).
Windup	b. A trend condition in which the process is in equilibrium and running smoothly producing a relatively straight line.
Offset	c. The amount of deviation of the controlled variable from the setpoint required to move the controller's output through its entire range (expressed as 10 percent of span).
Tuning	d. A plot of a process variable over time.
Trend	e. The difference between the setpoint and the controlled variable under steady state (not changing) conditions.
Hunting	f. Setting the controller's parameters of gain, reset, and rate (PID) to optimize the control of the process variable.
Steady state	g. The saturation of the controller output signal to its minimum or maximum value if the process variable does not return to the setpoint.

Student Activities

1. Given a set of trend charts, break into small groups and analyze the trend charts with respect to deviation and offset. Determine instances deviation and offset are clearly visible. Report your observations and conclusions to the class.
2. Given a computer program or process variable simulator, demonstrate the different behaviors of pressure, temperature, level, or flow control systems. Focus the discussion and lab activity on the resistance to change and the capacitive element of each system that you describe.
3. Given an analog pneumatic controller, perform a controller calibration and alignment procedure.
4. Given an analog electronic controller, perform a controller calibration and alignment procedure.
5. Given a single loop digital controller, perform a controller calibration and configuration procedure.
6. Given a control loop or a simulation of a control loop, conduct tuning operations and tune the loop.

7. Given a computer program or a process simulator, investigate the concepts of proportional band, proportional action, and proportional gain.
8. Given a computer program or a process simulator, investigate the concepts of integral action or reset.
9. Given a computer program or a process simulator, investigate the concepts of derivative or rate action. [Note: For clarity and understanding it is recommended that only gain, gain and integral, and gain and derivative be used to reinforce this session. Without a higher level of math the combination of all three (PID) may not serve any useful purpose other than to confuse the student.]
10. Given a drawing, picture, or actual trend, distinguish between the effects of proportional, integral, and derivative control. To do this: (a) study a chart with a measured variable plotted against time; (b) work in small groups to analyze the trend charts with respect to controller mode influences; (c) discuss with the group what you think is causing the trend to be the way it is; (d) report your group's conclusions to the class.
11. Conduct the following pressure lab:

Pressure Lab

This lab is designed to utilize a pneumatic controller on a simulator. (Note: A bench unit, using air for the process, could be used as well.) The instructor will change the tuning constants on the trainer controllers to demonstrate how a process system responds when the controller tuning constants are correct and incorrect.

Procedure

Note: If a controller begins to act up, it is probably because of a load change, a drastic upset, or someone has adjusted the tuning constants.

a. The instructor will demonstrate how a controller reacts to a load change or upset with the proper tuning constants, and will show how a controller reacts to a load change or upset when the tuning constants are incorrect.
   Note 1: If a drastic load change occurs, the tuning constants of a controller should be changed.
   Note 2: Some electronic controllers are self-tuning, such as Foxboro 761 and Taylor Mod 30.
b. The instructor will lower the gain to .5.
c. Raise the flow to 9 units and watch the process reaction, then discuss controller action with the instructor.
d. Lower the flow back to 4 units and let process stabilize.
e. The instructor will set gain back to normal and let process stabilize.
f. The instructor will then raise the gain to High.
g. Raise the flow to 9 units and watch the LRC-100 reaction, then discuss the controller action with the instructor.
h. Lower the flow back to 4 units and let process stabilize.
i. The instructor will set gain back to normal and let process stabilize, then lower reset to .5.
j. Raise flow to 9 units and watch process reaction, then discuss controller action with the instructor.
k. The instructor will lower flow back to 4 units and let process stabilize, then set reset back to normal and let process stabilize.
l. The instructor will raise reset to 8.
m. Raise flow to 9 units and watch the process reaction, then discuss the controller action with the instructor.
n. The instructor will lower the flow back to 4 units and let process stabilize, then set reset back to normal and let process stabilize.

CHAPTER 17

Control Schemes

Objectives

After completing this chapter, you will be able to:

- Identify and describe types of control schemes:
  - on/off
  - lead/lag
  - feedback
  - feedforward
- Describe a control scheme that allows the following modes:
  - local manual control
  - local automatic control
  - remote manual control
  - remote automatic control
  - cascading of the remote automatic control
- Match appropriate control schemes to a process.
- Given a process scenario, design an appropriate process control scheme.

Key Terms

Cascade control—employing two controllers so one process variable is controlled by controlling another; in this scheme one controller is the remote setpoint of another.

Lead/lag control—describes how a process reacts to a disturbance or other manipulating conditions.

Local manual control—the act of controlling a process variable by hand within the processing area.

Remote automatic control—the act of controlling an instrument loop remotely from a control room.

Remote manual control—the act of controlling a valve manually from a remote location, such as a control room.

Introduction

This chapter focuses on several different types of controls and their applications. These different control types fall into two categories: control schemes and control modes. "Scheme" refers to the approach used to control a particular process variable. Different process problems require different approaches for optimal control.

"Mode" refers to the agent or location of control (e.g., manual versus automatic or local versus remote). The term mode can be quite confusing. You may see "control modes" or "control actions" such as proportional (P), integral (I), and derivative (D) modes or actions stated as a three-mode control or a PID controller. On modern programmable controllers, controller modes may be stated as "Man," "Auto," "RSP," "Cas," "Sup," and "DDC."

Control Schemes

Process control involves many different types of control schemes. Some of the most common types of control schemes include on/off, lead/lag, feedback, and feedforward.

ON/OFF CONTROL

On/off control may be used when a controlled variable cycling above and below the setpoint is considered acceptable. It occurs when a final control element is moved from one of two fixed positions to the other with a small change of the controlled variable above or below the setpoint.

On/off control is the most common type of control used in our homes. Ovens, central air and heat, and water heaters all use on/off control schemes to operate. In industry on/off control is used in places where a fairly large control zone is not a problem. For example, a condensate collection vessel or other knockout pot typically uses two points for control: a high-level point and low-level point. At the high-level point, a pump is turned on to move the liquid out of the tank into another tank. At the low-level point the pump is turned off.

LEAD/LAG CONTROL

Lead/lag is a broadly used term in the processing industry that describes how a process reacts to a disturbance (e.g., a process reactor temperature change leads to a corresponding pressure change) or other manipulating conditions (a measurement change lags a controller output change).

Lead/lag is also used to describe how a control mode action reacts to an error (rate = lead; reset = lag), or which measured variable will be used to control a loop. It is even used to describe equipment utilization strategies.

FIGURE 17-1 Boiler

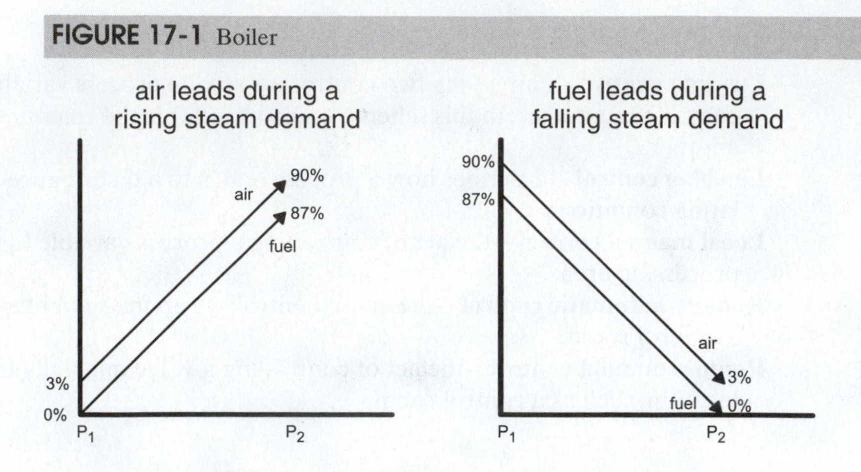

Examples of Lead/Lag Control

In a boiler system, **lead/lag control** specifically means to utilize both high and low selector relays to force the fuel to follow the airflow on a rising steam demand, and forcing the airflow to follow the fuel on a falling steam demand (see Figure 17-1).

Another lead/lag controller circuit is used to swap the lead role between two equal parallel components, such as compressors and/or evaporators in a refrigerator system like the one shown in Figure 17-2. This is done to provide a mechanism that will ensure equal wear to the two components while allowing the lagging component to be used in a supplementary capacity in the event of an unusually high demand cycle.

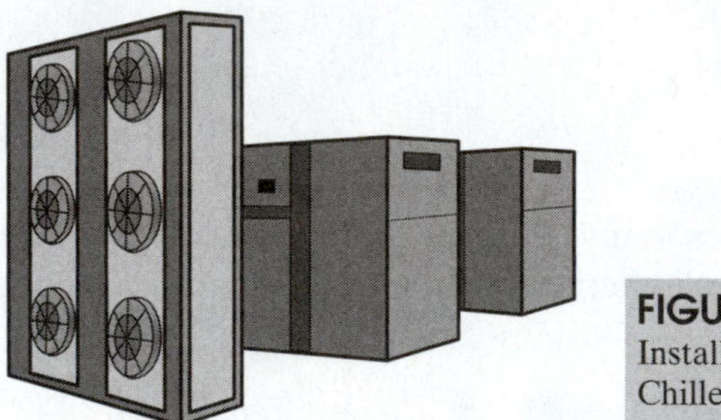

FIGURE 17-2 Free Cooling Installation (Two 75-kW Chillers and One RH Unit)

FEEDBACK CONTROL

Feedback control is a closed loop control strategy where the difference in the setpoint and measurement (called error) drives the controller output to change the position of the final control element to make PV = SP. There must be a difference between setpoint and measurement for this control action to operate.

Feedback control is designed to eliminate or minimize error. The fundamental concept that truly defines this control strategy is the feeding back of the measured variable to the controller.

FEEDFORWARD CONTROL

Feedforward control is an open loop control strategy where the controller output is based on knowledge (usually mathematical modeling) of the relationship between the output of the controller and the input received from the point in the process where the disturbance is measured (Figure 17-3).

Feedforward control is designed to prevent an error from occurring in the first place by changing the position of the final control element before the measured variable senses the change.

The fundamental concept that defines this control strategy is the open loop aspect of the measured variable in relationship to the controller. The controller output is changed by the magnitude of the disturbance rather than just the process variable of interest.

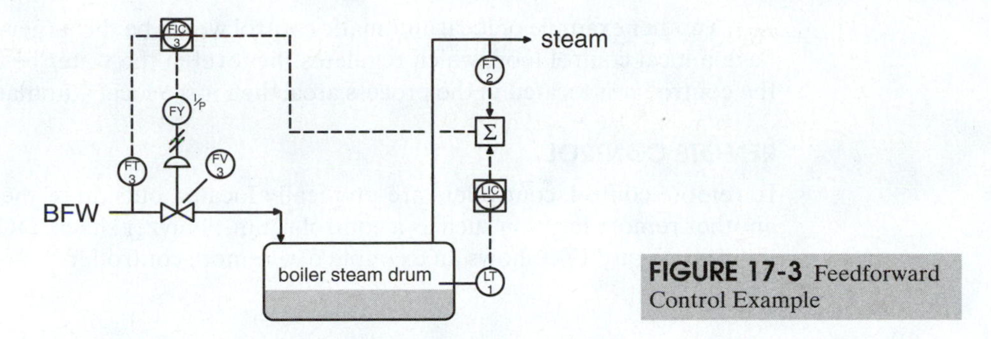

FIGURE 17-3 Feedforward Control Example

In many industrial applications feedforward and feedback control are used together. In high-pressure steam production (boiler) the steam demand signal is the feedforward input that is summed to the boiler level controller output that is the remote setpoint to the boiler feed water controller.

Control and Controller Modes

Controller mode refers to the agent and location of control. There is a difference between controller mode and control mode. The term controller mode refers to the status of a controller and whether it is located adjacent to (local) or removed from (remote) the process equipment area. Control mode or action refers to the proportional, integral, and derivative components or any other control algorithms of a controller.

LOCAL CONTROL (FIELD-MOUNTED)

In local control, the controllers are physically located in a process unit near a vessel, control valve, or transmitter. Figure 17-4 contains an example of a local controller.

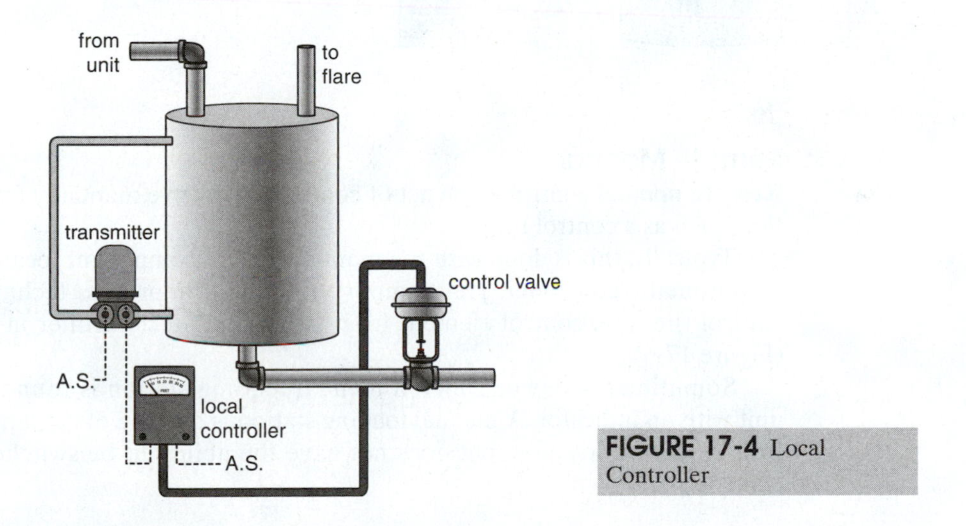

FIGURE 17-4 Local Controller

Local Manual

Local manual control involves a local controller in manual or the act of controlling a process variable by manually operating a valve within the processing area (e.g., a cooling tower may require the addition of makeup water only one time per shift, or less. If a process technician performs this duty by observing the basin water level and then regulating the makeup water valve by hand, this constitutes local manual control).

Local Automatic

With local automatic control, the automatic controller is also located in the processing area. An example is a steam letdown station consisting of a control valve and a pressure controller (integral mount) used to drop the main header steam pressure to a lesser pressure, making it appropriate for the steam requirements of a processing area.

Another example of local automatic control would be the larger cooling towers that have a local control loop, which regulates the level in the water basin automatically. If the controller is located in the process area, then it is a local automatic controller.

REMOTE CONTROL

In remote control, controllers are physically located outside of the process unit or in another remote location such as a control room, analyzer shack, DCS monitor, or PLC monitor. Figure 17-5 shows an example of a remote controller.

FIGURE 17-5 Remote Controller

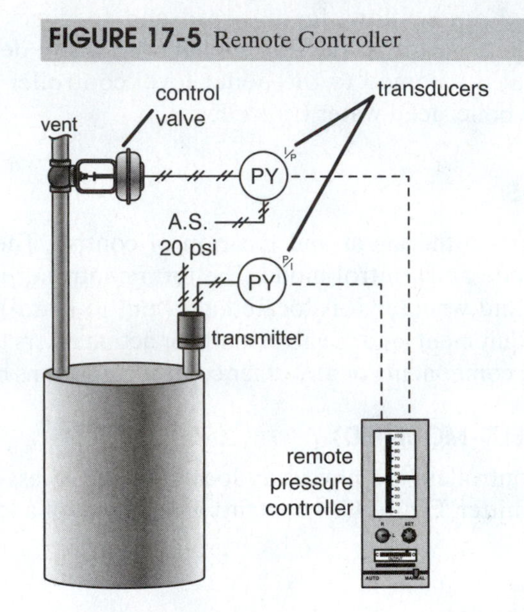

Remote Manual

Remote manual control is the act of controlling a valve manually from a remote location, such as a control room.

Typically, this is done with the manual control component located on the front of an automatic controller. This component allows the process technician to manually control the final control element from the automatic controller in the control room (Figure 17-6).

Sometimes, a device called a manual loading station is found as a stand-alone unit with an indicator. A manual loading station is capable of actuating a control valve from a remote location, but does not have the ability to be switched into automatic control.

Remote Automatic Control

Remote automatic control is the act of controlling an instrument loop remotely from a control room. Most controllers in the processing industry are remotely located. A remote automatic controller is defined as a controller that is located away from the processing area, usually in a centralized control room. Soon after the development of pneumatic transmitters, local controllers were moved to centralized control rooms where they could be monitored more efficiently.

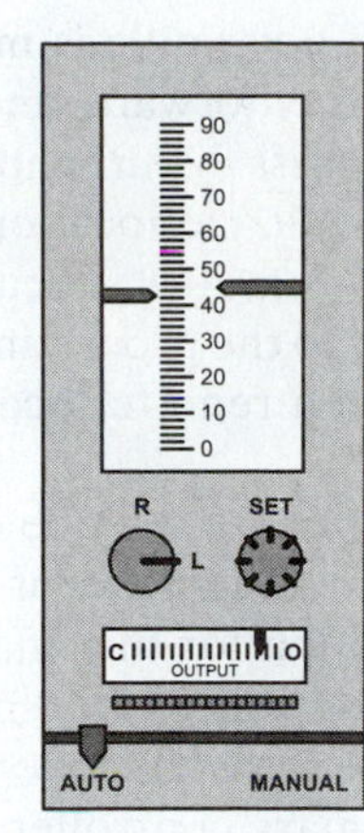

FIGURE 17-6 Local and Remote Settings

Cascading Of the Remote Automatic Control

In processes that exhibit excessive lag times or where the process variable has particularly tight constraints, it becomes necessary to control one process variable by controlling another. The control scheme that employs two controllers, where the output of one controller is the remote setpoint of another (e.g., a temperature controller output is the setpoint for a steam flow controller or slave) is called cascade control. An example of **cascade control** is shown in Figure 17-7.

As with most modern controllers, most cascaded controllers are remotely located away from the processing area.

The two cascaded controllers have controller modes as follows:

- Primary = AUTO/MAN
- Secondary = CAS/AUTO/MAN

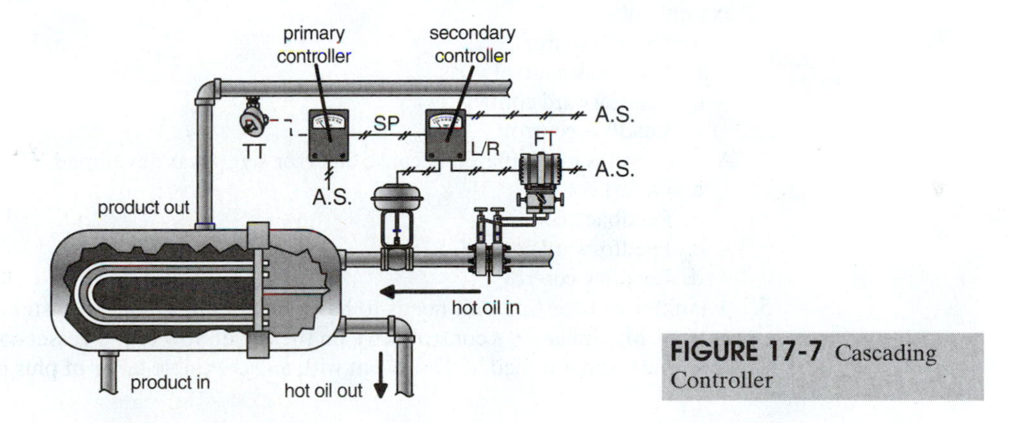

FIGURE 17-7 Cascading Controller

Summary

The control schemes selected to regulate the plant processes take into account many requirements such as quality, safety, economics, precision, production rates, and other factors. All plant processes are affected by many variables such as load changes, pressure swings, feedstock stream composition, and temperature variation. Selection of an appropriate control scheme is based on the acceptable level of process variability and the cost associate with reducing process fluctuations. Typically, precision control requirements mandate a more complex control scheme with more measurements and higher associated costs.

"What is the impact of variability in the control part of this process?" should be the primary question asked when selecting a control scheme, sensors, and final control elements for each area of the process. If the impact is extremely minimal, then simple on/off control may suffice. If the desired variability is somewhat less, then a feedback controller may be adequate. If precise control of a variable is mandated, then extra control hardware and combinations of feedback, feedforward, and/or lead/lag control maybe implemented to minimize the effects of process disturbances.

The mode of a control scheme refers to the agent and location of control (manual versus automatic or cascade and local versus remote). Local or field-mounted controllers are located within or in close proximity to the processing area. Remote controllers are just as the name implies, installed in a remote location away from the processing area (typically in a control room).

Most automatic controllers, whether local or remote, may be operated in multiple modes such as manual, automatic, or cascade. The manual mode is normally used when process conditions are abnormal and a fixed position for the final control element is highly desirable. The automatic mode allows a controller setpoint to be entered and achieved without further manual intervention. The cascade mode requires at least two controllers and two measured variables. The primary controller input is the slower-changing variable and the primary controller's output is the setpoint for the secondary controller. The secondary controller controls the faster-changing variable ultimately keeping the primary controller input variable on setpoint.

Checking Your Knowledge

1. A reactor's catalyst feed is manipulated due to a predicted load change. This is an example of:
   a. On/off control
   b. Feedback control
   c. Feedforward control
   d. Lead/lag control
2. The liquid in a tank is emptied once it reaches a specific level. This is an example of:
   a. On/off control
   b. Feedback control
   c. Feedforward control
   d. Lead/lag control
3. A parallel piece of equipment begins to operate due to an increase in process load. This is an example of:
   a. On/off control
   b. Feedback control
   c. Feedforward control
   d. Lead/lag control
4. A tank level is manipulated because an error signal has developed. This is an example of:
   a. On/off control
   b. Feedback control
   c. Feedforward control
   d. Lead/lag control
5. A tank in a processing unit needs to have an automatic control system installed to control its level. Manipulating a control valve on the outlet flow of the vessel controls the level. The level must be controlled at 93 percent with an acceptable error of plus or minus 2 percent.

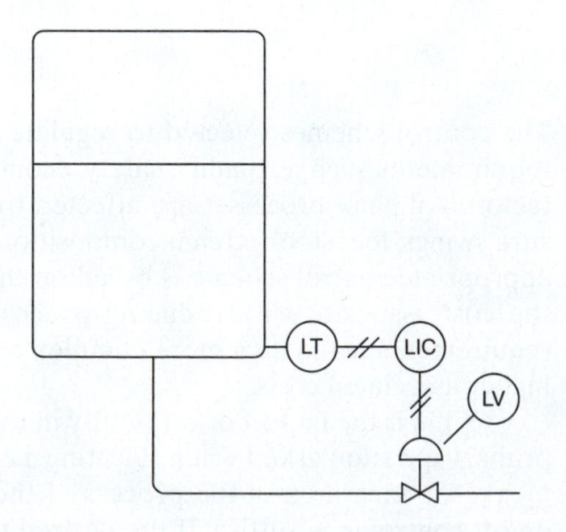

This is an example of:
  a. A feedback control loop
  b. A feedforward control loop

6. A tank in a processing unit needs to have an automatic control system installed to control its level. Manipulating a control valve on the outlet flow of the vessel controls the level. The level cannot tolerate even a 1 percent change. A single pipe emptying into the tank causes the only variable that can affect its level. Explain how feedforward can be used to accomplish this task?

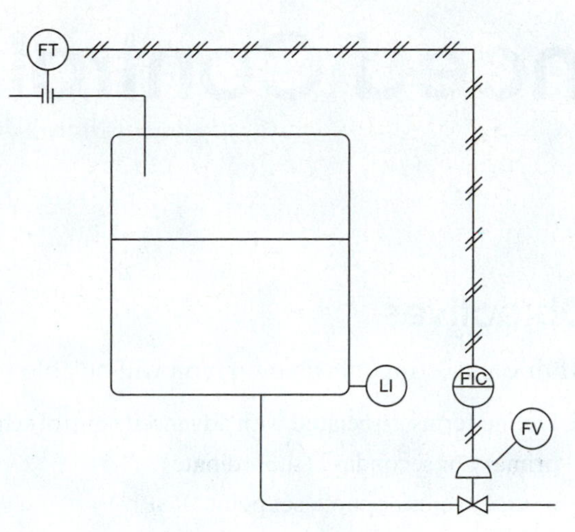

Student Activities

1. Given a description of a process, draw a graphic that represents the control scheme used in the process.
2. Given a control scheme scenario and illustrations, identify the appropriated control scheme. Discuss your reasons for selecting this scheme.
3. Given proper equipment:
   a. Manipulate a process variable using automatic mode and then using manual mode.
   b. Change the setpoint in automatic mode and observe the variable measurement change to match the new value.
   c. Change the control mode to manual and return the variable measurement to the original value.
   d. Draw a P&ID describing the loop.

CHAPTER 18

Advanced Control Schemes

Objectives

After completing this chapter, you will be able to:

- Define terms associated with advanced control schemes.
  - primary or secondary (subordinate)
  - cascade versus remote set point (RSP)
  - ratio
  - split range
  - multivariable input
  - override controllers
- Explain the purpose of a cascaded control scheme.
- Explain the function of a ratio (fractional) control scheme.
- Explain the purpose and function of a split-range control scheme.
- Explain the purpose and function of a multivariable control scheme.
- List steps required to change instrument controllers without bumping the process:
  - manual to automatic
  - remote to local setpoint (cascade)

Key Terms

Multivariable input—industrial processes that involve an interaction between two or more process variables.

Override controller—a device or program that remains inactive until a specific constraint (highest or lowest permitted extreme) on the measured variable is about to be reached.

Ratio—the proportion of flow between two separate flowing streams entering a mixing point.

Ratio control—control used in blending raw materials to make a final product and in proportioning the flow rates of reactants feeding into a reactor; also used to maintain the proper fuel-air ratio feeding into a furnace.

Ratio control loop—control loop designed to mix two or more flowing streams together while maintaining a quantitative ratio between them.

Introduction

Although single-loop, single-variable feedback controls are common, many applications require more complicated systems. Advanced control refers to those systems that use the three basic control modes: proportional, integral, and derivative, to yield control benefits not possible with simple loops. Such benefits include the following:

- reduced lag times
- faster response to process changes
- better control
- disturbance minimization
- high and/or low constant constraint controls
- ratio and mixing control

A single loop controller operates on the equilibrium theory that a control device will be configured to manipulate a variable to bring the controlled variable to a predetermined setpoint and maintain that value.

An advanced control scheme can be defined as a control system that goes beyond the single loop controller. There are many types of advanced control schemes and their configuration varies as do the processes they control.

Cascade Control

Cascade control is a control scheme in which the output of one controller becomes the setpoint for another. The two kinds of controllers found in cascade control are initiating controllers (referred to as primary controllers) and secondary or cascade controllers.

PRIMARY VERSUS SECONDARY CONTROLLER

The initiating or primary controller is a normal automatic controller with a setpoint adjustment located on the faceplate (local setpoint). This controller typically has controller modes of automatic and manual.

The secondary controller, which always controls the final control element, can be operated as a controller with a local setpoint or a remote setpoint (a setpoint received from an external source). Cascade or secondary controllers can have a controller mode of cascade (CAS) or remote setpoint (RSP).

Figure 18-1 shows an example of primary and secondary controllers.

The drawing in Figure 18-2 is a schematic of Figure 18-1. This schematic shows how the output of the primary controller becomes the remote setpoint (RSP) input to the secondary controller.

Regardless of whether the controller is locally or remotely operated, the controller will fundamentally operate the same as all other controllers, that is, operating to eliminate the difference between setpoint and measurement (error).

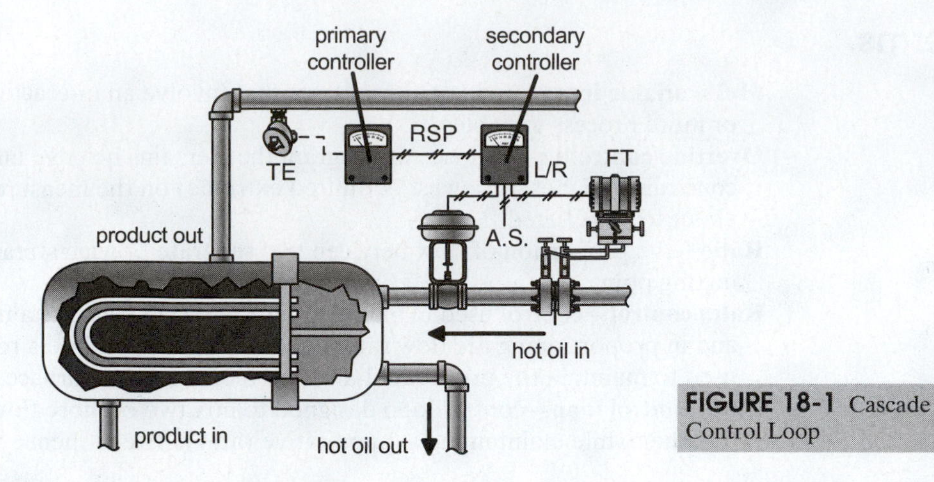

FIGURE 18-1 Cascade Control Loop

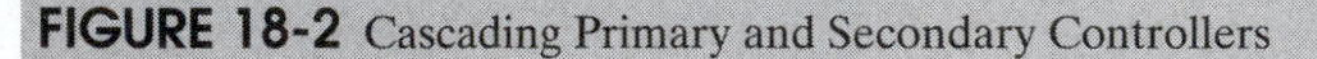

FIGURE 18-2 Cascading Primary and Secondary Controllers

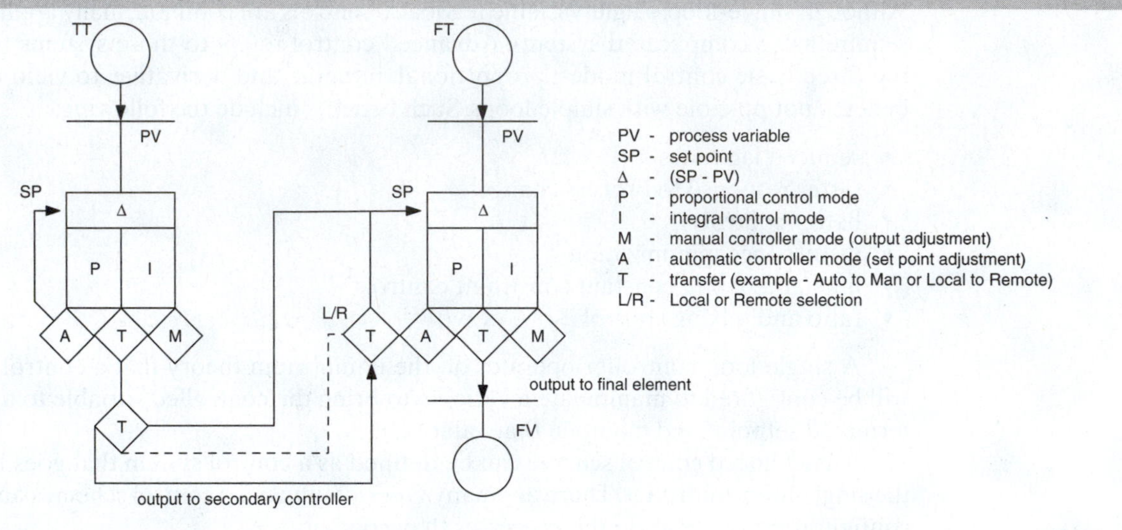

REASONS TO USE CASCADE LOOPS

Some important reasons for using a cascade loop include better control and reduced lag time.

Better Control

The secondary controller in many cascade systems controls flow rate. One problem with controlling a process variable with only a control valve is the potential that the upstream pressure to the valve may increase or decrease, causing the actual flow rate through the valve to vary as well. This may seem insignificant until you take a closer look at the system. Take as an example the valve that is used to supply heat to the process flowing through a heat exchanger. If the pressure in the line feeding this valve changes, the BTUs of heat flowing through the valve will change as well. Controlling the flow through the valve can eliminate this problem producing better control of the temperature.

Reduced Lag Times

In the previous example, a description of how upstream pressure can increase or decrease the flow rate through a control valve should be considered again. Consider this example: if the flow rate through the control valve diminishes because of a pressure drop, the temperature controller will not be able to react to the reduced flow rate until the change in temperature measurement is sensed. This problem can be solved if a flow loop is added to the temperature loop so that the primary controller can send an RSP to the flow controller. With this modification, the flow loop will immediately compensate for pressure changes in the steam line feeding the control valve and effectively eliminate the process lag time.

CASCADE CONTROL TIPS

When working with cascade control schemes, there are a few tips that should be considered. They are as follows:

- In order for a cascade system to function correctly, the secondary controller must respond more rapidly than the primary controller.
- The output from the primary controller should always follow the setpoint of the secondary controller when the secondary controller is not in RSP/CAS mode. This is sometimes called setpoint track or output track.
- The secondary controller should always setpoint track its process variable in manual mode. Some cascaded control schemes may use as many controllers as needed [e.g., an upper controller (% composition) should be cascaded to a middle controller (temperature) that is cascaded to a lower controller (flow)].
- The general procedure for a bumpless transfer in cascade control is as follows:
  1. Line out the primary variable with the secondary controller in local manual.
  2. With the secondary controller still in local (local setpoint), switch it from manual to automatic.
  3. With the primary controller in manual, adjust the output so the remote setpoint indicator on the secondary controller lines up with the local setpoint indicator.
  4. Toggle the secondary controller setpoint switch from local to remote.
  5. If the primary variable and the setpoint on the primary controller are the same, then switch from manual mode to automatic.

If the primary variable and the setpoint on the primary controller are not the same then adjust the primary controller setpoint so it equals the measurement and then switch it to automatic.

EXAMPLES OF CASCADE CONTROL

In order to better understand the concept of cascade control, it's important to look at several examples:

Example 1: Heat Exchanger

In Figure 18-3 the temperature transmitter located on the "shell-side" outlet pipe of the heat exchanger measures product temperature and sends it to the primary controller. The output signal from the primary controller becomes the external setpoint of the secondary controller. Recall that the secondary controller is capable of receiving an external setpoint. The secondary controller then controls the flow rate of the steam entering the heat exchanger (tube-side). As with all control loops, a relationship between the two process variables must be understood and quantified. In the case of this loop, the relationship between steam flow and the heating of the product through a specific heat exchanger is first established. Once this temperature-flow relationship is established, any disturbance, whether it be steam supply pressure or tube-side plugging, will trigger the secondary control loop to respond by repositioning the steam valve.

An important thing to note is an increase in steam supply pressure would cause the valve to close more whereas a decrease in steam supply pressure would cause the valve to open more. The flow loop will compensate for this problem. Level cascaded to flow is better than plain level control. The cascaded control can more tightly control the vessel level should the flow through the control valve change due to disturbances upstream of the control valve.

Example 2: Vessel "Bottoms" Level Control

Figure 18-4 the vessel is level controlled by resetting the flow rate of "bottoms" from the vessel. The level transmitter measures the vessel level and sends the measurement

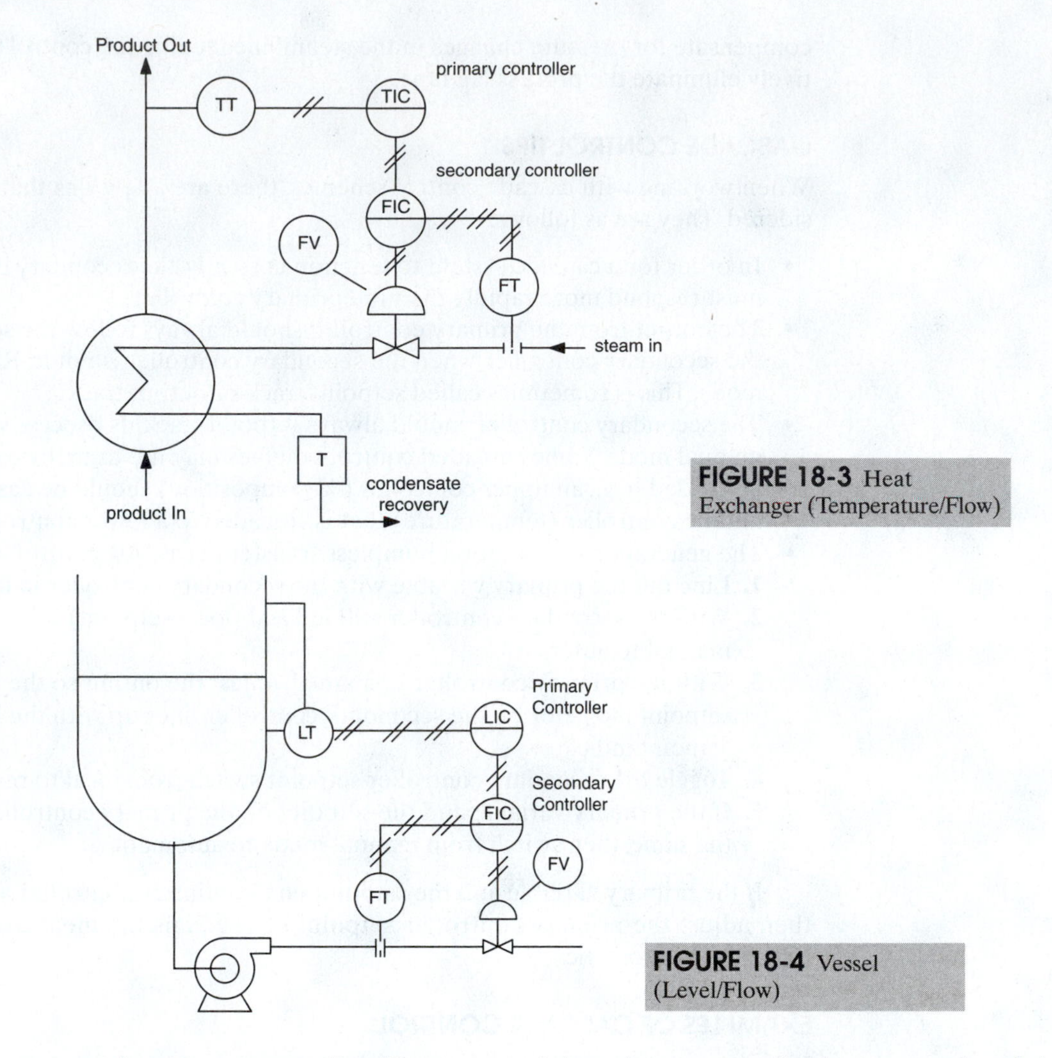

FIGURE 18-3 Heat Exchanger (Temperature/Flow)

FIGURE 18-4 Vessel (Level/Flow)

to the primary controller. The output of the primary controller is the setpoint of the secondary controller, which controls the flow rate from the vessel.

Example 3: Stirred Tank Reactor

In Figure 18-5 the temperature of the reactor is controlled by controlling the reactor jacket temperature. The temperature of the reactor contents can be controlled tighter and quicker by allowing the secondary temperature controller to adjust for cooling water supply changes rather than waiting for disturbances to be propagated through the jacket and reactor contents.

The temperature transmitter measuring the reactor temperature provides the input to the primary controller. The output of the primary is the external setpoint to the secondary controller. In this example, the secondary loop is also a temperature loop. The reactor jacket is a contact cooling system that is built around the entire reactor wall providing a large, consistent, and reliable direct contact cooling area. The reaction causes the reactor temperature to increase. The secondary temperature control loop responds by increasing the reactor cooling. If the reaction causes the reactor temperature to decrease, the reverse will happen.

Ratio Control

The term **ratio** means to proportion the rates of flow between two separate flowing streams entering a mixing point. The diagram in Figure 18-6 shows an example of **ratio control**.

Ratio control loops are designed to maintain a fixed ratio of two or more flowing streams regardless of flow changes in the uncontrolled stream. Ratio control is used

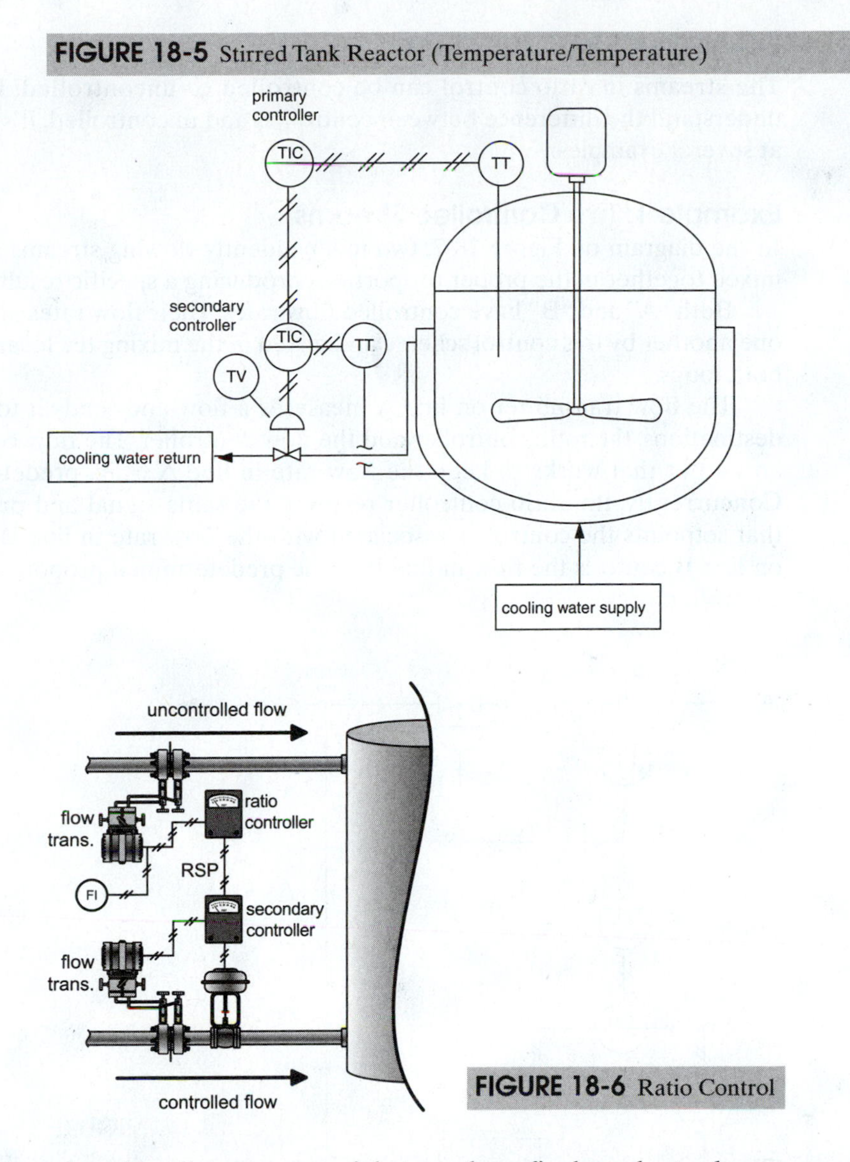

FIGURE 18-5 Stirred Tank Reactor (Temperature/Temperature)

FIGURE 18-6 Ratio Control

when blending raw materials to make a final product, when proportioning the flow rates of reactants feeding into a reactor, and to maintain the proper fuel-to-air ratio feeding into a furnace.

The hardware used to accomplish ratio control may look like a cascade loop between two flow loops. There are, however, significant differences between the two. In a cascade control loop the secondary loop operates in a subservient role to control the primary loop's process variable, whereas in a ratio control loop the secondary control loop simply proportions its flow rate to the primary loop. The bottom line is that the ratio loop doesn't receive feedback from the primary loop.

The ratio-producing component in a ratio loop may be a computing relay, such as a dividing relay, or it may be a specialized ratio controller. In either case, the ratio-producing component ratios the flow rates accordingly.

RATIO CONTROL TIPS

When working with ratio control schemes, there are a few tips that should be considered. They are as follows:

- Let the flow ratio controller calculate the ratio PV, but allow the process technician to adjust the ratio setpoint within certain allowable limits.
- Ratio control is not recommended on streams whose compositions may vary. Ratio control on streams whose compositions may vary must use a proper analyzer and cascade the setpoint control to the flow ratio controller.

EXAMPLES OF RATIO CONTROL

The streams in ratio control can be controlled or uncontrolled. In order to better understand the difference between controlled and uncontrolled, it's important to look at several examples:

Example 1: Two Controlled Streams

In the diagram on Figure 18-7, two independently flowing streams ("A" and "B") are mixed together in the proper proportions producing a specific resulting mixture.

Both "A" and "B" have controlled flow rates. Their flow rates are proportioned to one another by this control scheme and mixed in the mixing tee located downstream of both loops.

The flow transmitter on line A measures a flow and sends it to the two separate destinations, the ratio controller and the flow controller. The flow controller produces an output that works to keep the flow rate in line A at its predetermined setpoint. Concurrently, the ratio controller receives the same signal and produces an output that setpoints the controller associated with the flow rate in line B. The control loop on line B controls the flow in line B at the predetermined proportion set on the ratio controller.

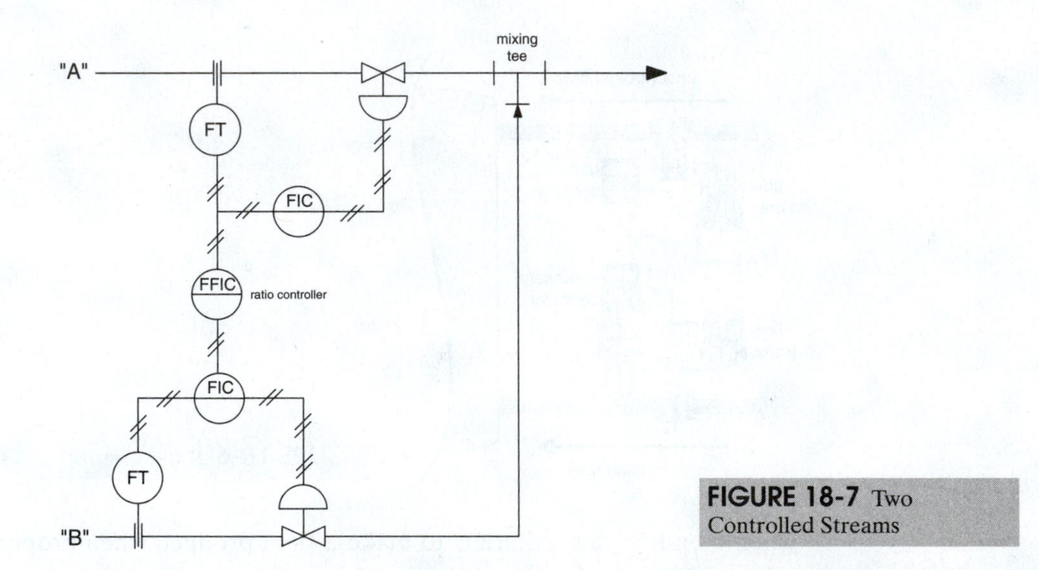

FIGURE 18-7 Two Controlled Streams

Example 2: One Controlled Stream and One Uncontrolled Stream

In the diagram on Figure 18-8, two independently flowing streams are mixed together in the proper proportions producing a specific resulting mixture.

Stream "A" is considered to be an uncontrolled flow, while "B" is considered to be the controlled flow. Again, their flow rates are proportioned to one another by this control scheme and mixed in the mixing tee located downstream of both loops.

The flow transmitter on line A measures a flow and sends it to the ratio controller only. The ratio controller then produces an output that provides a setpoint to the controller in the flow loop on line B. The control loop on line B then controls the flow in line B at the predetermined proportion set forth in the ratio controller.

Example 3: Feedforward Control

Figure 18-9 shows a cascade control loop with the secondary controller adjusting steam flow. This is an improvement over a single-loop control system, but the problem remains that the system cannot correct for load changes until they cause a change in the product temperature from the heat exchanger.

Figure 18-10 shows an improved scheme with a secondary loop that adjusts for product load change. This loop responds to changes in load rate and temperature. The improved *noiseless* response is shown in Figure 18-11.

FIGURE 18-8 One Uncontrolled Stream and One Ratio-Controlled Stream

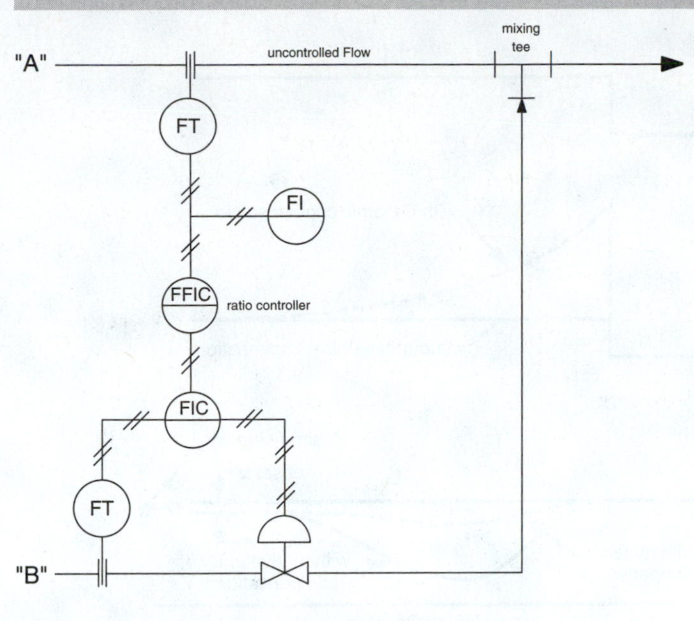

FIGURE 18-9 Cascade Control

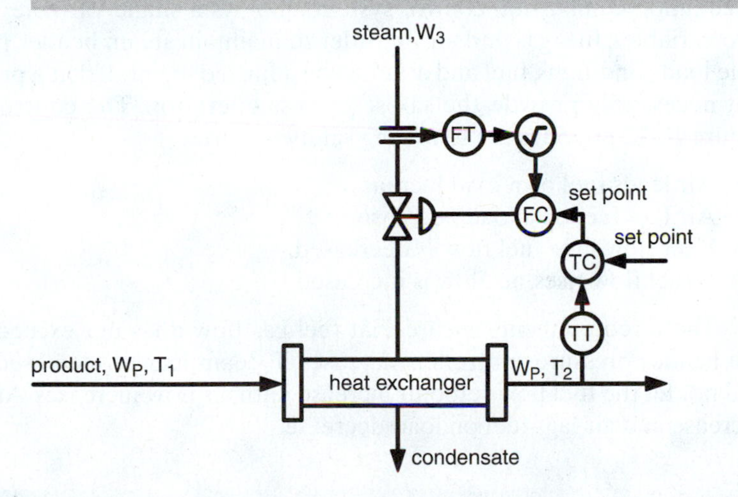

FIGURE 18-10 Feedforward Control

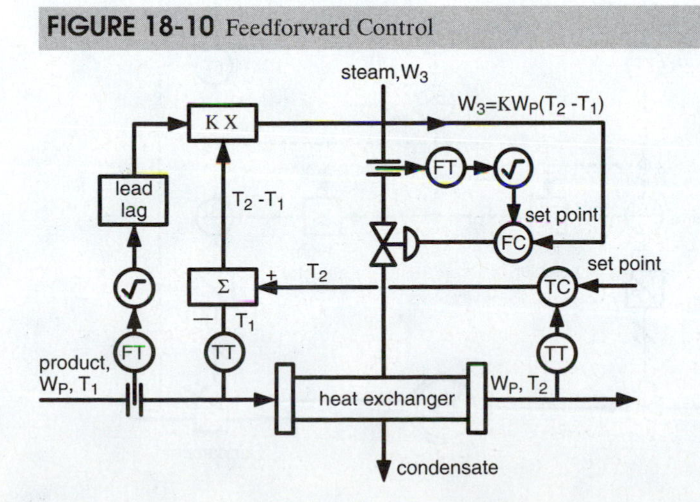

FIGURE 18-11 Feedforward Response

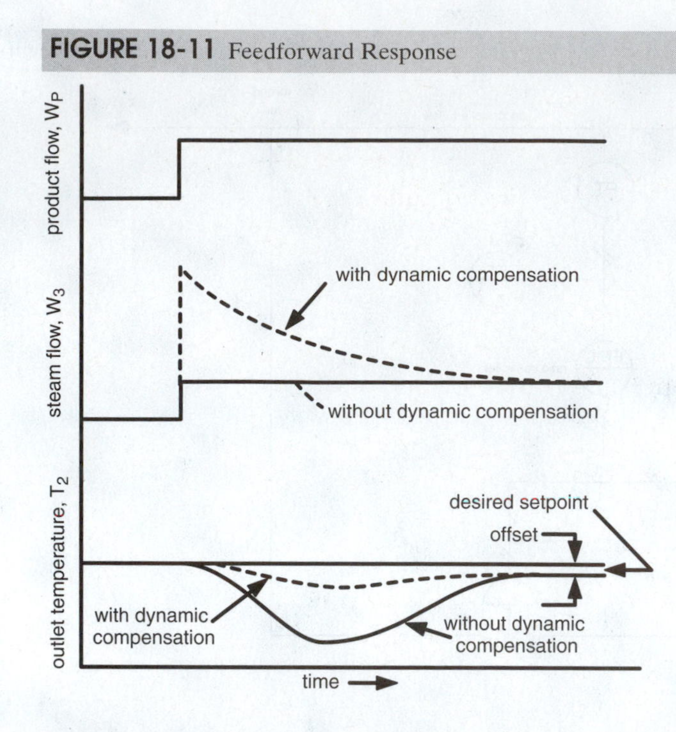

Example 4: Boiler Combustion Control

A common combustion control system employs a single variable, pressure to control two variables, fuel gas and air. In order to maintain steam header pressure under variable load conditions, fuel and air must be adjusted in parallel at a preset ratio. This does not necessarily provide the safest process operation. The control scheme shown in Figure 18-12 provides the following safety features:

- Air leads fuel on a load increase.
- Air lags fuel on a load decrease.
- If air flow falls, fuel flow is decreased.
- If fuel flow rises, air flow is increased.

These requirements ensure that fuel gas flow does not exceed air flow output of the header pressure controller increases if steam load is increased. The load selector will not let the fuel flow setpoint increase until air flow increases. Air leads fuel on load increase and air lags fuel on load decrease.

FIGURE 18-12 Boiler Combustion Control

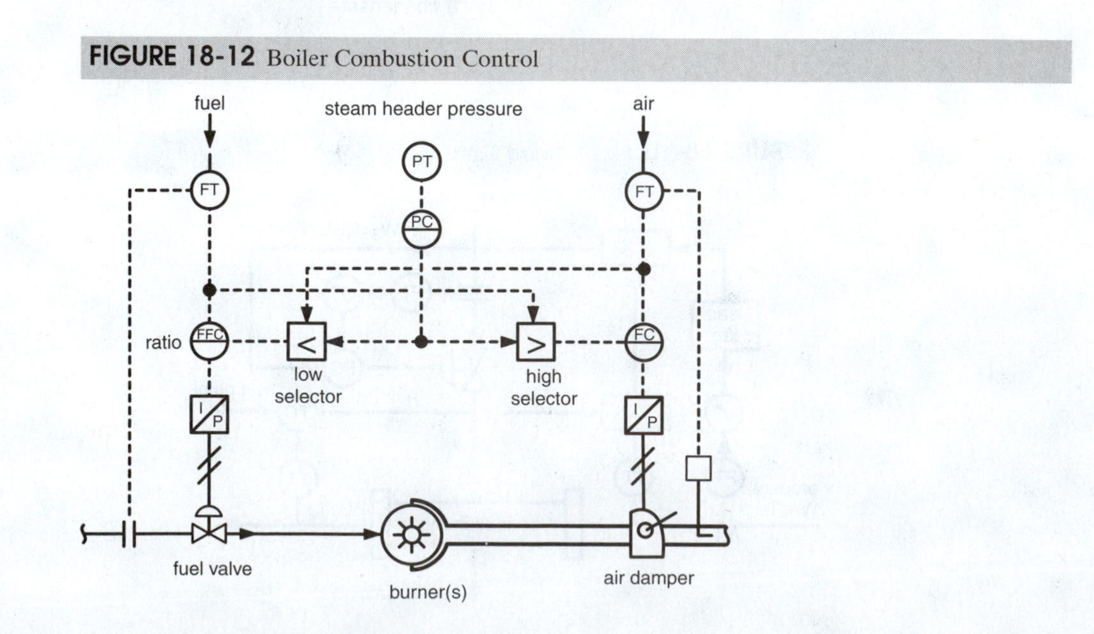

Split-Range Control

Split-range control is a control scheme in which the control signal is split into several parts, each of which is associated with one of the manipulated variables. The final outcome of split-range control is a single process that is controlled through the coordination of several manipulated variables, all of which have the same effect on the controlled output. The output of a split-range controller is divided between two final control elements, examples of which are shown in Figure 18-13.

In this example, output of a split-range controller is divided between two final control elements. The most common scenario involves one final control element (valve A) responding to the lower half of the output signal, while the other control element (valve B) responds to the upper half of the output signal. Consider the following example:

A processing tank requires a constant vapor space pressure (space above the liquid) of 10 psig. During charging (tank filling) and heating cycles, the pressure in the vapor space increases and a pressure release is required. After the heating cycle, the pressure drops so that increased pressure is required. To accommodate this, a controller with a split-range output applied to two separate control valves is required. One valve applies pressure to the tank, while another valve relieves pressure from the tank.

The most common way split-range valves would work together is to have one valve, valve B, respond by opening to an increasing signal from 9.5 psi to 15 psi and the other valve, valve A, respond by opening in an air-to-close arrangement from 8.5 psi to 3 psi. When the valves are receiving a signal of 8.5–9.5 psi, they are both closed. As the pressure drops in the tank, valve B opens allowing pressure to enter the tank, while the other remains closed. Conversely, when pressure rises in the tank valve A opens, allowing pressure to be relieved from the tank, while valve B remains closed.

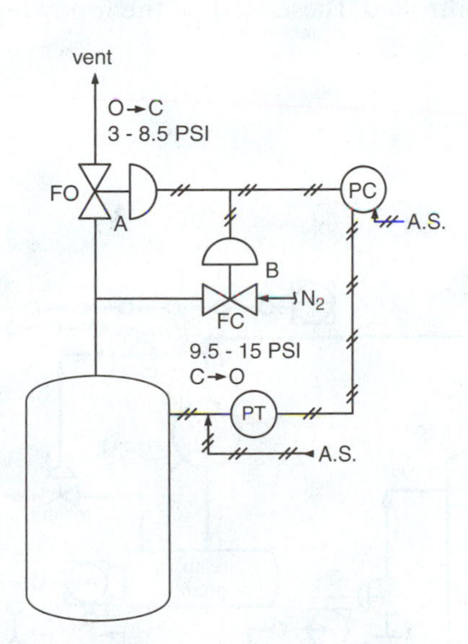

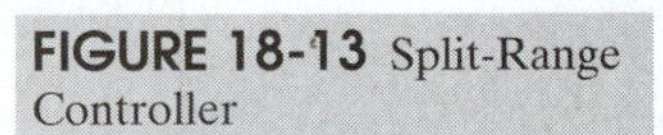
FIGURE 18-13 Split-Range Controller

Multivariable Control

Multivariable input refers to industrial processes that demonstrate an interaction between two or more process variables. When any one of these variables is changed, it affects the others in a swift and predictable manner. In some highly interactive processes, such as fractionator control, a multivariable controller with multiple inputs and outputs can produce a level of control that is impossible with single-input or single-output controllers. This is called multivariable control.

Multivariable control schemes are some of the most complicated and intricately designed control schemes in the processing industry. To fully understand and design a multivariable system requires extensive knowledge of process and control techniques.

Normally, single-input and output loops operate independently from other loops. When another variable in the same process changes, it may not affect other loops in a destabilizing manner.

In some cases, two or more loops may interact in a destabilizing manner in relationship to each other. If one manipulated variable changes it adversely affects one or more of the other variables. To illustrate this point, look at Figure 18-14, which contains two interacting loops.

Multivariable controllers are most generally model-based controllers. In a multivariable controller the inputs and outputs are custom "tailored" with a model (gain, time response, and delay) that is matched to the actual process being controlled. Being "model-based" versus standard PID–based is a distinguishing attribute.

Notice that a change in the output of Loop 1 affects the measurement of Loop 2, which in turn affects Loop 1 again. This usually triggers the loops to start cycling with steadily increasing amplitude.

The need to implement a multivariable control system increases proportionally with the complexity of the interactions between instrument loops. The more complex the loop interaction, the more necessary a multivariable control system becomes. In industry today, multivariable control solutions have become increasingly more common due to the reliability and programmability of modern computerized control systems.

Some examples of places where multivariable control schemes can be used include the following:

1. Chemical reactors
2. Distillation and fractionation towers
3. Heat exchangers

Figure 18-14 illustrates a depropanizer distillation system with all the variables that must be measured and controlled. These include the following:

- feed rate
- reflux rate

FIGURE 18-14 Multivariable Control Example

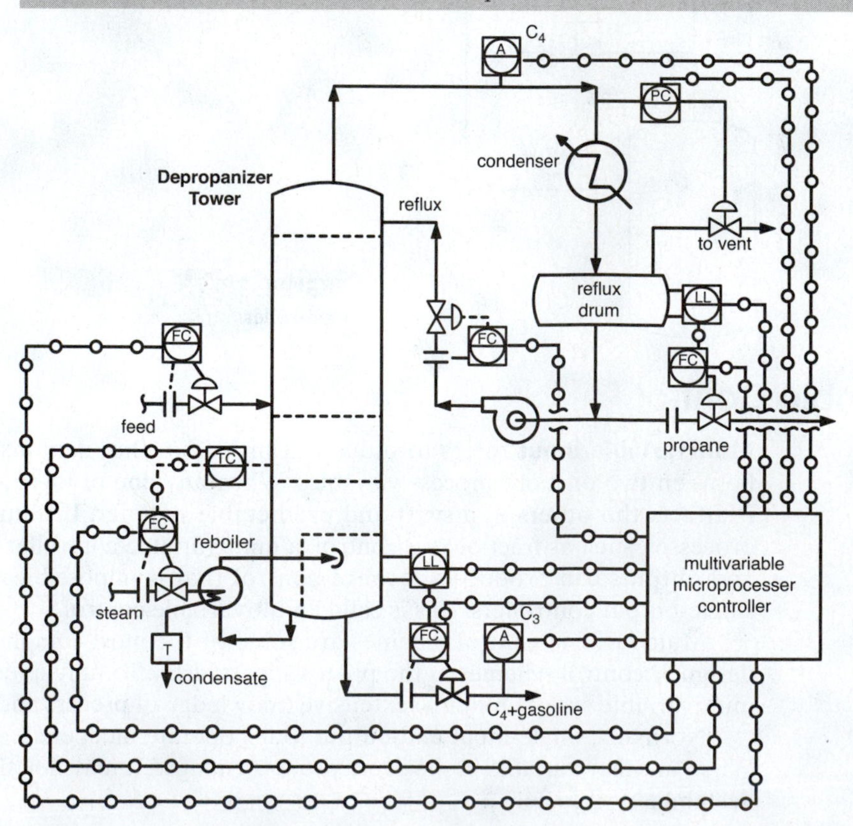

- pressure
- temperature
- reboiler
- overhead product (propane) rate, analysis
- bottom product (C_4 + gasoline) rate, analysis

Conventional digital instruments are provided primarily for startup, shutdown, and troubleshooting operations. The inputs and outputs to all instruments are connected to a microprocessor controller. Not only will a change in one input variable cause a change in several outputs, but changes in any output will cause input variable changes. For example, a feed rate or composition change will trigger potential changes in reboil, reflux, and product rates. The microprocessor controller continuously updates its data and makes a series of adjustments seeking the best attainable system performance. Continuous product analyses confirm the results of these adjustments. The computer controller adjusts all actuators in the same way as an analog controller.

The multivariable system allows continuous fine-tuning by the computer. In addition to maintaining all variables at conditions that provide desired results, the program can be modeled to perform helpful material balances and flag the approach to constraint conditions.

Override Controllers

An **override controller** is a device or program that remains inactive until a specific constraint (highest or lowest permitted extreme) on the measured variable is about to be reached. At that time the override mechanism takes control of the manipulated variable (the final control element) and prevents the constrained variable from exceeding its limit. Only after the measured variable returns to a predetermined "safe" level, does control of the manipulated variable returned to the normal controller.

In process refrigeration, override controllers are used to prevent an undesired condition. For example, in Figure 18-15 refrigerant liquid is admitted to a chiller so a desired process outlet temperature can be attained. However, if the level in the chiller increases above a certain point, there is danger of liquid being carried over into the compressor causing upset and potential damage to the compressor. Thus, the override controller limits the chiller level to a safe preset value (e.g., 80 percent) regardless of the desire of the process temperature controller.

FIGURE 18-15 Override Control Scheme Example

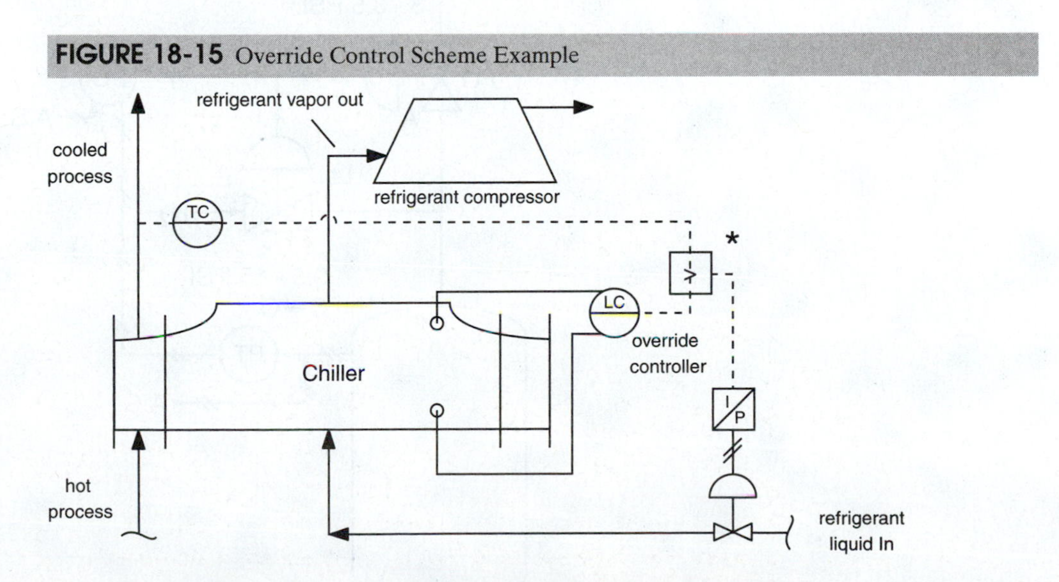

Summary

Many applications require systems that are more complicated than single-loop, single-variable feedback controls. These types of systems are referred to as advanced control systems.

The term advanced control refers to those systems that use the three basic control modes: proportional, integral, and derivative, to yield control benefits not possible with simple loops. Such benefits include reduced lag time, faster response to process changes, better control, disturbance minimization, high and/or low constant constraint controls, and ratio and mixing control.

Single loop controllers operate on the equilibrium theory that a control device will be configured to manipulate a variable to bring the controlled variable to a predetermined set point and maintain that value.

Advanced control schemes are defined as control systems that go beyond the single loop controller. There are many types of advanced control schemes and their configurations vary as widely as the processes they control.

Checking Your Knowledge

1. Identify the control scheme shown in the diagram below.
   a. Multivariable
   b. Split-range
   c. Override
2. A control system designed for a process with more than one input and more than one output is called:.
   a. Multivariable
   b. Split-range
   c. Override
   d. Cascade
3. Which of the following processes would be the most likely to require computer-assisted control?
   a. Multivariable
   b. Split range
   c. Override
   d. Cascade
4. *True or False* Switching a controller from automatic to manual requires adjusting the setpoint of the controller to the actual controlled point.

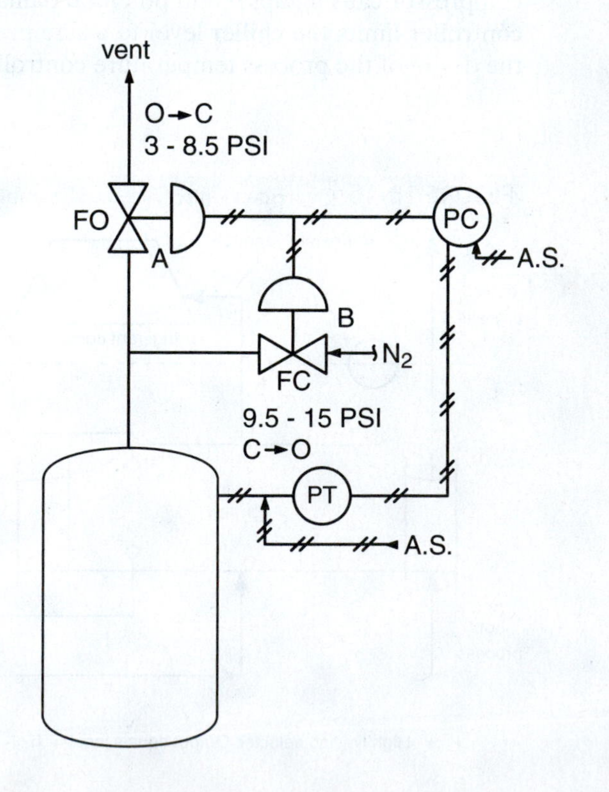

Student Activities

1. Define the following terms:
   a. Primary
   b. Secondary
   c. Cascade
   d. Remote setpoint (RSP)
   e. Ratio
   f. Split-range
   g. Multivariable input
   h. Override controllers
2. Explain the purpose and function of cascaded control schemes.
3. Explain the purpose and function of a split-range control scheme.
4. Explain the purpose and function of a multivariable control scheme.
5. List the steps required to change instrument controllers without pumping the process.
6. Draw a cascade control system as a P&ID and then describe how the cascade loop operates.

Given a cascade control system with both an inner (secondary) and outer (primary) loop:

1. Control the secondary (inner) loop at several different values in both manual and automatic control.
2. Switch between local setpoint and remote setpoint on the secondary controller without significantly bumping the process.
3. Control the process at several different setpoint values.
4. Given a ratio control setup configured according to the P&ID illustration used in the classroom, use a valve or rotameter with valve to adjust flow A and a ratio controller to adjust flow B.

CHAPTER 19

Introduction to Digital Control

Objectives

After completing this chapter, you will be able to:

- Explain the purpose of digital control.
- In general terms, describe the difference between analog and digital controllers.
- Define the following terms associated with process control schemes:
  - analog
  - discrete controllers
  - DCS
  - PLC
  - input
  - output
  - D to A
  - A to D
  - converter
  - multiplexer
  - demultiplexer
- In general terms, describe how a digital controller transmits an output signal to an analog final control element.
- In general terms, describe how an analog controller transmits to a digital controller.

Key Terms

Analog to digital (A/D)—the conversion of an analog signal to a digital signal so it can be processed by a computer; also referred to as "A to D."

Converter—a device that changes or converts a substance or a signal from one state to another (for example, changing AC current to DC current).

Demultiplexer (DEMUX)—a device that separates single stream, multiplexed signals back into their original multi-stream constituents.

Digital to analog (D/A)—the conversion of a digital signal to an analog signal so it can be interpreted and reported on an analog gauge or instrument (e.g., using an I/P converter to convert a digital signal to a pneumatic signal that triggers the opening or closing of a control valve); also referred to as "D to A."

Discrete controllers—digital controllers that can accept and produce only digital input and output.

Distributed control system (DCS)—a manufacturing system (usually digital) consisting of field instruments connected to multiplexers and demultiplexers and A/Ds (analog to digital) via wiring or busses and which terminate in a human-to-machine interface device or console.

Input—the process of entering data or information into a computer system or program.

Multiplexer (MUX)—a device that merges or interleaves multiple signals into one stream or output signal in such a way that each individual signal can be recovered or separated out using a demultiplexer.

Output—the number or value that comes out from a computer program or process.

Programmable logic controller (PLC)—a computer-based controller that uses inputs to monitor processes and outputs to control processes.

Introduction

In order to understand how information is transmitted and received, it is important to understand how digital control systems work and how they interrelate with analog components and converters.

Analog instruments, which were one of the earliest equipment types, uses static gauges with an indicating pen that must move smoothly across all values. Digital instruments, however, provide a more accurate and detailed way of collecting, sending, and interpreting data.

In order to transfer signals between analog and digital equipment, converters must be used. There are several different types of converters, although D/A (digital to analog) and A/D (analog to digital) are the most common.

Once signals are converted and are ready to be transmitted, a multiplexer (MUX) is used to merge or interleave the signals so they can all be transmitted. Once received, a demultiplexer separates the signal back out into its original multistream constituents. Both multiplexers and demultiplexers are key components of modern DCS systems.

A Brief History of Controllers

The concept of automated control is not new. As early as the mid-1920s, rudimentary forms of automatic control began appearing in industry, although a vast majority of processes were still controlled manually.

In these manually controlled systems process technicians were required to monitor gauges and make adjustments to variables in every tank throughout the workday. This was a very time and energy-consuming process.

By the 1950s and 1960s technology had advanced such that analog electronics began replacing manually controlled systems. This significantly improved the process and reduced the amount of time and energy required for monitoring.

In 1971 the Intel Corporation introduced the model 4004 microprocessor, the world's first "processor on a chip." With the advent of this new technology came a new type of control system: *digital control*.

Digital control began as direct control. In direct control, a computer was used to process measurement data, calculate the required control output, and control final control elements (usually valves). However, using a single computer to control a process had a major weakness. If the computer failed, the whole unit was shut down. Thus, improvements had to be made.

As part of the improvement process, direct control systems evolved into supervisory control, and then further evolved into the **distributed control systems (DCS)** we use today.

The first-generation DCS system was, basically, a series of microprocessor-based multiloop controllers built along a communications network called a data highway. Through this data highway and remote processing units (RPUs), distributed controllers were able to manage multiple process loops simultaneously and the digital signals associated with them.

Analog Versus Digital

Digital signals are characterized by data that are represented with coded information in the form of binary numbers (a series of 0's and 1's). Measurements taken from a process are continuously variable quantities (analog) that must be converted into binary numbers and entered into a computer for processing (digital). The next sections discuss the differences and similarities of analog and digital equipment.

ANALOG

Analog means analogous or similar to something. In the case of a measured variable, that something may be a real-world indication of a variable such as pressure or temperature. In an analog system, pressure and temperature measurements change in a continuous manner or *wave* mirroring the actual process variable.

In other words, while a digital readout devices jump from one numerical value to the next, analog pressure gauges with an indicating pen (like the one shown in Figure 19-1) move smoothly across all values in between.

DIGITAL

Digital is a term applied to a device that uses binary numbers to represent continuous values or discrete (individual or distinct) states. Binary numbers are first converted to decimal values before being displayed on a digital instrument like the ones shown in Figure 19-2. Unlike analog instruments, digital instruments can jump from one value to the next, skipping values in between.

Computers and computer-based instruments are characterized by digital **input** and **output**. There are many advantages to using computer-based instruments, including programmability, self-diagnostics, and the ability to recalibrate from a remote location.

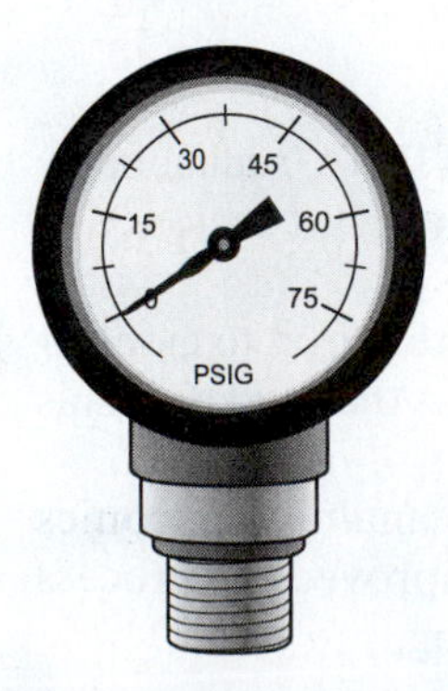

FIGURE 19-1 Analog Instrument

FIGURE 19-2 Digital Instrument

Industry uses both analog and digital systems because each has qualities desirable in specific situations. However, the true power of a digital system lies in the ability to manage information far beyond the capabilities of an analog system. Because of this, the use of digital instrumentation is on the rise, and analog systems are on the decline. As far as the near future is concerned, however, analog equipment will retain a significant role in the processing industry.

Other Components of a Digital Control System

REMOTE PROCESSING UNITS (RPUS)

Remote processing units (RPUs) are the heart of a distributed control system. An RPU is a microprocessor that is located close to the equipment that it monitors and controls. It is connected to workstations by the transmission subsystem or *data highway*. A large DCS can have hundreds of RPUs.

Originally an RPU contained only 8 field terminations or 8 *channels*. Therein lies the beauty of this type of system. If an RPU fails, only eight channels, out of potentially thousands, are affected. In addition, the abundance of channels makes it easy and economical to provide redundancy or other backup strategies.

PROGRAMMABLE LOGIC CONTROLLERS (PLCS)

Programmable logic controllers (PLCs), which are used mostly in discrete manufacturing, were designed to replace the switches and push buttons of motor control centers. The use of PLCs makes microprocessors and computers practical in industrial applications and enables the development of the DCS system.

While the original PLCs were limited to on/off operations, they did gain some analog capabilities over time. Current devices have complete digital and analog capabilities and are simply referred to as programmable controllers or PCs.

DISCRETE CONTROLLERS

Discrete controllers, which tend to be smaller in size than their analog counterparts, are digital controllers that can only accept digital input and provide digital output. All computers and PLCs are considered discrete controllers.

MULTIPLEXERS/DEMULTIPLEXERS (MUX/DEMUX)

The data highway in a plant DCS can accommodate multiple signals from field sources at one time. However, a controller must have a means to differentiate and process each individual signal.

A **multiplexer (MUX)** is a device that merges or interleaves multiple signals into one stream or output signal in such a way that each individual signal can be recovered or separated out later. Multiplexing can be done with mechanical relays; however, digital methods are more common and are less likely to develop problems.

A **demultiplexer (DEMUX)** is a device that separates single stream, multiplexed signals back into their original multistream constituents. In essence, demultiplexing is the opposite of multiplexing.

Multiplexers and demultiplexars are both key components of modern DCS systems. See Figure 19-3.

FIGURE 19-3 Multiplexer/Demultiplexer (MUX/DEMUX)

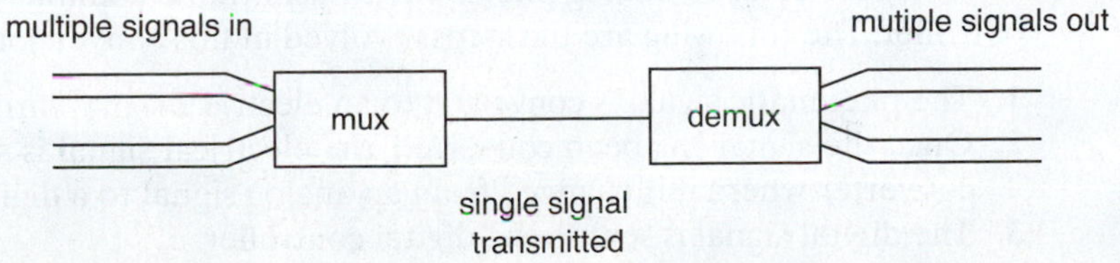

Signal Conversion and Transmission

As stated in the previous section, present-day process technology incorporates both analog and digital equipment. However, conversion is necessary for this equipment to interact. In D/A conversion, digital signals are converted into analog signals. In A/D conversion, the reverse is true. Most devices perform these A/D and D/A conversions internally.

While there are a number of different techniques used to convert a digital signal to its analog counterpart, the most common method is known as parallel-type conversion.

DIGITAL TO ANALOG (D/A)

There are several steps involved in a **digital to analog (D/A)** conversion. The following steps, which are diagrammed in Figure 19-4, are just one example:

1. The digital controller creates a digital (binary) signal.
   NOTE: This signal is proportional to the difference between the setpoint and the actual value of the process variable.
2. The digital signal is then converted to an equivalent analog signal with the help of a D/A converter.
3. The equivalent analog signal is then converted into a pneumatic (air) signal in an I/P converter.
4. The pneumatic signal is then sent to the control valve.
   NOTE: In some situations, there may be a positioner between the I/P and the control valve. The purpose of a positioner is to open the control valve very precisely. A positioner may be used when it is very critical that the control valve opening be very precise or very accurate.

In the example in Figure 19-4, the signal from the computer to the control valve requires two conversions before it can activate the control valve. Recent introduction of "digitized" control valves make such conversions unnecessary.

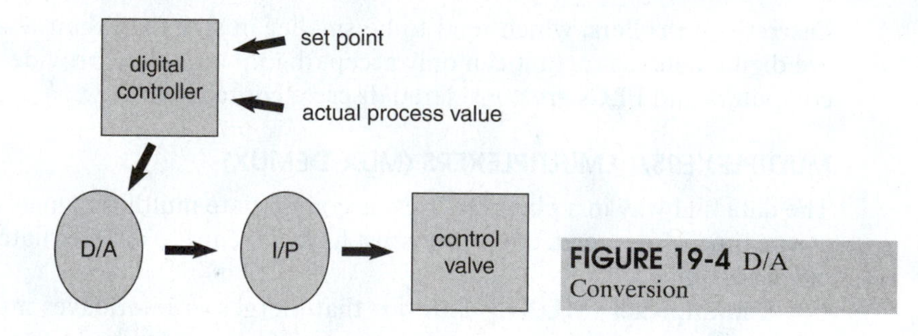

FIGURE 19-4 D/A Conversion

ANALOG TO DIGITAL (A/D)

Like D/A conversion, it is also important to convert signals the other direction, from **analog to digital (A/D)**. Analog instruments such as flow transmitters, pressure transmitters, level transmitters, or thermocouples produce a 4–20-mA analog signal. This analog signal must be converted into its digital equivalent before it can be processed by a computer. There are numerous techniques for accomplishing A/D conversion. These include parallel, cascade, and indirect conversions.

In the first diagram on Figure 19-5, an analog electrical transmitter is communicating with a discrete (digital) controller. An A/D converter is used to convert the signal.

In Figure 19-5, an analog pneumatic transmitter is communicating with a discrete controller. The following are the steps involved in this conversion:

1. The pneumatic signal is converted to an electrical signal through a P/I **converter**.
2. Once the signal has been converted, the electrical signal is sent through an A/D converter where it is changed from an analog signal to a digital signal.
3. The digital signal is sent to the digital controller.

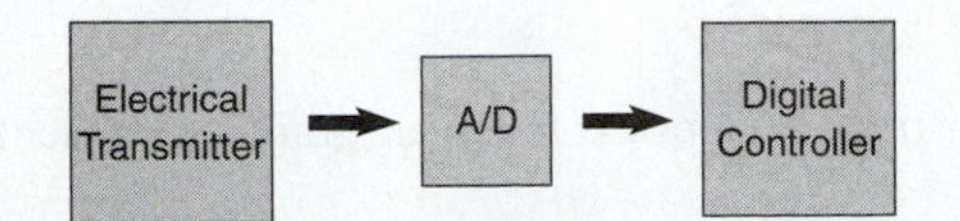

Digital Controller / Control Valve Controller

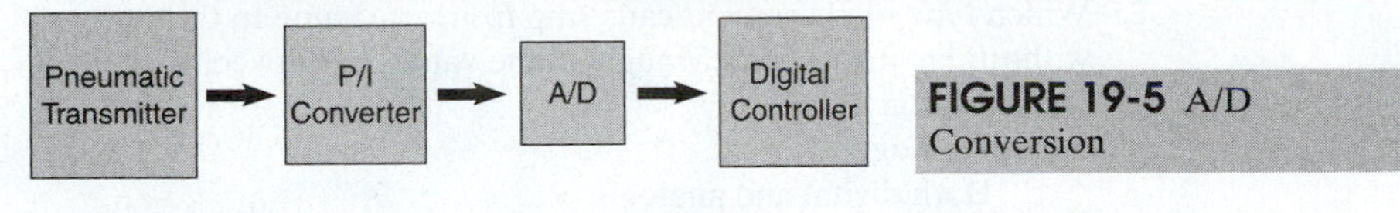

FIGURE 19-5 A/D Conversion

STAND-ALONE DIGITAL CONTROLLERS

Digital controllers are usually board-mounted with 120-volt AC power. These units are able to take many different types of inputs and convert them internally to digital information. Some of the multiple inputs include RTDs, thermocouples, 4–20 mA or 1–5 volt, and of course digital. Some of the outputs can be sent as analog or digital signals.

Digital controllers usually have self-diagnostic and auto-tune capability. They can be programmed with limited logic and several different control schemes such as cascade, ratio, feedforward, and feedback. The faceplate has a limited keypad for programming and shows SP, PV, output, mode alarms, and tag number.

SMART TRANSMITTERS

In an attempt at streamlining the conversion process, transmitters called *smart transmitters* (microprocessor-based instruments) have been created and are currently being used in industry. Smart transmitters have an A/D converter built into them so they can process information within the onboard microprocessor. It's important to note, however, that the output signal from a smart transmitter may or may not be digital or may be both. Some transmitters use high-frequency digital signals superimposed on the standard 4–20-mA output to communicate over the loop. The digital communication does not disturb the 4–20-mA signal since the net energy added to the loop is zero.

Summary

In this chapter you learned that there are two main types of equipment signals, analog and digital. Analog, one of the earliest equipment types, uses static gauges with an indicating pen that must move smoothly across all values between point A and point B.

Digital, which was not introduced until the 1970s, provides a more accurate and detailed way of collecting, sending, and interpreting data. Unlike analog equipment, digital instruments can jump from one point to the next without passing all the values in between.

Analog and digital systems are both used in industry because each has their own unique qualities that are desirable in certain situations. While analog instrumentation has been the most prevalent in the processing industry, the use of digital is on the rise because of the amount of flexibility and accuracy it provides.

In order to transfer signals between analog and digital equipment, converters must be used. There are several different types of converters. However, D/A (digital to analog) and A/D (analog to digital) are the most common ones encountered during digital to analog conversion.

Once signals are converted and are ready to be transmitted, a multiplexer (MUX) is used to merge or interleave the signals so they can all be transmitted simultaneously. Once received, a demultiplexer separates the signal back out into its original multistream constituents. Both multiplexers and demultiplexers are key components of modern DCS systems.

Checking Your Knowledge

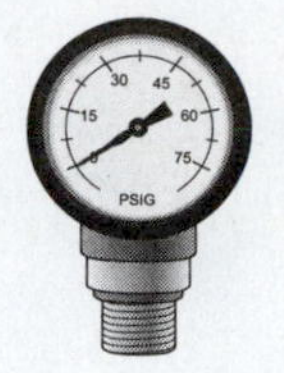

1. Which type of instrument does the following graphic represent?
   a. Digital
   b. Analog
   c. Both digital and analog
2. Which type of instrument can jump from one value to the next without having to pass through all the values in between?
   a. Digital
   b. Analog
   c. Both digital and analog
3. If you wanted to convert a digital signal to analog, which type of conversion would you use?
   a. A/D
   b. I/P
   c. D/A
4. You are a technician setting up a system and you have a pneumatic transmitter (A) and you need to send a signal to a digital controller (B). On the diagram below, identify and label all of the components (converters) that need to be added in between points A and B in order for the signal to be transmitted properly.

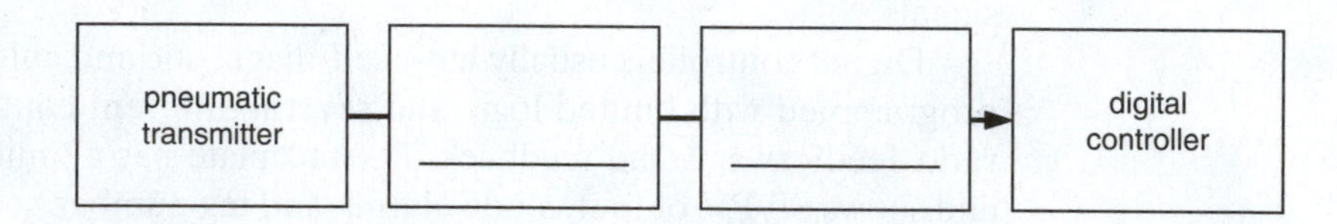

Match the following terms with their definitions:

Term	*Definition*
_____ Converter	a. A device that separates single stream, multiplexed signals back into their original multistream constituents.
_____ Demultiplexer	b. Computers, designed to replace switches and push buttons of motor control centers
_____ Discrete controllers	c. Digital controllers that can accept and produce only digital input and output.
_____ Distributed control systems	d. The process of entering data or information into a computer system or program.
_____ Input	e. A manufacturing system (usually digital) that consists of field instruments connected to multiplexer/demultiplexers and A/D's (analog to digital) via wiring or busses and that terminate in a human-to-machine interface device or console.
_____ Multiplexer	f. The number or value that comes out from a computer program or process.
_____ Output	g. A device that changes or converts signals from one state to another
_____ Programmable logic controllers	h. A device that merges or interleaves multiple signals into one stream or output signal

Student Activities

1. Draw a digital control loop and identify the parts. Discuss what each component is used for.
2. Given a digital control loop, identify the following components:
   a. Digital controller
   b. D/A
   c. Wiring
   d. I/P
   e. Positioner
   f. Control valve
3. Turn on power to the entire loop. Show the signal output from the controller and observe the response of the control valve. Record your observations.
4. Turn on power to the entire loop. Keeping the controller in manual, show the signal output from the controller and observe the response of the control valve. Record your observations.
5. Given a nonfunctioning D/A, determine what the problem is and how it will affect the control loop and the control valve.
6. Given an I/P with a failed air supply, determine the mode of the failure and the impact the failure has on the control valve action.

CHAPTER 20

Programmable Logic Controls

Objectives

After completing this chapter, you will be able to:

- Define terms associated with programmable logic control (PLC):
  ladder logic diagram
- Explain the purpose of PLCs:
  sequential control
  on/off
  emergency shutdown (ESD) systems
  integral to distributed control systems (DCS)
  stand-alone capability
- Explain how "ladder logic" applies to programmable logic control.

Key Terms

And/or—a conditional logic statement used to control a process.
If/then—a conditional logic statement used to control a paocess.
Ladder diagrams—diagrams that guide electricians in the fabrication process.
Motor control center (MCC)—hard-wired relays, switches, or contacts, housed in large metal cabinets that control the starting, stopping, and sequencing of motors and other devices.
On/off—a conditional logic statement used to control a process.
Shelf position—the contact position of an electrical device when de-energized (for example, a "normally open" switch is normally open when it is de-energized).

Introduction

Control engineering has made many changes and advancements over the years. In the past the main method for controlling a system was human intervention. Over time, however, fluid dynamics, electricity, and relays were employed to improve the process and reduce the dependence on manual intervention.

Relays were used to make simple logical control decisions and allow power to be switched on and off without a mechanical switch. However, once low-cost computers became readily available a new revolution occurred . . . the computer-based programmable logic controller (PLC).

Modern electronic PLCs were first introduced during the 1970s. Today PLCs represent the first and most widespread application of computers in process control today, mainly because of the numerous advantages they offer over previous hardwired systems. Those advantages include the following:

- cost-effectiveness (a relatively inexpensive way to control complex systems)
- more sophisticated control (computational abilities increase sophistication)
- flexibility (modular construction allows easy replacement and addition of units)
- improved troubleshooting (programming is easier and downtime is reduced)
- reliability (components tend to be more reliable and don't fail as frequently)

Historical Background of PLCs

In the pre-computer PLC era, logic was accomplished by hard-wired relays, switches, rotary contactors, or contacts, housed in relay control cabinets. These relay control cabinets were then connected to electric motor control circuits contained in **motor control centers (MCCs)** or to electric solenoid valves for **on/off** control of control valves. These MCCs, which are shown in Figure 20-1, controlled the starting, stopping, and sequencing

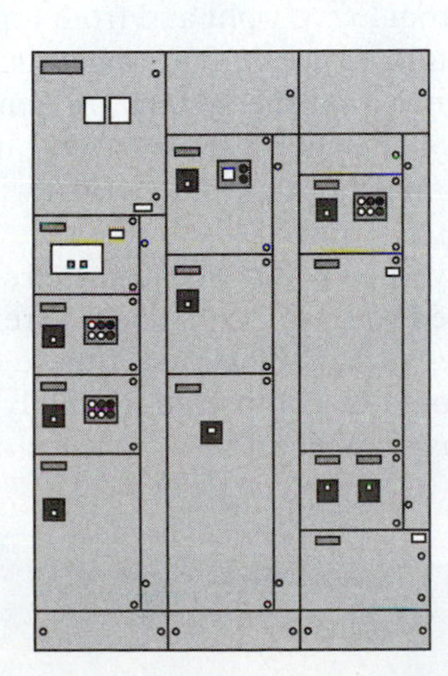

FIGURE 20-1 Motor Control Centers

FIGURE 20-2 Sample Ladder Diagram

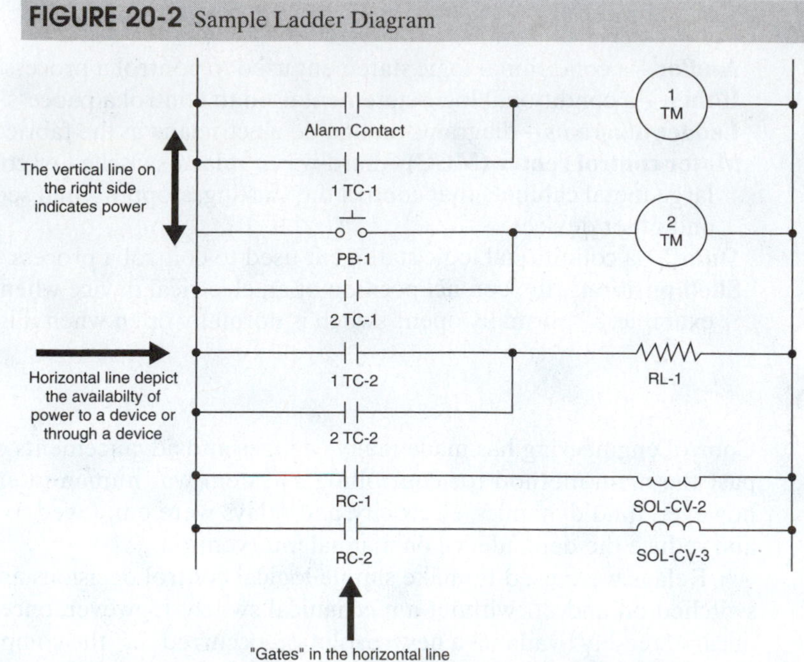
1
TM
Alarm Contact
The vertical line on
the right side
indicates power
1 TC-1
2
TM
PB-1
2 TC-1
Horizontal line depict
the availabilty of
power to a device or
through a device
1 TC-2
RL-1
2 TC-2
RC-1
SOL-CV-2
RC-2
SOL-CV-3
"Gates" in the horizontal line
indicate whether current is acually
flowing through a particular device

of motors and other devices. Computer-based PLCs now perform these functions faster and with greater reliability.

Because logic had to be hardwired into the older MCC-based controls, engineers were required to create diagrams to guide electricians in the construction process. These were called **ladder diagrams**. Figure 20-2 shows an example of a ladder diagram.

Having accurate ladder drawings was important because major modifications in control required time-consuming and expensive rewiring. Any time a change was made to the wiring or the control system the ladder diagrams had to be updated for future reference and troubleshooting purposes. This, too, was an expensive and time consuming process.

Reading Ladder Diagrams

Ladder logic diagrams derive their name from the ladder-like graphical symbolism used to depict PLC logic. The vertical lines, which look like upright ladder support, indicate power to an electrical system. The horizontal lines, which look like the rungs of a ladder, depict the availability of power to a device or through a device. "Gates" in the horizontal lines indicate whether current is actually flowing through a particular device.

Ladder logic diagrams are read from left to right and from top to bottom, just like a page of text would be read. Generally, devices are arranged in order of decreasing impact. Devices that have a larger impact on the system are generally shown above devices with relatively minor impact. Contacts, relays, and other devices are shown in their **shelf position**. For instance, a normally open *(NO)* contact will be shown as open. Consider switch A in Figure 20-3.

Switch A is considered normally open (NO) because there is no diagonal line intersecting it. When power is applied to this circuit relay, X remains de-energized because the circuit path is broken by switch A. However, if there had been a diagonal line like in Figure 20-4, the switch would be considered normally closed *(NC)*. When power is applied to this circuit relay, X is energized because the closed switch A enables current to flow in the circuit.

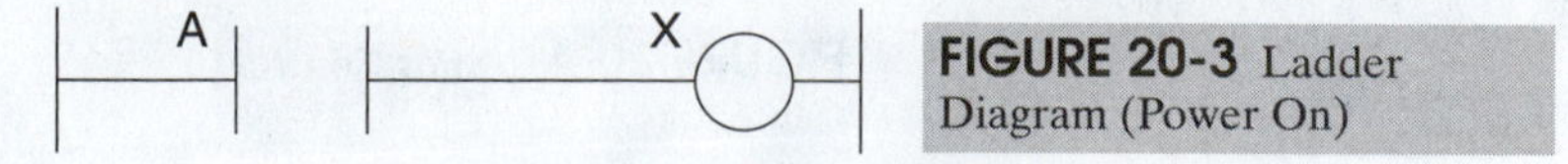
A
X

FIGURE 20-3 Ladder Diagram (Power On)

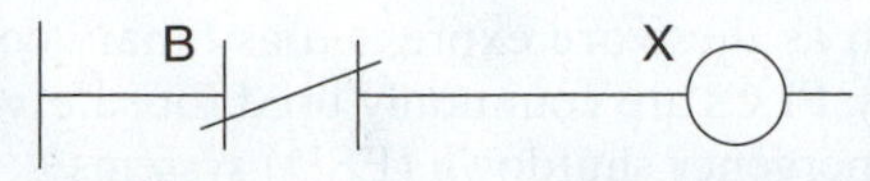

FIGURE 20-4 Ladder Diagram (Power Off)

Practical Tips

The following are practical tips:

- Instrument and electrical (I&E) technicians are typically quite good with PLCs and instrumentation. However, they may lack process knowledge. Process technicians provide a process liaison to work with the instrument technicians.
- Should it be necessary to temporarily bypass safety systems, management of change procedures must be strictly adhered to. A bypass, or "jumper" as it is called, should be documented and removed immediately after the completion of work.
- Before issuing a maintenance work permit to someone working on a PLC, plan the job thoroughly. It is not uncommon to see an entire plant shutdown accidentally because a PLC is being repaired or modified! Thus, process and maintenance technicians need to work together when planning PLC modifications.

PLCs Today

Modern PLCs allow logic changes to be easily programmed using desktop personal computers. In addition, these PLCs allow changes to ladder diagrams to be automatically recorded during the programming process, thus saving considerable amounts of time and money.

PLCs may be stand-alone units controlling only a single operation or they may be integrated into a broader control scheme such as modern distributed control systems (DCS).

Logic in PLCs

The main programming method used for PLCs is ladder logic. Ladder logic was intentionally developed to mimic relay logic. By using a familiar ladder logic structure, the amount of retraining needed for engineers and maintenance technicians was significantly reduced. While modern control systems do still use relays, they are seldom used for logic.

PLCs gained their name from the ability to program logic into these controls (e.g., programming **and/or, if/then**, or true/false functions like the one shown in Figure 20-5).

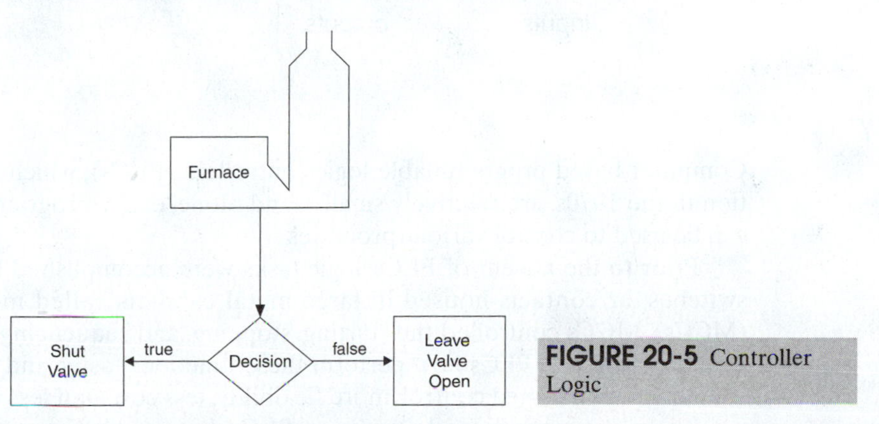

FIGURE 20-5 Controller Logic

In the following table, there are several decisions that can be made. These decisions, denoted here as "if" and "then" statements, could be as follows:

If...	*Then...*
The temperature in the furnace is <u>less than</u> 800 degrees	leave the fuel flow valve open.
The temperature in the furnace is <u>above</u> 800 degrees	shut the fuel flow valve off and shutdown the furnace.

On/off instructions such as these are expressed as binary code in the computer. Because of this functionality, PLCs are commonly used for safety-related applications and make up the heart of emergency shutdown (ESD) systems.

PLC functionality can be expressed as a three-part process:

1. Scan inputs from switches or transmitters.
2. Execute logic or control within the PLC.
3. Set outputs based upon results of logic.

Programming a PLC

There are a number of ways that a PLC can be programmed. These include wiring diagrams, Fortran, C++, Java, or other computer languages. However, ladder diagrams tend to be the method most plant personnel are familiar with and are, therefore, the most commonly used method for interfacing with a PLC. An example of ladder logic can be seen in Figure 20-6.

In Figure 20-6 the vertical lines on the left and right are the power source "rails." In between the left rail and the right rail are horizontal lines or "rungs." On these rungs are a combination of inputs and outputs. Inputs are entered into the system by sensors or switches. Outputs exit the system and are sent to devices such as lights and motors that are outside the PLC. If the inputs on the ladder diagram are opened or closed in the right combination, the power can flow from the left rail, through the inputs, to the power outputs, and finally to the right rail.

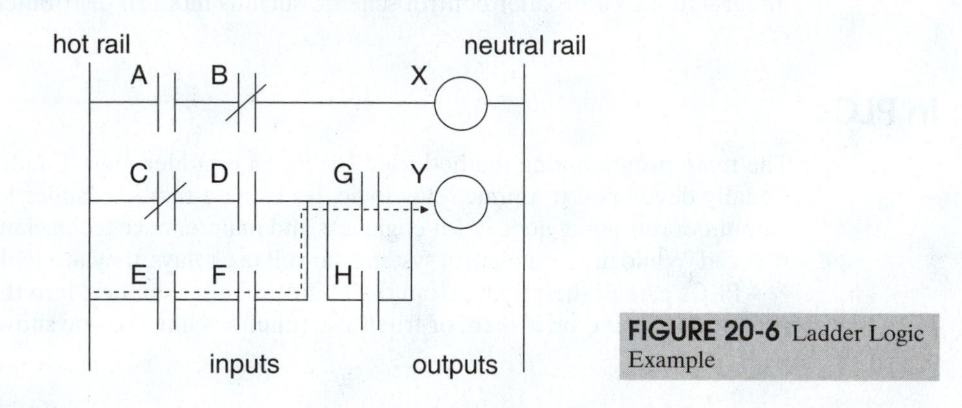

FIGURE 20-6 Ladder Logic Example

Summary

Computer-based programmable logic controllers (PLCs), which made their introduction in the 1970s, are relatively small, stand-alone, easily programmed computers that can be used to control various processes.

Prior to the advent of PLCs, logic tasks were accomplished by hard-wired relays, switches, or contacts housed in large metal cabinets called motor control centers (MCCs). MCCs controlled the starting, stopping, and sequencing of motors and other devices. However, PLCs now perform these functions faster and with greater reliability, more sophisticated control, more flexibility, less cost, and less downtime.

There are many ways to program PLCs; however, the most common method is through the use of ladder logic. Ladder logic, which was intentionally developed to mimic relay logic, is used because engineers and maintenance technicians were already familiar with it, and therefore required less retraining.

Ladder logic functions primarily as a series of "and/or," "if/then," or "true/false" statements and is represented through drawings called ladder diagrams. When reading ladder diagrams, it is important to note which devices are listed toward the top, since devices that have a larger impact on the system are generally shown above devices with relatively minor impact.

Checking Your Knowledge

1. *True or False* A PLC is easier and less expensive to install than a relay.
2. *True or False* In the following PLC diagram, switch "A" is considered "on."

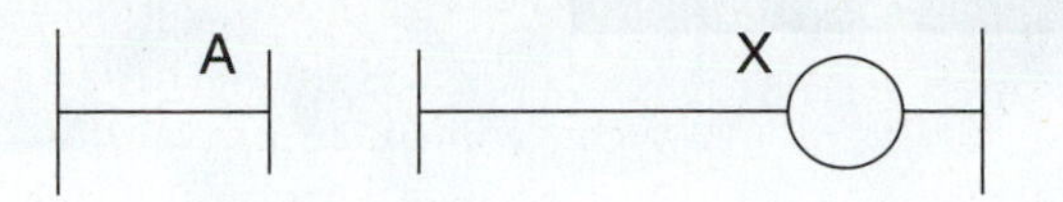

3. *True or False* In ladder diagrams like the one below, devices at the bottom have more impact on the system than the ones at the top.

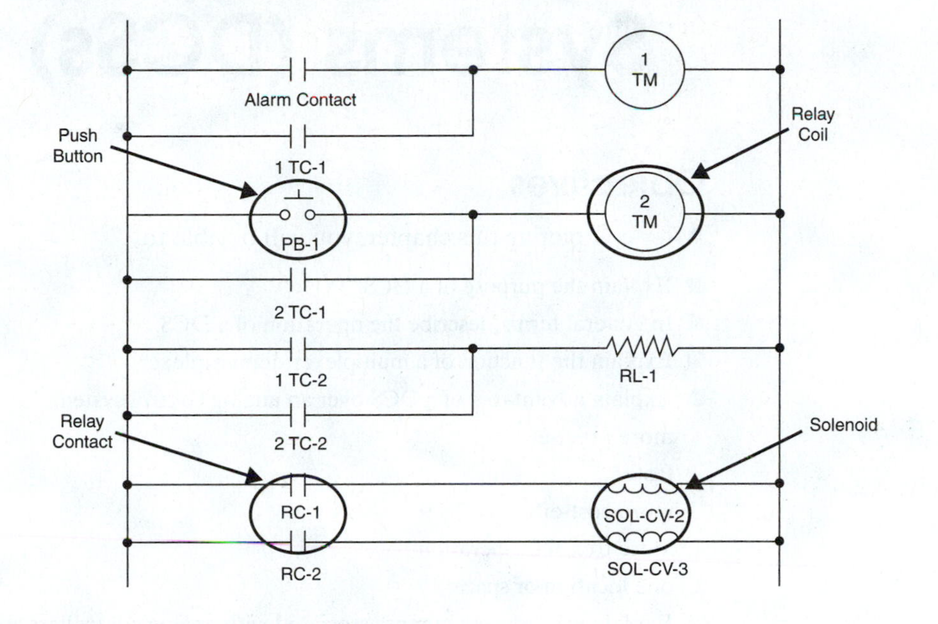

Student Activities

1. Describe what a PLC is.
2. Give an example of where you might use a PLC.
3. List the advantages of PLCs over relays.
4. Explain why you might use relays instead of a PLC.
5. Explain what a ladder logic diagram is and how "ladder logic" applies to programmable logic control.
6. Explain the purpose of programmable logic controllers (PLCs) with regard to the following:
    a. Sequential control
    b. On/off
    c. Emergency shutdown systems (ESDs)
    d. Integral to distributed control systems (DCSs)
    e. Stand-alone capability
7. Conduct a safety analysis of a simple process. Identify the type of ladder diagram that would be necessary to meet process safety requirements.
8. Consider effects of failure of a PLC on the plant safety. Emphasize the need for a separate power supply to such units.
9. Given a programmable logic controller (PLC):
10. Load a simple circuit into a PLC with the use of a PC. You may consider programming a START-STOP button to start up and shut down a system.
11. Change the input conditions and observe the response.

CHAPTER 21

Distributed Control Systems (DCSs)

Objectives

After completing this chapter, you will be able to:

- Explain the purpose of a DCS.
- In general terms, describe the operation of a DCS.
- Explain the function of a multiplexer/demultiplexer.
- Explain advantages of a DCS over an analog control system:

  more precise

  faster

  more cost-effective

  more trending operation

  one location or space
- Explain why process bumps associated with analog controllers are not a concern with digital controllers.
- Given a DCS diagram, identify the major components.

Key Terms

Data highway—a series of distributed computer processors or "nodes" that are all interconnected via a computer network.

Drift—a slow variation of a performance characteristic such as gain, frequency, or power output that can be attributed to factors like temperature and aging.

Node—a single computer terminal located on a computer network or "data highway."

Noise—disturbances that affect a signal and may distort the information carried by the signal.

Trending—the analysis of a change in measured data over at least three data measurment intervals.

Introduction

Distributed control systems (DCSs) are computer-based systems that are used to control processes (Figure 21-1). Because these systems are computer-based and interconnected, they are more fail-safe, reliable, and precise than traditional systems. In addition, they can process and store data more quickly and efficiently, are more cost-effective, and require less control room space than their analog counterparts.

Historical Background

In the 1970s the use of computers in process control became more prevalent than in years past. In these early systems, single computers were used to process measurement data, calculate the required control output, and control final control elements. This was known as direct digital control. It was called "direct" because a single computer was directly connected to a particular part of the process. While this type of control was effective, it had a major weakness . . . it lacked redundancy.

Over the years digital control evolved from direct digital control, to supervisory control, and then to the distributed control systems (DCSs) we use today.

In a direct control scenario there was only a single computer, so if the computer failed the entire unit was shut down. This translated to increased safety risks and decreased productivity. Thus, direct control was forced to evolve into a more fail-safe system.

The first-generation DCS was basically a series of microprocessor-based, multiloop controllers built along a communications network called a **data highway**.

A data highway is a series of distributed computer processors or **nodes** that are all interconnected via a computer network. Because of this multinode architecture, DCSs are able to manage multiple process loops at one time. Furthermore, they are more fail-safe than traditional systems. If one node becomes inoperable or goes down, signals can be rerouted to one of the other nodes and the process can continue. In risk management terminology this is referred to as distributed risk since the risk of control system failure is no longer concentrated at one point. Because it is unlikely that all points will fail simultaneously, the overall reliability of the control system is increased. This translates to increases in both safety and productivity.

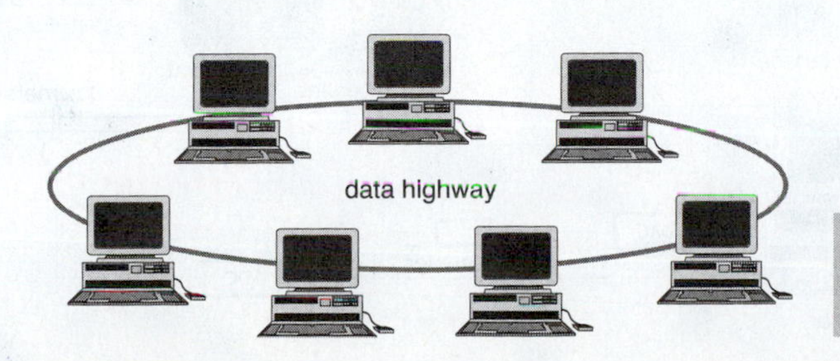

FIGURE 21-1 Distributed Control System (DCS) with Individual Nodes

DCS Operation

A DCS divides production control in a plant into several subparts or nodes. The exact manner in which it is divided depends on the degree of distribution desired and the cost and availability of the control systems.

Each node in a DCS is controlled by a local computer and is essentially independent of the other nodes (see Figure 21-2). A node may include a considerable control loop. The local computer for each node communicates with a central computer called the host computer. All nodes communicate with the host computer. Figure 21-3 shows an example of a typical DCS node.

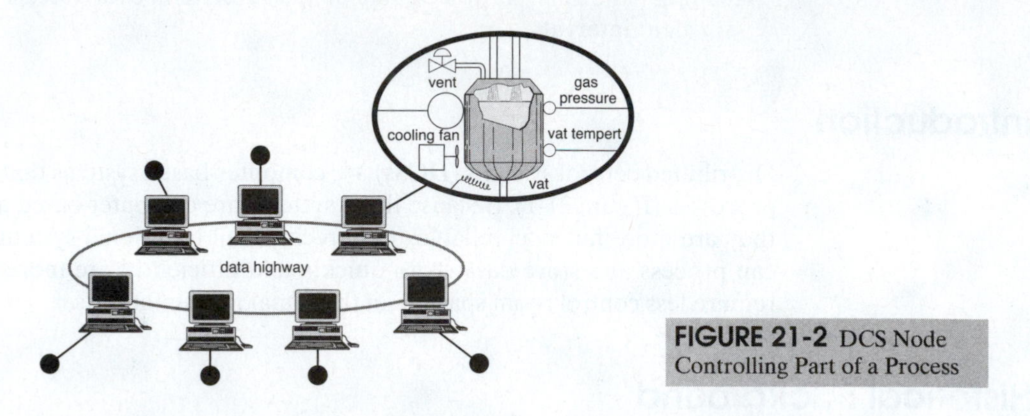

FIGURE 21-2 DCS Node Controlling Part of a Process

FUNCTION OF A TYPICAL DCS NODE

The functionality of a typical DCS node can be explained as a series of steps as follows:

1. The DCS node processes signals received from the control loop such as temperature, level, flow, pressure, and on/off signals (binary).
2. The signals processed by the DCS are then converted from analog to digital through an analog-to-digital (A/D) converter.
3. The digital signals produced by the A/D converter are then forwarded to a local computer by a multiplexer (MUX), a device that sends many signals down a data highway that can later be recovered or separated out into the original constituents by a demultiplexer (DEMUX).
4. The local computer then processes the incoming information and sends out control commands based on internally programmed logic and instructions.

FIGURE 21-3 An Individual Node

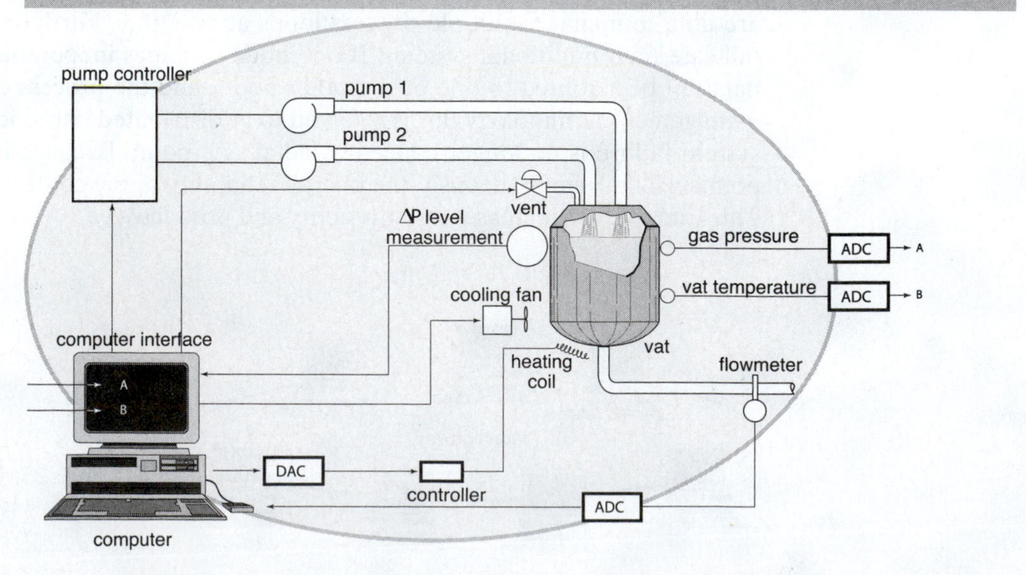

5. The control commands generated by the local computer are forwarded by the demultiplexer to a digital-to-analog (D/A) converter.
6. The analog signals produced by the D/A converter are then forwarded to the appropriate valve or control element and the host computer.

DCS Architecture

Let's take the above information and look at a typical distributed control system (DCS) installation. In a DCS the amount of distribution of control and its location is determined by vendor selection and user requirements. Let's build a DCS starting with field instruments, programmable controllers, operating stations, and accessory devices.

INPUT/OUTPUT (I/O) DEVICES

Field instruments are usually connected to some type of input or output device (electronic card) that resides on an input/output (I/O) highway. This I/O highway may be located in the process unit or in a remote terminal building. The connections to the field devices are made by conventional wiring or newer "bus" technology. Each field instrument is wired to its own or representative electronic card. There are several different kinds of I/O cards. For example, analog input (AI), analog output (AO), digital input (DI), digital output (DO), frequency input (v), thermocouple input (T/C), and field bus cards.

These cards contain programmable microprocessor chips and may perform many functions. For example, the AI card may perform A/D conversion and signal conditioning and may have current limiting in order to eliminate the use of fuses.

The power supply not only provides electrical power to the electronic cards, but also to field devices like transmitters and switches. The electrical power can be DC or AC depending on the type of card. Figure 21-4 gives an example of an I/O highway.

Usually each card has eight channels or eight input/output connections that are labeled 1 through 8. For example, the analog input card has eight input connections (+, –, and ground terminals), each wired to a field instrument that is given a unique tag number. For instance, in Figure 21-4, flow transmitter FT-1 is wired to an analog input (card 1, channel 2) and its output is wired to an analog output (card 2, channel 3).

The field bus card is connected to a field terminal block that can contain as many as eight field instruments. These instruments are "smart" digital devices that contain programmable microprocessors in newer d/p transmitters, mass flowmeters, vortex meters, digital valves, and more. The bus card can handle two terminal blocks. The wiring is single twisted pair from field instruments to the terminal block, and one pair wires to the bus card, which is an obvious construction cost advantage.

FIGURE 21-4 I/O Highway Example (1)

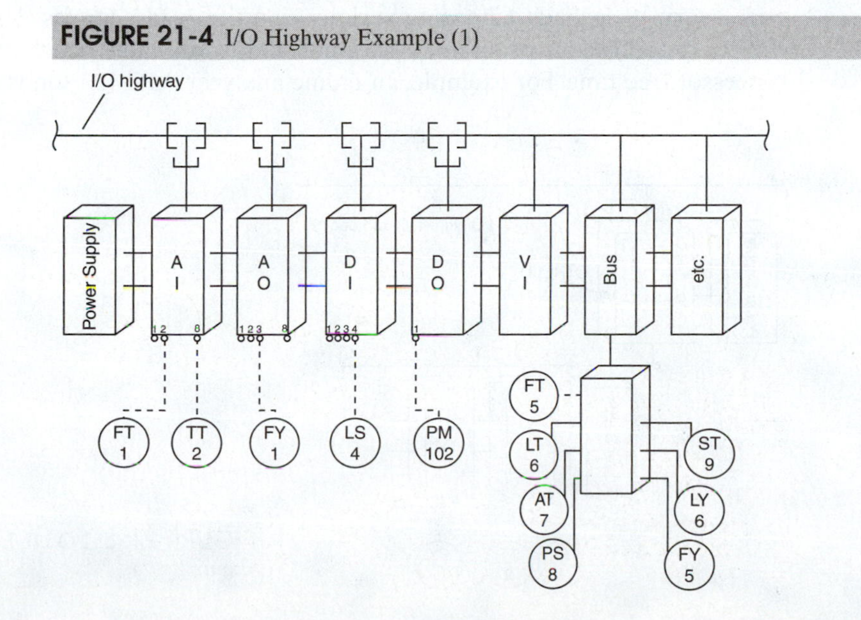

Communication is performed digitally with scan times of 100 milliseconds to 4 seconds depending on the number and type of instruments. The communication contains control data, instrument status, diagnostic, and other information. A field bus transmitter or valve positioner can be programmed to perform PID control functions.

It is important to note that the tag number is usually the same as the one shown on the unit P&ID. Each instrument tag number and where it is connected or located becomes a global database that is used by the DCS. This is configured (or programmed into the computer) when setting up the DCS architecture or when adding new controls.

DATA HIGHWAY

The I/O highway with its devices is connected to the programmable controller that lies on an Ethernet or TCP/IP highway (in the example in Figure 21-5, highway 1 or node 1).

Ethernet and Transmission Control Protocol/Internet Protocol (TCP/IP) are networks set up for transmitting information. Protocol is a set of rules that govern the transmission and makeup of binary data. Devices on the highway are assigned a unique address (or tag data). They communicate by sending and receiving information through data packets. An example of a packet might be start code, source address, control information, destination address, conformation handshake, or stop code. Devices on the highway are polled or broadcast when they need to send or receive information. Typical transmission speed can be faster or slower than 100 million bits per second.

The programmable controller is given a node address or tag number (e.g., CTRL-101, which translates to controller 01 on highway 1 or node one).

These programmable controllers can be configured to handle 500 to 1,000 loops. These loops can be a mixture of control loops, monitored inputs or outputs, startup or shutdown sequences, batch control, and logic functions. These controllers contain many microprocessors that can be configured to handle any type of control or unit functions. A loop could be input FT-1, output FY-1, and PID flow controller FIC-1 that is configured in CTRL-101. The system uses highway 1 or node 1/CTRL-101/FIC-1 for this control loop or a TCP/IP address. Most DCSs use tag number addresses to communicate with devices. Another loop shown might be column temperature indication TI-2 that is configured in CTRL-101 with its input TT-2: card 1, channel 8.

The programmable controllers have a built-in execution rate (usually 10 to 20 times a second or faster) to read all input signals, perform control functions, and send out output signals. These controllers do not make control changes (e.g., auto to manual, setpoint or output changes) unless requested by a supervisory computer or a process technician. In addition, each configured loop can be given its own scan time. For example, flow control loop FIC-1could be configured with a scan time of 0.5 seconds, whereas column tray temperature indication TI-2 could be selected with a scan time of once every 10 seconds. Choosing various scan times, fast for fast loops or slower for slow loops, allows the programmable controller to run more efficiently and with more processor-free time. For example, an online analyzer AIC-123 sends a new signal only

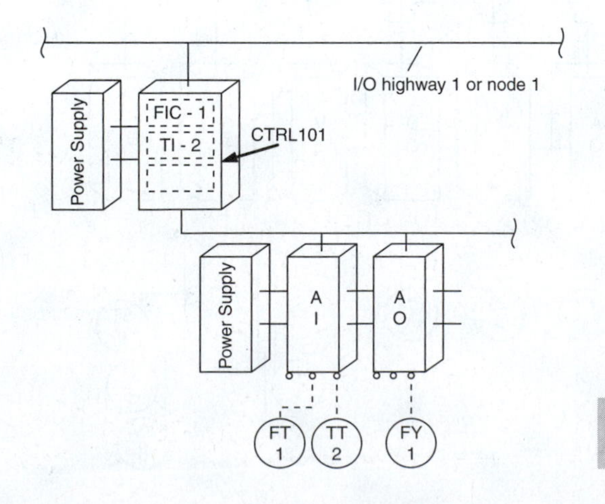

FIGURE 21-5 I/O Highway Example (2)

once per minute. If a 1-second scan time were configured, then for 59 scans or seconds the AIC-123 controller would calculate the same result and the output may only change on the 60th scan. This scan time should be equal to when the analyzer has a new result (in this case, 1 minute).

Data highway 1 can have several more programmable controllers connected to it, each with its own I/O highway (e.g., CTRL-102 (highway 1, controller 02) in Figure 21-6).

Thus, hundreds or thousands of loops can be connected to a single highway to control a large process unit. The programmable controllers and I/O may be located in several different places: in a control room, a remote terminal building, or in the field. The highways may be hundreds of feet long to miles long depending on the type and speed of the transmission system. The data highway from slowest to fastest transmission speed ranges from coaxial cable, to twisted pair shielded wire, to fiber-optic cable. Other transmission media may also be available such as wireless communications for remote transmission and control (e.g., remote wireless communication to supervise or run an offshore oil platform from shore).

There can be more than one data highway; usually there are 8 to 12. These highways can communicate with one another through devices such as network traffic directors, bridges, gateways, and other types of communication hardware.

FIGURE 21-6 I/O Highway Example (3)

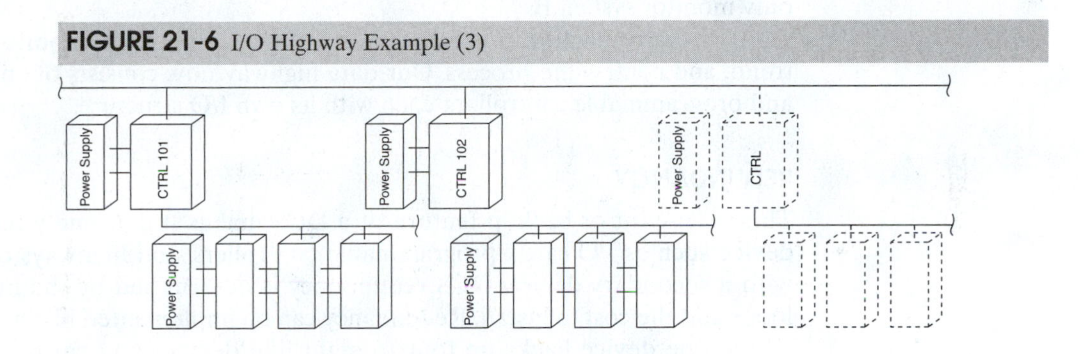

OPERATOR OR WORK STATIONS

So far we have not discussed how to control the process. Process control is performed by a device called a *work station* or *operator station*. An operator station can be thought of as a window to the process. The main function of a work station is to allow the operator to control the process. The station not only collects process data but also monitors the DCS hardware. This data can be presented to the operator in the form of alarm lists, real time or historical trends, unit graphics, material balance displays, interlock and help screens, and diagnostic information. Unit graphics have programmed control points that can be used to call up a controller faceplate. These stations can be connected to any data highway (e.g., highway 1 or node 1 in Figure 21-7).

This type of computer used to be a proprietary device furnished by the DCS vendor. Now it is an off-the-shelf computer that must meet the DCS vendor's specifications. Hardware may consist of several hard drives for data collection, several monitors

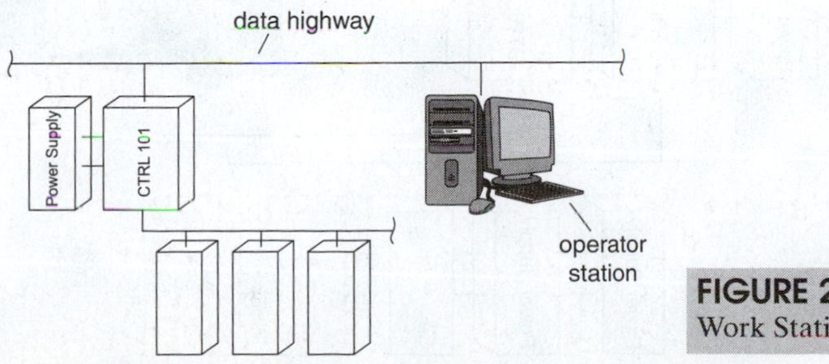

FIGURE 21-7 Operator and Work Stations

(with or without touch screens), keyboards, mice or trackballs, voice commands, and more. Software provided by the DCS vendor determines how the process data is displayed for the process technician.

Most process information can be configured as plant overview, unit graphics, faceplates, trends, alarm lists, interlock, and help displays. Special displays can also be configured. A digital picture of part of a process unit can be imported as a display with control points added. A mixture of these displays is configured to represent the process unit to the process technician. The number of displays that can be configured may number in the thousands.

The stations can be selected to have a periodic update time (or display refresh time) that can be as fast as 1 second or as slow as several seconds. Any loop can be configured to report periodically or by exception to the programmable controller and the operation station. A range of exception reporting values can be selected. For example, column temperature indication TI-2 can be selected to report if its PV changes from 1 percent of span. Alarms, however, are programmed to report when they occur and usually have a "first out" feature, which informs the user as to which alarm point was reached first.

Operator stations are usually located in main control rooms or satellite control rooms. There is considerable flexibility with regard to how these stations are configured. For example, a particular operator ID can be configured to control system A but only monitor system B.

The main function of an operator station is to collect, monitor, display, alarm, trend, and control the process. Our data highway now consists of an operator station and programmable controllers, each with its own I/O structure.

REDUNDANCY

The redundant or backup feature of a DCS makes it extremely reliable. Any DCS device such as I/O cards, programmable controllers, and highways can be backed up with a secondary device. DCS redundancy is determined by the number of critical loops and the cost to install. Redundancy can be implemented like a "spare tire" setup where one device backs up four to eight like devices (AI cards) or a one-for-one backup like a primary and secondary power supply.

It should be pointed out that any device on the highway can be made redundant or backed up with a secondary device (refer to Figure 21-8). If the primary device fails, the secondary takes over during the next scan and the process technician is alerted to the failure by an alarm. If more than one operator station is used then all are online at the same time. If power were lost to the operator station screen the programmable controllers would continue to run the unit until power was restored. Minimum redundancy should include primary and secondary highways, primary and secondary programmable controllers, primary and secondary power supplies, and the operator station.

FIGURE 21-8 Redundancy Example

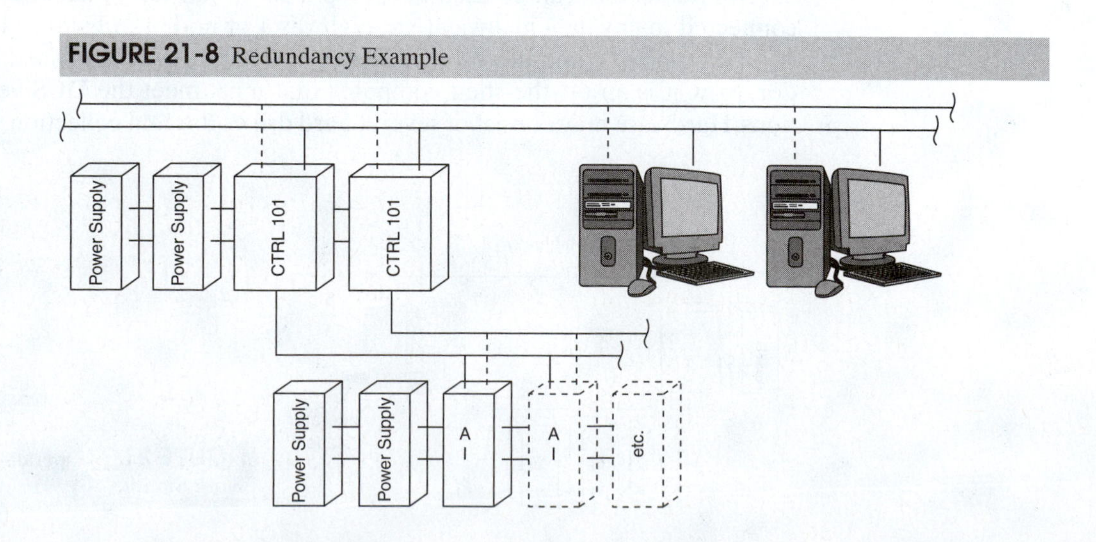

CONFIGURATION STATION

The primary function of a configuration station is to configure the DCS with all its hardware devices and control schemes. A configuration station may be connected to any data highway and may even be located in a separate building. Changes can be made offline and checked out. Following required coordination and approval (e.g., management of change) the changes can be downloaded to the unit control system. The changes can also be downloaded to a computer simulator. Simulators may be used to train process technicians in the new control schemes prior to implementation on the unit.

The configuration station stores the complete plant configuration on its hard drive. Backup copies of the plant configuration should be stored on tape, DVD, CD, or other computers. A good idea is to store a copy onsite and another copy offsite.

ACCESSORY DEVICES

Some accessory devices connected to the data highway include printers, hard pen recorders, process information computers, data loggers, and advanced process control (APC) computers. Printers are used to print unit graphics and morning reports. Hard pen recorders are used to trend troublesome loops. Data loggers are used to store alarms, sequential events, keystrokes, and more. A separate process information system may be connected to the DCS highway to collect and store process data for historical trending. The APC computer usually performs advanced control schemes and model predictive control. Some programmable controllers can perform the APC functions, while the work station contains the process information system and some of the data logger functions.

PLANT NETWORK

A local area network (LAN) is usually connected to the DCS through some type of gateway or security device that allows read-only data to be available to engineers, maintenance, and operational personnel. The LAN may contain a telephone, business computers, personal computers, and a web server. The operator should have access to a separate PC for items such as e-mail, work orders, and Internet. The operator should never have access to the Internet from the work station.

DCS Advantages

Computer-based distributed control systems (DCSs) are preferred over analog-based systems for a number of reasons including: repeatability, accuracy, speed, cost-effectiveness, improved data management, space efficiency, and increased safety and reliability.

REPEATABILITY AND ACCURACY

The overall repeatability and accuracy of a DCS is greater than that of the analog alternative. Digital systems can process more information involving signal checking and reduction of **noise**, and distinguish erroneous signals caused by **drift**.

SPEED

Analog systems require a multitude of individual components to perform complex computations. Because of this, data processing speeds are slower. DCSs have the ability to process complex calculations quickly, thereby increasing the output speed.

From a process technician's perspective, it also means the speed of response to an event is greatly improved because data is more readily accessible.

COST-EFFECTIVENESS

The initial installation cost for a DCS may be less than or equal to that of an analog system. The cost of maintenance for a digital system, however, is considerably lower than for an analog system since digital components are significantly more reliable than their analog counterparts.

DCSs allow a more convenient means for configuring and changing the system to accommodate the needs of the given operation. Software programming changes are much more cost effective and efficient than hardware changes. In addition, access to historical data, improved control schemes, graphics, complex calculations, predictive maintenance information, and more are all inherent in the system and significantly less expensive than hardware devices.

IMPROVED DATA MANAGEMENT

DCSs offers superior data storage, archiving, retrieval, and analysis capabilities. For example, if an alarm is indicated, a diagnostic message indicating which part of the system is not functioning properly may be displayed and an alarm log created. These logs, along with the archiving capability and the ability for statistically analyzing, tracking, and **trending** data, are immensely valuable when it comes to troubleshooting process deviations and ensuring regulatory compliance.

SPACE EFFICIENCY

DCS displays require less physical space than their analog counterparts. This allows the size of the control room to be reduced while increasing operational productivity and troubleshooting capability, and reducing maintenance costs.

SAFETY AND RELIABILITY

DCSs now incorporate functions that were traditionally handled by separate, hard-wired PLCs for safety trips and interlocks. Because a DCS is computer-based it allows for safer operation of a unit than analog counterparts. The redundancies inherent to this type of system ensure that signals can be rerouted and operations can continue should a specific local computer (node) fail, thereby increasing reliability and mean time between failures and reducing the potential for unsafe conditions.

BUMPLESS TRANSFER

In DCSs, configuration choices (e.g., setpoint or PV tracking) eliminate bumpless transfer issues.

Summary

Distributed control systems (DCSs) are computer-based systems that are used to control processes. These computer-based systems first entered into process technology in the 1970s. Prior to the 1970s, processes were controlled by analog controllers, which were effective but did not contain the same level of functionality as their digital counterparts.

A DCS consists of a series of components that are interconnected via a computer network or "data highway." DCS systems are more fail-safe than traditional systems because of the redundancy built into the systems and the reliability of digital components (e.g., if a node in a DCS becomes inoperable or goes down, data signals are automatically rerouted to a different node and the process can continue). This translates to increases in both safety and productivity.

In addition to being safer and more reliable, distributed control systems are also more user-friendly, more accurate, and provide more flexibility in choosing control station locations. Furthermore, they can process and store data more quickly and efficiently, are more cost-effective, and require less control room space than their analog counterparts.

Checking Your Knowledge

1. A single computer failure using DCS would likely cause:
    a. Environmental violation
    b. Catastrophic fires
    c. No operational interruption
    d. Unit shutdown

2. *True or False* With digital controllers, it is more difficult to execute a bumpless transfer from manual to automatic.
3. A node in a DCS system consists of:
   a. A single control loop
   b. A host computer and several local computers
   c. A local computer and several control loops
   d. A host computer and several control loops
4. *True or False* A DCS reduces the noise and drift that can make an analog system less accurate.
5. The reduced space requirement for DCS controls may result in:
   a. Greater productivity in operation
   b. All controls now being located in the field
   c. Greater productivity in maintenance and troubleshooting
   d. Greater flexibility in control station locations

Student Activities

1. Given a diagram of a DCS system, identify and label the components and describe the function of each.
2. Draw a schematic of a typical DCS and identify parts.
3. Given a series of "what-if" analysis of failures:
   a. Explain which sections will be affected.
   b. Identify the steps that will ensure that the affected parts will shut down safely.
   c. Discuss the effect of noise (crosstalk) or loose connections on the safety and performance of a DCS system.
4. Given a DCS system:
   a. Identify the components and describe the function of each.
   b. Trace the system from the sensor to computer (node), from the computer to the host, and then back to the final control element.
   c. Deliberately cause failure of a thermocouple and observe the results.

CHAPTER 22

Instrumentation Power Supply

Objectives

After completing this chapter, you will be able to:

- Define terms associated with instrumentation power supply:
  - automatic transfer switches
  - battery and charger
  - inverter
  - emergency generator
- Explain the purpose of uninterruptible power supply (UPS) systems.
- Given a diagram, identify components in a UPS system.

Key Terms

Alternating current (AC)—a type of electrical current that reverses direction at regular intervals or cycles (as opposed to direct current that does not reverse).

Automatic transfer switch (ATS)—an electromechanical device that switches an electric load between two separate power sources (primary and secondary) depending on power availability from these sources; it is typically configured to be in the normal position when primary power is available.

Battery—an electrical device consisting of one or more cells that converts chemical energy into electrical energy and provides a steady-state of DC voltage.

Breaker—a switchlike device in electrical panel boxes used to keep electrical current from exceeding the recommended load.

Charger—a device used to restore batteries to a proper electrical charge.

Direct current (DC)—electrical current that flows in only one direction (as opposed to alternating current, which reverses directions at regular intervals).

Generator—a device that uses fuel or other mechanical energy to create electrical energy.

Inverter—a device used to change direct current (DC) into alternating current (AC).

Uninterruptible power supply (UPS)—a backup power unit, usually consisting of large batteries, a rectifier, inverter, battery charger, and static switch that provides continuous auxiliary power when the normal power supply is interrupted.

UPS alarm—an audible or visual signal used to draw attention to problem situations.

Introduction

Electrical power is a very important part of industrial processes. It is used to provide light, drive motors, power computer equipment, and much more. However, in order for us to use electrical power, we must harness it, convert it, and then route it so it is useable. Harnessing, converting, and routing can be accomplished using devices such as automatic transfer switches, batteries and battery chargers, inverters, generators, and uninterruptible power supplies (UPSs).

Plant Power Supply Overview

Most high-voltage power lines that you see stretching across the landscape carry 138 kilovolts (138 kV) of **alternating current (AC)**. Alternating current is used in industrial applications because it can be stepped up or stepped down to meet different voltage requirements with the use of transformers (Figure 22-1).

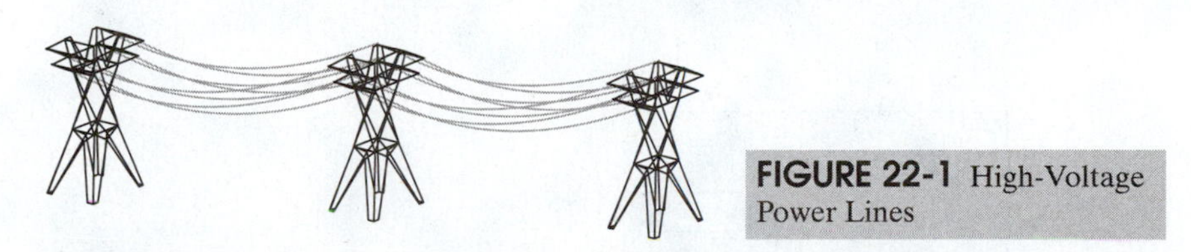

FIGURE 22-1 High-Voltage Power Lines

When power reaches a plant substation, transformers step the voltage down from 138 kV to 13.8 kV prior to it entering the substation. Once inside the substation power is routed to the various units within a plant. Figure 22-2 shows an example of power routing.

Once power reaches the unit, it is then routed to the unit's main **breaker** and disconnect switches located in an electrical equipment room like the one shown in Figure 22-3. Power passes though the motor control centers (MCCs) and disconnect switches to several individual transformers and their respective MCCs.

Motor control centers (MCCs), which are shown in Figure 22-4, are rows of steel cubicles that contain electrical switching equipment. This equipment supplies the

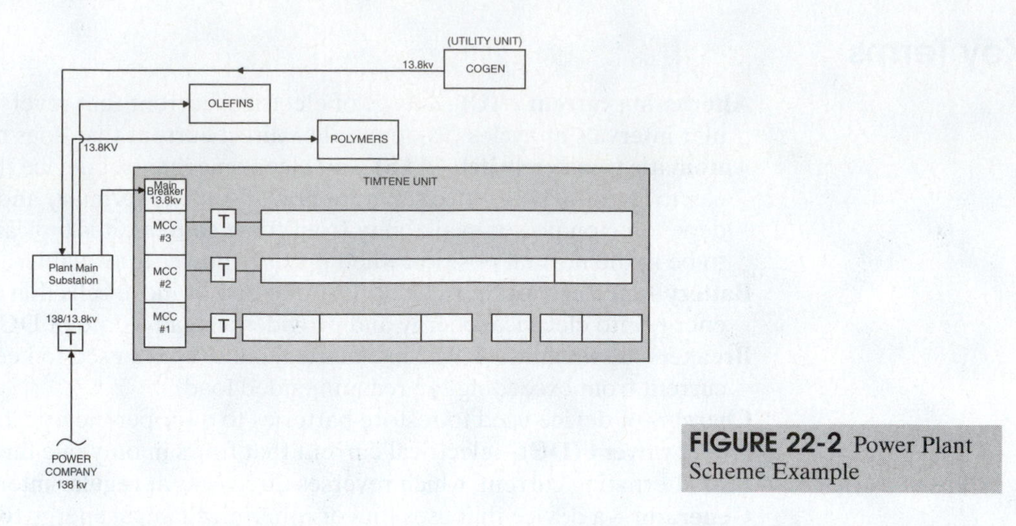

FIGURE 22-2 Power Plant Scheme Example

FIGURE 22-3 Power Routing Example

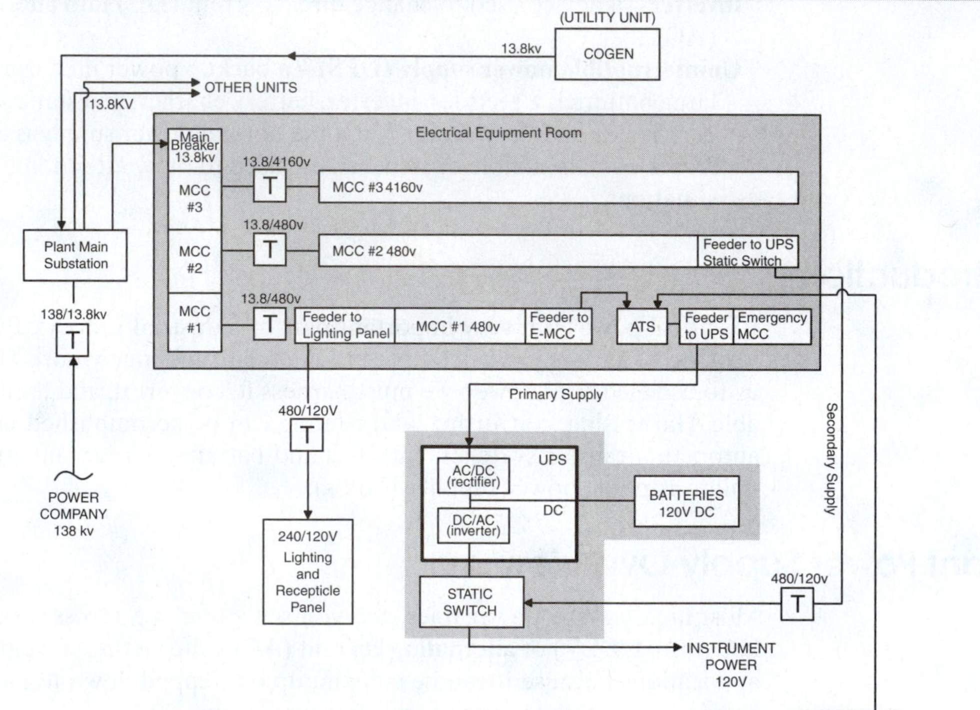

FIGURE 22-4 Circuit Breakers in a Motor Control Center (MCC)

FIGURE 22-5 Sample Power Flow

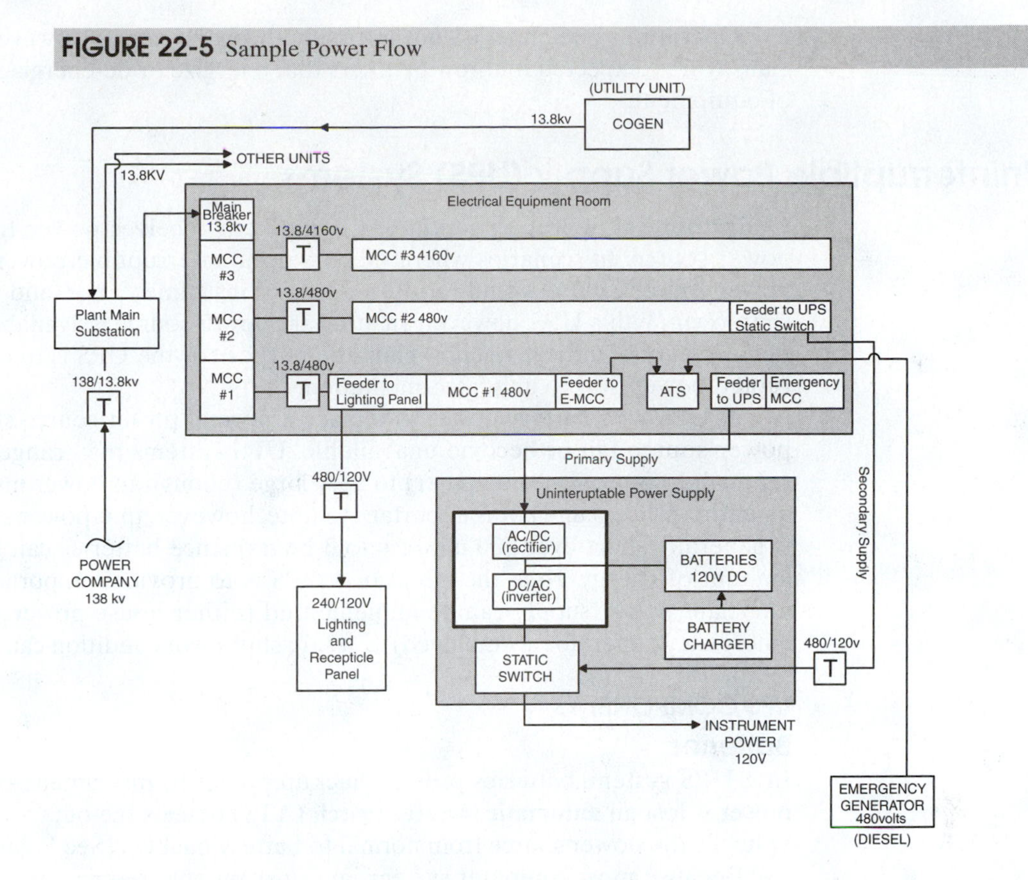

power needed by the many motors and devices that are used in the unit, as well as the power needed to run the emergency motor control centers (E-MCCs) and the **uninterruptible power supply (UPS)**.

Figure 22-5 shows an example of power flow from the main 138-kV power line, through the breakers and MCCs to the UPS.

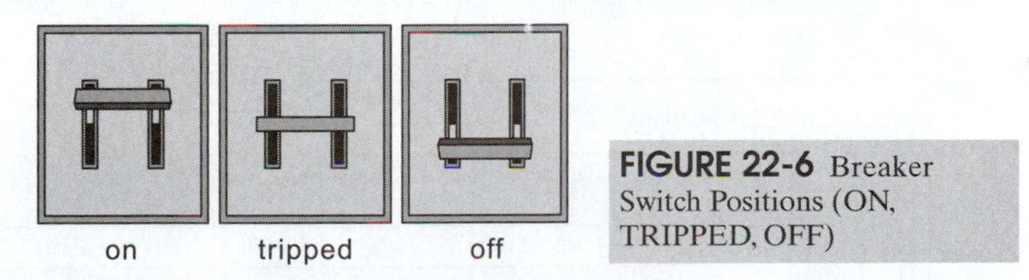

FIGURE 22-6 Breaker Switch Positions (ON, TRIPPED, OFF)

When studying electricity it is important to know that every piece of electrical equipment in the unit is supplied by a cubicle in one of the MCCs. Each cubicle has its own breakers. Breakers are the primary point in which to connect or disconnect power from a piece of equipment (Figure 22-6). The following are the three states of a breaker and their descriptions.

Breaker Handle Position	*Description*
ON	The handle is in the "up" position; power to the equipment is on and conditions are normal.
TRIPPED	The handle is in the "middle" position; the breaker has been tripped and power to the equipment is off; the breaker will have to be reset in order to restore power to the equipment. **Note:** To reset a breaker, pull the handle to the down (off) position and then return it to the up (on) position.*
OFF	The handle is in the "down" position; power to the equipment is off and conditions are normal.

*Different plants will have different regulations on the procedures for resetting tripped breakers.

Operations personnel do not normally throw breakers. However, process technicians will be expected to throw breakers that energize or de-energize individual pieces of equipment.

Uninterruptible Power Supply (UPS) Systems

Under normal operating conditions equipment receives power from a traditional power source. In scenarios where there is no uninterruptible power supply (UPS), a power source failure would result in a loss of instrumentation and an abrupt system shutdown. With a UPS, however, an auxiliary power source is available to maintain the functionality of critical systems. Thus, the purpose of the UPS is to ensure an uninterrupted power flow to unit instrumentation.

A UPS uses **battery** power to replace a normal power source should the normal power source fail or become unavailable. UPS systems may range from very small (enough to power one computer) to very large (enough to power instrumentation for an entire process unit). It is important to note, however, that power supplied by a UPS is generally short-lived (30 minutes to 3 hours) since batteries can hold only a finite amount of energy. Thus, the intent of a UPS is to provide temporary power until an alternate power supply can be implemented (either house power is restored, or an emergency generator is employed) or a safe shutdown condition can be achieved.

UPS COMPONENTS

Batteries

In a UPS system, batteries provide backup power to instrument systems. If normal power is lost an **automatic transfer switch (ATS)** detects the outage and automatically switches the power source from normal to battery backup. (See Figure 22-7.)

Because most computer systems and instruments require AC power, the power coming from the battery system must be converted from **direct current (DC)** to AC by an **inverter** located within a UPS. Once normal power is restored the ATS switches the

FIGURE 22-7 UPS Block Diagram

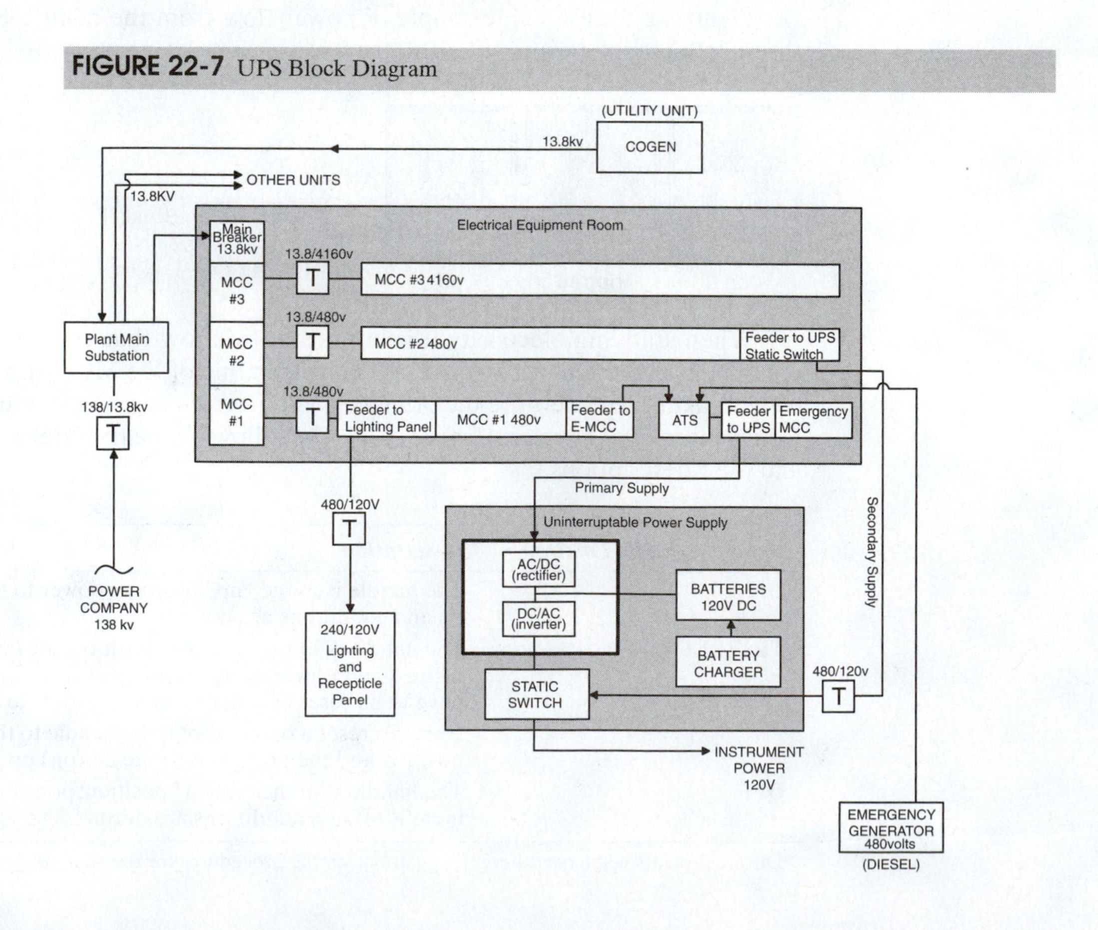

power source from battery backup to normal power. A battery **charger** is an integral part of the battery backup system.

Automatic Transfer Switches

During normal operation of the unit electrical system the automatic transfer switch (ATS) plays no special role. Uninterrupted and unaltered electricity flows from the motor control center (MCC), through the ATS, to the emergency MCC (E-MCC).

The ATS monitors the presence of power from an MCC feeder. If that power fails, the ATS will start an emergency generator. The purpose of the ATS is to always supply 480-volt power to the emergency MCC. However, the ATS cannot maintain continuous power to the E-MCC. There will be a temporary loss of power from the moment that the ATS senses a loss of feed and provides a start signal to the **generator** until the generator gets up to speed and the ATS switches to it.

Emergency Generator

The primary purpose of emergency generators is to provide electrical power at the desired voltage, phase, and frequency. The emergency generator system consists of a driver (diesel engine or gas turbine), AC generator, and necessary switches and instruments. Generation capacity design is based on the power demand of the electrical load. Many of these systems are designed to come on automatically with a loss of normal power. To ensure operability of these systems when needed, they should be operated and tested at prescribed intervals.

UPS Alarms and Indicators

Due to the critical roll played by the UPS, process technicians need to be familiar with **UPS alarms** and indicators, especially those used to indicate abnormal conditions. The following is a list of typical status indicators that might be found on a small computer UPS system:

Indicator	*Description*
Online	An LED that illuminates when the unit is running nominally online power. Note: This indicator will be off if the "On Battery" LED is on.
On battery	An LED that illuminates when the UPS is running on battery power. Note: This indicator will be off if the "online" LED is on.
Overload	A light that illuminates if someone tries to power up more equipment than the UPS system can accommodate. Note: If this light appears, the amount of equipment attached to the unit needs to be reduced or the size of the UPS needs to be increased.
Site wiring fault	An LED, often located on the back of the unit, which illuminates if there is a problem in the circuit feeding the UPS (not the UPS itself). Note: A qualified electrician should examine the circuit if this LED illuminates.
Replace battery	An LED that illuminates if the battery is not charging properly or staying within normal operating parameters. Note: Battery charging and operating parameters are determined automatically by the UPS through periodic internal tests.
Low battery	An LED that illuminates when the UPS detects the battery is almost exhausted and is close to a shutdown. Note: UPS systems will normally shut down before the battery is completely drained as a failsafe measure, since a complete battery discharge could damage the unit.
Battery status	A series of LEDs, often in a vertical "bar graph" configuration, that indicates the amount of battery power still remaining. Note: These types of indicators may not be found on lower-end (i.e., less expensive) UPS systems.
Load status	A series of LEDs, similar to battery status LEDs, that indicate how much of the unit's capacity is currently being drawn by the equipment attached to it. Note: These LEDs are good indicators as to whether or not the UPS can support additional equipment.

Be aware that, in addition to these visual indicators, some UPS systems also have audible indicators or "alarms" that are used to draw attention to problem situations.

To learn more about indicators, especially those specific to your UPS system, refer to the user manual that came with the equipment.

Troubleshooting Tips and Warnings

Few if any process technicians will be responsible for servicing or maintaining a UPS system. However, process technicians may be asked to assist an I/E technician with preventative maintenance activities, so it is important to be aware of some of the features and hazards associated with UPS systems:

- Risk of Electrical Shock—UPS systems produce live AC current at the output even if the unit is disconnected from the primary power source. Therefore, personnel should always be aware of the presence of AC current and the potential for electrical shock.
- Battery Maintenance—The status of the battery charger and the batteries must be checked on a routine basis. It is important to maintain the proper operating conditions specified by the manufacturer.

Checking Your Knowledge

1. The _______ will signal the generator to turn on in the event of power failure from the primary power supply.
   a. Main breaker
   b. Static switch
   c. Automatic transfer switch
   d. Backup relay switch
2. *True or False* During normal operation, power will not pass through the ATS.
3. To reset a tripped breaker:
   a. Pull the breaker handle down.
   b. Pull the breaker handle up.
   c. Pull the breaker handle down and then up.
   d. Pull the breaker handle up and then down.
4. The_________ ensures clean and uninterrupted power to instrumentation.
   a. Motor control center
   b. Uninterruptible power supply
   c. Static safety switch
   d. Emergency generator
5. AC voltage becomes DC voltage as it passes through the _____________.
   a. Rectifier
   b. Converter
   c. Inverter
   d. Feeder

Student Activities

1. Given a UPS or a drawing of a UPS, identify the location of each of the following indicators, explain when each one would be illuminated, and explain what each one indicates.
   a. Battery status
   b. Load status
   c. Low battery
   d. Replace battery
   e. Online
   f. On battery
   g. Overload
   h. Site wiring fault
2. Given a breaker or a picture of a breaker, identify the position of the switch (on, off, or tripped).
3. Given a "tripped" breaker, reset the breaker to a position that would allow current flow to resume through a device.
4. In the failure scenario diagrammed below, power to motor control center (MCC) #1 has failed.
   a. Describe the problem(s) this scenario will cause.
   b. Explain why this could mean major trouble for the entire unit.

c. Explain how this failure might impact the UPS.
d. Explain why you should check to see that generator has started and that the UPS is not running off batteries.

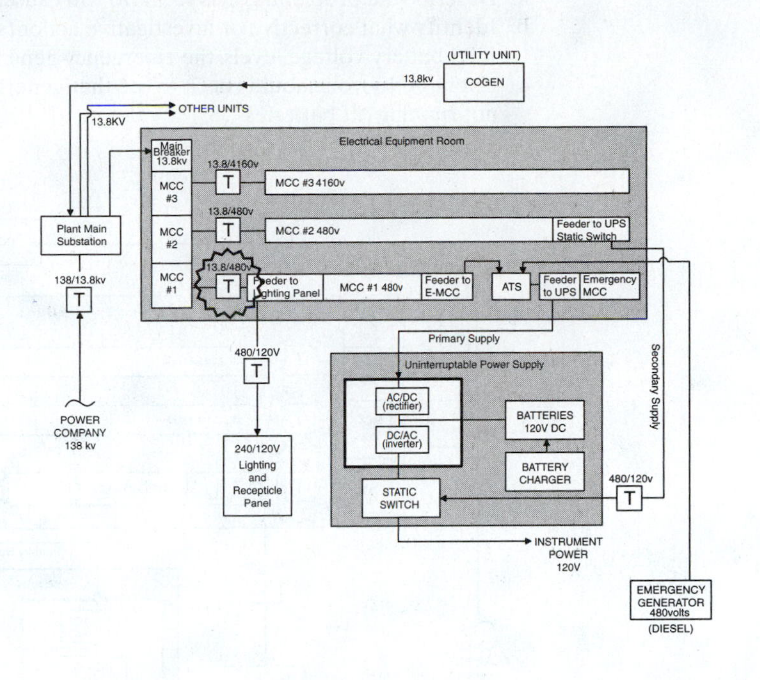

5. In the failure scenario diagrammed below, a breaker trips in the MCC #1 feeder to the ATS.
   a. Describe the problem(s) this scenario will cause.
   b. Identify what corrective action you should take if the breaker is tripped.
   c. Explain why you should check to see that generator has started and that the UPS is not running off batteries.

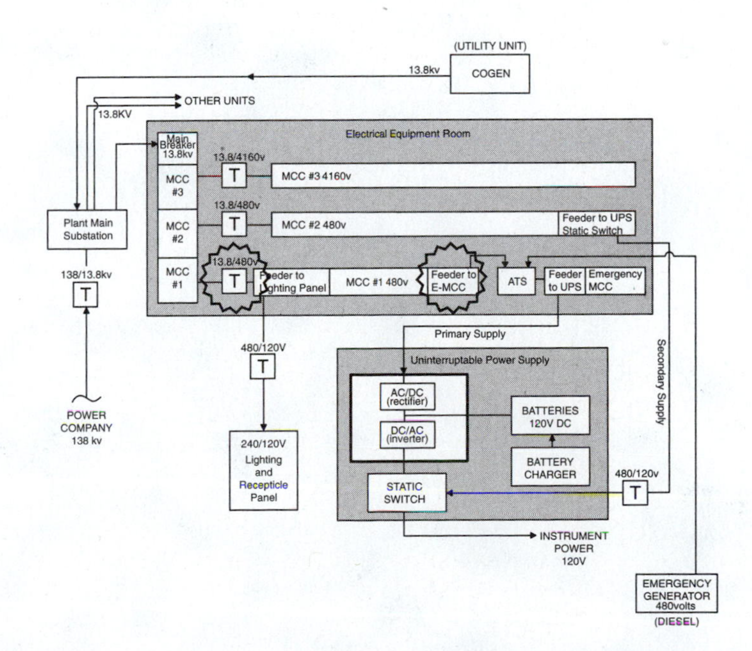

6. In the failure scenario diagrammed below, either power to motor control center (MCC) #1 has failed or a breaker has tripped in the MCC #1 feeder to the ATS and the generator has not started.
   a. Describe the problem(s) this scenario will cause.
   b. Identify what corrective or investigative action(s) you should take with regard to UPS battery voltage levels, the emergency generator, and all UPS alarms.
   c. Explain why you should check to see that generator has started and that the UPS is not running off batteries.

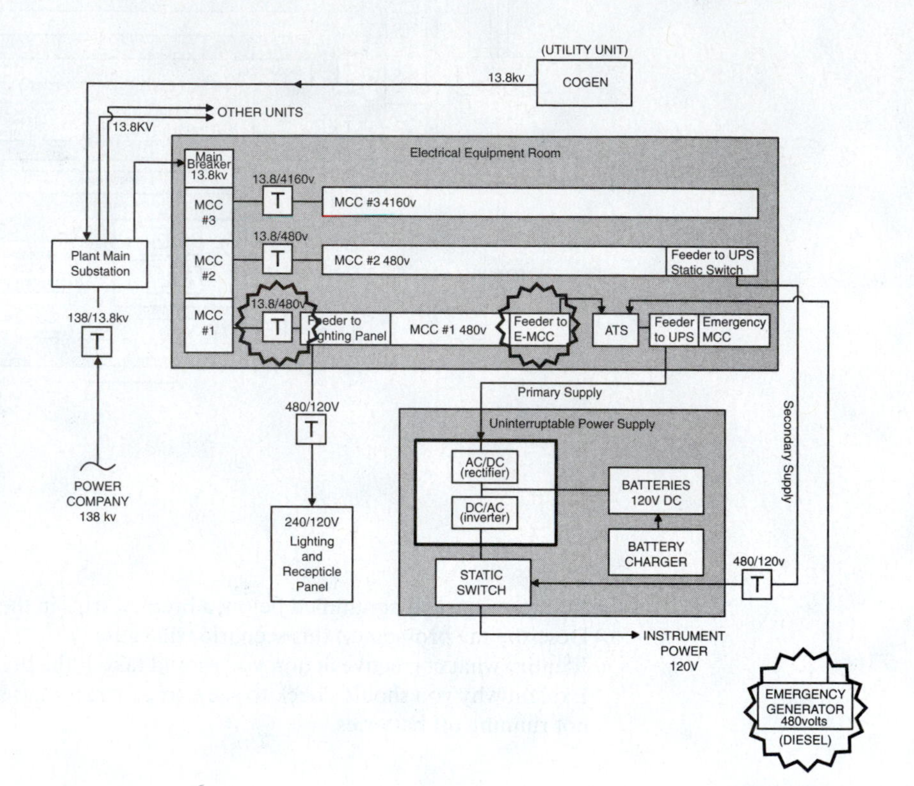

CHAPTER 23

Emergency Shutdown (ESD), Interlocks, and Protective Devices

Objectives

After completing this chapter, you will be able to:

- Define the following terms associated with emergency shutdown systems, interlocks, and alarms:
  - startup permissives
  - high alarm
  - high high alarm shutdown
  - low alarm
  - low low alarm shutdown
  - permissive
  - interlock
  - emergency shutdown (ESD)
- Given a drawing, picture, or actual device, identify and describe basic shutdown devices.
- Recall the use of programmable logic controllers in ESD systems.
- Given a PFD and/or P&ID and a legend, locate and identify emergency shutdown devices.
- Given a basic description of a process and a P&ID of its ESD system, describe what a specific ESD is designed to protect.
- Describe common types of alarms:
  - deviation
  - vibration
  - gas detect
  - defeat
  - deluge system
  - redundancy
- Explain voting logic.
- Explain methods for testing and resetting ESDs.
- Discuss issues associated with bypass of an ESD.
- Discuss how redundancy devices are used to prevent nuisance shutdowns.

Key Terms

Bypass—a short patch cable or wire used to complete an electrical circuit, usually temporarily, for testing or diagnostic purposes (also called a *jumper*).

Defeat—*See* bypass.

Deluge system—an emergency shutdown (ESD) system component that, when a gas leak is detected, sprays the area of leakage with large quantities of water to dilute the concentration of vapors so they are below the lower explosive limit (LEL), thereby reducing the potential of fire.

Deviation alarm—an alarm that is activated when the difference between two variables exceeds a set limit (for example, in a particular catalyst bed, the difference or deviation between the outlet temperature and the inlet temperature should not exceed 50 degrees F; in this example the actual temperature value is not important, what is important is the difference between the two values).

Emergency shutdown (ESD) system—a system consisting of interlocks, breakers, sensors, and other equipment responsible for shutting down equipment in an extremely abnormal situation in order to protect the equipment and reduce the risk of catastrophe.

High (H) alarm—the first alarms triggered when a process variable (e.g., fluids in a tank) rises above a predetermined high level; the purpose of this alarm is simply to notify an operator that the level is abnormal.

High high (HH) alarm—alarm triggered when a process variable (e.g., fluids in a tank) continues to rise above a pre-determined maximum level after a high (H) alarm has already been triggered; the purpose of this alarm is to notify a process technician that the level is becoming increasingly abnormal and to initiate a shutdown or other corrective action.

Interlock—a shutdown circuit designed to detect dangerous conditions and safely shut down a process or a piece of equipment.

Jumper—*See* bypass.

Low (L) alarm—the first alarm triggered when a process variable (e.g., fluids in a tank) drops below a pre-determined low level; the purpose of this alarm is to notify a process technician that the level is abnormal.

Low low (LL) alarm—alarm triggered when a process variable (e.g., fluids in a tank) continues to drop below a predetermined minimum level after a low (L) alarm has already been triggered; the purpose of this alarm is to notify a process technician that the level is becoming increasingly abnormal and to initiate a shutdown or some other corrective action (e.g., shutting off power to a pump to prevent cavitations, which could result in pump damage).

Permissives—a set of conditions designed to ensure safe operations that must be met before a piece of equipment can be turned on.

Redundancy—a design feature that provides more than one function for accomplishing a given task so the failure of one function does not impair the system's ability to operate.

Vibration alarm—an alarm designed to protect vibrating equipment, which is triggered if the level of vibration increases above an acceptable level.

Voting logic—computer logic that analyzes the signals from several devices, all of which are monitoring the same condition; it will initiate shutdown if a majority of the monitoring devices signal a dangerous condition.

Introduction

The process industry utilizes equipment and processes that can be extremely hazardous. In order to ensure a safer work environment and reduce the likelihood of equipment damage or catastrophe, procedural and equipment safeguards must be implemented. These safeguards include safe startup procedures, equipment designed to detect and warn of dangerous conditions, and equipment and procedures necessary for shutdown in the event dangerous conditions occur.

Some of the main devices responsible for ensuring system safety include permissives, interlocks, sensors, and alarms. In addition to these devices, specific operating procedures, such as those used when installing a bypass or jumper, are also employed.

Permissives and Interlocks

When operating equipment it is imperative that safeguards be implemented prior to start-up and during operation. Two of the most common types of safeguards include **permissives** and **interlocks**.

PERMISSIVES

Before a piece of equipment can be turned on, a set of conditions must be met to ensure safe operation. This set of conditions is called the startup permissives for that piece of equipment.

For example, startup permissives for a furnace may include verification of fuel pressure, LEL check (a check for the presence of hydrocarbons), and a steam flow purge. Permissives may be checked manually in a step-by-step manner or by programmed computers. The primary purpose of start-up permissives is to ensure that it is safe to put a particular piece of equipment into operation (Figure 23-1).

Permissives should be clearly documented in plain, easy-to-understand language. In addition, they should never be bypassed, tampered with, or their duration shortened without written management approval, as doing so could jeopardize safety.

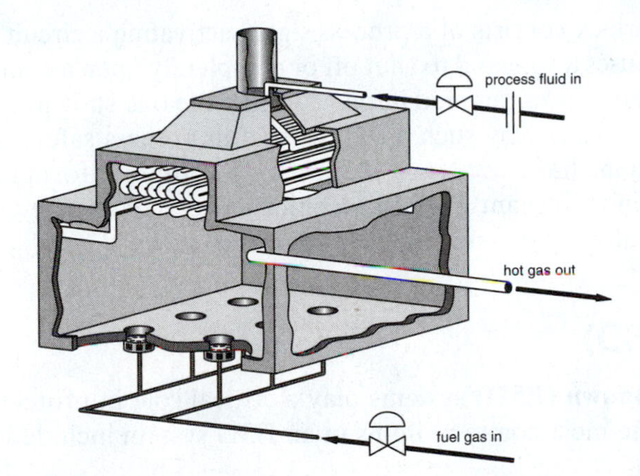

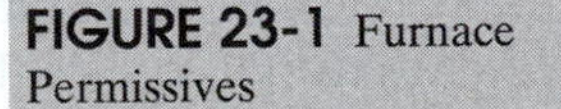
FIGURE 23-1 Furnace Permissives

INTERLOCKS

Although process emergencies are rare, the potential for danger to equipment and lives is high. Therefore, a means for detecting dangerous conditions and for safely shutting down a process or a piece of equipment is imperative. All critical equipment is designed with a shutdown circuit for accomplishing this task. This shutdown circuit is called an interlock. Figure 23-2 shows an example of an interlock.

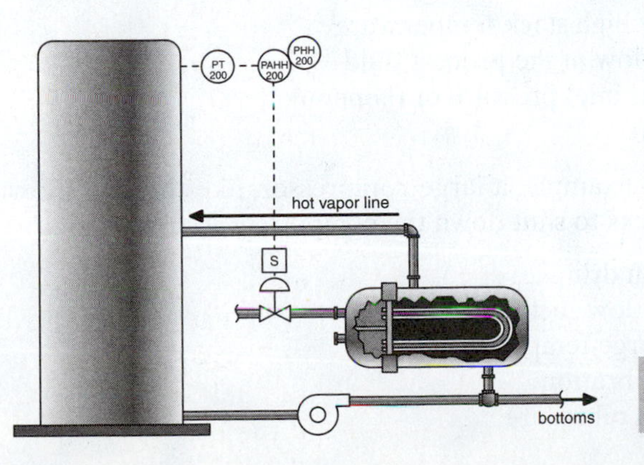

FIGURE 23-2 Distillation and Reboiler Interlocks

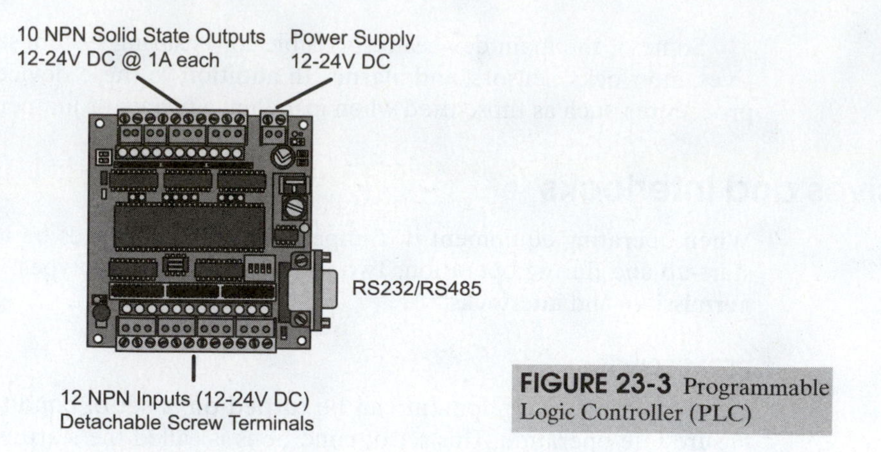

FIGURE 23-3 Programmable Logic Controller (PLC)

In this example, an interlock prevents the pressure within a distillation column from exceeding a safe limit by executing the steps below:

1. A monitoring device signals that a dangerous pressure has been reached.
2. The interlock action closes the steam valve going into the reboiler.
3. The product pressure from the reboiler into the distillation column begins to drop.
4. Once pressure has fallen to a safe level, the steam valve may be manually reset.

A basic interlock consists of a process signal activating a circuit (including relays or contacts) that causes a solenoid to shut off or completely open a control valve. Interlocks may be hardwired or programmed into a PLC like the one shown in Figure 23-3.

Because interlocks play such a critical role in process safety it is imperative that process technicians have access to the most up-to-date information on the specific interlocks used in their plant. Process technicians need to also be aware that bypassing interlocks typically requires management approval.

Emergency Shutdown (ESD)

Emergency shutdown (ESD) systems play a critical role in protecting equipment and lives. Some of the most common items in an ESD system include interlocks, breakers, and sensors.

INTERLOCKS IN EMERGENCY SHUTDOWN

An emergency shutdown (ESD) for a piece of equipment or process may involve a number of related interlocks.

For example, a furnace, like the one shown in Figure 23-4, could be equipped with interlocks designed to shutdown fuel flow in response to any one of several conditions such as:

- excessively high stack temperature
- low or no flow of the process fluid
- low fuel gas inlet pressure or flameout
- loss of draft

In another example, a large compressor, like the one shown in Figure 23-5, may include interlocks to shut down the compressor in the event of:

- high suction drum level
- excessively low suction pressure
- high discharge temperature
- excessive vibration
- lube or seal oil failure

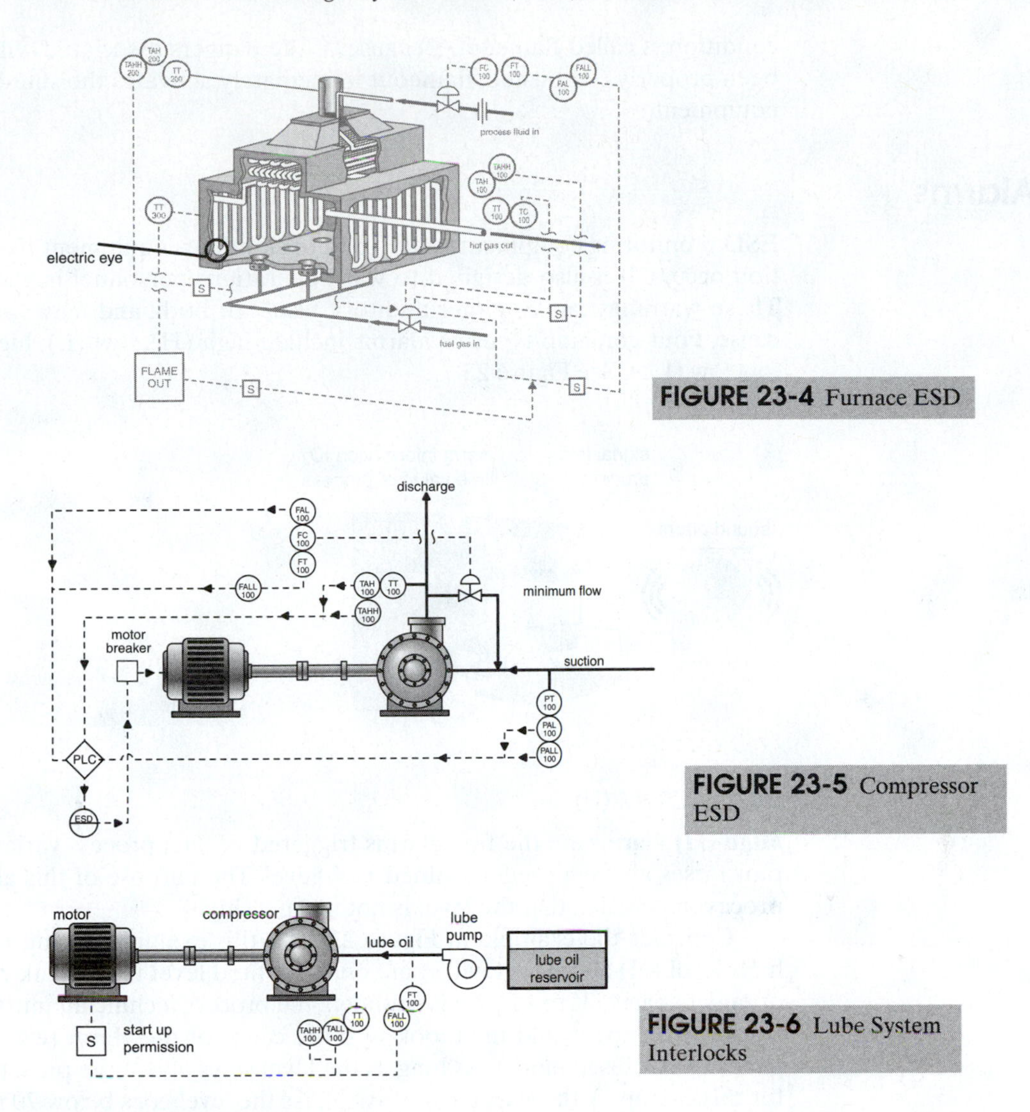

FIGURE 23-4 Furnace ESD

FIGURE 23-5 Compressor ESD

FIGURE 23-6 Lube System Interlocks

Finally, a third example of emergency interlocks is a lube oil/seal systems like the one shown in Figure 23-6.

In many types of equipment lube oil must achieve a certain minimum temperature to ensure the viscosity range appropriate for lubrication. Conversely, if the lube oil temperature exceeds a certain value, the oil will disintegrate and "carbon up." Therefore, lube oil systems contain both high and low-temperature alarms.

Lubrication is critical for rotating or reciprocating equipment. Lube oil system permissives or interlocks are designed to ensure that the equipment will not operate unless the lube oil system is functioning properly.

BREAKER SWITCHES IN EMERGENCY SHUTDOWN

A breaker switch is a safety device used to provide protection against overcurrent. In the event of high temperature a bimetallic strip expands inside the breaker causing the breaker to trip. Once a breaker is tripped it must be manually reset (note: many large motors have restrictions regarding how many times they can be started).

ELECTRIC EYE SENSORS IN EMERGENCY SHUTDOWN

Some boilers and furnaces are equipped with sensors called *electric eyes* (also referred to as *flame scanners*) that monitor the flame burning in the combustion zone. Absence of flame means that the fuel is pouring out of the burner and is not being burned completely. If all burners, or at least a large number of them, have ceased burning, the condition is called flameout. Because of the dangers associated with gas that has not been properly combusted, flameout immediately activates the shutdown of fuel gas to equipment.

Alarms

ESD monitoring equipment is designed to shutdown equipment if a dangerous condition occurs. It is also designed to warn operations personnel in such a circumstance. These warnings can be either auditory, visual or both, and may vary by severity and cause. Four common types of alarms include high (H), low (L), high high (HH), and low low (LL). See Figure 23-7.

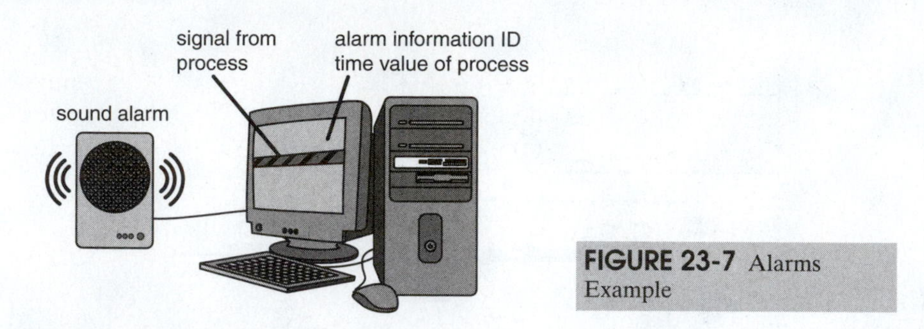

FIGURE 23-7 Alarms Example

HIGH ALARM (H)

High (H) alarms are the first alarms triggered when a process variable (e.g., fluid in a tank) rises above a predetermined high level. The purpose of this alarm is to notify a process technician that the level is not normal (high).

Consider the example in Figure 23-8. In this example, a tank is equipped with a high-level (H) alarm that will come on if the fluid level in the tank reaches 70 percent of tank capacity. If this alarm is sounded, the process technician must acknowledge the alarm to silence it, and then look into the cause of the alarm (e.g., the control valve needs to be closed more). As long as the level stays above the preset value (70 percent for this example), the alarm light stays lit. As the level goes below 70 percent, the alarm light turns off.

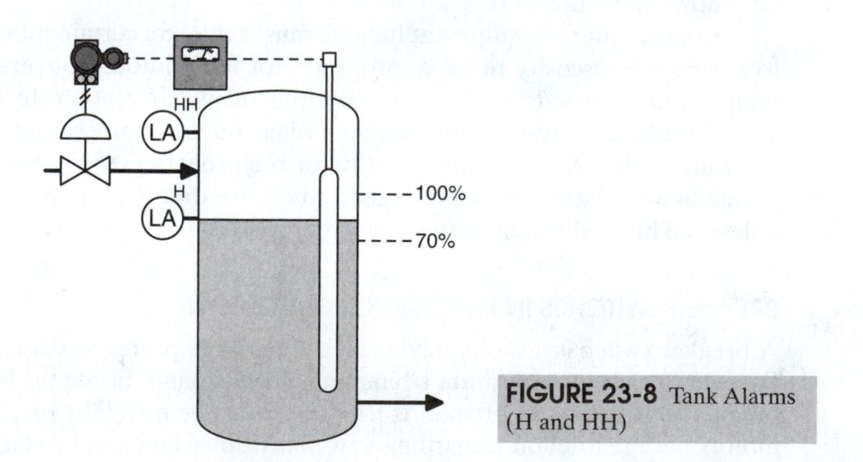

FIGURE 23-8 Tank Alarms (H and HH)

HIGH HIGH ALARM SHUTDOWN (HHSD)

High high (HH) alarms are triggered when a process variable (e.g., fluid in a tank) continues to rise above a predetermined maximum level after a high (H) alarm has already been triggered. The purpose of this alarm is to notify a process technician that a shutdown has been initiated.

Again, consider the example in Figure 23-8. In this example a high (H) alarm would have been triggered when the fluid level reached 70 percent. If the tank level continued to rise above 90 percent, the controller would provide another audiovisual alarm (an HH alarm) and would initiate a safe shutdown of the system (typically the actuation of a solenoid that shuts off the inlet valve). Once shutdown is initiated, safety mechanisms prevent the system from being restarted until the level has fallen below the high high (HH) preset values determined by process needs and safety considerations.

LOW ALARM (L)

Low (L) alarms are the first alarms triggered when a process variable (e.g., fluid in a tank) drops below a predetermined low level. The purpose of this alarm is to notify a process technician that the level is not normal (low). Whether a given process needs a low (L) alarm depends on the process requirements. A tank, like the example shown in Figure 23-9, may include a low-level (L) alarm because a very low fluid level may cause cavitation problems in the outlet pump. Similarly, certain heat exchangers may also need a low-temperature (L) alarm if there is a risk that low temperatures may lead to solidification or freezing of certain chemicals.

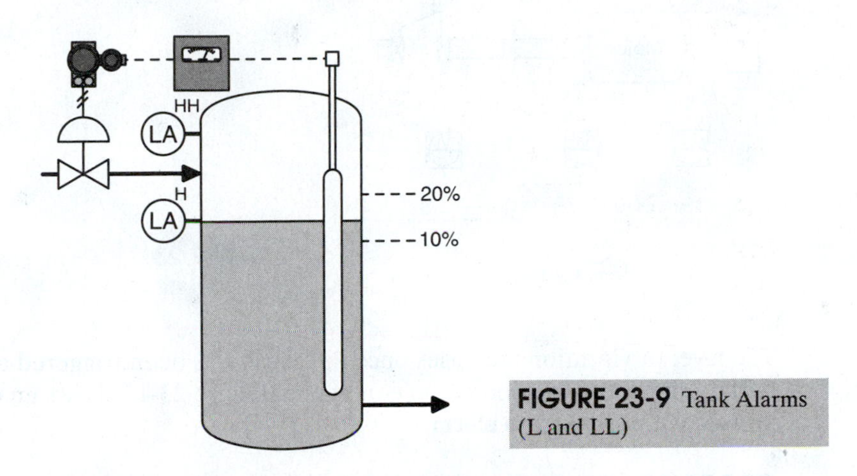

FIGURE 23-9 Tank Alarms (L and LL)

LOW LOW ALARM SHUTDOWN (LLSD)

Low low (LL) alarms are triggered when a process variable (e.g., fluid in a tank) continues to drop below a predetermined minimum level after a low (L) alarm has already been triggered. The purpose of this alarm is to notify a process technician that the level is becoming increasingly abnormal (low) and initiates a shutdown or some other corrective action (e.g., shutting off power to a pump to prevent cavitations, which could result in pump damage).

DEVIATION ALARM

In certain processes, like the one shown in Figure 23-10, the absolute value of a variable may not be important but a certain differential is important. For instance, in a particular catalyst bed the difference (deviation) between the outlet temperature and the inlet temperature should not exceed 50 degrees F. In this case, signals from two thermocouples are sent to a different unit or a computer. If the difference is greater than 50 degrees F an audiovisual **deviation alarm** is activated. This type of alarm can also be used with other variables such as the pressure found in a distillation column.

VIBRATION ALARM AND SHUTDOWN

Large rotating machinery is often equipped with vibration monitoring since running such machinery with high vibrations could damage it.

Typically radial and axial vibration monitoring is performed by instruments known as proximity probes located at several strategic points (e.g., radial and thrust bearings and gear units). If vibrations exceed a certain value an audiovisual alarm is activated.

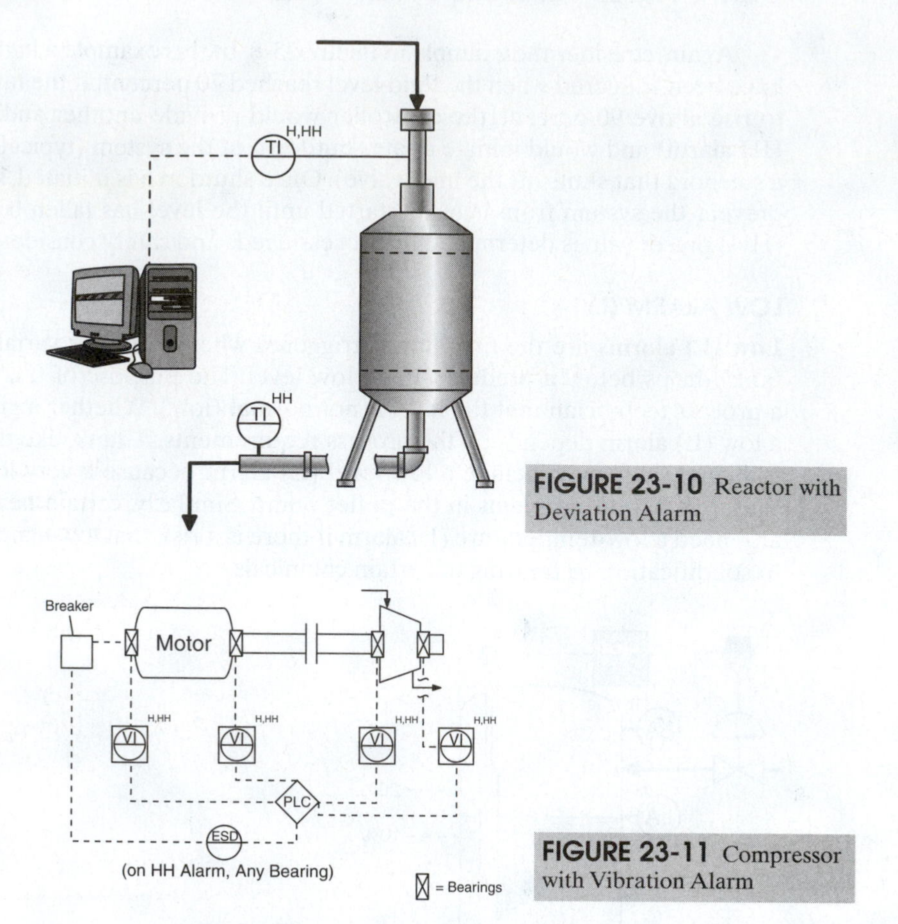

FIGURE 23-10 Reactor with Deviation Alarm

FIGURE 23-11 Compressor with Vibration Alarm

The level of vibration increases once the alarm has been triggered a second alarm will initiate shutdown to protect the equipment. Figure 23-11 shows an example of a compressor with a **vibration alarm.**

GAS DETECTION ALARM AND SUPPRESSION SYSTEMS

Although modern pumps and sealing systems are designed to minimize leaks, the possibility of leaks still exists. A large spontaneous leak of hydrocarbons, or even a small leak over an extended period of time, can lead to fire or explosion and threaten the loss of equipment and lives. Therefore, facilities that handle flammable materials almost always employ several gas detection alarms located strategically throughout the plant. Depending on the type of flammable vapors involved, the monitors may employ several different gas-detection techniques including infrared (IR), ultraviolet (UV), or cloud chamber detection.

Typically, large units have several gas detection monitors that provide an audiovisual alarm to the control room. The alarm from each of these monitors identifies the area or location of the leak.

Many gas-detection alarms are also tied to deluge systems, like the one shown in Figure 23-12. Deluge systems spray the area with large quantities of water to dilute the

FIGURE 23-12 Compressor with Gas Detection Alarm

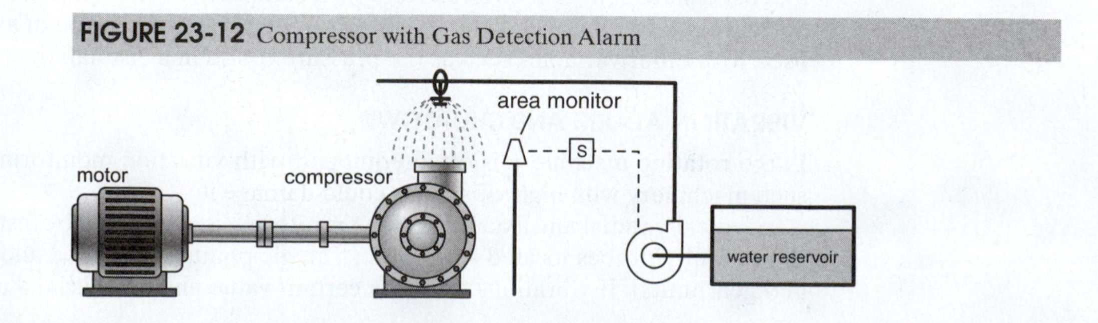

concentration of vapors so they are below the lower explosive limit (LEL), thereby reducing the potential of fire and also protecting equipment.

To ensure continual reliability and avoid nuisance tripping of deluge systems, monitor systems should be checked periodically.

Avoiding Unnecessary Shutdowns

Unnecessary shutdowns are both time consuming and expensive. In order to reduce the likelihood of a shutdown a variety of hardware, software, and techniques are employed. Two of the most common techniques used are the incorporation of redundant hardware and computer logic called voting logic.

REDUNDANCY

Previous chapters have used the term **redundancy** to describe functional duplication of plant components to assure uninterrupted operations. Two examples of redundancy are found in distributed control systems (DCSs) and uninterruptible power supplies (UPSs).

In a DCS, more than one computer is capable of handling a particular operation. Therefore, if one computer fails, the data can be rerouted and another DCS can assume the task.

In a plant UPS system, batteries are used as a backup power source in the event of a power failure or power decrease. This ensures that power flow to critical instrumentation will continue, thereby reducing the likelihood of unnecessary shutdown.

VOTING LOGIC

Redundancy can be designed into an alarm system to minimize nuisance or spurious shutdowns. A common method of redundancy involves a computer function called **voting logic**, which is shown in Figure 23-13.

With voting logic a computer program analyzes the signals from several devices, all of which are monitoring the same condition, and will initiate shutdown only if a sufficient number of the monitoring devices signal a dangerous condition.

Consider temperature control in a reactor. In this scenario thermocouples are used to monitor reactor temperature. If the temperature exceeds a certain value, shutdown is activated. However, if a faulty thermocouple is installed, the result might be a costly and unnecessary shutdown. Therefore, a voting logic approach is often the best solution.

In a voting logic approach, several thermocouples are installed with each sending readings to a microprocessor. The microprocessor then compares the incoming readings and initiates a shutdown only if several of the thermocouples are registering unacceptable values.

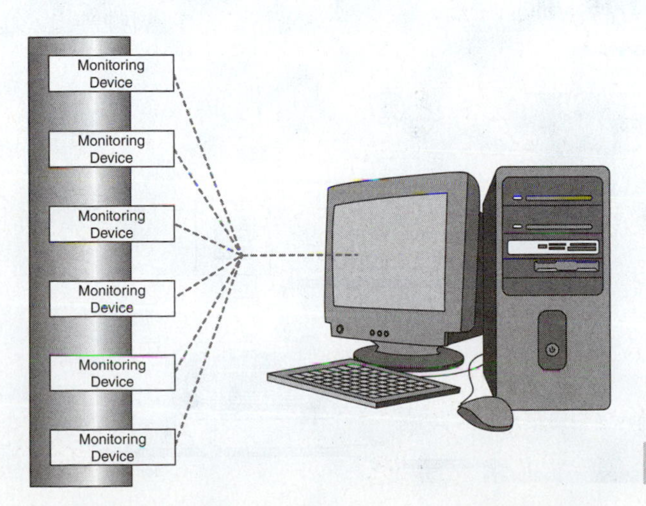

FIGURE 23-13 Voting Logic

ESD Resetting and Testing

The purpose of any ESD system is to shut down a system or piece of equipment in the event of an unsafe condition. Because it is a safety shutdown, an automatic reset of an ESD system is not be feasible. Thus, ESD systems must be reset manually. Resets can be either simple (single push-button type) or multistage and may include a manual relatch of the field device.

SIMPLE RESET

Relatively simple ESD systems, such as those involving only one or two related loops, can be reset by simply pushing a button (e.g., some lube or burner management systems are designed with a single push-button configuration). Figure 23-14 shows an example of a simple reset.

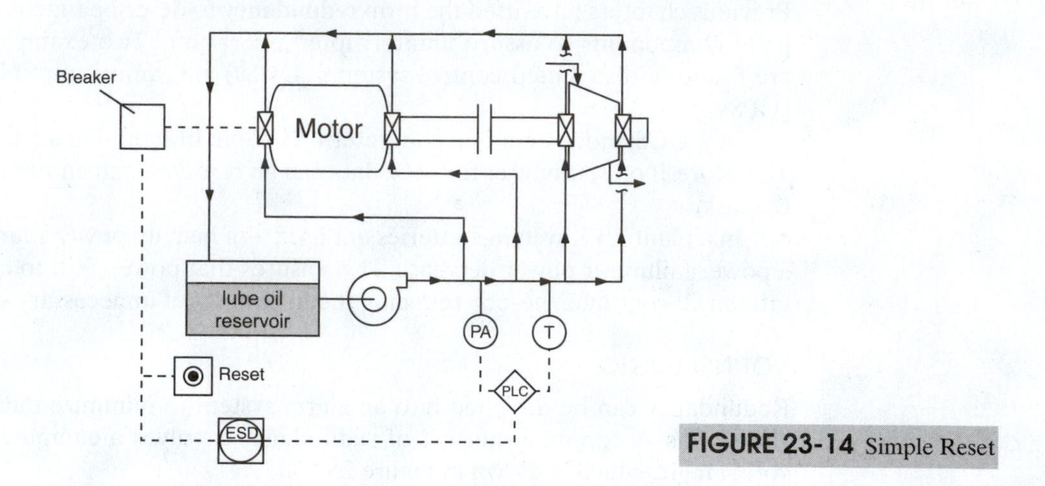

FIGURE 23-14 Simple Reset

MULTISTAGE RESET

Typically, ESD systems that involve many loops also utilize multistage resets like the one shown in Figure 23-15. In this example an ESD system on a distillation column has shut down product feed and steam to the reboiler. In order to get the system up and working again, each component must be reset individually.

Reset of a multiple-stage system should take place only after each component of the system is carefully checked and the cause of the shutdown has been determined, since knowing the cause of the shutdown may be useful in avoiding a reoccurrence.

While ESDs require manual reset, some alarms can be reset automatically. This is allowed because efficiency concerns outweigh the minimal risk involved in the auto reset of an alarm.

FIGURE 23-15 Multiple Reset

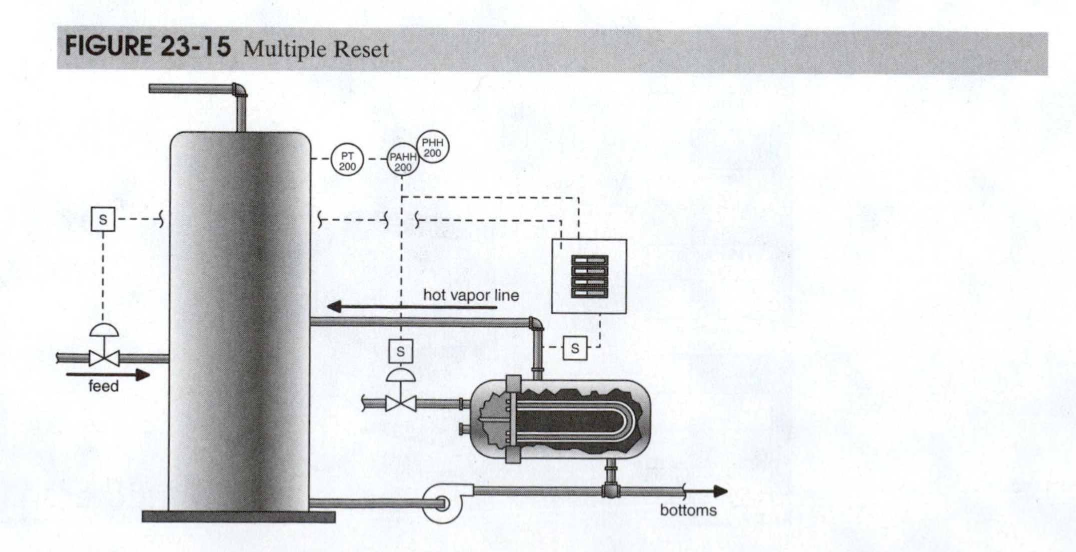

TESTING AN ESD

ESD testing is necessary to ensure the system will function properly when needed. A typical test involves the testing of a measurement instrument in order to activate the ESD system and alarms. Once an instrument has been tested one of several results may occur: (1) the system will either perform an actual shutdown of a valve or a final control element with all systems activated (full test); or (2) the system will perform a shutdown with some systems modified or disengaged (dry testing).

FULL TESTING

Full testing, which is almost always done when a plant or process is being commissioned for the first time or following a turnaround, involves the application of a specific electrical charge to an ESD instrument in order to induce the shutdown of a valve or final control element (Figure 23-16).

While full testing provides a high degree of confidence in the system, it does have a downside, and that is the potential for equipment damage. For this reason, many individuals opt to perform a dry test, which is less conclusive, but is also less likely to damage critical and expensive equipment.

FIGURE 23-16 Full Test

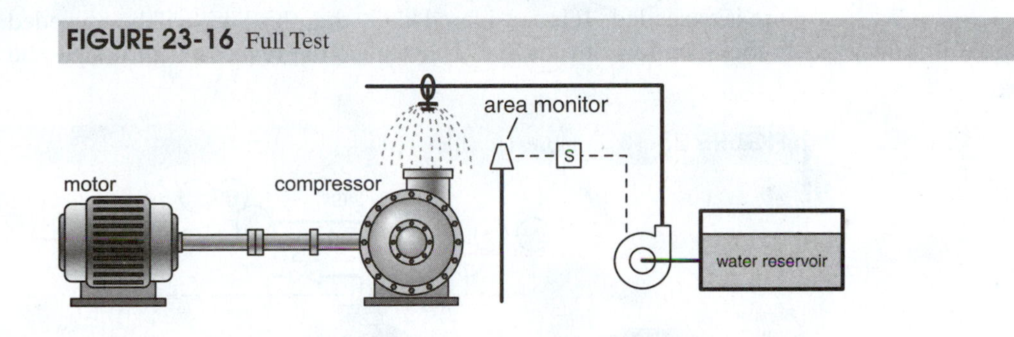

DRY TESTING

Dry testing is similar to full testing in that the alarm and shutdown devices are activated to induce a shutdown. However, in dry testing the actual shutdown system is disengaged so the test cannot be fully executed (Figure 23-17).

Consider the full test example shown in Figure 23-16. In this example the **deluge system** is activated, and water is dispensed. If this same system were to be tested using a dry test method, the signal would still be sent to the deluge system, but the water would not be dispensed.

While it is possible to observe whether or not a deluge system mechanism is activated, dry testing cannot give an assurance that the system will function as specified in the design since the system is unable to execute completely.

FIGURE 23-17 Dry Test

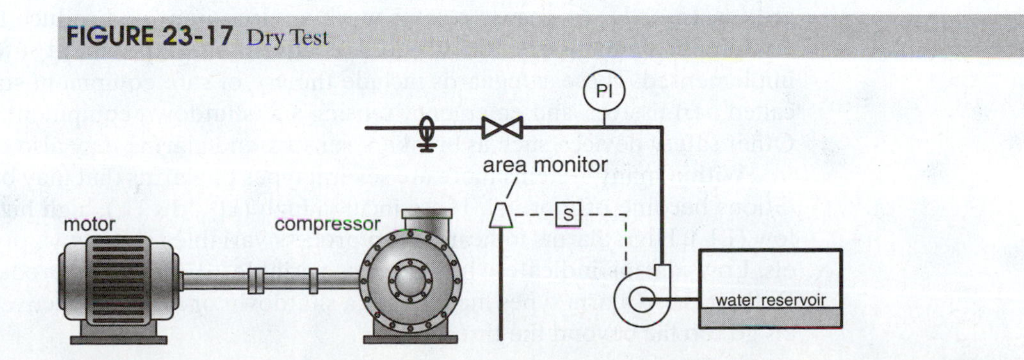

PRACTICAL TIPS

ESD system testing and calibration should be done when the plant, or section of the plant, is down. Close coordination between process technicians and instrumentation personnel is essential. All test findings should be carefully documented for regulatory, legal, and technical reasons.

ESD Bypassing

There may be times when it is necessary to override an interlock. In order to perform this override, a piece of electrical wire called a **bypass** (also referred to as a **jumper** or **defeat**) can be installed. However, while bypasses can be a convenient and necessary expedient, misuse may result in damage to expensive equipment. Therefore, top management approval is mandatory before bypasses can be installed.

Bypass systems can be designed to permit testing of alarms and trips during operation. Close coordination between process technicians and instrument personnel is essential in order for these tests to be successful. Upon completion of the work, the bypass system must be deactivated and documented for regulatory, legal, and technical reasons.

Consider the example shown in Figure 23-18. In this example, it has become necessary to perform repairs on a compressor while the compressor is running. However, the compressor is equipped with an interrupt designed to initiate a shutdown if a certain temperature is exceeded. It is considered likely that this limit will be exceeded during the repair. In this situation a bypass may be installed to prevent a high-temperature shutdown.

FIGURE 23-18 Compressor with Interlock Defeats

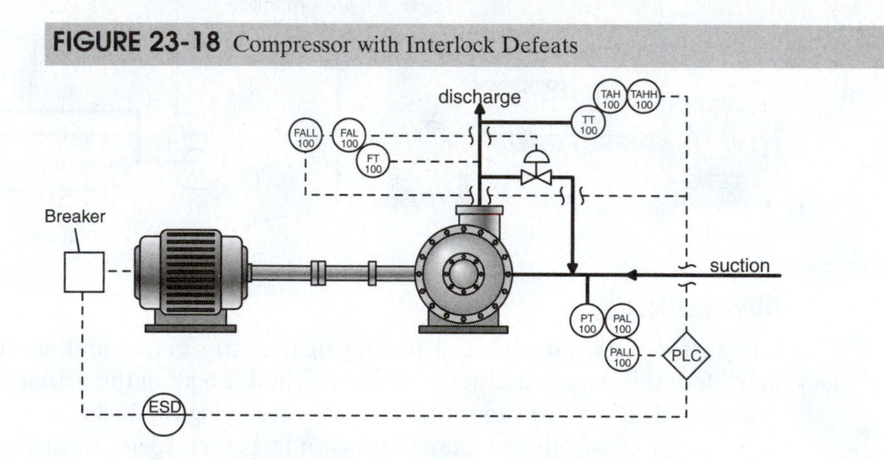

ESD BYPASS MANAGEMENT

ESD systems are designed to protect lives and equipment. In the event that a bypass is necessary, companies must have established procedures that are followed carefully in order to minimize risk.

Summary

The process industries utilize equipment and processes that can be extremely hazardous. In order to ensure a safer work environment and reduce the likelihood of equipment damage or catastrophe, procedural and equipment safeguards must be implemented. These safeguards include the use of safe equipment startup procedures called permissives, and emergency sensing and shutdown equipment called interlocks. Other safety devices, such as breakers, sensors, and alarms, may also be incorporated.

Within many systems there are several types of alarms that may be sounded if conditions become off normal. These include high (H), low (L), high high (HH), and low low (LL). High alarms indicate when process variables rise above predetermined levels. Low alarms indicate when process variables drop below predetermined levels. Both of these alarm types may trigger a shutdown or other corrective action if the levels go too far beyond the threshold.

In addition to high (H), low (L), high high (HH), and low low (LL) alarms, many systems also contain deviation, vibration, and gas detection alarms.

Deviation alarms notify the process technician when the difference between two variables exceeds acceptable levels. Vibration alarms notify the process technician if the amount of vibration generated by rotating equipment exceeds acceptable limits (excessive vibration can damage equipment). Gas detection alarms notify process technicians of gas leaks and may trigger various suppression systems, such as a deluge system which douses the area with water in order to dilute the concentration of vapors so they are below the lower explosive limit (LEL).

In addition to alarm and suppression systems, many systems also include components that help protect the equipment and the process by avoiding unnecessary shutdowns. Two of these components include system redundancy (e.g., a UPS system supplying backup power to critical systems during a power failure, and computer logic that measures inputs from multiple sensors and only initiates a shutdown if a majority of the sensors are registering unacceptable values).

If shutdown is initiated, many systems require a manual reset in order to make them operational again. Resets can either be simple (e.g., a single push button) or multistage (in which each component must be individually checked and reset before the system will be operational).

In order to ensure an alarm or suppression system will function properly during abnormal conditions, performance testing is needed. Testing can either be full testing or dry testing. Full testing is almost always done when a plant or process is being commissioned for the first time or following a major turnaround. During a full test, the emergency shutdown system (ESD) is triggered and suppression system (e.g., a deluge system) is activated. While this is more accurate than a dry test, there are limitations, namely the likelihood of equipment damage. Dry testing, is similar to full testing in that the emergency shutdown system (ESD) is triggered and a signal is sent to the suppression systems. However, the suppression system in this case is disengaged so it does not dispense.

Finally, when working with systems there may be times when it is necessary to override an interlock using a bypass or a jumper. If a bypass is to be used, management must first approve the use of the bypass and document their approval on a bypass form. Once the work is complete, bypasses must be deactivated and the removal documented for regulatory, legal and technical reasons.

Checking Your Knowledge

1. A_____________ is an instrument used to detect vibration.
   a. Proximity probe
   b. Vibratect
   c. Vibro-indicator
   d. Contact probe
2. In order to restart equipment after ESD, ______ will be necessary.
   a. Interlock
   b. Bypass
   c. Reset
   d. Defeat
3. *True or False* Multiple reset means that a single button reset of each piece of equipment will be necessary.
4. *True or False* Dry test verifies the effectiveness of an ESD system.

Student Activities

1. Given a diagram of a relatively simple loop, identify the instrumentation and the alarm and shutdown systems.
2. Take a relatively simple loop (one loop) and identify the instrumentation and the alarm and shutdown systems.
3. Provide students with a P&ID illustrating several pieces of common plant equipment. Have students identify each interlock system along with its components. Have students discuss what each ESD is designed to protect and to protect against.

4. For example:
   - Level applications: Can the tank overflow? What are the safety and environmental consequences of such an overflow? In many applications, you may provide an H-level alarm and a HH-level shutdown.
   - Pressure applications: When would you use a high-pressure alarm, followed by an HH pressure shutdown? Although relief valves (RV) provide a protection against overpressure, a HH pressure shutdown gives an added assurance of protection.
   - Temperature applications: If a potential exists for a runaway reaction (where the temperature goes out of control), an HH temperature shutdown is worthy of consideration.
5. Flow applications: Some situations may require you to shut down the process. A severe flood in a distillation column would necessitate that the feed be shutdown.
6. Take a single interlock system and identify components that can fail and can cause spurious shutdown.

CHAPTER 24

Instrumentation Malfunctions

Objectives

- Upon completion of this chapter, you will be able to:
- Recall the methods used for determining if a sensing or measuring device is malfunctioning:
  - remote versus local indication
- Describe the failure modes of the following:
  - temperature elements
  - thermocouples
  - RTDs
  - level floats
  - flow elements
  - pressure elements
  - analytical elements
- Explain how a control loop will respond to typical malfunctions in the following:
  - primary sensing elements
  - transmitters
  - controllers
  - final control elements

Key Terms

Noisy signal—a signal that fluctuates dramatically that is most likely the result of a lose sensor connection.

Introduction

Process technicians rely upon instrumentation to monitor all process variables and ensure safety. Therefore, it is important that all instrumentation be accurate and functioning properly, since malfunctioning instruments may fail to indicate an actual problem or may indicate process problems where none exist.

In the event of an instrument malfunction, process technicians need to know how to identify the malfunction, isolate it, and determine the possible cause. In order to do this, technicians must be familiar with various types of instrumentation (e.g., pressure, temperature, level, flow, and analytical elements) and the symptoms associated with their failure.

Identifying Instrument Malfunction

Process technicians work closely with instrumentation technicians during instrumentation troubleshooting. Therefore a basic understanding of typical instrumentation problems and troubleshooting methods is imperative. Most importantly, technicians need to know how to isolate and identify the source of the problem, and identify the types of failures and their causes.

ISOLATING THE PROBLEM SOURCE

When troubleshooting, technicians should always use the process of elimination to systematically identify and isolate faulty components (Figure 24-1). Take, for example, a control loop.

A control loop is an integrated system consisting of a sensor or transmitter, a controller, and a final control element. A malfunction in any of these components affects the loop performance and stability. Therefore, it may be necessary to use the process of elimination to verify signals into and out of each instrument in turn in order to isolate the faulty component.

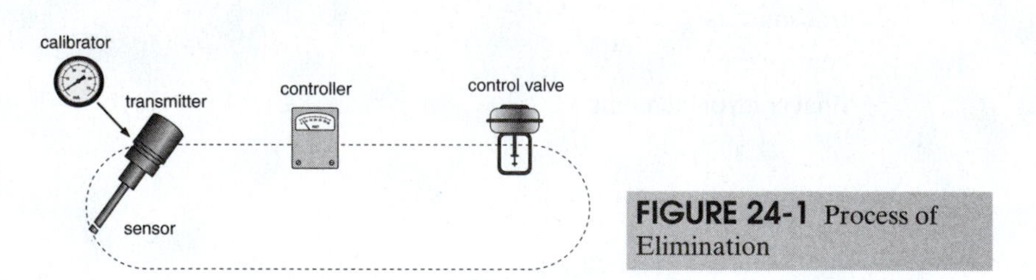

FIGURE 24-1 Process of Elimination

CALIBRATION

One of the first things that should be investigated when troubleshooting instrumentation is calibration (Figure 24-2). Calibration is a common method used to determine if a sensor is working. Take, for example, a sensor in a control loop.

In a control loop a sensor may be removed or isolated from the process and subjected to rigorous testing. For example, a thermocouple can be placed in boiling water and the output observed with the use of a handheld calibrator to see if the output temperature is within the manufacturer's published accuracy for that thermocouple at the boiling point of water. If the output is inaccurate, one might assume the sensor (thermocouple) is faulty. This method can be applied to both local and remote indications. Be aware, however, that, while it may be practical to calibrate a temperature-measuring instrument as described in this example, it may be difficult to apply this method with other types of sensors (e.g., those that measure flow, level, and pressure).

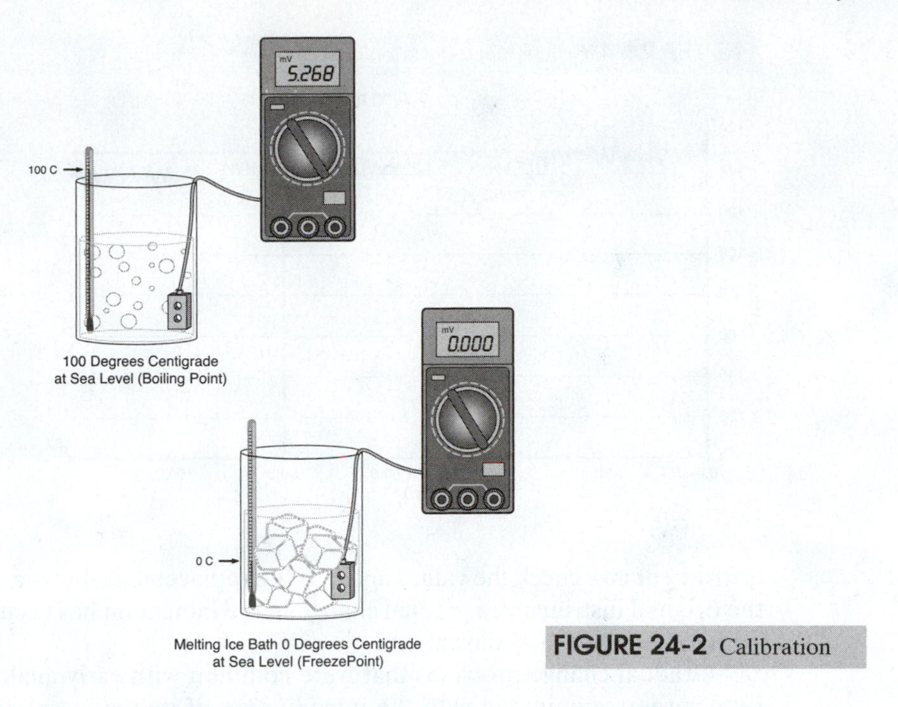

FIGURE 24-2 Calibration

Once it has been determined that an instrument is faulty, the next step is to determine the cause of the failure. When the faulty component is identified, it must be repaired or replaced.

FAILURE TYPES AND POSSIBLE CAUSES

Instrument failures can be of many types and causes. The change in output, which could be attributed to signal noise, may be abrupt, gradual, or a fixed deviation.

Abrupt Change

Abrupt changes in the indicated value of a process variable with no corresponding change in the process are reliable indications that the sensor or transmitter is faulty (Figure 24-3). For example, if a level indicator registers an abrupt change in a tank level, but the sight glass does not show a corresponding change, a transmitter wire with a short in it may be the problem.

Simple observation may identify the most likely source of the problem in both remote and local indications.

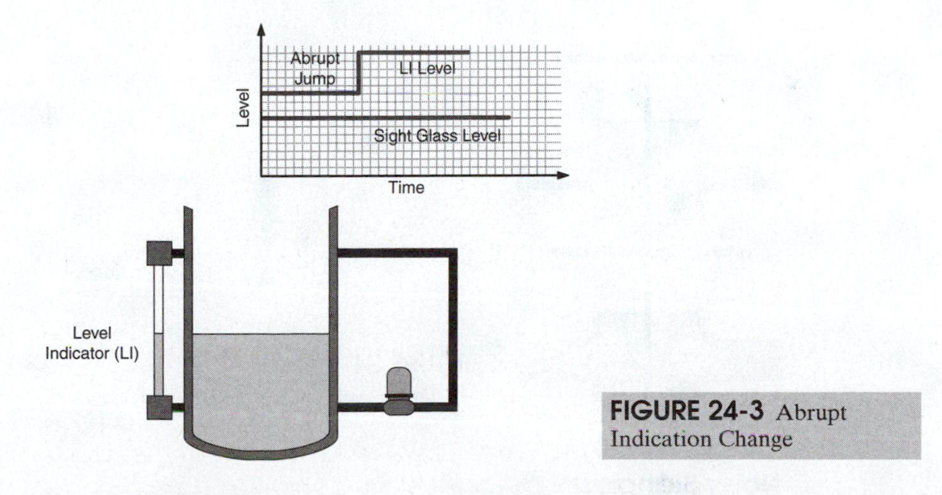

FIGURE 24-3 Abrupt Indication Change

Gradual Change

Gradual change in instrument readings is somewhat harder to diagnose (Figure 24-4). However, if there is gradual change in the indication, install a freshly calibrated

FIGURE 24-4 Gradual Deterioration and Drifting Measurements

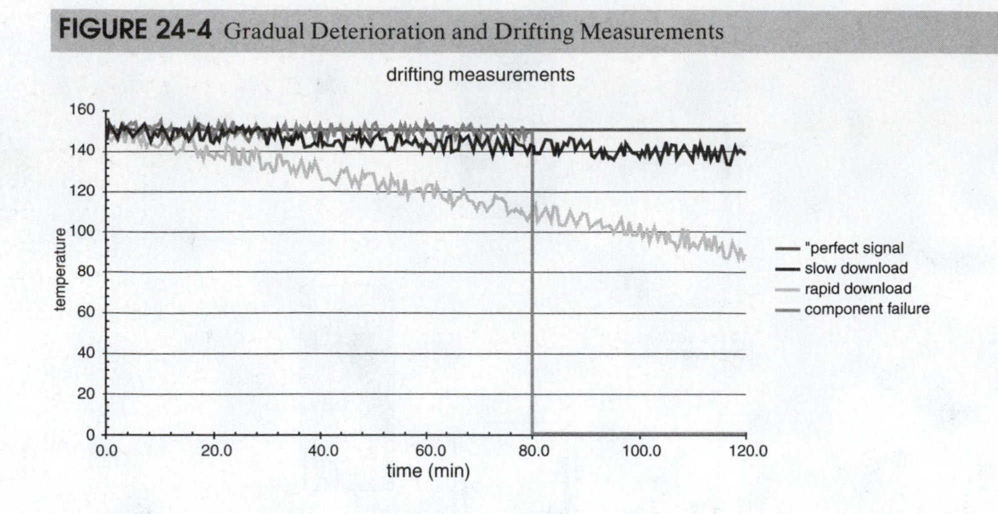

instrument and check the values again. If the replacement shows a different value from the original instrument, a gradual change in the indication has occurred. This is true for both remote and local indications.

Gradual change problems that were common with early analog instruments have been largely eliminated with the introduction of digital instrumentation. However, most thermocouples do show a gradual change with age. Plugged lines to a d/p cell on a flowmeter can show gradually increasing or gradually decreasing signals depending on which leg is plugging faster (the responses are discussed in more detail later on in this chapter). Gradual changes in flow and level indications may be caused by leaks in the impulse lines to the transmitter. If the impulse lines are fluid filled and the fluid is not replaced, or is replaced with a different fluid, such as process fluid, the transmitter output may show a gradual change (e.g., the time scale in Figure 24-4 could also be in days or weeks). The slower the drift, the more difficult the problem is to detect and analyze.

Fixed Deviation

A fixed deviation may be more difficult to recognize and troubleshoot than abrupt or gradual changes. For example, the improper installation of an orifice plate, like the one installed backwards in Figure 24-5, will give flow indications that are in error by a fixed margin. Troubleshooting this type of problem requires the use of engineering calculations. However, this technique can be applied to both remote and local indications.

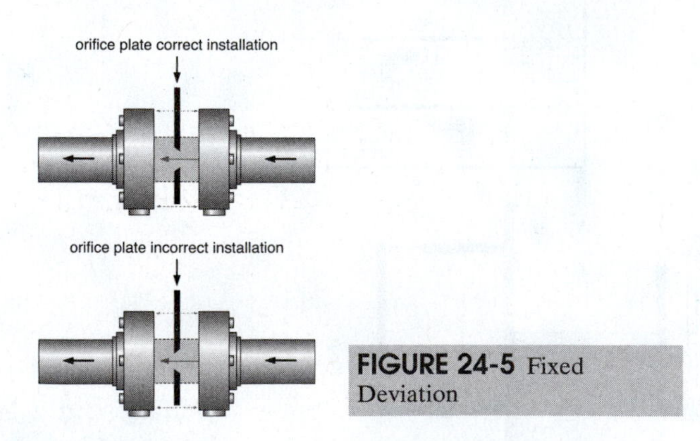

FIGURE 24-5 Fixed Deviation

Noisy Signals

Most process measurements have a small amount of inherent noise, which does not signify a problem. Excessive noise on signals that are normally "clean," however, does indicate a problem that must be fixed. Excessive *noise* (signals that fluctuate

FIGURE 24-6 Noisy Signal Measurements

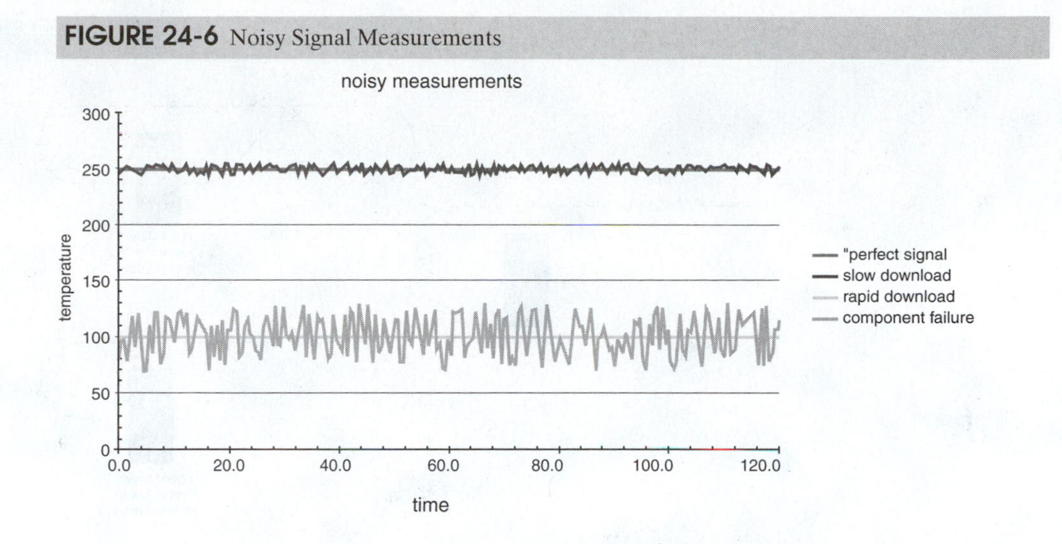

dramatically), like that shown in Figure 24-6, can be generated when signal lines are not properly shielded or are placed too close to electrically noisy equipment such as large motors or high-voltage electrical lines. For this reason, avoid running temporary power and extension cords in the vicinity of instrument signal wiring. A **noisy signal** may indicate instrumentation failure. The source of this noise may be loose sensor connections. If that is the case, tightening the connections should eliminate the problem. This technique is applicable to both remote and local indications.

No Indicated Change

The fact that a process variable does not change at all can often be an indication of a failure. Rarely do process variables go for very long completely unchanged. This is particularly true for faster processes like flow. The visibility of this type of failure is more evident with digital readouts, which often include one or two decimal points. For example, the process technician should become suspicious of an absolutely steady flow rate reading of 52.46 gpm. This could well be a false reading caused by a transmitter failure or a software calculation that has "locked up" and stopped functioning.

Instrument Failure Modes

Different types of sensor or transmitter elements have different failure modes. This means that each type of equipment will exhibit particular behaviors and characteristics when they begin to fail. The following paragraphs describe some of the different failure modes for various types of instrumentation.

TEMPERATURE ELEMENTS

A resistance temperature device (RTD), also called a resistance temperature detector, is composed of both a resistance source and a voltage source. The current allowed to flow through the circuit is inversely proportional to the temperature being measured. An open RTD would be a high resistance, which would provide a false indication of a high temperature. A shorted RTD would be a low resistance, which would provide a false indication of a low temperature.

Environmental conditions may also cause the gradual deterioration of an RTD and, consequently, a gradual decline in the RTD output signal (Figure 24-7). In addition, RTD circuits can show fluctuating output if the wiring connections are loose or if there is cross-talk along the signal path. Although twisted pair wires, shields, and grounding are effective means for preventing the interference from cross-talk, defective workmanship in any of theses items can result in a fluctuating signal.

Thermocouple failure modes are comparable to those of an RTD at least from an electrical point of view. Although a thermocouple may be protected by a thermowell,

FIGURE 24-7 RTD Failure

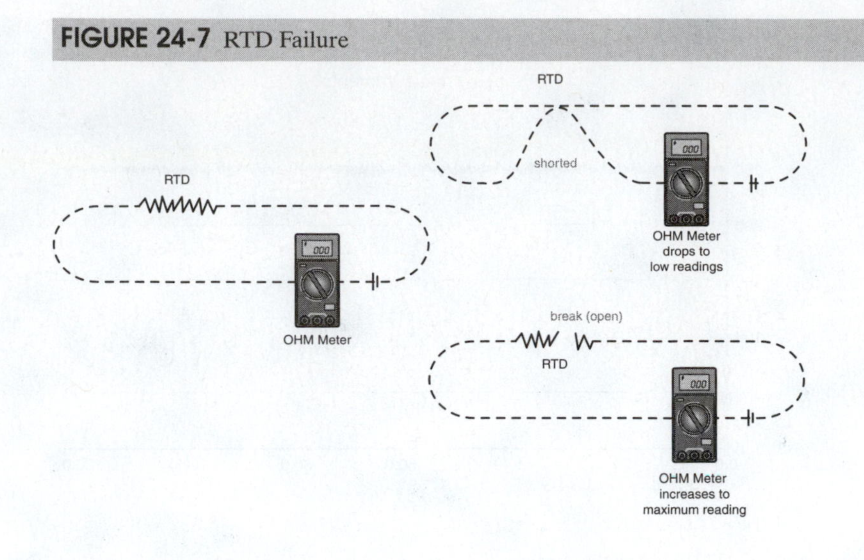

an improper environment could cause instrument failure. While RTDs and thermocouples are the most common types of temperature sensors, other types are sometimes used. These include thermal bulb, bimetallic, and infrared.

Examples of thermocouple materials are: Type T (copper constantin), Type J (iron constantin), Type E (chromel constantin), and Type K (chromel alumel). These thermocouple types are arranged in order of increasing temperature applications (Type T being the lowest, and Type K being the highest).

FLOW ELEMENTS

Flow may be measured by a variety of devices and methods including: orifice, venturi, pitot, Doppler, transit time, vortex, magnetic meter, coriolis, turbine meter, elbow meter, and thermal. Use of an orifice meter is the most common of these methods (Figure 24-8). An orifice meter typically includes an orifice plate connected to a differential pressure (d/p) cell. An orifice meter may malfunction as a result of incorrect installation, orifice plate corrosion, plugged pressure lines, incorrect valve position, or component failure.

Incorrect installation of the orifice plate can cause problems. If an orifice plate is installed backwards, the flow indication will be inaccurate by a fixed or constant margin. Similarly, if an orifice plate is installed without a sufficient length of straight pipe upstream and downstream from the plate, a consistently erroneous signal will result. These errors can be diagnosed rather quickly simply by checking the field installation of the orifice plate.

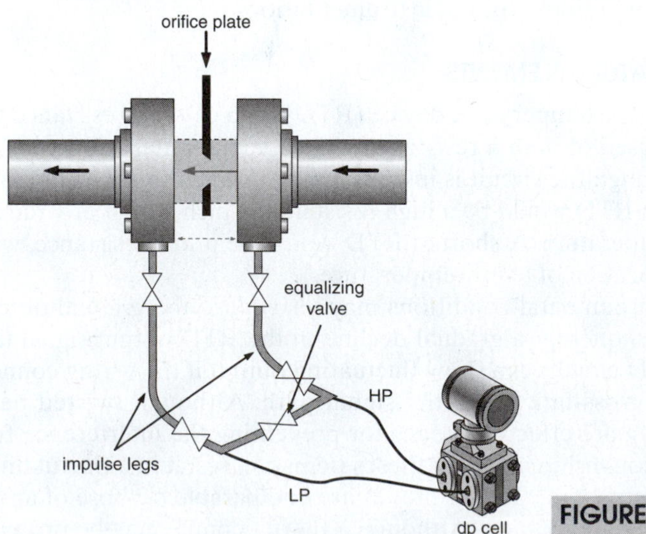

FIGURE 24-8 Orifice Meter

Corrosion of the orifice plate will result in a gradually declining flow indication. This type of failure is rare and should not occur if the plate is constructed from the proper material.

The impulse legs connecting the flow pipe with the d/p cell may be the source of instrument failure. Impulse legs are short lengths of metal tubing. One is called the high-pressure leg (HP) and the other is called the low-pressure leg (LP). If either of the HP or LP taps becomes plugged, it will result in an erroneous flow indication. Failure to winterize (freeze preparation) the impulse legs, when necessary, can also result in an erroneous flow indication. If the equalizing valve is inadvertently left open, the flow indication will continue to stay at zero. If the equalizing valve is opened during normal operation, the freeze protection fluid will be flushed from the HP leg to the LP leg and could cause an imbalance and an erroneous output. Failure of any component of the d/p cell will also result in erroneous or zero-flow indications.

PRESSURE ELEMENTS

If a d/p cell is used as a pressure sensor or transmitter, error can occur. Some pressure transmitters contain diaphragms to isolate the process fluid from the internal transmitter components. If the pressure transmitter is used in a viscous service or slurry service, the diaphragm may become coated with thick deposits. This will dampen the response of the pressure sensor. In extreme cases, the sensor will register a zero reading. Rapidly pulsating pressures can destroy the pressure sensor. This can often be corrected with the use of a restrictive device (pulsation dampener).

Analytical Elements

It may be necessary to test the composition of the product at a given stage in the process. To accomplish this, a variety of analysis techniques may be used including: flame ionization detection (FID), thermal conductivity (TC), photo ionization detection (PID), infrared (IR), and ultraviolet (UV). Although analyzers associated with each technique have their own specific problems, all types share common problems related to sample conditioning. See Figure 24-9.

Sample conditioning may utilize strainers, filters, and guard beds (adsorbents) to remove moisture. Samples containing impurities in the form of particulates are passed through a fine filter. A heavy load of trash may plug the filter and result in zero output from the analyzer. For this reason, filters must be changed at regular intervals.

Excessive moisture can quickly saturate a guard bed and then seep into the analyzer. Most moisture traps have a color indication. When a color change occurs, it is necessary to change out the guard bed.

The sensor for in-situ analyzers may become plugged by dirt or trash in the process rendering the analyzer unusable.

FIGURE 24-9 Analyzer

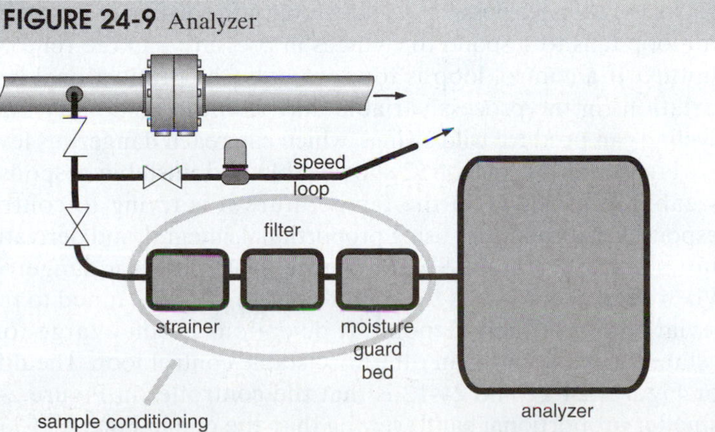

Loop Response to Instrumentation Failure

So far in this chapter, failures in different components of a control loop have been discussed in isolation. These components interact making it difficult to determine which component is responsible for a particular failure. Furthermore, more than one component may contribute to the problem. For example, the unstable behavior of a control loop may be the result of an improperly tuned controller, a noisy signal from the transmitter, a hesitation in valve response (e.g., the packing may be too tight) or a combination of all of these causes. See Figure 24-10.

FIGURE 24-10 Loop Response

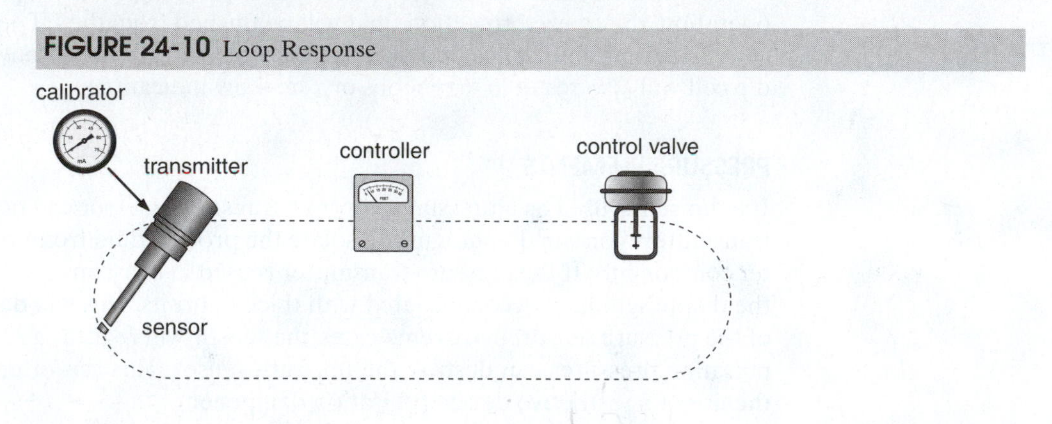

PRIMARY SENSING ELEMENTS

Primary sensing elements may affect a loop in different ways. A gradual signal deviation, or a fixed signal deviation, may result in a corresponding deviation in the process variable. For example, an incorrectly installed orifice plate may result in a consistently low flow indication. This will not make a loop unstable. If the orifice plate shows fluctuating flow due to corrosion, the controller response will also be unstable and the valve will constantly open and close. Similarly, a poor ground on a thermocouple may produce a fluctuating temperature signal, which will make a temperature loop unstable.

TRANSMITTERS

Transmitters convey information from a sensor to a controller. A faulty transmitter may be responsible for a signal that is consistently in error. In addition, a fluctuating transmitter signal (due to loose wires or interference) may cause a control loop to be unstable.

CONTROLLERS

The most common problem associated with controllers stems from improper tuning. A sluggish controller (one with a proportional band that is too high) will fail to respond to process changes possibly leading to unsafe conditions. For example, if a sluggish control loop fails to respond to changes in pressure and the relief valve fails, a vessel can rupture. If a control loop is too sensitive (the proportional band is too small) small variations in the process variable may result in instability that is known as cycling. Cycling can produce wild swings, which can reach dangerous levels.

Figures 24-11 and 24-12 show stable and unstable responses of a controller to a sizeable step change in the temperature it is trying to control. Both figures show responses of a controller using proportional, integral, and derivative modes. Figure 24-11 shows a stable response while Figure 24-12 shows a dangerous unstable response. When controllers are set for "tight" control, they are tuned to respond rapidly to small deviations from their setpoint. If deviations become large for unexpected reasons, instability can occur in an otherwise stable control loop. The difference in the settings for Figures 24-11 and 24-12 is that the controller in Figure 24-12 has a higher gain (smaller proportional band) setting than the one in Figure 24-11.

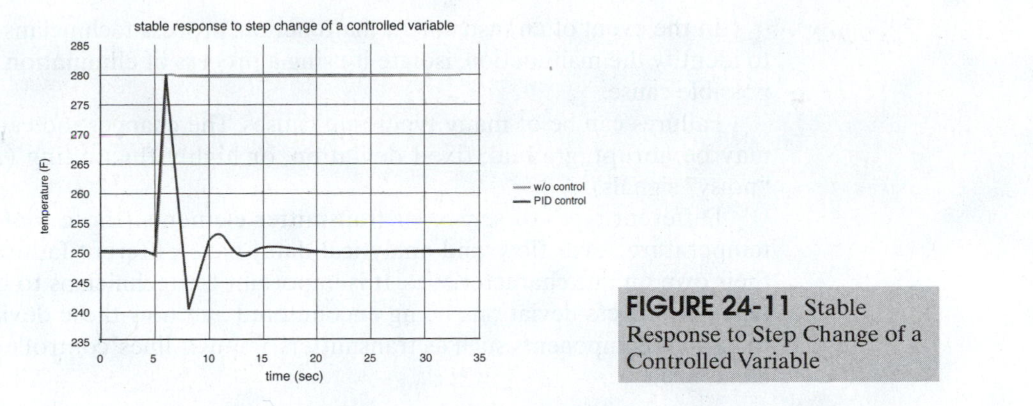

FIGURE 24-11 Stable Response to Step Change of a Controlled Variable

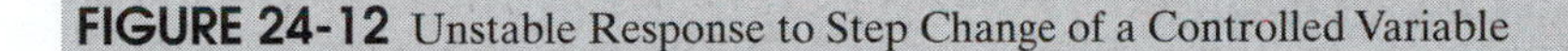

FIGURE 24-12 Unstable Response to Step Change of a Controlled Variable

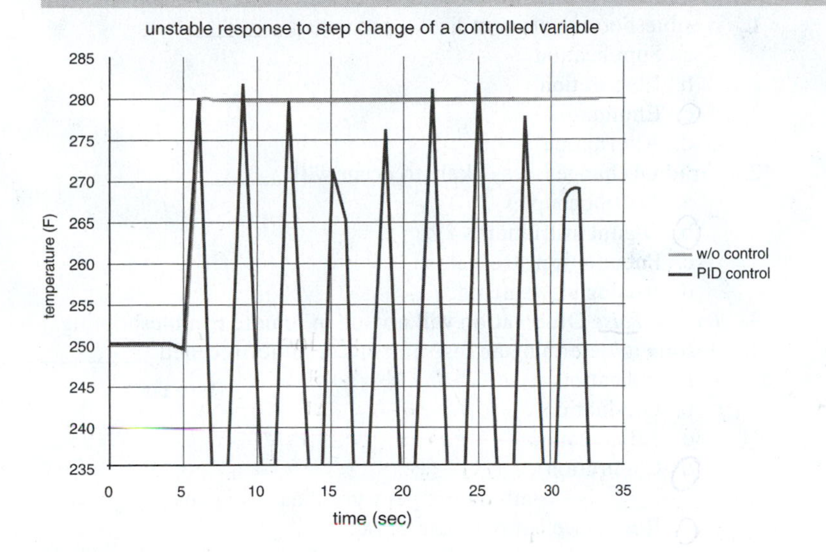

CONTROL VALVE

A control valve may be the source of loop instability. Highly viscous liquid service or slurry service may cause plugging problems. This may force the control valve to open more than would be required under normal conditions. Severe plugging may create the illusion that a control valve is too small. This will not normally affect the stability of the control loop. However, in extreme cases, a control valve may not be able to respond adequately to the needs of the control loop. In such cases, the loop will be off setpoint all the time and the controller will be unable to control the process. In addition, if plugging or overtightened packing causes the valve stem to stick, the valve will constantly be sticking and freeing itself. Air leaks on the pneumatic valve operators or positioners often cause the valve to be erratic. Any of these problems could cause the control loop to be unstable.

A useful rule of thumb says that control valves should normally operate between 15 percent and 85 percent open. If a valve is constantly at 90 percent open it is plugged or it is too small for the service. At the other extreme, if a valve is constantly at 10 percent open, it is too large for the service or the trim has worn severely. In both cases, the valve flow characteristics should be changed. This may require a complete new valve or possibly a simpler (and cheaper) change of the internal trim package.

Summary

Process technicians rely upon instrumentation to monitor and control all process variables to maintain desired targets and ensure safety. Improperly functioning instrumentation may fail to indicate a problem when one actually exists, or may send up false indications when no problems exist.

In the event of an instrument malfunction, process technicians need to know how to identify the malfunction, isolate it using a process of elimination, and determine the possible cause.

Failures can be of many types and causes. The changes indicated during a failure may be abrupt, gradual, fixed deviation, or highly fluctuating (as is the case with "noisy" signals).

Different types of sensor or transmitter elements (those that measure pressure, temperature, level, flow, and analytical data) have different failure modes, each with their own unique characteristics. It is important for technicians to be familiar with the types of process deviations being encountered and how these deviations might relate to various components such as transmitters, impulse lines, controllers, and final control elements.

Checking Your Knowledge

1. Troubleshooting often utilizes a process of ________.
   a. Supplication
   b. Elaboration
   c. Elimination
   d. Alternation
2. Gradual change is less likely to occur with ________.
   a. Thermocouples
   b. Digital instruments
   c. Pneumatic instruments
   d. Analog instruments
3. *True or False* Observation will not aid in remote troubleshooting.
4. Testing to determine an instrument's accuracy is called ____________.
   a. Indication
   b. Graduation
   c. Adjudication
   d. Calibration
5. A ________ is a common temperature-measuring device.
   a. Resistance temperature device
   b. Temperature transmitter
   c. Heat indication device
   d. Temperature variance recorder
6. The backwards installation of an orifice plate will result in:
   a. A gradual deviation in the meter output signal
   b. An erratic output signal
   c. No change in the meter output signal
   d. A fixed error in the meter output signal
7. If the diaphragm in a pressure transmitter becomes severely coated, a sensor may register:
   a. A zero reading
   b. A high reading
   c. An erratic output signal
   d. A gradually increasing out signal
8. Controller failure most often stems from:
   a. Improper tuning
   b. Improper atmosphere
   c. Electrical spikes
   d. A plugged HP leg
9. *True or False* A sticking control valve can cause instability in a loop.

Student Activities

1. Look around your home or workplace and think of all the different instrument and control malfunctions you have experienced in either of these places (e.g., heating or air conditioning system failures, clothes dryer failures, and automobile system failures such as the exhaust gas sensor or oxygen sensor). Record these failures on a piece of paper and provide a description of each.

2. Utilizing a flow control loop, artificially cause the orifice plate to show erroneous flow. With the controller in AUTO, observe the valve response.
3. Discuss what could go wrong with a displacer.
4. Using the control loop in lab activity 1, disconnect the air supply and observe the loop response (output from the controller).
5. Loosen the wiring connections to and from a transmitter and observe the loop response.

Glossary

And/or A conditional logic statement used to control a process.

Accelerometer A vibration measuring device (e.g., Piezoelectric type).

Acceptable limits The operating range within which a piece of rotating equipment can operate without causing excessive wear to the bearings or other types of catastrophic failure.

Accuracy How close a measurement corresponds to its true value.

Actuator A device that provides motion to a valve for controlling purposes.

Alarm switch Used to notify an operator when a process variable enters an abnormal range (e.g., high or low) by triggering an alarm (a light or annunciator).

Algorithms Preset mathematical functions calculated in a controller that can be mechanical, analog, or digital. The three most common output functions deal with proportional (P), integral (I), and derivative (D) tuning.

Alternating current (AC) A type of electrical current that reverses direction at regular intervals or cycles (as opposed to direct current that does not reverse).

Analog Analogous or similar to something else; in an analog system, process variable measurements change in a continuous manner, mirroring the actual process variable.

Analog to digital (A/D) The conversion of an analog signal to a digital signal so it can be processed by a computer; also referred to as A to D.

Analytical The use of a logical technique to perform an analysis; in instrumentation, a measurement of a physical or chemical property.

Analyzer measurements Qualitative and quantitative.

Analyzer probe A device that is placed in a process stream to obtain specific data regarding the stream contents.

Annubar® tube A trademark name for one manufacturer's multiport tube having four impact points spaced across the pipe and facing the flow with another tube sensing static pressure; measures the average pressure produced by the four impact points.

Atmosphere(s) The pressure at any point in the atmosphere due solely to the weight of the atmospheric gases above the point concerned; 14.7 psia is at sea level.

Atmospheric pressure Weight of the air comprising the atmosphere; 14.7 lb per square inch is the basic reference point for pressure gauges.

Auto/manual switch A switch that allows a process technician to select either automatic or manual control from the front of a controller.

Automatic to manual and manual to automatic switching When controller action is moved from automatic to manual or vice versa by adjusting the setpoint of the controller to the actual controlled point and then switching the mode; or switching from automatic to manual by simply repositioning the mode selector.

Automatic transfer switches (ATS) An electromechanical device that switches an electric load between two separate power sources (primary and secondary) depending on power availability from these sources; it is typically configured to be in the normal position when primary power is available.

Autostart switch Where a predetermined process condition activates a switch that, in turn, triggers an autostart (turns the process to the ON position automatically) sequence.

Back pressure regulator A device used to regulate and/or control the pressure of a process fluid upstream of the device location.

Balloon The basic instrumentation symbol.

Bars Measurement of pressure equal to 0.987 atmospheres.

Basic control functions Sensing, measuring, comparing, calculating, correcting, and manipulating.

Basic equipment symbol Common equipment such as pumps, towers, furnaces, etc. are basic pieces of equipment for most processing facilities and have commonly recognizable, or basic, equipment symbols.

Battery An electrical device consisting of one or more cells that converts chemical energy into electrical energy and provides a steady-state of DC voltage.

Bimetallic strip Two dissimilar strips of metal bonded together that expand and contract at different rates when exposed to temperature change, causing a blending or rotating effect; used as the primary element in a temperature gauge or bimetallic thermometer.

Block flow diagram (BFD) A flow scheme in a simple sequential block form.

Body The housing component of a valve.

Bonnet The top portion of the valve that connects the valve to the actuator; it can be removed to allow entry into the valve body cavity; usually contains the packing box and stem mechanism.

Breaker A switchlike device in electrical panel boxes used to keep electrical current from exceeding the recommended load.

British thermal unit (BTU) The amount of heat required to raise the temperature of a pound of water 1 degree Fahrenheit.

Bubbler A special kind of head pressure measuring method allowing the measurement of head pressure in a liquid without the pressure sensor coming in contact with the process fluid.

Bump A process upset occurring when a controller is switched from auto to manual mode.

Bumpless transfer The act of changing the controller from manual to automatic (or vice versa) without a significant change in controller output.

Bypass A short patch cable or wire used to complete an electrical circuit, usually temporarily, for testing or diagnostic purposes (also called a "jumper").

Bypass switch Used to override the normal operation of a system or device.

Calibration The act of applying a known input span to an instrument and adjusting the device so that it provides an indication or output corresponding to the known values.

Capacitance transducer A device that contains a measurement diaphragm and capacitor plates; changes pressure measurement to an electronic signal.

Cascade control Employing two controllers so one process variable is controlled by controlling another; in this scheme one controller is the remote setpoint of another.

Cascade control loop Where the control loop is characterized by the output of one controller becoming the setpoint of another.

Celsius/centigrade Scale of measurement to determine temperature; freezing point of water is 0 degrees C and boiling point of water is 100 degrees C.

Charger A device used to restore batteries to a proper electrical charge.

Chart recorder A device that physically or electronically records data that process and instrument technicians can use to troubleshoot process trends and instrumentation problems.

Chromatogram A graphic record of the separated components.

Chromatography A process where molecular components of a liquid or gas are separated and identified by means of a tube called a column.

Closed control loop When a control loop has feedback (e.g., controller in automatic mode).

Color/optical analyzer A photometer that operates in the visible light spectrum (400–800 mμ); two types include either the visual color analyzer and the photometer or spectral color analyzer.

Combustion The rapid consumption of fuel resulting in its conversion to heat, light, and gases.

Comparator A component of a controller that compares the measurement to a predetermined setpoint.

Comparing, calculating, and correcting element The control loop component that receives the appropriate signal from the transmitter and compares the signal to a desired value (setpoint); if there is a difference, then the output of the comparison causes a calculation to be performed to cause a corrective response by the controller output signal to the final control element.

Conduction The transfer of heat through solid matter by moving from one molecule to the next.

Conductivity A measurement of the ability of a material to conduct an electrical current; the measure of the ability of a liquid (or any solution) to conduct electricity.

Conductivity analyzer A device that measures all ions in an aqueous solution.

Conductivity meter A device that measures the conductivity of a process sample by comparing it to a known value or standard cell.

Continuous level measurement Monitors all level points in the tank from the zero percent level (bottom of the measuring device) to 100 percent full.

Control loop A group of instruments working together to control a single process variable such as pressure, temperature, level, or flow.

Control mode The control action or the control algorithm response such as PID or a programmed function.

Control valve The most common final control element in the processing industry; has an actuating device mounted to it; drives the flow controlling mechanism (the plug or disc) in a valve.

Controlled variable A process variable that is sensed to initiate the control signal.

Controller An instrument that receives a signal from the transmitter and compares it to a setpoint, and produces an output to a final control element.

Controlling Keeping a variable at a specific quantity.

Convection The transfer of heat through the circulatory motion occurring in a fluid when there is a difference in temperature from one region to another; density differences between the hot and cold molecules causes convection motion in fluids.

Converter A device that changes or converts a substance or a signal from one state to another (for example, changing AC current to DC current).

Converting and transmitting element The control loop component that converts the sensed process variable and transmits the measured signal.

Converting device A device that receives information in one form of an instrument signal and changes it into another form of an instrument signal.

Cycling The moving or shifting of a process variable above and/or below the setpoint.

Data highway A series of distributed computer processors or "nodes" that are all interconnected via a computer network.

Dead time The elapsed time between the initiation of an input (measured, controlled, or manipulated variable), change, or other stimulus and the point at which the resulting response can be measured or observed.

Defeat *See* bypass.

Deluge system An emergency shutdown (ESD) system component that, when a gas leak is detected, sprays the area of leakage with large quantities of water to dilute the concentration of vapors so they are below the lower explosive limit (LEL), thereby reducing the potential of fire.

Demultiplexer (DEMUX) A device that separates single-stream, multiplexed signals back into their original multistream constituents.

Density Mass per unit volume.

Depth A distance from the surface of a liquid that extends downward.

Derivative action A controller output response that is proportional to the rate at which the controlled variable deviates from the setpoint.

Deviation The difference between the setpoint and the controlled variable under dynamic (changing) conditions.

Deviation alarm An alarm that is activated when the difference between two variables exceeds a set limit (for example, in a particular catalyst bed, the difference or deviation between the outlet temperature and the inlet temperature should not exceed 50 degrees F; in this example the actual temperature value is not important, what is important is the difference between the two values).

Device error The accuracy of the instrument (+/–) full scale.

Differential pressure (Δp, d/p) The difference between two related pressures; usually used in measurements of process variables (pressure, temperature, level, and flow).

Differential pressure (d/p) cell An instrument that measures the difference between two pressure points and produces a corresponding output signal for tank level.

Differential pressure (d/p) transmitter A device that responds to the pressure created by the primary device and then converts the pressure signal into a standard output signal that can represent either differential pressure or flow rate.

Differential pressure cell A simple mechanical type sensor found in pneumatic transmitters or a more sophisticated transducer type found in electronic transmitters; used to measure the difference between two pressure points and produces a corresponding output signal; may be used to measure level, flow rates, or differential column pressure.

Digital A transmission method that employs discrete electrical signals as opposed to continuous signals or waves.

Digital signal Characterized by data that is represented as coded information in the form of binary numbers; used to transmit data to and from field transmitters on a twisted pair of wires; may also be between computers and computer components.

Digital to analog (D/A) The conversion of a digital signal to an analog signal so it can be interpreted and reported on an analog gauge or instrument (e.g., using an I/P converter to convert a digital signal to a pneumatic signal that triggers the opening or closing of a control valve); also referred to as D to A.

Digital transmission A signal transmission method that uses digital devices to create a signal.

Direct acting controller A controller whose output signal value increases or decreases as the controller input signal value increases (increased input = increased output) or decreases.

Direct current (DC) Electrical current that flows in only one direction (as opposed to alternating current, which reverses directions at regular intervals).

Direct level measurement Measures the process variable directly in terms of itself (e.g., a sight glass or dipstick).

Disc or plug The only moveable component in the valve that is actuated to open or close the flow path through the valve.

Discreet sensing element A sensing element that can stand alone or is individually distinct; connected to the transmitter by sensor wires.

Discrete controllers Digital controllers that can accept and produce only digital input and output.

Displacer A sealed cylindrically shaped tube used to measure buoyancy.

Distributed control system (DCS) A manufacturing system (usually digital) consisting of field instruments connected to multiplexers and demultiplexers and A/Ds (analog to digital) via wiring or busses and which terminate in a human-to-machine interface device or console.

Drift A slow variation of a performance characteristic such as gain, frequency, or power output that can be attributed to factors like temperature and aging.

E/F conversion The conversion of a signal from voltage to frequency.

E/I conversion The conversion of a signal from voltage to current.

E/P conversion The conversion of a signal from voltage to pneumatic.

Electrical transmission A signal transmission method that uses electrical devices to create a signal.

Electrolyte A nonmetallic substance that is an ionic conductor; the greater a substance conducts an electrical charge, the more ionic the substance becomes.

Electromagnetic flowmeter (Magmeter) A magnetic flowmeter designed to determine volumetric flow of electrically conductive liquids, slurries, and corrosive and/or abrasive materials.

Electronic Powered by electricity.

Electronic signal Either an analog or digital signal; current or voltage signal.

Emergency shutdown (ESD) system A system consisting of interlocks, breakers, sensors, and other equipment responsible for shutting down equipment in an extremely abnormal situation in order to protect the equipment and reduce the risk of catastrophe.

F/E conversion The conversion of a signal from frequency to voltage.

F/I conversion The conversion of a signal from frequency to current.

Fahrenheit Scale of measurement to determine temperature; freezing point of water is defined as 32 degrees F and the boiling point of water is 212 degrees F.

Fail last (fail-in-place) The condition of the valve seat upon loss of instrument air or power failure.

Feedback control A closed loop control strategy where the difference in the setpoint and measurement drives the output of the controller to the final control element.

Feedback loop The most common type of control loop where the change caused by the output of the controller is fed back to the process providing a self-regulating action.

Feedforward control An open loop control strategy where the controller's output is based on knowledge (usually mathematical modeling) of the relationship between the output of the controller and its input received from the point in the process where the disturbance occurs.

Filled thermal system A temperature-sensing bulb filled with a liquid, vapor, or gas and connected by means of a capillary tube to a pressure-measuring element.

Final control element The last active device in an instrument control loop; directly controls the manipulated variable; usually a control valve, louver, or an electric motor.

Flat-glass level gauge There are three types of flat-glass level gauges: reflex, transparent, and welded pad. The heavy, thick-bodied glass promotes safety in high temperature and high-pressure hydrocarbon service.

Flatline A slang term sometimes used to describe the trend condition in which the process is in equilibrium and running smoothly, producing a relatively straight line.

Float Gauge An instrument that uses cables, pulleys, levers, or any other mechanism to convey the position of a float to a liquid level.

Flow A fluid in motion.

Flow measurement Flow rate (an instantaneous flow measurement) and total flow (a summation of instantaneous flow rates over a time interval or an accumulation of counts provided by a positive displacement device); measured in volume or mass units without respect to time.

Flow nozzle Similar to a venturi tube but has an extended tapered inlet commonly installed in a short piece of pipe called a spool piece.

Flow rate A measure of how much fluid moves through a pipe or channel within a given period of time.

Fluid Flowing liquids and gases that mix easily because of the continual movement between molecules.

Force A push or pull exerted on an object that causes the object to change direction.

Gain Change in output divided by the change in input.

Gas chromatograph analyzer An analyzer that provides the necessary means to accomplish the chromatographic separation and analysis.

Gas chromatography (GC) An analytical method that can provide a molecular separation of one or more individual components in a sample.

Generator A device that uses fuel or other mechanical energy to create electrical energy.

Glass-stem thermometer A temperature-measuring device constructed of a glass bulb and a capillary (small) tube made of glass with a numbered scale; the bulb and the capillary tube contain a liquid such as mercury or colored alcohol that expands whereby a reading can be made from the etched scale on the length of the tube.

Handwheel An actuator accessory used to manually override the actuator or to limit its motion.

Heat The energy that flows between bodies of differing temperatures.

Heat tracing The technique of adding heat to process piping, instrument piping or an instrument in order to keep the process fluids at a constant temperature so signal transmission is not impacted.

Height A distance from a zero reference point to the surface.

High (H) alarm The first alarms triggered when a process variable (e.g., fluids in a tank) rises above a predetermined high level; the purpose of this alarm is simply to notify an operator that the level is abnormal.

High high (HH) alarm Alarm triggered when a process variable (e.g., fluids in a tank) continues to rise above a pre-determined maximum level after a High (H) alarm has already been triggered; the purpose of this alarm is to notify a process technician that the level is becoming increasingly abnormal and to initiate a shutdown or other corrective action.

Hunting A condition that exists when a control loop is improperly designed, installed or calibrated, or when the controller itself is not properly aligned; this condition will be observed by the process technician as a random behavior or cycling above and/or below the setpoint (desired control point).

I/E conversion The conversion of a signal from current to voltage.

I/F conversion The conversion of a signal from current to frequency.

I/P conversion The conversion of a signal from current to pneumatic.

I/P transducer (current-to-pneumatic transducer) A device that converts a milliampere signal into a pneumatic pressure.

If/then A conditional logic statement used to control a process.

Impulse tubing A tube that is usually made of stainless steel and allows the process variable to be sensed by the sensor located in the transmitter.

Inches of mercury (Hg) Most common measurement scale for a manometer; measures the level of mercury in pounds per cubic inch.

Inches of water (H_2O) A very small measurement of pressure equal to 0.036 pounds per square inch per inch at 4 degrees Celsius.

Inches of water column Liquid head pressure measurement expression; also expressed as pounds per square inch (psi).

Indicating The showing of a current condition via a readout (e.g., digital or analog).

Indirect flow measurement Measuring one variable of a process to infer another.

Indirect level measurement Measures another process variable (e.g., head pressure or weight) in order to infer level.

Innage The measurement from the bottom of a tank to the surface of the product.

Input The process of entering data or information into a computer system or program.

Instrument scale A range of ordered marks that indicate the numerical values of the process variable.

Instrumentation, Systems and Automation Society (ISA) A global, nonprofit technical society that develops standards for automation, instrumentation, control, and measurement.

Insulation A protective covering placed around the steam and electric tracing tubes to prevent heat loss and to protect personnel from contact with hot surfaces.

Integral action (or reset) A controller output response that is proportional to the length of time the controlled variable has been away from the setpoint.

Integrally mounted sensing element Where the sensing element is a physical part of the transmitter.

Interface level A plane where two materials meet; can be between two liquids, a liquid and a slurry, or a liquid and foam.

Interlock A shutdown circuit designed to detect dangerous conditions and safely shut down a process or a piece of equipment.

Inverter A device used to change direct current (DC) into alternating current (AC).

Jumper *See* bypass.

Kelvin Scale of measurement that is sometimes called an absolute scale because 0 K is the point at which no heat exists; freezing point of water is 273 K and boiling point of water is 373 K.

Ladder diagrams Diagrams that guide electricians in the fabrication process.

Lag time A relative measure of the elapsed time between two events, states, or processes.

Laminar flow Smooth-flowing fluid.

Lead/lag control Describes how a process reacts to a disturbance or other manipulating conditions.

Legend An explanation of what the symbols and codes on a drawing represent; usually located on an individual drawing in a framed area or on a page within a set of drawings.

Level The height of a liquid in reference to a zero point.

Level measurement The act of establishing the height of a liquid surface in reference to a zero point.

Limit switch Used to verify the state or presence of a condition that exists in the process.

Line symbols Connectors between the basic pieces of equipment without which process streams could not be moved.

Linear scaling A linear relationship between two scales (input versus output).

Liquid (hydrostatic) head pressure The pressure exerted by the height of a column of liquid; the most common indirect level measurement.

Live zero A standard bias has been added to the instrument signal (e.g., pneumatic 3–15 psig or electronic 4–20 mA); instead of reading zero the reading is 3 psig or 4 mA.

Load cell A device comprised of a strain gauge bonded to a robust support column called a force beam; together these are capable of supporting the entire weight of a vessel and its contents; a transducer that measures force or weight.

Load change A change in any variable in the process that affects the value or state of the controlled variable.

Local Located at or near the point of measurement.

Local controller When the controller is physically mounted in the processing area near the other instruments in the loop.

Local indicator Instruments that are placed on equipment in the field only to be read in the field; may be used for comparison with transmitted instrumentation readings or, used on noncritical processes to indicate pressure values.

Local manual control The act of controlling a process variable by hand within the processing area.

Loop error The accumulated error of each device in the loop; calculated as the square root of the sum of the sum of the squares of individual device accuracy.

Louvers (dampers) Devices used to control airflow.

Low (L) alarm The first alarm triggered when a process variable (e.g., fluids in a tank) drops below a predetermined low level; the purpose of this alarm is to notify a process technician that the level is abnormal.

Low low (LL) alarm Alarms triggered when a process variable (e.g., fluids in a tank) continues to drop below a predetermined minimum level after a low (L) alarm has already been triggered; the purpose of this alarm is to notify a process technician that the level is becoming increasingly abnormal and to initiate a shutdown or some other corrective action (e.g., shutting off power to a pump to prevent cavitations, which could result in pump damage).

Lower range value (LRV) The number at the bottom of the scale

Manipulating element The final control element (e.g., control valve) is manipulated by the corrective response of the controller output so that the process variable is maintained at the appropriate setpoint value.

Manometer A gravity-balanced pressure-measuring device with two fluid chamber tube gauges connected by a U-shaped tube so fluid (a liquid or mercury) flows freely between the chambers.

Mass The amount of matter in a body or object; mass has to do with the amount of matter in a molecule rather than its size.

Mass flowmeter (coriolis) The most common type of true mass flowmeter; a meter that eliminates the need to compensate for typical process variations such as temperature, pressure, density and even viscosity.

Mass flow rate Where a solid material is weighed as it is conveyed on a moving belt (conveyor) and an instantaneous weight measurement is taken and the rate of motion of the belt is known.

Mass flow units Measure for the weight being passed through a certain location per unit of time; usually expressed in pounds per unit of time as in pounds per minute.

Mass spectrometer (MS) A device capable of separating a gaseous stream into a spectrum according to mass and charge.

Measured variable A process variable that is measured.

Mechanical link A way of mechanically transmitting the motion of a primary sensor to a controlling mechanism; conveys linear or rotary motion by using a pivoting crank.

Mechanical transmission A signal transmission method that uses mechanical methods to create a signal.

Meniscus The curved upper surface of a column of liquid having either a convex or concave shape.

Molecular speed Molecular speed changes as a result of numerous factors. Energy, normally in the form of heat energy, may change, which will increase or decrease molecular speed. Speed can also be affected by collisions with other molecules. The collisions may be with molecules from the material (for example, a product being agitated) or from other molecules (for example, the walls of a vessel).

Motor control centers (MCC) Hardwired relays, switches, or contacts, housed in large metal cabinets that control the starting, stopping, and sequencing of motors and other devices.

Multiplexer (MUX) A device that merges or interleaves multiple signals into one stream or output signal in such a way that each individual signal can be recovered or separated out using a demultiplexer.

Multivariable input Industrial processes that involve an interaction between two or more process variables.

Node A single computer terminal located on a computer network or "data highway."

Noise Disturbances that affect a signal and may distort the information carried by the signal.

Noisy signal A signal that fluctuates dramatically that is most likely the result of a lose sensor connection.

Nuclear device Uses a tightly controlled gamma radiation source with a detector to infer a level in a tank and used when other technologies are unsuccessful; located on the outside of a tank and impervious to the effects of adverse process conditions; common gamma radiation sources are the radioisotopes cobalt 60 and cesium 137.

Offset The difference between the setpoint and the controlled variable under steady-state (not changing) conditions.

On/off A conditional logic statement used to control a process.

On/off control Control used when controlled variable cycling above and below the setpoint is considered acceptable, or when the process quickly responds to an energy change, when the energy change itself is not too quick.

Inline analyzer An analyzer (either continuous or contiguous) installed directly in a process line that may be either a sample system or of the newer probe type.

Opacity analyzer An optical analyzer used to determine how much particulate matter is in a gas sample.

Open control loop When a control loop does NOT have feedback (e.g., controller in manual mode).

Operating range One number that is the difference between the upper and lower range values (URV and LRV) on a scale.

Optical measurement A measurement that uses reflection, refraction, or absorption properties of light to measure the chemical or physical properties of a sample.

Optical meter Any or all of the following: turbidity meters, opacity meters, and color meters.

Orifice flange A flange with holes drilled in the flange through to the pipe to allow pressures upstream and downstream of an orifice plate to be measured.

Orifice plate A piece of 1/8-inch to 1/2-inch-thick metal with a calibrated hole drilled (or cut) through; types of orifice plates include the concentric plate, eccentric bore, and segmental plate.

ORP meter An electrochemical analyzer that measures a small electromotive force (EMF) across a hydrogen-ion-sensitive glass bulb; an ORP (oxidation reduction potential) meter also measures a small voltage (EMF) across its electrode and has the capability of detecting all oxidizing and reducing ions in the solution.

Output The number or value that comes out from a computer program or process.

Over range The condition that exists when the signal value of a device or system exceeds the maximum allowable value.

Override controller A device or program that remains inactive until a specific constraint (highest or lowest permitted extreme) on the measured variable is about to be reached.

Overspeed A dangerous condition that can occur in a turbine or other type of equipment that moves too fast.

Oxidation reduction potential (ORP) Measures redox potential created by the ratio of reducing agents to oxidizing agents present in the sample.

P/I conversion The conversion of a signal from pneumatic to current.

Parallel data communication One wire per bit or 64+ wires for a 64-bit binary word; used primarily in short distances (a few feet).

Percentage flow rate A common way to indicate a flowing process with 100 percent equating to an actual quantity such as gallons per minute (gpm).

Percent level measurement A measurement of level based on percentage with zero percent level at the bottom of the measuring device and 100 percent level at the top.

Permissives A set of conditions designed to ensure safe operations that must be met before a piece of equipment can be turned on.

pH A measurement of the hydrogen ion concentration of a solution indicating how acidic (below 7.0) or basic (above 7.0) a substance is from neutral (7.0).

pH meter An electrochemical instrument that measures the acidity and alkalinity of a solution; an analyzer that measures a specific ion (e.g., hydrogen) concentration in a solution.

Photometer A color analyzer that operates in the visible light spectrum (400–800 μm).

Photometry system An automated system used to determine the color of a sample.

Piping and instrumentation diagram (P&ID) Contains more detail than a PFD to include piping and instrumentation details and the entire control system.

Pitot tube An L-shaped tube that is inserted into a pipe with its open end facing the flow and another tube sensing static pressure.

Pneumatic Powered by a gas.

Pneumatic signal An instrument communication with a range of 3–15 psig; must have an air supply; has a lag time associated with the signal; relatively short transmission distances.

Pneumatic transmission A signal transmission method that uses compressed air/gas to create a signal.

Point level measurement Level is measured at one distinct point in a tank.

Positive displacement flow measurement When flow if measured in absolute volumes where the flowing materials is admitted into a chamber of known volume and then transferred to a discharge point; a counter registers the number of times the chamber fills and discharges.

Positive displacement meter Piston, oval-gear, nutating-disk, and rotary-vane types of positive displacement flow meters.

Pounds per square inch (psi) Liquid head pressure measurement expression; also expressed as inches of water column.

Precision How close repeated measurements are versus the action; reproducibility; the closeness of repeated measurements of the same quantity; the agreement between the numerical values of two or more measurements made in the same way and expressed in terms of deviation.

Pressure A force per unit of area (force ÷ area = pressure).

Pressure gauge The most common instrument for measuring pressure; the three primary types of pressure gauge measuring scales are absolute, gauge, and vacuum.

Pressure measurement (bars) Measurement of pressure equal to 0.987 atmospheres.

Pressure-reducing regulator A device used to regulate and/or control the pressure of a process fluid downstream of the device location.

Primary controller Perceives the secondary loop as a separate entity; responds to one process variable while the secondary controller responds to another.

Process analyzer An unattended analytical instrument that is capable of continuously monitoring a process stream.

Process control The act of regulating one or more process variables so that a product of a desired quality can be produced.

Process equilibrium The condition that exists when there is a balance of material and energy within a given system or process; also referred to as "steady state."

Process error The difference between setpoint and process variable (SP – PV).

Process flow diagram (PFD) A pictorial description of an actual process including the major process equipment while providing process information including the heat and material balances; usually developed when initiating the design of a new plant.

Process variable (PV) Measured property of a process.

Process variable switch A type of switch that actuates when a predetermined value of a process variable (e.g., pressure, temperature, level, flow, and analytical) is present.

Programmable logic controller (PLC) A computer-based controller that uses inputs to monitor processes and outputs to control processes.

Properties Characteristics of a substance such as pH, ORP, conductivity, optical measurement, composition, and/or combustion capability.

Proportional action A controller output response that is proportional to the amount of deviation of the controlled variable from the setpoint.

Proportional band (proportional gain) The amount of deviation of the controlled variable from the setpoint required to move the output of the controller through its entire range (expressed as a percent of span).

Proximity switch A type of switch that requires the presence of an object or device to facilitate its operation such as a magnetic coupling or decoupling.

psia Pounds per square inch absolute; the absolute scale at zero would be no pressure at all.

psig Pounds per square inch gauge; pressure measurement that references 14.7 psia as its zero point.

Qualitative A measurement of the properties of a substance.

Quantitative A measurement of the amount of properties within a substance.

Radiation The transfer of heat through emitting radiant energy in the form of waves or particles.

Rankine An absolute scale of measurement to determine temperature; freezing point of water is 492 degrees R and boiling point of water is 672 degrees R.

Rate action The faster the rate of change of the process variable, the greater the output response (derivative action).

Ratio The proportion of flow between two separate flowing streams entering a mixing point.

Ratio control Control used in blending raw materials to make a final product and in proportioning the flow rates of reactants feeding into a reactor; also used to maintain the proper fuel-air ratio feeding into a furnace.

Ratio control loop Control loop designed to mix two or more flowing streams together while maintaining a quantitative ratio between them.

Ratio controllers Controllers that are designed to ratio (or proportion) flow rate between to separate flows entering a mixing point; designed so that its output represents the exact flow rate needed by the controlled flow loop to remain in alignment with the desired ratio to the uncontrolled flow; may also be capable of receiving two separate flow inputs and ratioing its output to the control valve located in the control line.

Recording The keeping of historical data by making a physical record.

Rectilinear speed Linear speed expressed in distance per unit of time (e.g., feet/second).

Redundancy A design feature that provides more than one function for accomplishing a given task so the failure of one function does not impair the system's ability to operate.

Reflex level gauge Has a single flat-glass panel capable of refracting light off a prismlike backside creating a silvery contrast above the liquid level that allows it to be seen from a distance.

Regulator A self-contained and self-actuating controlling device used to regulate variables such as pressure, flow, level, and temperature in a process.

Relay A device that "boosts," maintains, or controls the flow of a signal so it can be properly received.

Remote Located away from the point of measurement (usually in a control room).

Remote automatic control The act of controlling an instrument loop remotely from a control room.

Remote controller Any controller that is not located in the processing area with the transmitter and control valve.

Remote manual control The act of controlling a valve manually from a remote location, such as a control room.

Remote setpoint (RSP) A setpoint received from an external source.

Representative sample A sample that contains portions of the process stream combined together over time or distance.

Resistance temperature device (RTD) Primary element that measures temperature changes in terms of electrical resistance.

Reverse acting controller A controller whose output signal value decreases as the controller input signal value increases (i.e., increased input = decreased output).

Reynolds number A mathematical computation describing the flow of fluids numerically.

Rotameter A direct-read variable area (tapered) flow tube where the fluid enters through the bottom, then flows upward, lifting a free-floating indicator plummet (float); the position of the float references to the calibrated marks on the glass tube to indicate flow rate.

Rotational speed Number of revolutions per unit of time (e.g., revolutions per minute or rpms).

Sample system The various components of a sampling apparatus that obtains, transports and returns the sample with the analyzer itself conditioning and analyzing the sample.

Scaling The act of equating the numerical value of one scale to its mathematically proportional value on another scale.

Seat The stationary part of the valve trim connected to the body that comes in contact with the valve plug; when the valve plug is fully seated, the flow through the valve ceases.

Secondary controller A special type of controller sometimes called a cascade controller or remote setpoint controller; a standard controller with the added capability of choosing to receive a remote setpoint from an external source or a local (internal) setpoint.

Sensing The act of detecting.

Sensing element The control loop component that detects, or senses, the process variable.

Sensor Detects the process variable; can be an integral part of a transmitter.

Serial data communication Two data wires; the most common means of communication used between plant equipment.

Setpoint The desired process value.

Setpoint knob The mechanism by which a technician could manipulate the setpoint.

Shelf position The contact position of an electrical device when de-energized (for example, a "normally open" switch is normally open when it is de-energized).

Shielding A technique used to control external electromagnetic interference (EMI) by preventing transmission of noise or static signals from the source to the receiver; consists of foil, mesh, or woven wire.

Shutdown switch Used to actuate a circuit that shuts down a process.

Signal converter transducer Part of a transmitter that effectively converts the process variable into a standard instrument signal; a device that converts one energy form into another.

Smart instruments Instruments that have one or more microprocessors or "smart chips" included in their electronic circuitry so they may be programmed and have diagnostic capability.

Span The algebraic difference between the URV minus the LRV of a scale; expressed as one number.

Spanning The process of setting or calibrating the lower and upper-range values of a device; also the process of testing a device or system by traversing up and down the scale through the entire calibrated or operational range.

Specific gravity The density of a substance relative to the density of water.

Spectrogram A graph where transmittance is plotted with respect to wavelength.

Spectrometer analyzer An analyzer used to detect chemical components in a process sample by measuring variations in transmittance (or absorption) of a spectrum of light passed through the sample; may use visible light (VIS), ultraviolet light (UV), or infrared light (IR and NIR) as a source and detection measurement.

Speed The distance traveled per unit of time irrespective of direction (e.g., feet/second).

Speed monitor A device that measures speed; comprised of a speed sensor and a readout and receiving device.

Split-ranged controller Where the output signal is divided between two final control elements.

Spring The device that provides the energy to move a valve in the opposite direction of the diaphragm loading motion so that the valve can be opened and closed proportionally with the instrument signal; also provides energy to return the valve back to its fail-safe condition.

Standard signal The language that instruments use to communicate between one another (4–20 mA, 3–15 psif, or digital).

Static switch A switch within an uninterruptible power supply (UPS) that automatically switches a piece of equipment from primary (line) power to secondary (UPS battery backup) power in the event of a power failure or voltage decrease.

Steady state A trend condition in which the process is in equilibrium and running smoothly producing a relatively straight line on a graph or chart.

Stem The pushing rod that transfers the motion of the actuator to the valve plug..

Strain gauge transducer A pressure-measuring device consisting of a group of wires that stretch when pressure is applied, creating resistance, thereby changing the process pressure into an electronic signal.

Switch A mechanical or electrical device that is used to operate or energize mechanical or electrical circuits for alarm, shutdown, or control purposes using a predetermined operating point or setpoint.

Symbology Various graphical representations used to identify equipment, lines, instrumentation, or process configurations.

Tape A narrow strip of calibrated ribbon (steel) used to measure length.

Tape gauge A level-measuring device consisting of a metal tape that has one end attached to an indicator and the other end attached to a float.

Temperature A specific degree of hotness or coldness as indicated on a calibrated scale; the measurement of the average kinetic energy of the molecules of the substance being measured.

Temperature differential (delta) Difference between two temperatures, normally across two points in a control loop.

Temperature gauge An independent device with a sensing element such as a bimetallic strip, bourdon tube, or bellows that is linked to a pointer displaying the temperature on a calibrated face.

Thermistor A type of resistor used to measure temperature changes, relying on the change in its resistance with changing temperature.

Thermocouple Primary element consisting of two wires of dissimilar metals connected at one end.

Thermowell A thick-walled, typically stainless steel device shaped like a tube that is inserted into a hole in piping or equipment specifically prepared to house a temperature-sensing and measuring element.

Total carbon analyzer An analyzer used to determine how much carbon is in a sample; used to detect carbon-based contaminants in steam condensate and wastewater.

Transmitter A device that transmits a signal from one device to another.

Transmitting Communicating via a signal from one place to another.

Transparent level gauge Has two transparent flat-glass panels, front and back, that form a vertical chamber in conjunction with the metal sides where the process level can be seen and measured using this vertical chamber.

Trend The plotting of a process variable over time.

Trending The analysis of a change in measured data over at least three data measurement intervals.

Troubleshooting The process of systematically examining, localizing, and diagnosing equipment malfunctions or anomalies.

Tubular-type sight glass A reflex or clear glass tube open into a vessel on top and bottom.

Tuning Adjusting the control action settings so that they produce an appropriate dynamic response to the process resulting in good control.

Turbidity analyzer An optical analyzer used to determine the cloudiness of a liquid.

Turbidity and opacity meters Optical meters; measure the transmittance or absorption of light passing through a sample.

Turbine meter A flow tube containing a free-spinning turbine (fan) wheel where the revolutions per minute (rpm) are proportional to flow rate; the faster the flow, the faster the turbine spins; the output is fed directly into an instrument to indicate either flow rate or total flow.

Turbulent flow Fluid flowing with turbulence.

Ullage (outage) The measurement from the surface of the product to the top of the tank.

Ultrasonic and radar device An accurate distance measuring instrument usually inserted into the top of a vessel; emits a pulse of energy that is reflected off the surface of the material back to the receiver; may be either a single unit (transponder) or two separate units (transmitter and receiver).

Uncontrolled flow Flow in a process line where there is no control valve.

Uninterruptible power supply (UPS) A backup power unit, usually consisting of large batteries, a rectifier, inverter, battery charger, and static switch, that provides continuous auxiliary power when the normal power supply is interrupted.

Upper range value (URV) The number at the top of the scale.

UPS alarm An audible or visual signal used to draw attention to problem situations.

Vacuum Where the pressure measured is less than atmospheric pressure; usually measured in inches of mercury (Hg).

Valve positioner A device used to make the valve position match the controller output signal by positioning the moving parts of a valve in accordance to a predetermined relationship with the instrument signal received from the loop controller; may also be used to adjust the position of the valve according to the specific needs or change the amount of signal needed to fully stroke the valve as in a split-range application.

Velocity Speed with a specific direction.

Venturi tube A primary element used in pipelines to create a differential pressure such that when converted to flow units (e.g., gpm), the flow in the line is measured.

Vibration The periodic motion of an object.

Vibration alarm An alarm designed to protect vibrating equipment, which is triggered if the level of vibration increases above an acceptable level.

Vibration meter A device used to measure displacement, velocity, or acceleration due to vibration;

consists of a pickup device, an electronic amplification circuit, and an output meter.

Vibration sensor or monitors A device used to sense the effects of vibration by sending a signal to a meter or monitor, or to shut down a device if operating limits are exceeded.

Vibration switch Used to determine the velocity, acceleration, displacement or any combination of these characteristics, for the purpose of predicting wear or impending failure.

Visual color analyzer Compares a physical sample with a standard by shining visible light through the sample; the color comparison is made by the human eye.

Volumetric flow units Measure for the volume being passed through a certain location in a pipe or channel per unit of time; usually measured in gallons per minute and cubic feet per minute.

Voting logic Computer logic that analyzes the signals from several devices, all of which are monitoring the same condition; it will initiate shutdown if a majority of the monitoring devices signal a dangerous condition.

Welded pad gauge Generally made of flat glass and integrally mounted to the vessel by either a welded or flanged connection with threaded pipe; extremely rugged to prevent vessel drain out.

Windup The saturation of the controller output signal to its minimum or maximum value if the process variable does not return to the setpoint.

Zeroing out Adjusting a measuring instrument to the proper output value for a zero measurement signal.

Index

G

H

I

K

L

M

S

T

U

V

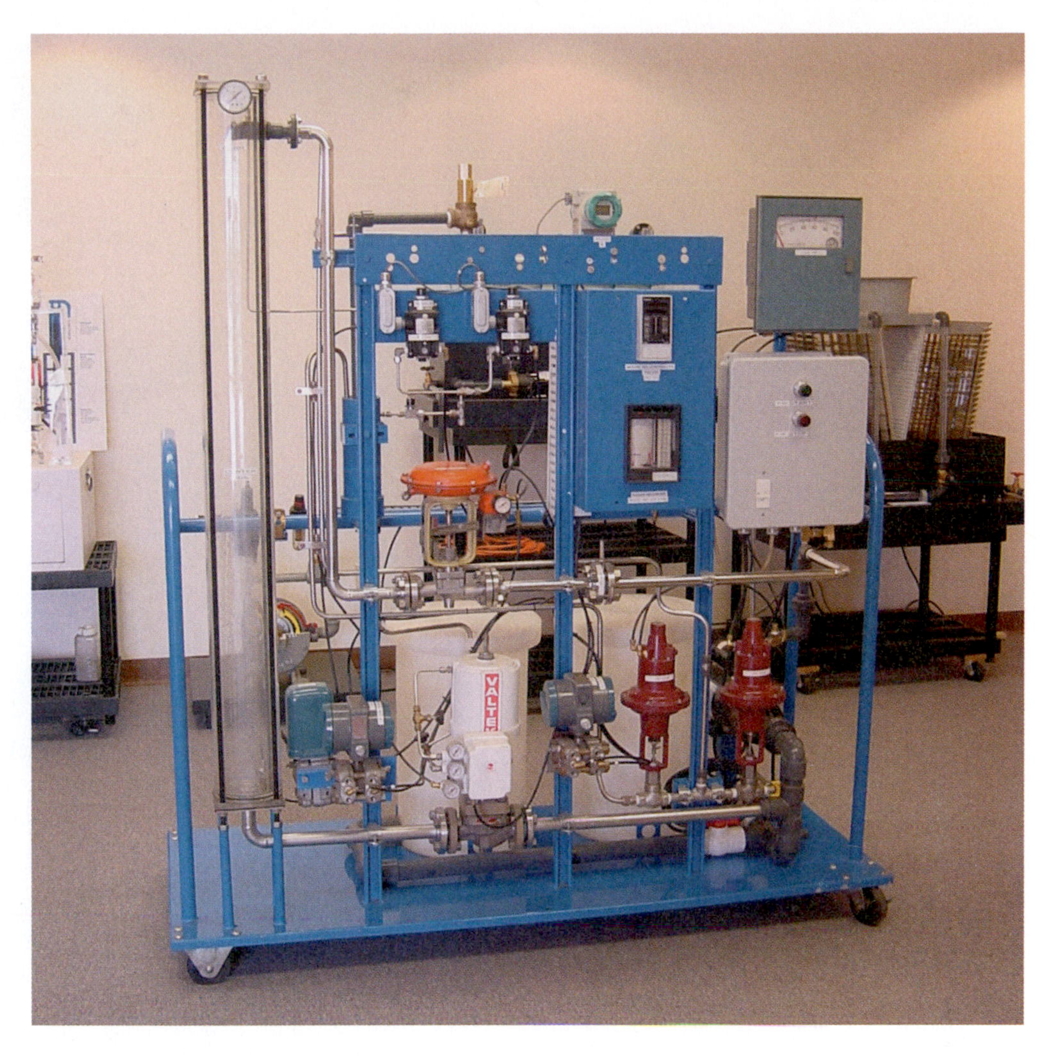

Basic instrument training unit
Courtesy of Bayport Training and Technical Center

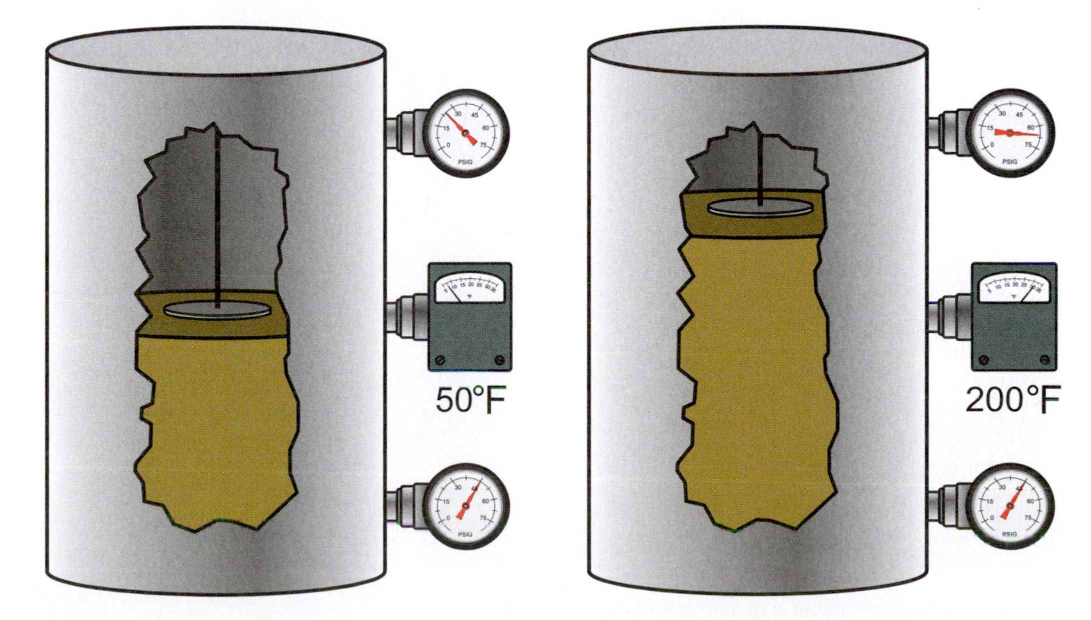

Measurement accuracy

Dual gauge in bars

Pressure gauge

Pressure gauge used in a sanitary process
Courtesy of Winters Instruments

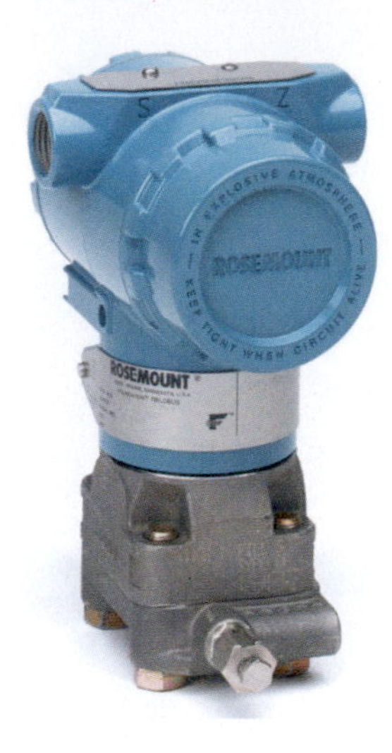

Pressure transmitter
Courtesy of Emerson Process Management

Local pressure display with transmitter and digital display
Courtesy of Emerson Process Management

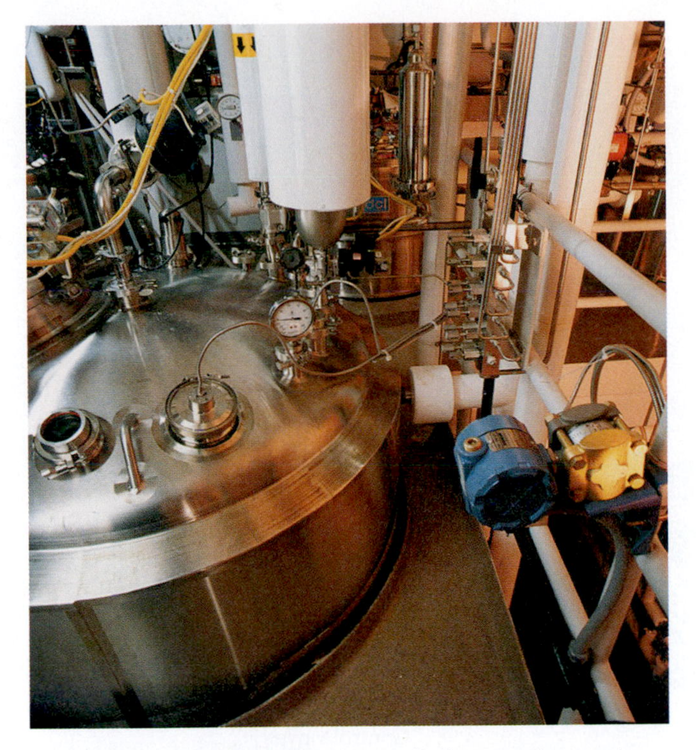

Batch reactor with associated pressure indicator and transmitter
Courtesy of Emerson Process Management

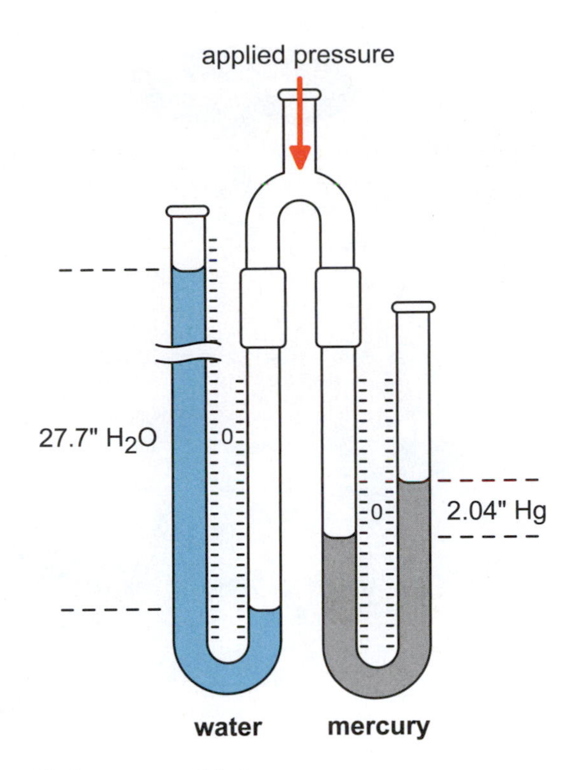

Inches water and inches mercury pressure

Digital temperature display
Courtesy of Emerson Process Management

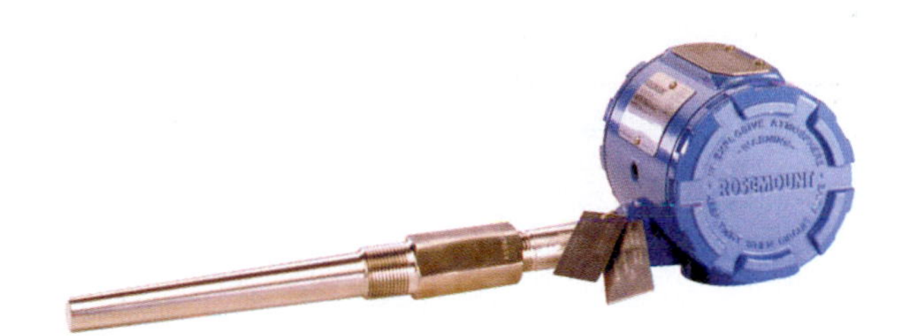

Multivariable temperature transmitter
Courtesy of Emerson Process Management

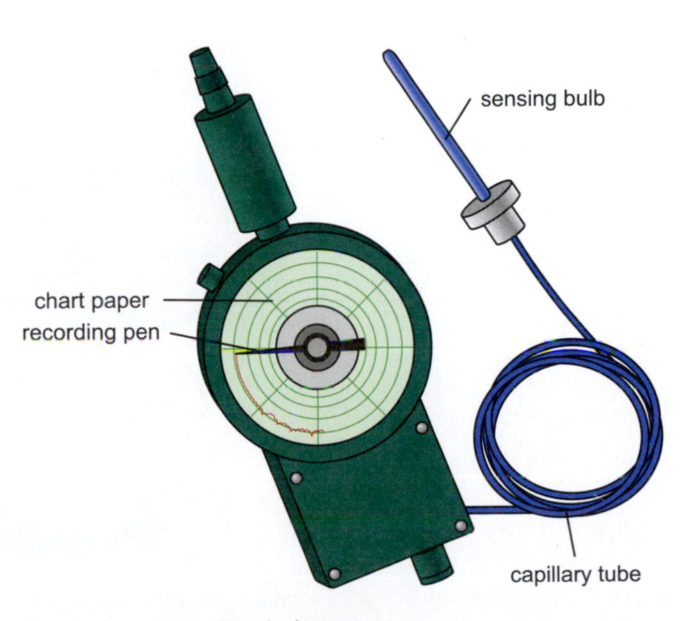

Temperature recording device

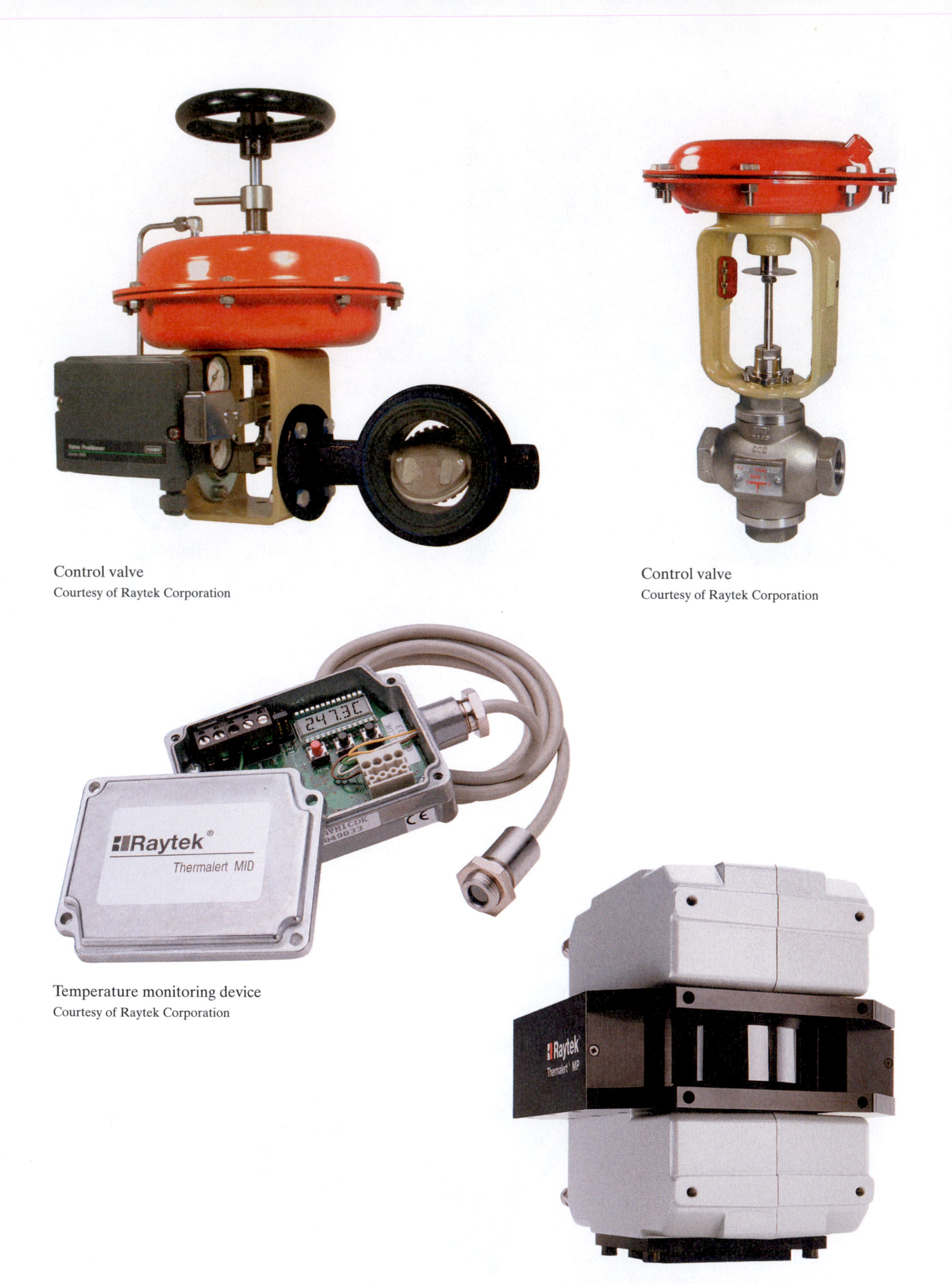

Control valve
Courtesy of Raytek Corporation

Control valve
Courtesy of Raytek Corporation

Temperature monitoring device
Courtesy of Raytek Corporation

Temperature monitoring device
Courtesy of Raytek Corporation

Level transmitter
with cable sensor
Courtesy of SOR, Inc.

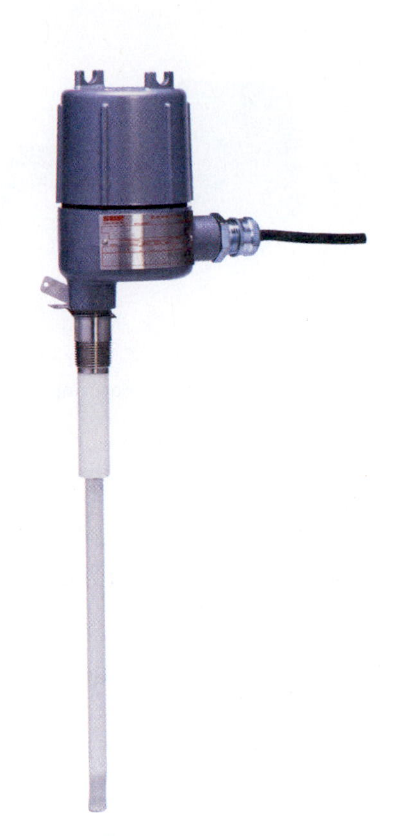

Level transmitter with
sheathed sensor
Courtesy of SOR, Inc.

Single point ultrasonic
level switch
Courtesy of SOR, Inc.

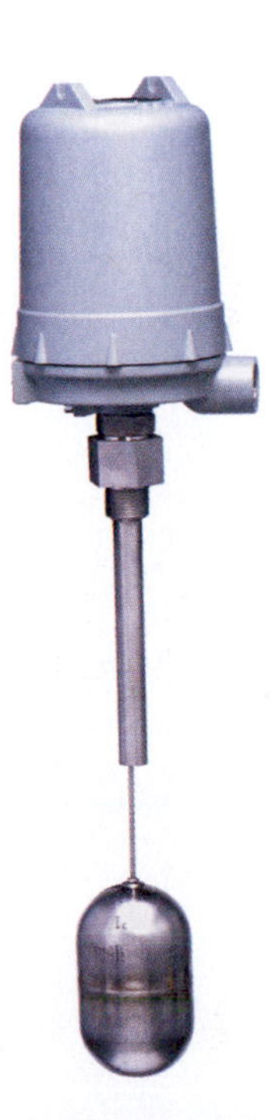

Top mounted float level switch
Courtesy of SOR, Inc.

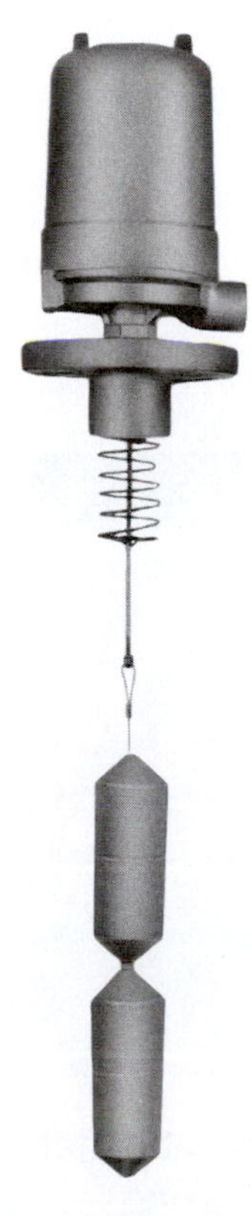

Top mounted displacer switch
Courtesy of SOR, Inc.

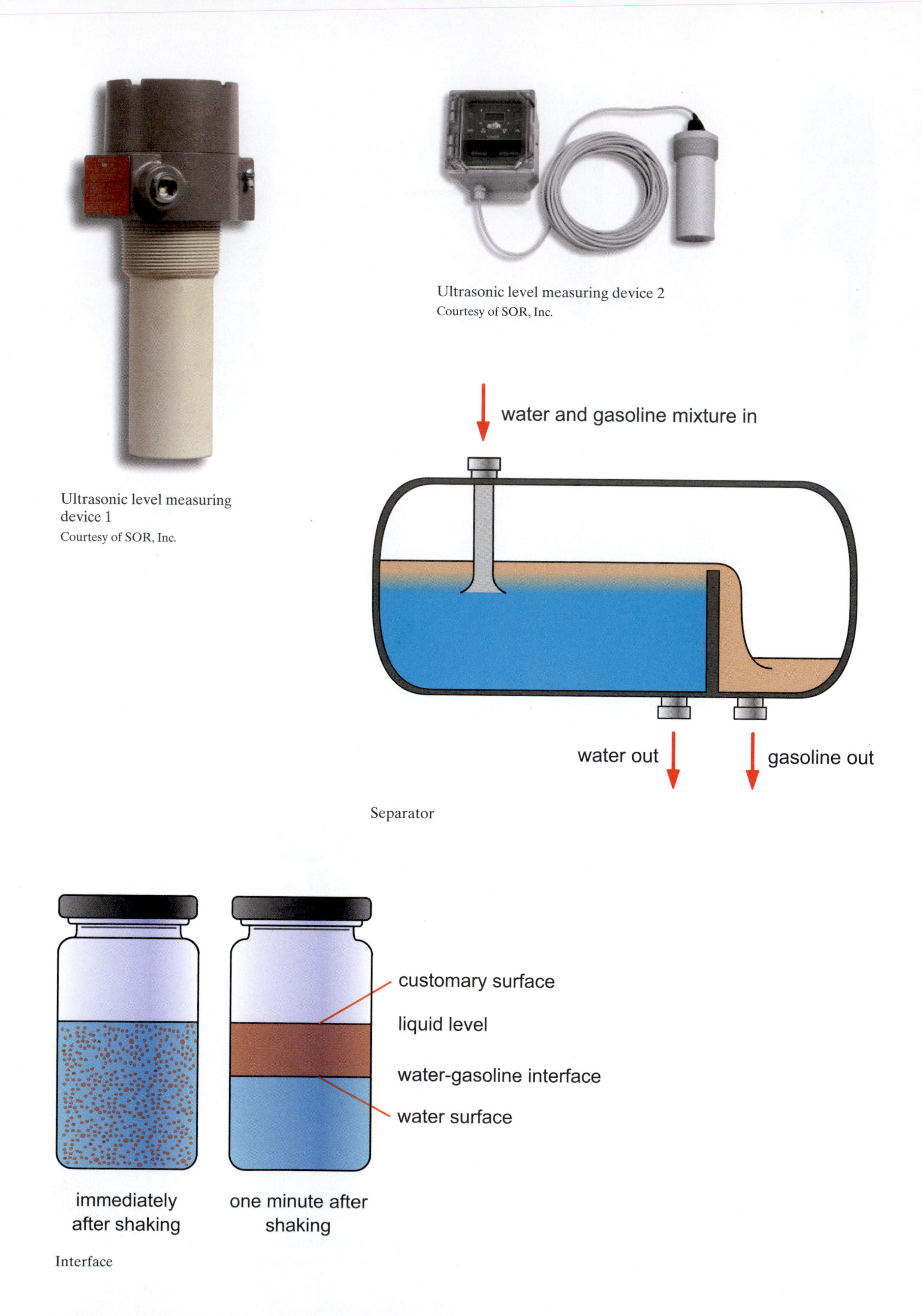

Ultrasonic level measuring device 2
Courtesy of SOR, Inc.

Ultrasonic level measuring device 1
Courtesy of SOR, Inc.

Separator

Interface

Rotameter

Flow meters and control valves on process streams
Courtesy of Lamar Institute of Technology

Signal Converter
(analog or voltage output)
Lock Nut
Pressure
Port Adapter
Temperature
Port Adapter
Housing
Retainer
Rings
Rotor
Supports
Turbine
Rotor

Flow meter with signal converter
Courtesy of Hedland Division of Racine Federated Inc.

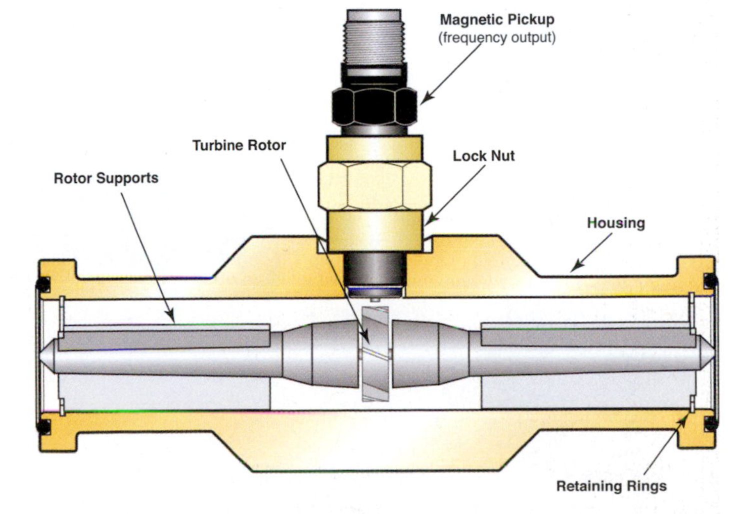

Flow meter with magnetic pickup
Courtesy of Hedland Division of Racine Federated Inc.

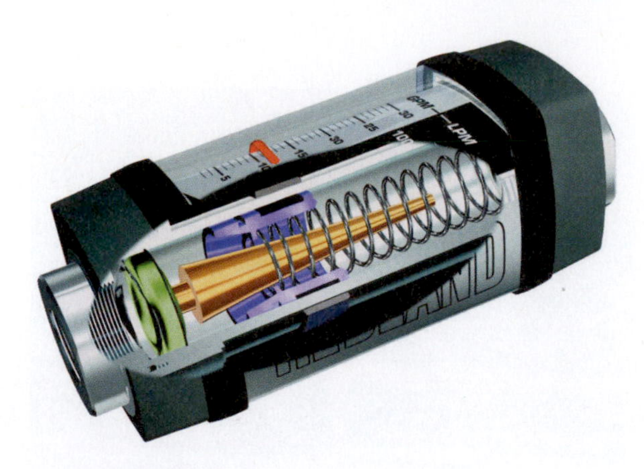

Flow meter
Courtesy of Hedland Division of Racine Federated Inc.

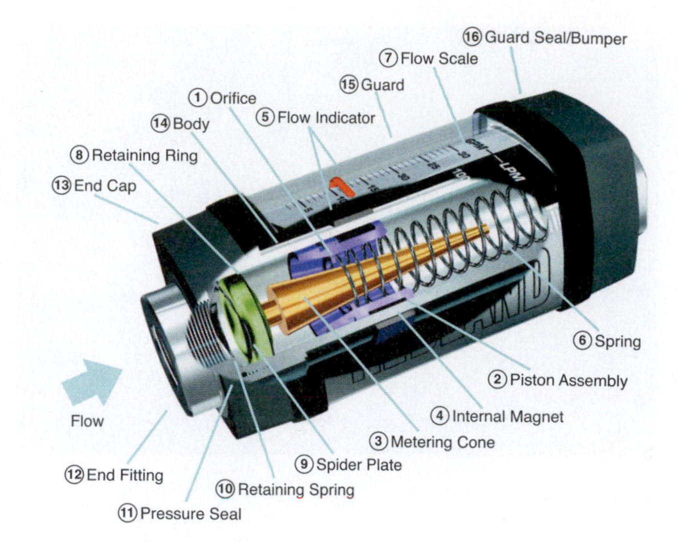

Flow meter cutaway
Courtesy of Hedland Division of Racine Federated Inc.

Magnetic flow meter
Courtesy of Emerson Process Management

Mass flow meter
Courtesy of Emerson Process Management

Flow meter with digital display
Courtesy of Emerson Process Management

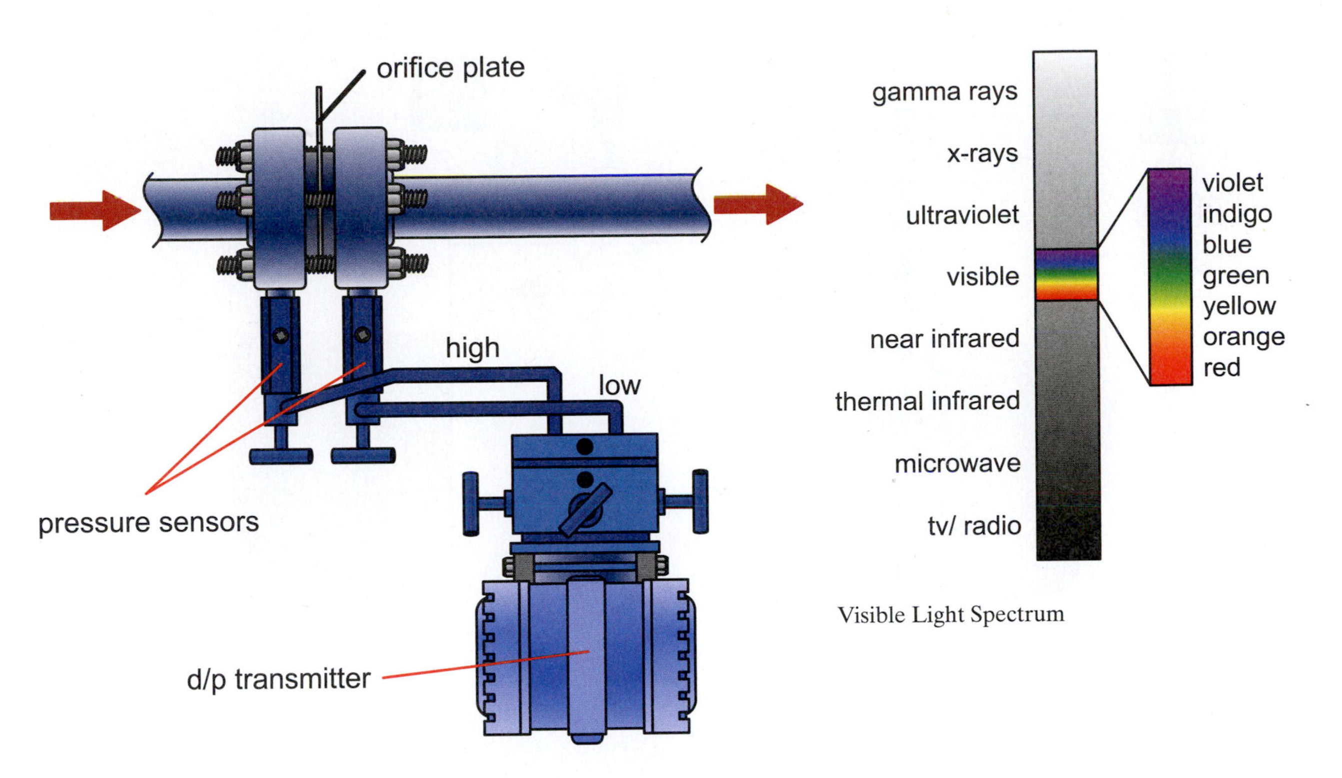

Visible Light Spectrum

Differential Pressure (D/P) Transmitter

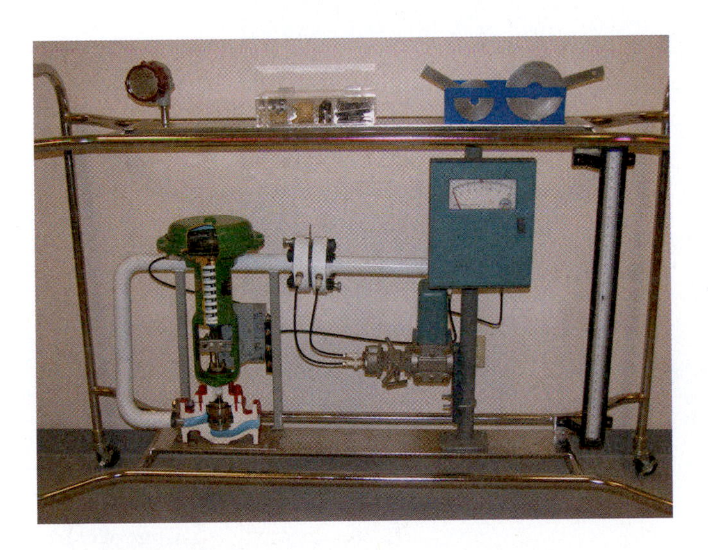

Control loop training unit
Courtesy of Bayport Training and Technical Center

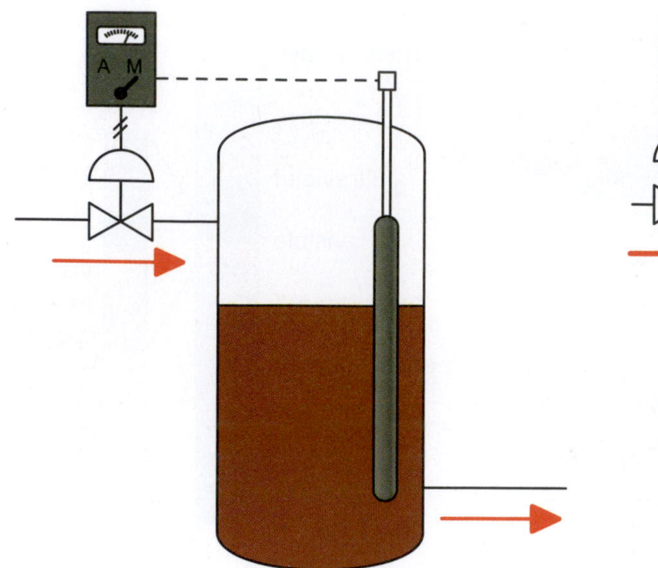

Open-loop control

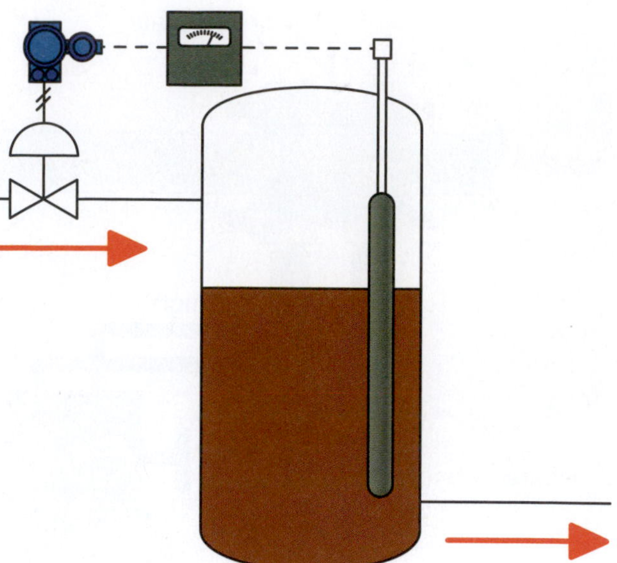

Closed-loop control (feedback)

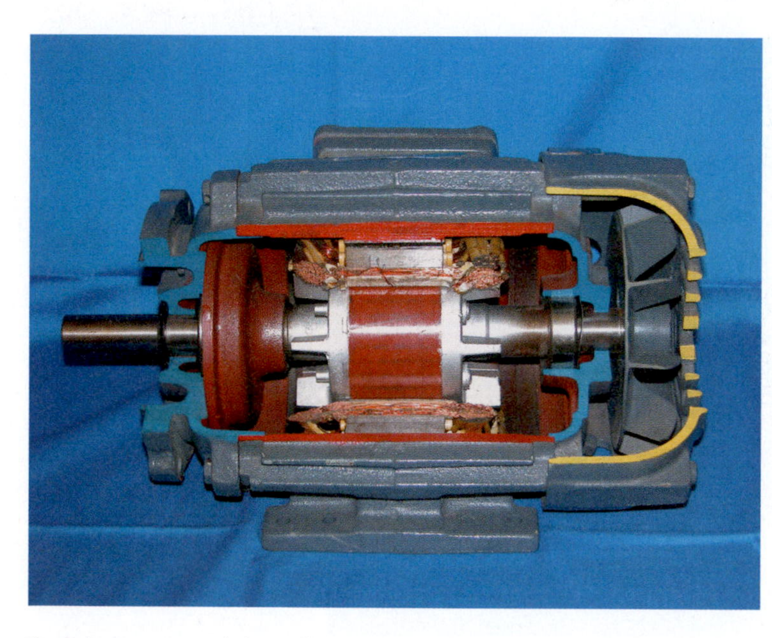

Basic instrument training unit
Courtesy of Bayport Training and Technical Center

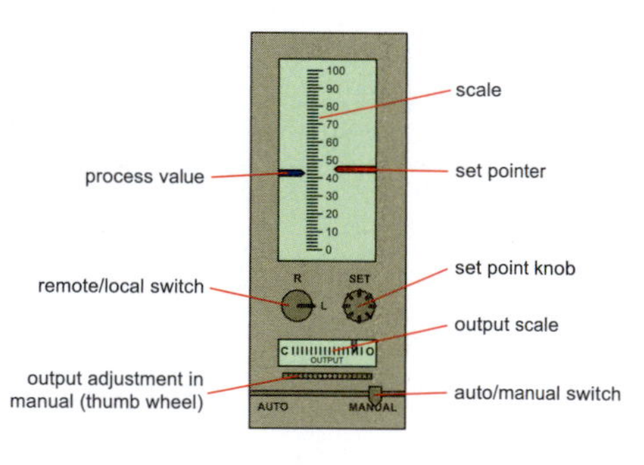

Controller front panel

Rotary plug control valve
Courtesy of Emerson Process Management

Control valve with valve positioner 2
Courtesy of Emerson Process Management

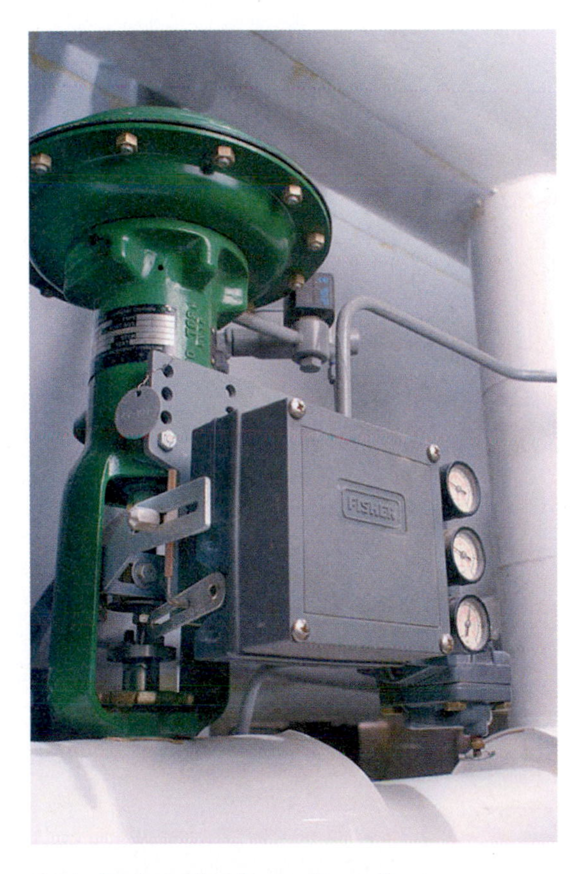

Control valve with valve positioner 1
Courtesy of Emerson Process Management

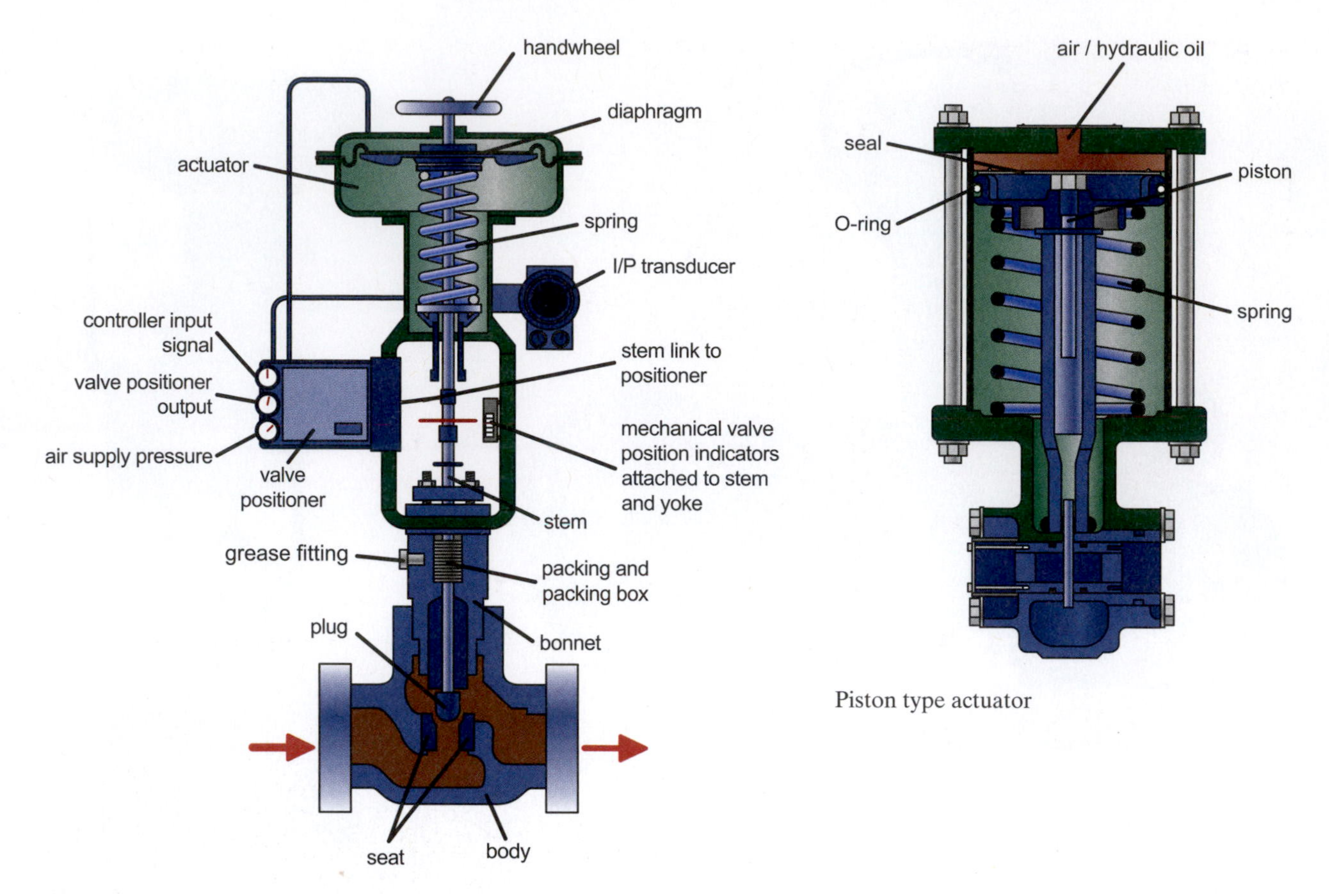

Piston type actuator

Control valve key components

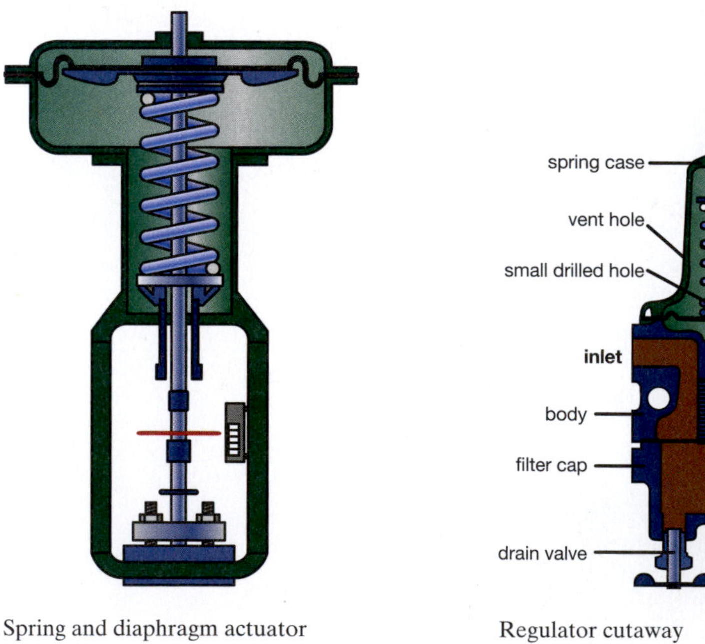

Spring and diaphragm actuator

Regulator cutaway

MPC Operate - 40D2-MPC/MPC1
File View Trend Help
Contents of '40D2-MPC/MPC1'
Mode: AUTO MAN
Control: MPC Local
AUTO
AUTO MPC
Bottom Temperature
°F
250.0
77.6
0.0 Not Limited
Detail
17 Wed May 2000

Parameter Reference	Descriptor	Value	Units	Y-Scale Low	Y-Scale High	Timestamp
40D2-MPC/MPC1/CNTRL1.CV	Top Temperature	61.7	°F	-7.98	192.02	5/17/00 9:27:16 AM
* 40D2-MPC/MPC1/CNTRL2.CV	Bottom Temperature	100.2	°F	0.00	250.00	5/17/00 9:27:16 AM
40D2-MPC/MPC1/CNSTR1.CV	Top Pressure	29.3	psig	0.00	50.00	5/17/00 9:27:16 AM
40D2-MPC/MPC1/MNPLT1.CV	Reflux Flow	35.0	lb/hr	0.00	100.00	5/17/00 9:27:16 AM
40D2-MPC/MPC1/MNPLT2.CV	Steam Flow	40.0	no ...	0.00	100.00	5/17/00 9:27:16 AM

Mpc Block Status/Error Summary
Good
Top Temperature 0.0 200.0
Reflux Flow 0.0 100.0
Top Pressure 0.0 50.0
Feed Flow 0.0 20000.0
Bottom Temperature 0.0 250.0
Steam Flow 0.0 100.0
For Help, press F1
History: ADVCNT1

DCS Display
Courtesy of Emerson Process Management

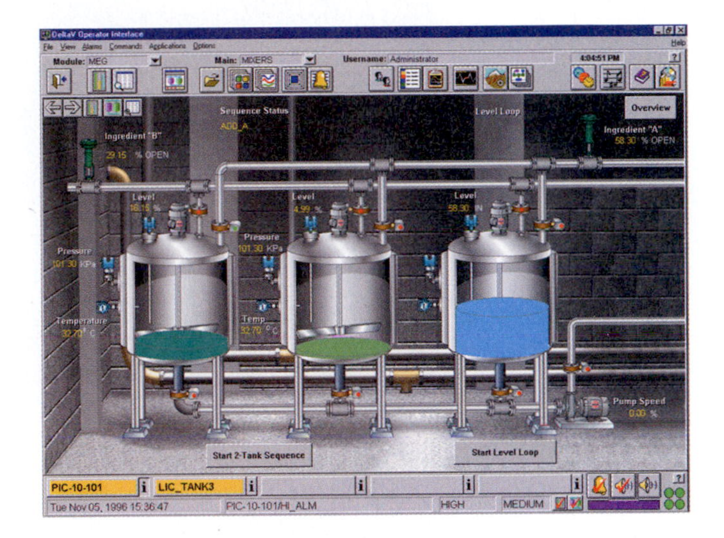

Screen display of mixing tanks
Courtesy of Emerson Process Management

Specific instrument location in the field
Courtesy of Emerson Process Management

PLC Box 1 Inside View
Courtesy of Bayport Training and Technical Center

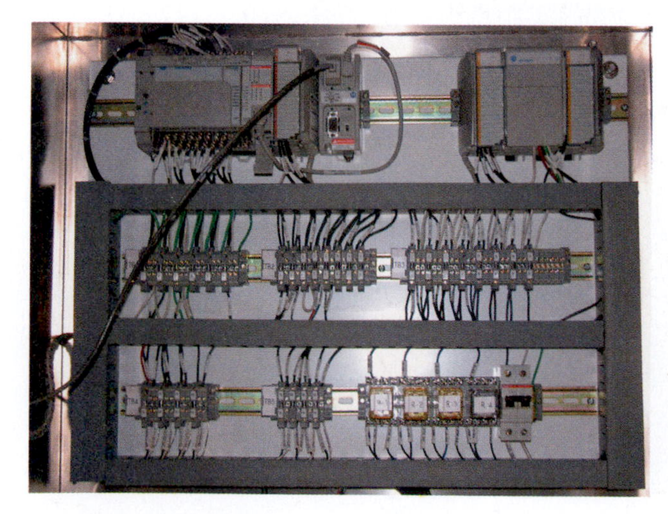

PLC Box 2 Inside View
Courtesy of Bayport Training and Technical Center

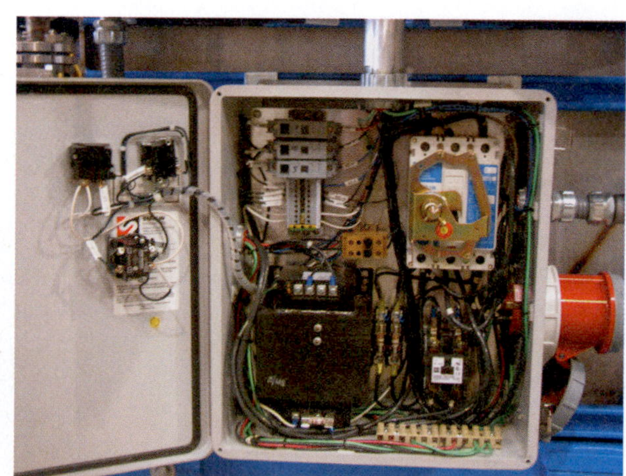

PLC Box 3 Inside View
Courtesy of Bayport Training and Technical Center

PLC Box 4 Inside View
Courtesy of Bayport Training and Technical Center

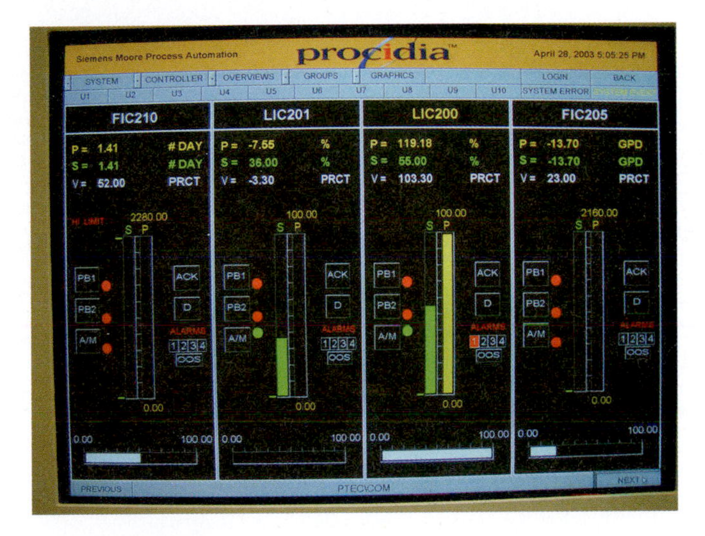

DCS screen display
Courtesy of Lamar Institute of Technology

Motor control center
Courtesy of Bayport Training and Technical Center

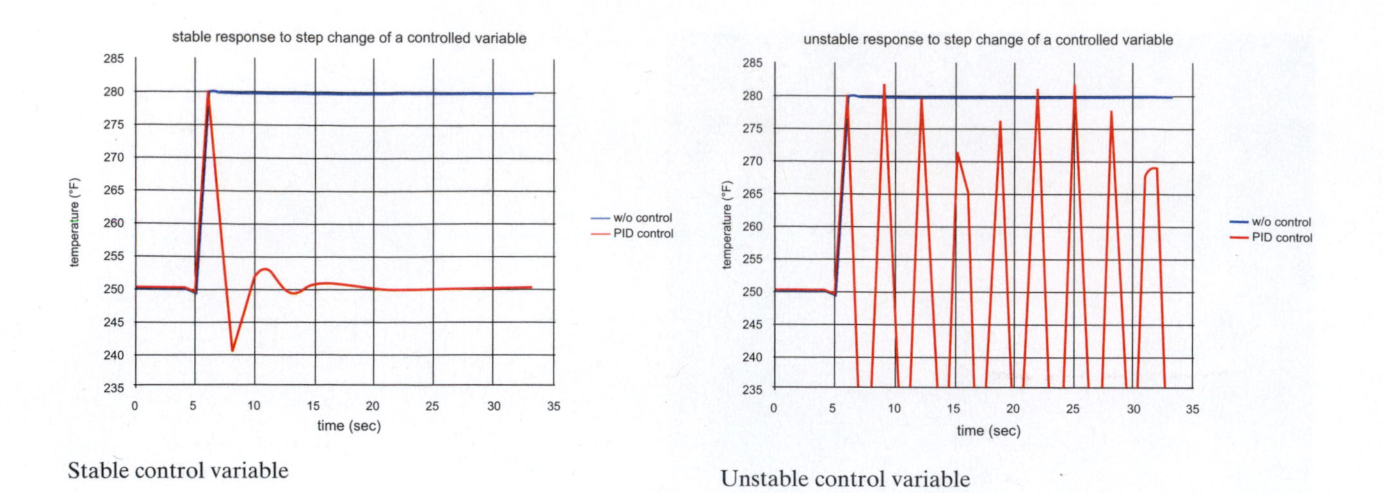

Stable control variable

Unstable control variable

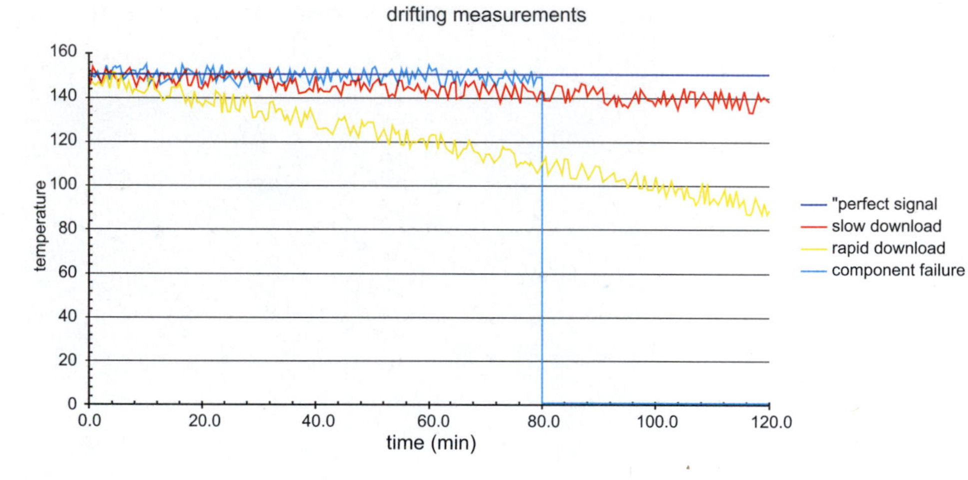

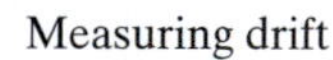

Measuring drift

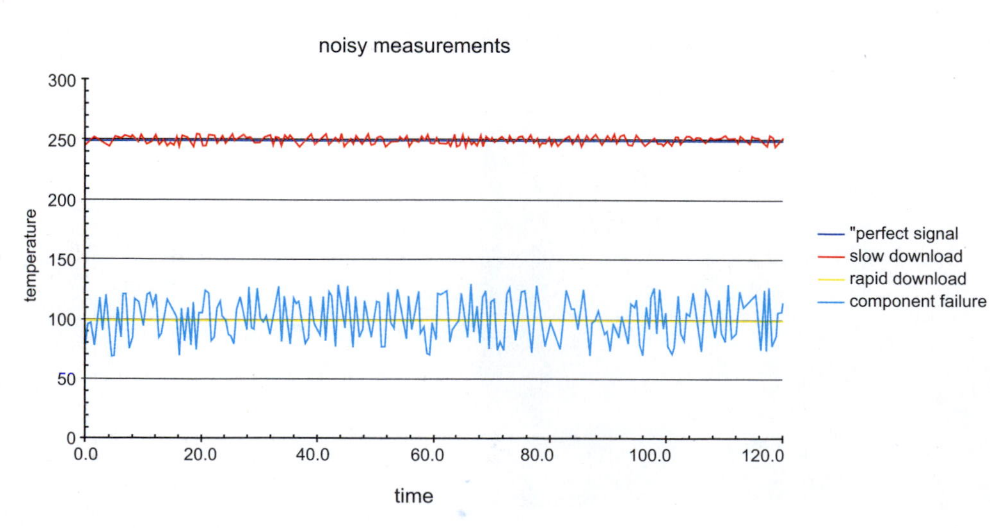

Noisy measurements